Methods in Enzymology

Volume 358
BACTERIAL PATHOGENESIS
Part C
Identification, Regulation, and Function of Virulence Factors

METHODS IN ENZYMOLOGY

EDITORS-IN-CHIEF

John N. Abelson   Melvin I. Simon

DIVISION OF BIOLOGY
CALIFORNIA INSTITUTE OF TECHNOLOGY
PASADENA, CALIFORNIA

FOUNDING EDITORS

Sidney P. Colowick and Nathan O. Kaplan

Methods in Enzymology

Volume 358

Bacterial Pathogenesis

Part C

*Identification, Regulation, and Function
of Virulence Factors*

EDITED BY

Virginia L. Clark

DEPARTMENT OF MICROBIOLOGY AND IMMUNOLOGY
SCHOOL OF MEDICINE AND DENTISTRY
UNIVERSITY OF ROCHESTER
ROCHESTER, NEW YORK

Patrik M. Bavoil

DEPARTMENT OF ORAL AND
CRANIOFACIAL BIOLOGICAL SCIENCES
UNIVERSITY OF MARYLAND
BALTIMORE, MARYLAND

ACADEMIC PRESS

An imprint of Elsevier Science

Amsterdam   Boston   London   New York   Oxford   Paris
San Diego   San Francisco   Singapore   Sydney   Tokyo

Academic Press
An imprint of Elsevier Science.
525 B Street, Suite 1900, San Diego, California 92101-4495, USA
http://www.academicpress.com

Academic Press
84 Theobalds Road, London WC1X 8RR, UK
http://www.academicpress.com

International Standard Book Number: 0-12-182261-3

PRINTED IN THE UNITED STATES OF AMERICA
02 03 04 05 06 07 MM 9 8 7 6 5 4 3 2 1

Table of Contents

Section I. Alternatives to Mammalian Model Systems

Section II. Virulence and Essential Gene Identification

Section III. Global Gene Expression: Microarrays and Proteomics

Section IV. Bacterial Perturbations of Eukaryotic Cell Cycle and Apoptosis

Section V. Bacterial Modification or Exploitation of Eukaryotic Signal Transduction

Section VI. Type III Secretion Systems

Section VII. Quorum Sensing and Gene Regulation

Contributors to Volume 358

Article numbers are in parentheses following the names of contributors.
Affiliations listed are current.

MARGARETA AILI (26), *Department of Molecular Biology, Umeå University, SE-901 87 Umeå, Sweden*

SHIN-ICHI AIZAWA (28), *Department of Biosciences, Teikyo University, Toyosatodai, Utsunomiya 320-8551, Japan*

BRIAN J. AKERLEY (6), *Department of Microbiology and Immunology, University of Michigan Medical School, Ann Arbor, Michigan 48109*

DARRIN AKINS (12), *Department of Microbiology and Immunology, University of Oklahoma Health Sciences Center, Oklahoma City, Oklahoma 73104*

REGINA L. BALDINI (1), *Department of Surgery, Harvard Medical School and Massachusetts General Hospital, Boston, Massachusetts 02114*

ALAN BARBOUR (12), *Department of Microbiology and Molecular Genetics, University of California, Irvine, California 92697*

STOYAN S. BARDAROV (5), *Department of Microbiology and Immunology, Albert Einstein College of Medicine, Bronx, New York 10461*

JORGE BENACH (12), *Center for Infectious Diseases, State University of New York, Stony Brook, New York 11794*

SVEND BIRKELUND (19), *Department of Medical Microbiology and Immunology, University of Aarhus, and LOKE Diagnostics ApS, DK-8000 Aarhus C, Denmark*

JAMES B. BLISKA (24), *Department of Molecular Genetics and Microbiology, Center for Infectious Diseases, State University of New York, Stony Brook, New York 11794*

ROY J. M. BONGAERTS (4), *Institute of Food Research, Norwich Research Park, Norwich NR4 7UA, United Kingdom*

ELISE BOREZÉE (14), *INSERM U570, Faculté de Médecine Necker-Enfants Malades, 75730 Paris Cedex 15, France*

MIRIAM BRAUNSTEIN (5), *Department of Microbiology and Immunology, University of North Carolina, Chapel Hill, North Carolina 27599*

CHAD BROOKS (12), *Department of Microbiology and Immunology, University of Oklahoma Health Sciences Center, Oklahoma City, Oklahoma 73104*

DIRK BUMANN (21), *Department of Molecular Biology, Max Planck Institute for Infection Biology, D-10117 Berlin, Germany*

MARTIN K. R. BURNHAM (8), *Department of Microbiology, GlaxoSmithKline Pharmaceuticals Research and Development, Collegeville, Pennsylvania 19426*

MELISSA CAIMANO (12), *Center for Microbial Pathogenesis and Departments of Medicine and Genetics and Developmental Biology, University of Connecticut Health Center, Farmington, Connecticut 06030*

SHERWOOD CASJENS (12), *Department of Pathology, University of Utah Medical Center, Salt Lake City, Utah 84132*

GUNNA CHRISTIANSEN (19), *Department of Medical Microbiology and Immunology, University of Aarhus, DK-8000 Aarhus C, Denmark*

JOSEPHINE E. CLARK-CURTISS (7), *Departments of Biology and Molecular Microbiology, Washington University, St. Louis, Missouri 63130*

STUART J. CORDWELL (15), *Australian Proteome Analysis Facility, Macquarie University, Australia 2109*

FRANCE DAIGLE (7), *Department of Microbiology, University of Montreal, Montreal, Quebec, Canada H3C 3J7*

DAYLE A. DAINES (11), *Seattle Biomedical Research Institute, Seattle, Washington 98109*

REBEKAH DEVINNEY (27), *Department of Microbiology and Infectious Diseases, University of Calgary Health Sciences Centre, Calgary, Alberta, Canada T2N 4N1*

MARIA DIMITROVA (11), *INSERM U74, Institute of Virology, University of Strasbourg, 67000 Strasbourg, France*

ABDALLAH ELIAS (12), *Rocky Mountain Laboratories, National Institute of Allergy and Infectious Diseases, National Institutes of Health, Hamilton, Montana 59840*

STEPHEN K. FARRAND (32), *Departments of Crop Sciences and Microbiology, University of Illinois, Urbana, Illinois 61801*

MICHAEL FOUNTOULAKIS (20), *Center for Medical Genomics, F. Hoffmann-La Roche Ltd., CH-4070 Basel, Switzerland*

GAD FRANKEL (25), *Department of Biological Sciences, Center for Molecular Microbiology and Infection, Imperial College of Science, Technology, and Medicine, London SW7 2AZ, United Kingdom*

FRANCISCO GARCÍA-DEL PORTILLO (29), *Department of Microbial Biotechnology, Centro Nacional de Biotecnología–CSIC, 28049 Madrid, Spain*

MATHIEU GISSOT (23), *Unit of Molecular Biology of the Gene, INSERM U277 and Université Paris 7, Institut Pasteur, 75724 Paris Cedex 15, France*

MICHÈLE GRANGER-SCHNARR (11), *Délégation Rhône-Alpes-Site Vallée du Rhône, 69609 Villeurbanne, France*

LEANDER GRODE (17), *Department of Immunology, Max Planck Institute for Infection Biology, D-10117 Berlin, Germany*

BENGT HALLBERG (26), *Department of Pathology, Umeå University, SE-901 87 Umeå, Sweden*

ISABELLE HAUTEFORT (4), *Institute of Food Research, Norwich Research Park, Norwich NR4 7UA, United Kingdom*

JAY C. D. HINTON (4), *Institute of Food Research, Norwich Research Park, Norwich NR4 7UA, United Kingdom*

LES M. HOFFMAN (9), *Epicentre Technologies, Madison, Wisconsin 53713*

JOAN Y. HOU (7), *Department of Biology, Washington University, St. Louis, Missouri 63130*

WILLIAM R. JACOBS, JR. (5), *Howard Hughes Medical Institute, Departments of Microbiology and Immunology and Molecular Genetics, Albert Einstein College of Medicine, Bronx, New York 10461*

ALGIS JASINSKAS (12), *Department of Microbiology and Molecular Genetics, University of California, Irvine, California 92697*

JERRY J. JENDRISAK (9), *Epicentre Technologies, Madison, Wisconsin 53713*

YINDUO JI (8), *Department of Veterinary Pathobiology, College of Veterinary Medicine, University of Minnesota, St. Paul, Minnesota 55108*

PETER R. JUNGBLUT (17, 21), *Core Facility Protein Analysis, Max Planck Institute for Infection Biology, D-10117 Berlin, Germany*

LAURA KATONAH (12), *Center for Infectious Diseases, State University of New York, Stony Brook, New York 11794*

STEFAN H. E. KAUFMANN (17), *Department of Immunology, Max Planck Institute for Infection Biology, D-10117 Berlin, Germany*

CHRISTOPHER KIRKPATRICK (16), *Department of Biology, Kenyon College, Gambier, Ohio 43022*

STUART KNUTTON (25), *Institute of Child Health, Clinical Research Block, Birmingham B4 6NH, United Kingdom*

LEON F. KUBENA (10), *U.S. Department of Agriculture, Agricultural Research Service, Food and Feed Safety Research Unit, College Station, Texas 77845*

YOUNG MIN KWON (10), *Center of Excellence for Poultry Science, University of Arkansas, Fayetteville, Arkansas 72701*

DAVID J. LAMPE (6), *Department of Biological Sciences, Duquesne University, Pittsburgh, Pennsylvania 15282*

ROBERT A. LaROSSA (13), *DuPont Central Research and Development, Wilmington, Delaware 19880*

GEE W. LAU (1), *Division of Pulmonary and Critical Care Medicine, University of Cincinnati College of Medicine, Cincinnati, Ohio 45267*

JENS MATTOW (17), *Department of Immunology, Max Planck Institute for Infection Biology, D-10117 Berlin, Germany*

HANS-JOACHIM MOLLENKOPF (17), *Department of Immunology, Max Planck Institute for Infection Biology, D-10117 Berlin, Germany*

STEVEN F. MOSS (22), *Department of Medicine, Division of Gastroenterology, Rhode Island Hospital, Brown University, Providence, Rhode Island 02903*

XAVIER NASSIF (14), *INSERM U570, Faculté de Médecine Neckers-Enfants Malades, 75730 Paris Cedex 15, France*

DAVID J. NISBET (10), *U.S. Department of Agriculture, Agricultural Research Service, Food and Feed Safety Research Unit, College Station, Texas 77845*

PHILIPPE OGER (32), *Earth Sciences Laboratory, École Normal Supérieure de Lyon, Lyon Cedex 07, France*

CAROLINE OJAIMI (12), *Department of Microbiology and Immunology, New York Medical College, Valhalla, New York 10595*

DAVID M. OJCIUS (23), *Institut Pasteur, INSERM, Université Paris, 75724 Paris Cedex 15, France*

LUCIANO PASSADOR (31), *Department of Microbiology and Immunology, University of Rochester Medical Center, Rochester, New York 14642*

JEAN-LUC PERFETTINI (23), *Unit of Molecular Biology of the Gene, INSERM U277 and Université Paris 7, Institut Pasteur, 75724 Paris Cedex 15, France*

AGNÈS PERRIN (14), *INSERM U570, Faculté de Médecine Necker-Enfants Malades, 75730 Paris Cedex 15, France*

LUU PHAN-THANH (18), *Infectious Disease and Immunology Unit, Institut National de la Recherche Agronomique, 37380 Nouzilly-Tours, France*

STEPHEN K. PICATAGGIO (13), *DuPont Central Research and Development, Wilmington, Delaware 19880*

M. GRACIELA PUCCIARELLI (29), *Department of Microbial Biotechnology, Centro Nacional de Biotecnología–CSIC, 28049 Madrid, Spain*

YINPING QIN (32), *Department of Crop Sciences, University of Illinois, Urbana, Illinois 61801*

JUSTIN RADOLF (12), *Departments of Medicine and Genetics and Developmental Biology, Center for Microbial Pathogenesis, University of Connecticut Health Center, Farmington, Connecticut 06030*

LAURENCE G. RAHME (1), *Department of Surgery, Harvard Medical School and Massachusetts General Hospital, Boston, Massachusetts 02114*

RENATE REIMSCHUESSEL (3), *Center for Veterinary Medicine, Food and Drug Administration, Laurel, Maryland 20708*

STEVEN C. RICKE (10), *Department of Poultry Science, Texas A&M University, College Station, Texas 77843*

PATRICIA ROSA (12), *Rocky Mountain Laboratories, National Institute of Allergy and Infectious Diseases, National Institutes of Health, Hamilton, Montana 59840*

MARTIN ROSENBERG (8), *Department of Microbiology, GlaxoSmithKline Pharmaceuticals Research and Development, Collegeville, Pennsylvania 19426*

ROLAND ROSQVIST (26), *Department of Molecular Biology, Umeå University, SE-901 87 Umeå, Sweden*

EDWARD G. RUBY (30), *Pacific Biomedical Research Center, Kewalo Marine Laboratory, University of Hawaii at Manoa, Honolulu, Hawaii 96813*

KRISTIN M. RULEY (3), *Center for Vaccine Development, University of Maryland School of Medicine, Baltimore, Maryland 21201*

CHIHIRO SASAKAWA (28), *Department of Microbiology and Immunology, Division of Bacterial Infection, Institute of Medical Science, University of Tokyo, Minato-ku, Tokyo 108-8639, Japan*

ULRICH E. SCHAIBLE (17), *Department of Immunology, Max Planck Institute for Infection Biology, D10117 Berlin, Germany*

IRA SCHWARTZ (12), *Department of Microbiology and Immunology, New York Medical College, Valhalla, New York 10595*

ROBERT SHAW (25), *Institute of Child Health, Clinical Research Block, Birmingham B4 6NH, United Kingdom*

JULIE M. SIDEBOTHAM (4), *Institute of Food Research, Norwich Research Park, NR4 7UA Norwich, United Kingdom*

RICHARD P. SILVER (11), *University of Rochester Medical Center, Rochester, New York 14642*

SIMON SIMS (12), *Sigma-Genosys, The Woodlands, Texas 77380*

JON SKARE (12), *Department of Medical Microbiology and Immunology, Texas A&M University Health Science Center, College Station, Texas 77843*

JOAN L. SLONCZEWSKI (16), *Department of Biology, Kenyon College, Gambier, Ohio 43022*

DANA R. SMULSKI (13), *DuPont Central Research and Development, Wilmington, Delaware 19880*

EMILIA MIA SORDILLO (22), *Department of Pathology and Laboratory Medicine, College of Physicians and Surgeons of Columbia University, St. Luke's–Roosevelt Hospital Center, New York, New York 10025*

PHILIPPE SOUQUE (23), *Unit of Molecular Biology of the Gene, INSERM U277 and Université Paris 7, Institut Pasteur, 75724 Paris Cedex 15, France*

ERIC V. STABB (30), *Department of Microbiology, University of Georgia, Athens, Georgia 30602*

KRISTEN SWINGLE (12), *Department of Medical Microbiology and Immunology, Texas A&M University Health Science Center, College Station, Texas 77843*

BÉLA TAKÁCS (20), *Biological Technologies, F. Hoffmann-La Roche Ltd., CH-4070 Basel, Switzerland*

KOICHI TAMANO (28), *Department of Microbiology and Immunology, Division of Bacterial Infection, Institute of Medical Science, University of Tokyo, Minato-ku, Tokyo 108-8639, Japan*

MAN-WAH TAN (2), *Departments of Genetics, and Microbiology and Immunology, Stanford University School of Medicine, Stanford, California 94305*

LORI J. TEMPLETON (13), *DuPont Central Research and Development, Wilmington, Delaware 19880*

COLIN R. TINSLEY (14), *INSERM U570, Faculté de Médecine Necker-Enfants Malades, 75730 Paris Cedex 15, France*

MICHELE TRUCKSIS (3), *Division of Infectious Diseases and Immunology, University of Massachusetts Medical School, Worcester, Massachusetts 01605*

BRIAN BERG VANDAHL (19), *Department of Medical Microbiology and Immunology, University of Aarhus, and LOKE Diagnostics ApS, DK-8000 Aarhus C, Denmark*

GLORIA I. VIBOUD (24), *Department of Molecular Genetics and Microbiology, Center for Infectious Diseases, State University of New York, Stony Brook, New York 11794*

HANS WOLF-WATZ (26), *Department of Molecular Biology, Umeå University, SE-901 87 Umeå, Sweden*

GARY WOODNUTT (8), *Department of Microbiology, GlaxoSmithKline Pharmaceuticals Research and Development, Collegeville, Pennsylvania 19426*

Preface

The recent and rapid expansion of information on mechanisms of bacterial pathogenesis has led to this third volume in the *Methods in Enzymology* series on this topic. Part A (Volume 235) covered methods for the isolation and identification of bacterial pathogens and their associated virulence factors, while Part B (Volume 236) dealt with interactions between bacterial pathogens and their hosts. Since then new methods have been developed to identify virulence factors and to assess their mode of action. In addition, it has become apparent that most virulence factors are coordinately regulated. Thus, the methods described in this volume represent new approaches to identify virulence factors, to determine the spectrum of gene regulation under different environmental conditions, and to identify the mechanism of pathogenesis.

The first section of Part C deals with the use of nonmammalian hosts to assess virulence of a pathogen. It has been demonstrated that many pathogens are able to infect organisms of different kingdoms, which allows the more rapid identification of bacterial factors important in virulence by the use of high throughput screens using genetically manipulable hosts such as plants, nematodes, or insects. The remarkable conservation of many host defense mechanisms also allows the identification of genes important in preventing bacterial pathogenesis. The second section of this volume describes new genetic methods to identify bacterial genes essential and/or involved in virulence. This is followed by a section detailing the use of microarrays and proteomics to determine differences in global gene expression under different environmental conditions. The next three sections describe methods used to identify the function of bacterial virulence factors in pathogenesis. The final section is a brief introduction to methods involved in investigations of the role of gene regulation by quorum sensing. For a more complete description of this topic we refer you to Volumes 310, 336, and 337 of *Methods in Enzymology* that deal with various aspects of bacterial quorum sensing and biofilm formation.

We are greatly indebted to all the contributors for sharing their research expertise and making this volume possible.

VIRGINIA L. CLARK
PATRIK M. BAVOIL

METHODS IN ENZYMOLOGY

VOLUME 229. Cumulative Subject Index Volumes 195–198, 200–227

VOLUME 230. Guide to Techniques in Glycobiology
Edited by WILLIAM J. LENNARZ AND GERALD W. HART

VOLUME 231. Hemoglobins (Part B: Biochemical and Analytical Methods)
Edited by JOHANNES EVERSE, KIM D. VANDEGRIFF, AND ROBERT M. WINSLOW

VOLUME 232. Hemoglobins (Part C: Biophysical Methods)
Edited by JOHANNES EVERSE, KIM D. VANDEGRIFF, AND ROBERT M. WINSLOW

VOLUME 233. Oxygen Radicals in Biological Systems (Part C)
Edited by LESTER PACKER

VOLUME 234. Oxygen Radicals in Biological Systems (Part D)
Edited by LESTER PACKER

VOLUME 235. Bacterial Pathogenesis (Part A: Identification and Regulation of Virulence Factors)
Edited by VIRGINIA L. CLARK AND PATRIK M. BAVOIL

VOLUME 236. Bacterial Pathogenesis (Part B: Integration of Pathogenic Bacteria with Host Cells)
Edited by VIRGINIA L. CLARK AND PATRIK M. BAVOIL

VOLUME 237. Heterotrimeric G Proteins
Edited by RAVI IYENGAR

VOLUME 238. Heterotrimeric G-Protein Effectors
Edited by RAVI IYENGAR

VOLUME 239. Nuclear Magnetic Resonance (Part C)
Edited by THOMAS L. JAMES AND NORMAN J. OPPENHEIMER

VOLUME 240. Numerical Computer Methods (Part B)
Edited by MICHAEL L. JOHNSON AND LUDWIG BRAND

VOLUME 241. Retroviral Proteases
Edited by LAWRENCE C. KUO AND JULES A. SHAFER

VOLUME 242. Neoglycoconjugates (Part A)
Edited by Y. C. LEE AND REIKO T. LEE

VOLUME 243. Inorganic Microbial Sulfur Metabolism
Edited by HARRY D. PECK, JR., AND JEAN LEGALL

VOLUME 244. Proteolytic Enzymes: Serine and Cysteine Peptidases
Edited by ALAN J. BARRETT

VOLUME 245. Extracellular Matrix Components
Edited by E. RUOSLAHTI AND E. ENGVALL

VOLUME 246. Biochemical Spectroscopy
Edited by KENNETH SAUER

VOLUME 247. Neoglycoconjugates (Part B: Biomedical Applications)
Edited by Y. C. LEE AND REIKO T. LEE

Section I

Alternatives to Mammalian Model Systems

[1] Use of Plant and Insect Hosts to Model Bacterial Pathogenesis

By REGINA L. BALDINI, GEE W. LAU, and LAURENCE G. RAHME

Introduction

It has become acutely apparent that as the sequencing of the genomes of many human bacterial pathogens is near completion, the relevance of thousands of bacterial genes to human pathogenesis needs to be determined. The paucity of vertebrate animal models to systematically screen for virulence-related mutants creates an obstacle to the identification of novel virulence factors and the elucidation of mechanisms of pathogenesis. Fortunately, some bacterial pathogens that cause disease in humans are capable of establishing themselves and causing disease in phylogenetically diverse hosts. Studies demonstrate that this ability depends on the remarkable conservation in virulence mechanisms and factors required by the pathogen to infect organisms of different kingdoms.[1–4] The ability of a human pathogen to infect nonmammalian, genetically tractable hosts such as plants, nematodes, or insects allows for high-throughput screens to be performed in which bacterial mutants with reduced virulence in these hosts can be identified and the host responses during infection can be studied. Once bacterial mutants of the human pathogen of interest that are attenuated in virulence in one of these hosts have been identified, their relevance to mammalian pathogenesis can be validated using a mammalian model. The use of invertebrate hosts to model bacterial pathogenesis as adjuncts to mammalian models not only is ethical but offers numerous advantages. The genetic tractability that hosts such as plants, nematodes, or insects provide, the availability of their genome sequences, the fast generation time, the ease of handling, and low cost are some of the many benefits that the use of these model hosts can bring.

Human bacterial pathogens that infect one or more of the nonmammalian hosts mentioned above include *Pseudomonas aeruginosa*,[1,5–8] *Salmonella typhimurium*,[2]

[1] L. G. Rahme, E. J. Stevens, S. F. Wolfort, J. Shao, R. G. Tompkins, and F. M. Ausubel, *Science* **268,** 1899 (1995).

[2] A. Aballay, P. Yorgey, and F. M. Ausubel, *Curr. Biol.* **10,** 1539 (2000).

[3] D. Schneider and M. Shahabuddin, *Science* **288,** 2376 (2000).

[4] A. Basset, R. S. Khush, A. Braun, L. Gardan, F. Boccard, J. A. Hoffmann, and B. Lemaitre, *Proc. Natl. Acad. Sci. U.S.A.* **97,** 3376 (2000).

[5] G. Jander, L. G. Rahme, and F. M. Ausubel, *J. Bacteriol.* **182,** 3843 (2000).

[6] M.-W. Tan, S. Mahajan-Miklos, and F. M. Ausubel, *Proc. Natl. Acad. Sci. U.S.A.* **96,** 715 (1999).

[7] L. G. Rahme, F. M. Ausubel, H. Cao, E. Drenkard, B. C. Goumnerov, G. W. Lau, S. Mahajan-Miklos, J. Plotnikova, M.-T. Wa, J. Tsongalis, C. Walendziewicz, and R. G. Tompkins, *Proc. Natl. Acad. Sci. U.S.A.* **97,** 8815 (2000).

[8] S. Mahajan-Miklos, L. G. Rahme, and F. M. Ausubel, *Mol. Microbiol.* **37,** 981 (2000).

Enterococcus,[9] *Erwinia* spp.,[10] *Serratia marcescens,*[11] and *Proteus vulgaris.*[11] In this article we will focus on the methodologies used to screen for the identification of previously unknown *P. aeruginosa* virulence-related genes relevant to mammalian pathogenesis. None of the currently available animal models simulate all aspects of any human disease caused by *P. aeruginosa.* This pathogen is a leading cause of nosocomial infections in cystic fibrosis patients, burn victims, and other immunocompromised individuals.[12] The ability of *P. aeruginosa* to infect plants (lettuce and *Arabidopsis thaliana*),[1] insects (*Galleria mellonella* and *Drosophila melanogaster*),[5,11,13,14] nematodes (*Caenorhabditis elegans*),[6] and mice permitted the identification of a large number of "universal" virulence factors and mechanisms of pathogenesis[7,8] required to cause disease in mammals. Here we will describe the methodology used to identify *P. aeruginosa* "universal" virulence factors using lettuce and *Arabidopsis thaliana* as well as *Galleria mellonella* and *Drosophila melanogaster* as model hosts. We will also describe the protocols used to infect *D. melanogaster* with *Erwinia carotovora* using natural modes of infection. The methodology used to model *P. aeruginosa, Salmonella typhimurium,* and *Enterococcus* pathogenesis in *C. elegans* will be described in another article in this volume by Tan *et al.*[14a]

General Considerations

The first step in the infection of any host is the preparation of the bacterial inoculum. The growth phase of the bacterial cells to be used for infection is crucial to the outcome of the experiment because some bacterial virulence factors can be expressed at certain stages of bacterial growth and not at others. Although the cells are physiologically more active during the exponential phase, virulence factors that are cell-density regulated are highly expressed in the early stationary phase.[12] Therefore, for bacteria in which cell density constitutes an important mechanism of regulation of virulence-related factors, it is important to allow cells to grow up to the stage at which most of their virulence factors are expressed. Using bacterial inoculum from various growth stages to infect the desired host will help in determining the optimum growth phase of the bacterial cells to be

[9] D. A. Garsin, C. D. Sifri, E. Mylonakis, X. Qin, K. V. Singh, B. E. Murray, S. B. Calderwood, and F. M. Ausubel, *Proc. Natl. Acad. Sci. U.S.A.* **98,** 10892 (2001).

[10] S. S. Schneierson and F. J. Bottone, *CRC Crit. Rev. Clin. Lab. Sci.* **4,** 341 (1973).

[11] J. S. Chadwick, *Fed. Proc.* **26,** 1675 (1967).

[12] C. Van Delden and B. H. Iglewski, *Emerg. Infect. Dis.* **4,** 551 (1998).

[13] G. W. Lau and L. G. Rahme, unpublished (2002).

[14] D. A. D'Argenio, L. A. Gallagher, C. A. Berg, and C. Manoil, *J. Bacteriol.* **183,** 1466 (2001).

[14a] M.-W. Tan, *Methods Enzymol.* **358,** [2], 2002 (this volume).

used for infection. In the case of *P. aeruginosa,* disease in plants, insects, and animals is reproducibly caused when early stationary phase cells are used for infections.

The optimum titer of the bacterial inoculum for infection should also be determined. In determining this, one should keep in mind that the concentration of the inoculum to be used should allow for identification of slightly attenuated in virulence mutants. The genotype of the host is an additional concern. For example, some ecotypes of *Arabidopsis thaliana* or species of *Drosophila* are more susceptible to *P. aeruginosa* strain PA14 infection than others.

Although some of the considerations listed here may be valid for various pathogens, the conditions described here are ideal for studying virulence factors in *P. aeruginosa,* and the protocols must be optimized for the specific pathogen to be tested.

Plants as Model Hosts

Plants can be used either for assessing whether a particular strain is capable of causing disease (qualitative assay) or to measure the virulence (degree of symptoms) of such a strain by determining the bacterial growth within the host tissues (quantitative assay). Often qualitative assays are only the first step in characterizing or screening for the virulence potential of a bacterial strain, whereas quantitative assays can compare strains and thus more accurately determine differences in the colonization and invasion abilities among them.

Using Lettuce to Screen for Bacterial Mutants Attenuated in Virulence Qualitative Assay

In order to perform a high-throughput qualitative screen for mutant clones that are attenuated in virulence, a *P. aeruginosa* transposon mutant library should be screened in lettuce (variety Romaine or Iceberg) plants. A flow chart describing the various steps involved in this screen is presented in Fig. 1 (modified from Rahme et al.[7]). The detailed procedure is described below.

1. Individual mutant clones are grown overnight in Luria–Bertani (LB) broth in 96-well microtiter plates at $37°$, subsequently diluted $1:10$, and allowed to grow to the early stationary phase. The bacteria are then pelleted, washed once with 10 mM $MgSO_4$, and resuspended in 10 mM $MgSO_4$. After dilution to an optical density (OD) of 0.2 at 600 nm with 10 mM $MgSO_4$, two serial 10-fold dilutions are made in 10 mM $MgSO_4$. The resulting dilutions correspond to approximately 10^6 and 10^5 cfu (colony-forming units)/ml.

2. Romaine lettuce plants are grown in MetroMix (Scotts-Sierra Horticultural Co., Marysville, OH) potting soil in a greenhouse at $26°$, 12 hr light/12 hr

(A)   **Generate bacterial mutant library**

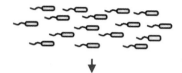

↓

(B)   **Screen for mutants attenuated in virulence
(qualitative assay)**

↓

(C)   **Measure growth in *Arabidopsis thaliana*
(quantitative assay)**

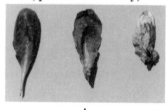

↓

(D) **Test selected nonpathogenic mutant in mice**

FIG. 1. Strategy used to screen for *P. aeruginosa* mutants attenuated in virulence. (A) Bacterial mutant clones were prepared. (B) Midribs of lettuce leaves were inoculated with the wild-type *P. aeruginosa* strain PA14 (left) or isogenic, nonvirulent mutants (right) and photographed 5 days postinoculation. Arrows indicate the inoculation sites. (C) Infected *Arabidopsis thaliana* leaves (from left to right): mock treatment, PA14 at day 1 showing weak symptoms, and severe symptoms at day 5. (D) Mutants were tested in mice.

dark. The older midribs of the 12-week-old plants are inoculated with 10-μl aliquots of 10^7, 10^6, and 10^5 cells/ml using a Pipetman (without forcing the tip all the way through), leaving about 0.5 cm between each point of inoculation (Fig. 1). Alternatively, detached leaves are washed with 0.1% bleach and placed in 15-mm diameter petri dishes containing one Whatman (Clifton, NJ)

#1 filter paper soaked in 10 mM MgSO$_4$.[15] The midrib of each leaf is inoculated as above. Each midrib should be inoculated with the wild-type strain as a control.

3. Following inoculation, plants or petri dishes containing the inoculated midribs should be incubated in a growth chamber at 30° with a relative humidity of 90–100%, and a 12-hr photoperiod. If the detached leaf method is used,[15] it is important to maintain the Whatman #1 filter paper soaked in 10 mM MgSO$_4$ throughout the duration of the experiment.

4. The symptoms must be evaluated daily for 4–5 days. Signs of symptoms with the wild-type strain will first appear as light brown spots at the site of inoculation approximately 30 hr postinfection. The expected full-blown symptoms 4–5 days following infection consist of strong brown and soft-rot areas surrounding the site of inoculation that could spread to the entire midrib, resulting in complete maceration and collapse of the tissue (Fig. 1).

This method could yield mutants that elicit null, weak, or moderate rotting symptoms as compared to symptoms caused by the parental strain.[15] The midribs of the lettuce leaves can also be inoculated by simply "tooth-picking" single bacterial colonies directly from agar plates. This approach can be used for the identification of mutants that cause null or weak soft-rot symptoms on lettuce. Using the "tooth-picking" as the method of inoculation, mutants that may cause moderate symptoms will be missed.

Using Arabidopsis thaliana Plants for Qualitative Assays or Quantitative Measurement of Virulence

Arabidopsis thaliana plants can be used for both qualitative and quantitative measurements of virulence. However, in high-throughput screens for qualitative analyses, it is more practical to use lettuce plants and subsequently verify the results obtained by using the *Arabidopsis* leaf infiltration model.[15] As mentioned above, the choice of the host genotype is crucial in obtaining the optimum symptoms and growth of the pathogen of choice. Therefore, infecting several *Arabidopsis* ecotypes increases the likelihood of identifying a highly susceptible ecotype. Various *Arabidopsis* ecotypes can be obtained from the *Arabidopsis* Biological Resource Center (Columbus, OH). The identification of cultivars with various degrees of susceptibility or resistance to the pathogenic strain used will also provide valuable information regarding the host-ecotype specificity of the strain.

For example, when the *Arabidopsis* ecotype Llagostera (Ll-O) is infected with *P. aeruginosa* strain PA14, severe soft-rot symptoms are developed (Table I and

[15] L. G. Rahme, M.-W. Tan, L. Le, S. M. Wong, R. G. Tompkins, S. B. Calderwood, and F. M. Ausubel, *Proc. Natl. Acad. Sci. U.S.A.* **94**, 13245 (1997).

TABLE I

SYMPTOMS ELICITED BY *P. aeruginosa* STRAIN PA14 AND ISOGENIC MUTANTS IN LEAVES
OF *Arabidopsis thaliana* ECOTYPE LLAGOSTERA[a]

Symptom	Description
Severe	Soft-rotting of the entire leaf characterized by a water-soaked reaction zone and chlorosis around the inoculation site at 2–3 days postinfection. Characteristic of the wild-type strain PA14 and other pathogenic *P. aeruginosa* strains
Moderate	Moderate water-soaking and chlorosis with most of the tissue softened around the inoculation site
Weak	Localized water-soaking and chlorosis of tissue circumscribing the inoculation site
None	No soft-rot symptoms, only chlorosis circumscribing the inoculation site

[a] 5 days postinoculation.

Fig. 1) whereas less severe symptoms are observed in the Columbia (Col-O) ecotype and no symptoms are observed in the Argentat (Ag) ecotype. Such findings are valuable for the study of host responses during infection.

Arabidopsis Leaf Inoculation Assay

1. *Arabidopsis* seeds are soaked in water and incubated at 4° for 48 hr. Subsequently, seeds are mixed with 0.01% (w/v) agarose (to keep them separate) and spread on MetroMix 200 soil using a Pasteur pipette.

2. Flats containing the planted seeds are transferred to a growth chamber at $20 \pm 2°$ with a 12-hr photoperiod and 70% (v/v) humidity. Two to 3 weeks later, the germinated seedlings are individually transplanted in new trays containing MetroMix 200 soil, 3 cm apart.

3. *P. aeruginosa* cultures from the early stationary phase are prepared as described above for the lettuce assays. Following the washing of the bacteria in 10 mM MgSO$_4$, bacteria are resuspended in 10 mM MgSO$_4$ to give an OD$_{600}$ of 0.2. The suspension is diluted 1 : 100 and 1 : 1000 and injected into leaves of 6-week-old *Arabidopsis* plants. For each bacterial sample, approximately 12 plants are used. Three to four leaves are inoculated per plant, because only the older leaves, which are more susceptible to infection, are used for inoculation.

4. The leaves are infiltrated by forcing a bacterial suspension through the leaf stomata using 1-ml disposable syringe without a needle and pressing it against the abaxial side of the leaf (Fig. 2).

5. Immediately following inoculation, the plants are transferred to a growth chamber at 30° in an atmosphere of 90–100% relative humidity and kept there during the course of the experiment. All trays containing plants are covered with clear plastic covers to maintain humidity at a high level. Small holes are made in the cover to allow evaporation. Symptoms and growth are monitored daily for 5 days.

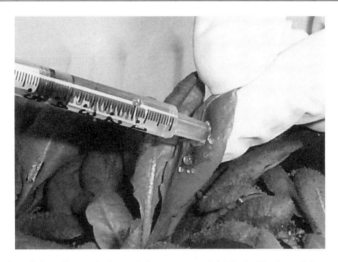

FIG. 2. Inoculation of an *Arabidopsis thaliana* ecotype Col-0 leaf with a bacterial suspension.

6. To assess bacterial proliferation in *Arabidopsis* leaves, two 0.28-cm^2 leaf disks are obtained from four different leaves using a #2 cork cutter (Fisher Scientific, Pittsburgh, PA). Leaf intercellular fluid containing bacteria is harvested by grinding the two leaf disks in 300 μl of 10 mM MgSO$_4$ using a plastic pestle. Serial dilutions of the homogenate are plated on LB agar plates to determine bacterial concentrations. Data can be reported as means and standard deviations of the log (cfu/cm^2) for each data point as described in Rahme *et al.*[1]

An example of the type of symptoms that can be observed in *Arabidopsis* leaves following infection with *P. aeruginosa* mutant strains identified as attenuated in virulence in the lettuce screen is listed in Table I.

Standard microscopic techniques can also be used to determine the ability of strains to attach and colonize leaves as well as to assess morphological changes in the plant tissues. Examples of how these techniques can be applied to the PA14-*Arabidopsis* model are described in Plotnikova *et al.*[16]

Insects as Model Hosts

Bacterial Infection of Drosophila and Galleria mellonella

Although plants can be used as genetically tractable host systems to dissect host responses to infection, parallels in innate immunity against pathogens in insects

[16] J. M. Plotnikova, L. G. Rahme, and F. M. Ausubel, *Plant Physiol.* **124,** 1766 (2000).

and mammalian hosts render insects unique hosts for dissecting diseases–host response pathways triggered by a pathogen. *Drosophila* has been used as a host to study bacterial virulence and host immune responses with various bacterial, fungal, and protozoan pathogens of mammalian hosts.[3,4,13,14,17,18]

As stated above, the ability of bacteria to cause infection is strongly linked to the growth phase of the bacterial population/culture. As in the case of plant infections, the *P. aeruginosa* growth phase most suitable for *Drosophila* infections is the early stationary phase. The bacterial cultures of *P. aeruginosa* for *Drosophila* infections are prepared as follows:

1. Grow *P. aeruginosa* strains in LB broth with appropriate antibiotic supplements to early stationary phase, corresponding to about 5×10^9 cfu/ml.

2. Make serial 10-fold dilutions in 10 mM MgSO$_4$ and plate on LB agar to determine the concentration of viable bacteria in the original culture. Use the 100-fold dilution for infection.

Preparing Drosophila for Infection. In general, the most widely used method to anesthetize flies is with CO_2. If CO_2 is not available, placing flies on a chilled glass plate ($4°$) on a bed of ice will also work well.

1. Anesthetize flies with CO_2. Place flies on a flat CO_2 stage and, if necessary, separate males from females. Use 25–30 flies for mortality assessment, or as many flies as necessary for the determination of bacterial growth (see step 7 in the following section).

2. As excessive exposure to CO_2 may cause the flies to die prematurely and give false results following infection, inject the flies (see below) as quickly as possible to minimize their exposure to CO_2.

Injection of Bacteria into Adult and Third Instar Drosophila Hemolymph. There are two general protocols used in the infection of *Drosophila* by direct introduction of bacteria into hemolymph[13,14,17–19]: simple pricking with a 10-μm tungsten needle previously dipped into a suspension of bacterial cells, or injection of bacterial suspension with a PL1 100 microinjector (Harvard Apparatus, Inc., Holliston, MA) into adult flies or larvae.[20] Both injection protocols may be used when infecting *Drosophila* adults or larvae with concentrated fungal spores as well.[17] The pricking protocol is described below.

[17] B. Lemaitre, E. Nicolas, L. Michaut, J. M. Reichhart, and J. A. Hoffmann, *Cell* **86,** 973 (1996).

[18] B. Lemaitre, J. M. Reichhart, and J. A. Hoffmann, *Proc. Natl. Acad. Sci. U.S.A.* **94,** 14614 (1997).

[19] L. P. Wu and K. V. Anderson, *Nature* **392,** 93 (1998).

[20] M. Elrod-Erickson, S. Mishra, and D. Schneider, *Curr. Biol.* **10,** 781 (2000).

1. Dip a 10-μm tungsten needle (Ernst Fullam, Inc., Latham, NY) halfway into the bacterial suspension.

2. Using the needle, prick flies in the dorsal half thorax (halfway through the fly) avoiding the wings. The needle should be as close to the cuticle as possible so as not to injure the internal organs of the fly.

3. Set flies aside on the CO_2 stage and repeat dipping and infection until all flies are infected.

4. Place the flies in a vial and leave the vial on its side so that the flies will not get stuck to the food.

5. For mortality studies, record how many flies have died because of injuries 12 hr after the infection procedure. These flies should be excluded from mortality determinations.

6. Record the number of dead flies at various time points: 18, 24, 30, 36, 42, 48, and 72 hr postinfection.

7. To assess bacterial proliferation in flies, collect five flies per time point, homogenize each one in 300 μl of 10 mM $MgSO_4$ using a plastic pestle, and plate serial dilutions on LB agar plates.

8. To study host immune responses over the course of the infection, collect flies at various time points, anesthetize them with CO_2, and store them at $-80°$ until RNA or protein extractions are performed.

Natural Infection of Drosophila. Natural infections of *Drosophila* with bacteria and fungi have been described.[4,21] Natural infection with a fungal pathogen is carried out by shaking anesthetized flies for a few minutes in a petri dish containing actively sporulating fungal cultures such as *Aspergillus fumigatus* or *Fusarium oxysporum.* Flies covered with spores are then transferred to fresh *Drosophila* medium and set at 29°. This article is focusing on bacterial pathogens; a comprehensive methodology of natural infections of *Drosophila* with fungi can be found in Nicolas *et al.*[21]

The plant pathogen *Erwinia carotovora* has been reported to infect humans.[4,10] *Erwinia* spp. belong to the family of Enterobacteriaceae and cause important plant diseases. Although strains of *Erwinia* spp. have been recovered from humans, better medical care and the use of antibiotics have largely eliminated the threat of infections caused by this pathogen. Basset *et al.*[4] have reported a natural infection model that involves the feeding of *Drosophila* larvae with a mixture of overnight *Erwinia carotovora* pellet and crushed banana. Fly larvae feeding on this mixture produce an immune response. Therefore, *Drosophila* can be used as an adjunct host to plants to identify virulence factors of *Erwinia* and host responses to it that are conserved through evolution. The protocol described below may be adapted

[21] E. Nicolas, J. M. Reichhart, J. A. Hoffmann, and B. Lemaitre, *J. Biol. Chem.* **273**, 10463 (1998).

for infections of *Drosophila* larvae with other bacterial species. Natural infection of *Drosophila* larvae with *Erwinia carotovora* is performed as follows:

1. Thoroughly mix 200 μl of concentrated bacterial pellet from an overnight culture with 400 μl of crushed banana in a 2-ml microfuge tube. Add 200 third-instar larvae, mix well, and cover the top of the tube with a foam plug. Incubate at room temperature for 30 min.

2. Transfer the mixture to standard fly medium and incubate at 29°.

3. To determine bacterial colony-forming units, collect five larvae per time point (up to 48 hr) and rinse them in water, followed by 3 × 5 sec changes in 70% (v/v) ethanol for external sterilization. The larvae are then homogenized and the bacteria counts are determined as described in the previous section.

Larvae can also be collected at different time points for analysis of host immune responses during infection. Infected larvae can be stored at −80° until use.

Infection of Bacteria into Hemolymph of Galleria mellonella. Galleria mellonella has been used as a model host for other bacterial and fungal pathogens besides *P. aeruginosa.* Indeed, some of them are opportunistic human pathogens such as *Serratia marcescens, Proteus vulgaris, Proteus mirabilis, Aspergillus fumigatus,* and *Fusarium oxysporum.*[11]

The protocol described here is used in *P. aeruginosa* infections. If a different pathogen is used, conditions may need to be optimized.

1. Dilute overnight PA14 cultures grown in LB 1 : 100 in the same medium. Grow the culture to an OD_{600} of 0.3–0.4, pellet the cultures, and resuspend in 10 mM MgSO$_4$ at OD_{600} 0.1. In contrast to plant and *Drosophila* infections, the infection of *G. mellonella* is successful with log-phase-grown bacterial cultures.[5]

2. Make 10-fold serial dilutions in 10 mM MgSO$_4$ with the appropriate antibiotics. The presence of the antibiotic will prevent infection by bacteria occurring naturally on the surface of the larvae. Use a 10-μl Hamilton syringe to inject 5-μl aliquots into *G. mellonella* larvae via the hindmost left proleg. About 1% of control larvae injected with the MgSO$_4$ die and are excluded from the mortality assessment.

3. Monitor the mortality up to 60 hr postinfection. Infected *G. mellonella* should behave and move normally until roughly 48 hr postinfection. Thereafter, melanization quickly sets in followed by rapid death of the larvae within 2 hr. Larvae that survive beyond 60 hr postinfection usually clear the bacteria from hemolymph and show no sign of infection.

4. To determine bacterial growth in infected larvae, harvest 10 groups of 10 larvae at 8 hr intervals. Homogenize each group in 200 ml 10 mM MgSO$_4$ in a Waring blender for 1 min at maximum speed, make serial dilutions, and plate in LB agar.

The relatively large size of *G. mellonella* enables the rapid injection of defined doses of bacteria and thus makes this insect an attractive model system to conduct high-throughput screens for the identification of bacterial mutants that are attenuated in virulence. Using this approach, Jander et al.[5] found a significant correlation between virulence of *P. aeruginosa* mutants in insects and mammals.

Conclusion

The approach and methodologies described in this article provide the opportunity to bypass the inherent problems with vertebrate animal models and to perform a genome-wide screen for bacterial virulence factors required for pathogenesis in mammalian hosts. However, to take advantage of this approach, the pathogen of interest should contain virulence factors required for pathogenesis in both invertebrates and mammalian hosts. Such a large-scale screening using invertebrate hosts not only will be an efficient method for identifying unknown virulence-associated genes relevant to mammalian pathogenesis but also will provide important information regarding the evolution of pathogenesis.

[2] Identification of Host and Pathogen Factors Involved in Virulence Using *Caenorhabditis elegans*

By Man-Wah Tan

Introduction

Functional interactions between pathogen and host are crucial to the process of pathogenicity and by identifying and characterizing genes involved in host defense mechanisms and the pathogen response to these mechanisms, we can understand pathogenicity more fully. To identify the complex cascades of events that are triggered in the host and the pathogen during an infection, ideally both the host and pathogen should be genetically tractable. A *Caenorhabditis elegans–Pseudomonas aeruginosa* pathogenesis model has been developed that allows us to tap into the multifaceted power of functional genomics and genetics to systematically and comprehensively dissect both the virulence determinants of the pathogen and innate immune system of the host in a single experimental system.[1] By studying pathogenesis in a genetically tractable host such as *C. elegans,* both the animal

---

[1] M.-W. Tan and F. M. Ausubel, *Curr. Opin. Microbiol.* **3,** 29 (2000).

host and pathogen can be genetically altered and the effects of these alterations on pathogenesis/host immunity can be readily tested. Moreover, complete genome sequences are available for both *C. elegans* and *P. aeruginosa*,[2,3] thus bringing to bear numerous genomics resources and technologies to the study of host–pathogen interactions. Comparison to the completed human genome has revealed that 43% of the *C. elegans* proteins have sequence similarities to predicted human proteins,[4] suggesting that *C. elegans* may be a valid model for studies of numerous disease processes, including the innate immune response to infectious agents. The use of *C. elegans* as a model host to study host–pathogen interactions has been extended to include several other human bacterial pathogens, such as gram-negative bacteria *Salmonella enterica*,[5,6] *Serratia marcescens*,[7] *Burkholderia pseudomallei*,[8] and gram-positive bacteria *Enterococcus faecalis, Streptococcus pneumoniae,* and *Staphylococcus aureus*.[9]

In the following sections I will discuss methods used to (1) identify virulence factors and (2) identify host defense factors using *C. elegans*. I will focus primarily on the interactions between *C. elegans* and *P. aeruginosa*. Detailed protocols for killing assays using other pathogens, such as *S. enterica* and *E. faecalis,* are given in Tan and Ausubel.[10] For readers new to *C. elegans* as a model system, some of the literature and web-based resources pertinent to this discussion are listed at the end of this article.

1. Identification of Virulence Factors

C. elegans is a bacteria-feeding nematode that is found naturally in the soil, which is also the natural habitat of many human pathogens, including *P. aeruginosa*. *P. aeruginosa* kills *C. elegans* by at least three largely distinct mechanisms that are dependent on growth conditions and the genotype of the bacteria. Strain PA14, when grown in low salt medium (SKM), kills worms over a period of 2–3 days ("slow killing") by an infection-like process that correlates with the

[2] *C. elegans Consortium, Science* **282,** 2012 (1998).
[3] C. K. Stover, X. Q. Pham, A. L. Erwin, S. D. Mizoguchi, P. Warrener, M. J. Hickey, F. S. Brinkman, W. O. Hufnagle, D. J. Kowalik, M. Lagrou, R. L. Garber, L. Goltry, E. Tolentino, S. Westbrock-Wadman, Y. Yuan, L. L. Brody, S. N. Coulter, K. R. Folger, A. Kas, K. Larbig, R. Lim, K. Smith, D. Spencer, G. K. Wong, Z. Wu, and I. T. Paulsen, *Nature* **406,** 959 (2000).
[4] G. M. Rubin, *Nature* **409,** 820 (2001).
[5] A. Aballay, P. Yorgey, and F. M. Ausubel, *Curr. Biol.* **10,** 1539 (2000).
[6] A. Labrousse, S. Chauvet, C. Couillault, C. L. Kurz, and J. J. Ewbank, *Curr. Biol.* **10,** 1543 (2000).
[7] C. L. Kurz and J. J. Ewbank, *Trends Microbiol.* **8,** 142 (2000).
[8] A. L. O'Quinn, E. M. Wiegand, and J. A. Jeddeloh, *Cell. Microbiol.* **3,** 381 (2001).
[9] D. A. Garsin, C. D. Sifri, E. Mylonakis, X. Qin, K. V. Singh, B. E. Murray, S. B. Calderwood, and F. M. Ausubel, *Proc. Natl. Acad. Sci. U.S.A.* **98,** 10892 (2001).
[10] M.-W. Tan and F. M. Ausubel, *in* "Molecular Cellular Microbiology" (P. Sansonetti and A. Zychlinsky, eds.), Vol. 31, p. 461. Academic Press, London, 2002.

accumulation of bacteria in the worm gut.[11] When PA14 is grown in a high-salt medium (PGS), it kills worms within 4–24 hr ("fast-killing") by the production of low molecular weight toxin(s), including a redox-active compound pyocyanin.[12] Another strain of *P. aeruginosa*, PA01, kills rapidly by a third distinct mechanism. When PA01 is grown on brain–heart infusion (BHI) agar, worms become paralyzed within 4 hr upon contact with the bacterial lawn and the lethal effect is mediated by cyanide.[13,14] When grown on BHI, strain PA14 does not cause the level of paralysis seen with strain PA01.[13]

The majority of *P. aeruginosa* mutants isolated using *C. elegans* as host also displayed reduced virulence in a burned mouse model.[11,12,15] These studies illustrate the extensive conservation in the virulence mechanisms used by *P. aeruginosa* to infect evolutionarily diverged hosts and validate the use of a nonvertebrate host to screen for virulence determinants relevant to mammalian pathogenesis.

A. Isolation of P. aeruginosa Mutants Defective in C. elegans Killing

Generation of Bacterial Mutants. A library of *P. aeruginosa* mutants can be constructed using a transposon insertion strategy. Several Tn5-based transposons have been used successfully in generating PA14 mutants. These include Tn*phoA*[16,17] and Tn5-B30(Tc^r).[18] Other mutagens that have been used successfully to generate random mutants of PA01 include transposon IS*phoA*/hah,[19,20] mTn5-Tc,[14] and a *mariner* transposon.[21] Although transposon mutagenesis has proved to be a powerful tool in dissecting virulence, care should be taken to overcome some of the potential pitfalls. First, insertion of a transposon in an operon is often polar on a downstream gene expression, which may cause an incorrect assignment of phenotype if the transposon is inserted upstream from the actual virulence gene. One strategy to circumvent this problem is to create a nonpolar deletion of the gene. Second, it is important to determine that no secondary transposition events have occurred in the mutant that result in insertional inactivation of two or more unlinked genetic loci.

[11] M.-W. Tan, S. Mahajan-Miklos, and F. M. Ausubel, *Proc. Natl. Acad. Sci. U.S.A.* **96,** 715 (1999).

[12] S. Mahajan-Miklos, M.-W. Tan, L. G. Rahme, and F. M. Ausubel, *Cell* **96,** 47 (1999).

[13] C. Darby, C. L. Cosma, J. H. Thomas, and C. Manoil, *Proc. Natl. Acad. Sci. U.S.A.* **96,** 15202 (1999).

[14] L. A. Gallagher and C. Manoil, *J. Bacteriol.* **183,** 6207 (2001).

[15] M.-W. Tan, L. G. Rahme, J. A. Sternberg, R. G. Tompkins, and F. M. Ausubel, *Proc. Natl. Acad. Sci. U.S.A.* **96,** 2408 (1999).

[16] M. R. Kaufman and R. K. Taylor, *Methods Enzymol.* **235,** 426 (1994).

[17] L. G. Rahme, M.-W. Tan, L. Le, S. M. Wong, R. G. Thompkins, S. B. Calderwood, and F. M. Ausubel, *Proc. Natl. Acad. Sci. U.S.A.* **94,** 13245 (1997).

[18] G. A. O'Toole and R. Kolter, *Mol. Microbiol.* **28,** 449 (1998).

[19] C. Manoil, *Methods Enzymol.* **326,** 35 (2000).

[20] D. A. D'Argenio, L. A. Gallagher, C. A. Berg, and C. Manoil, *J. Bacteriol.* **183,** 1466 (2001).

[21] S. M. Wong and J. J. Mekalanos, *Proc. Natl. Acad. Sci. U.S.A.* **97,** 10191 (2000).

Identification of Bacterial Mutants Defective in "Slow Killing."[15] Inoculate individual PA14::Tn*phoA* clones into 200 μl King's B medium in microtiter plates containing rifampicin and neomycin at 100 and 200 μg/ml, respectively. Spread 10 μl of the overnight culture on slow killing medium (SKM: 3 g NaCl, 3.5 g Bacto-peptone, 1 ml 5 mg/ml cholesterol in ethanol, 17 g Bacto-agar, 975 ml H_2O; add sterile 1 ml 1 M CaCl$_2$, 1 ml 1 M MgSO$_4$, and 25 ml 1 M KH$_2$PO$_4$ at pH 6 after autoclaving) in 5.5-cm petri plates and incubate at 37° for at least 24 hr. With shorter incubation period, we often observe significant reduction in bacterial virulence. After 8–24 hr at room temperature (23–25°), seed each plate with two L4-stage hermaphrodite *C. elegans* strain N2 (Bristol). Incubate seeded plates at 25° and examine for live worms after 5 days. On plates seeded with a nonpathogenic mutant, thousands of progeny worms can be seen at day 5 and the bacterial lawn is completely consumed, whereas very few or no live worms are found on plates seeded with the wild-type strain and the bacterial lawn remains intact. Putative nonpathogenic or attenuated mutants identified in the preliminary screen can be retested and subjected to a virulence assay to determine the kinetics of *C. elegans* killing.

Identification of Bacterial Mutants Defective in "Fast Killing."[12] Spread 5 μl of individual transposant strain grown overnight in King's B medium on 3.5-cm plates containing PGS medium (1% w/v Bacto-peptone, 1% w/v NaCl, 1% w/v glucose, 0.15 M sorbital, 1.7% w/v Bacto-agar). Incubate plates at 37° for 24 hr and then place at room temperature for 8–12 hr. Transfer 5 L4 worms to each plate. Putative bacterial mutants that are defective in fast killing are identified as those that have three or more surviving worms after 24 hr. These mutants can be subjected to the fast killing assay and those that consistently gave a lower rate of killing relative to the parental PA14 can be further characterized.

Identification of Bacterial Mutants Defective in Causing Worm Paralysis.[14] Grow individual PA01 transposon insertion mutants in BHI broth until the broth is visibly turbid. Spread 150 μl of each suspension onto a 3.5-cm-diameter BHI agar plate. After incubating the plates for 24 hr at 37°, spot 50-μl aliquot (containing 20–200 adult nematodes) from a suspension of N2 nematodes in M9 buffer onto the PA01 lawn. Incubate the seeded plates with the lid on for 4 hr at room temperature. Keep and retest bacterial strains that exhibited a minimum of 10% reduction in killing relative to the wild type.

B. *P. aeruginosa Pathogenesis Assays*

Slow Killing Assay

1. Seed 3.5-mm-diameter SKM agar plates with 10 μl of an overnight bacterial culture in King's B or LB medium with appropriate antibiotics. When spreading the bacteria, care should be taken not to break the agar surface of the plate as this will result in worms burrowing into the agar and present a significant problem in scoring that plate for worm mortality.

2. Incubate the plates at 37° for at least 24 hr. Equilibrate plates to ambient temperature (23–25°) for several hours prior to adding 20–30 worms at a specific developmental stage. Use 1-day-old hermaphrodites or the final stage larvae (L4). For statistical analysis, have 3–4 replicates per trial. Use *Escherichia coli* OP50 as a negative control.

3. Incubate plates at 25° and score for the number of dead worms every 4–6 hr (after the initial 24 hr) until all the worms exposed to the wild-type pathogen are dead. A worm is considered dead if it does not respond to touch. Exclude from the analysis worms that die as a result of getting stuck to the wall of petri plates; this can be minimized by ensuring that there is no condensate on the walls prior to the addition of worms.

4. The LT_{50} (time required to kill 50% of the nematodes) can be determined as follows: For each replicate, a log-logistic model can be fitted to the data with the aid of a computer program such as SYSTAT, using the equation

$$P_i = A + \frac{(1 - A)}{1 + e^{B - (G \times \ln T_i)}}$$

where P_i is the proportion of worms killed at each time T_i, A is the fraction of worms that died in a OP50 control experiment, and B and G are curve fitting parameters which are chosen for optimal fit of the curve to the data points. Once B and G have been determined, LT_{50} can be calculated by the formula

$$LT_{50} = e^{(B/G)} \times (1 - 2A)^{(1/G)}$$

Note: There are several host-associated factors that affect the susceptibility of nematodes to bacteria-mediated killing. Gravid adult hermaphrodites are more susceptible than adult males because of embryos hatching from within the gravid adults.[11] To eliminate mortality caused by internal hatching, use either *C. elegans* males or temperature-sensitive mutants, such as *fer-1 (hc1)* or *glp-4 (bn2),* which are sterile at 25°. The *C. elegans* strains can be obtained from the *Caenorhabditis* Genetics Center (CGC) at the University of Minnesota.

Fast Killing Assay

1. Inoculate 5 ml of King's B or LB medium with single colony of *P. aeruginosa* PA14 and grow at 37° overnight.

2. Spread the center of a 3.5-cm-diameter plate containing 5 ml PGS agar with 5 μl of the overnight culture. Incubate plates for 24 hr at 37°.

3. After equilibrating the plates to room temperature, add 30 L4-stage hermaphrodite *C. elegans* to each plate and incubate at 25°.

4. Assay for worm mortality at 4-hr intervals.

Note: For this assay, it is important that fresh PGS plates (1–7 days after pouring and kept at 4°) be used. Worms used for this assay must be cultivated on pure culture of OP50. Occasionally, OP50 strain may adopt a "slimy" or mucoid appearance. Do not use worms grown on slimy bacteria as they give inconsistent results.

2. Identification of Host Factors

The ability to detect an invading pathogen and activate an instantaneous defense is crucial for survival. This first line of defense is performed by the innate immune system, which acts effectively without previous exposure to a pathogen and confers broad protection. The innate immune system also controls and assists the acquired immune system, which takes days or weeks to develop maximum efficacy. Many human diseases result from a failure of the innate immune system. Yet, the precise molecular mechanisms of how an invader is recognized, what the pathogen-specific signaling pathways are, how these pathways are triggered, and what pathogen-specific response are activated remain to be elucidated. In order to fully understand the innate immunity, we need to know the various components that make up the innate immune system and their respective functions. It has not been feasible to systematically dissect components of the mammalian innate immune response because mammalian models are not genetically facile. Compelling evidence has been presented that the host innate immune responses to pathogen are conserved from invertebrates to humans at the molecular level.[22] This suggests that genetically tractable model invertebrates can be used to identify components of the innate immune system, many of which will also be conserved in mammals.

C. elegans is an attractive model to study host defense response because the unprecedented combined power of genetic and functional genomic approaches can be applied to comprehensively dissect the innate immune system of a eukaryote host. In addition to its ease of genetic analysis, extensive molecular genetic toolbox, and fully sequenced genome, developments in rapid mapping, genome-wide expression using high-density arrays, and inhibition of gene function by RNA interference (RNAi) have made the genome-wide identification of host defense factors very accessible.[23] In the following, I will discuss several genetic and functional genomics approaches that can be used to identify host defense factors: (1) genetic screens, (2) genome-wide gene expression, and (3) gene inactivation by double-stranded RNA-mediated interference (RNAi).

A. Genetic Screens

A direct approach to identify host factors that are involved in defense response to pathogens is by screening for *C. elegans* mutants that are either more resistant or more susceptible to pathogen attack. This screen does not rest on any assumptions

[22] D. A. Kimbrell and B. Beutler, *Nat. Rev. Genet.* **2**, 256 (2001).
[23] P. W. Sternberg, *Cell* **105**, 173 (2001).

about which signaling pathway or downstream components are important for defense against pathogens, thus allowing the identification of novel innate immune mechanisms. Darby and colleagues screened F_2 progeny of ethyl methane sulfonate (EMS) mutagenized *C. elegans* for resistance to *P. fluorescens* SE59-mediated paralysis. They identified two alleles of *egl-9* in the screen and showed that *egl-9* mutants are also resistant paralysis induced by *P. aeruginosa* PA01.[13] It was later shown that paralysis was caused by cyanide.[14] Using sequence profile analysis, Aravind and Koonin[24] showed that EGL-9 belongs to a conserved family of enzymes of the 2-oxoglutarate and Fe(II)-dependent dioxygenase superfamily and predicted that the EGL-9 family proteins are prolyl hydroxylases that modify intracellular proteins. This prediction was proved correct when Epstein and colleagues showed biochemically that EGL-9 is indeed a 2-oxoglutarate-dependent dioxygenase that regulates the hypoxia inducible factor (HIF) by prolyl hydroxylation.[25] Interestingly, an *egl-9* homolog, PA4515, is present in *P. aeruginosa* strain PA01.[24] What is the connection between PA4515 and the action of the *P. aeruginosa* toxin? Could the bacterial EGL-9-like proteins modify host proteins in a manner that favors the survival and spread of the pathogen? Further genetic analyses on both the host and pathogen, such as identification of genetic suppressors of the *egl-9*-mediated resistance in *C. elegans* together with functional characterization of PA4514 and the toxin(s) of *P. aeruginosa,* will likely provide insights into the role EGL-9 proteins in *P. aeruginosa* pathogenesis.

The analysis of EGL-9 is likely to reveal the specific interactions between pathogen-derived toxin(s) and the host protein(s). To identify defense pathway(s) in the host in response to pathogen, we have carried out genetic screens to identify mutant worms that are either more susceptible or resistant to *P. aeruginosa*-mediated slow killing. In characterizing mutants that showed increased susceptibility it is important to note some of the nonspecific factors that contribute to the susceptibility phenotype. *C. elegans* mutants with defects in feeding and defecation are also more susceptible to *P. aeruginosa* slow killing. *C. elegans* is a filter feeder, taking in liquid with bacteria and then spitting out the liquid while retaining and grinding up the bacteria when they reach the terminal bulb.[26] *C. elegans* mutants defective in grinding, such as *eat-13* and *phm-2,* receive a higher bacterial inoculum than wild type because they allow the entry of more intact pathogens into the intestines. Consequently these mutants are more sensitive to pathogen-mediated killing (M.-W. Tan, G. Alloing, and F. M. Ausubel, unpublished data, 1999). *C. elegans* defecates at regular intervals by a series of sequential muscle contractions. A defect in any of these steps leads to failure to remove intestinal contents at regular intervals and

[24] L. Aravind and E. V. Koonin, *Genome Biol.* **2,** 0007.0001 (2001).

[25] A. C. Epstein, J. M. Gleadle, L. A. McNeill, K. S. Hewitson, J. O'Rourke, D. R. Mole, M. Mukherji, E. Metzen, M. I. Wilson, A. Dhanda, Y. M. Tian, N. Masson, D. L. Hamilton, P. Jaakkola, R. Barstead, J. Hodgkin, P. H. Maxwell, C. W. Pugh, C. J. Schofield, and P. J. Ratcliffe, *Cell* **107,** 43 (2001).

[26] L. Avery, *Genetics* **133,** 897 (1993).

causes a "constipated" phenotype. Constipated mutants, such as *aex-2* and *unc-25*, which retain pathogens in the intestines longer are also more sensitive than wild-type worms to pathogen-mediated killing (M.-W. Tan, unpublished data, 1998). Mutants displaying these phenotypes would also be isolated from increased susceptibility screens and should be discarded from further analyses. Analysis of EMS-induced worm mutants that showed enhanced susceptibility to *P. aeruginosa* PA14, but without grinding or defecation defects, revealed that *C. elegans* uses the Sma pathway, a homolog of the mammalian transforming growth factor β (TGFβ) pathway, to transduce signals required for antibacterial defense (M.-W. Tan, in preparation).

C. elegans Mutagenesis. EMS is the favored mutagen used in generating mutations in *C. elegans* because of its low toxicity and its ability to induce mutations at high frequencies. It is important to note that approximately 92% of EMS-induced base pair changes are G/C to A/T transitions, which favor the production of A/T-rich stop codon mutations but very rarely mutate amino acids Asp (AAY), Ile (AUY, AUA), Lys (AAR), Phe (UUY), and Tyr (UAY) to other residues.[27] For genetic studies in which a variety of amino acid changes are desired, such as suppressor screens, *N*-ethyl-*N*-nitrosourea (ENU) is a superior mutagen because it does not produce the EMS-induced bias. The following EMS mutagenesis protocol is adapted from Hodgkin.[28]

1. Remove worms grown on a 9-cm plate by washing with 5 ml M9 buffer (3 g KH_2PO_4, 6 g Na_2HPO_4, 5 g NaCl, 1 ml 1 M $MgSO_4$, H_2O to 1 liter). Ensure that the majority of the worms are at the L4 stage.

2. Wash the worms by successive steps of centrifugation (500g, 30 sec) and resuspension, the first time in 10 ml M9 buffer, the second time in 1 ml M9 buffer.

3. In a fume hood, mix thoroughly 20 μl EMS in 3 ml of M9 buffer in a 2-cm-diameter glass test-tube. Glassware is recommended over plasticware because worms tend to stick to plastic, resulting in injury to the worm and increased susceptibility to direct toxicity of EMS.

4. Add the worm suspension from step 2 to the EMS solution, making the final concentration of EMS to 47 mM. Mix the contents by swirling and incubate at room temperature for 4 hr. Worms tend to settle at the bottom of the tube. To prevent the worms from becoming anoxic, swirl the tube every 30 min.

5. After 4 hr, remove as much supernatant as possible. Transfer the supernatant to an access volume of 2 M NaOH to inactivate the mutagen. Wash worms three times, each time with 10 ml M9 buffer, and transfer the supernatant to excess 2 M NaOH.

---

[27] P. Anderson, *in "Caenorhabditis elegans:* Modern Biological Analysis of an Organism" (H. F. Epstein and D. C. Shakes, eds.), Vol. 48, p. 31. Academic Press, San Diego, 1995.

[28] J. Hodgkin, *in* "*C. elegans:* A Practical Approach" (I. A. Hope, ed.), p. 245. Oxford University Press, New York, 1999.

6. Transfer mutagenized worms, suspended in 0.2 ml or less M9 buffer, to a 9-cm plate containing a NGM/OP50 lawn. The recipe for NGM is the same as that for SKM, except that NGM contains only 2.5 g Bacto-peptone.

7. After the plate has dried, transfer actively moving L4 worms to a fresh NGM/OP50 plate and allow them to recover and mature into young adults for about 24 hr.

8. From this population, choose healthy adults, which form the parental generation (P_0) for the mutant screen. The nature of the screen will dictate the subsequent manipulations beyond this step.

For example, to isolate mutants that show enhanced susceptibility to *P. aeruginosa,* transfer 45 P_0 gravid adults per plate into 20 9-cm NGM/OP50 plates. Move the P_0 adults to new plates every 4 hr in 3 successions, yielding approximately 1000 F_1 eggs per plate. When the F_1 develop into gravid adults, keep these worms as separate pools and isolate F_2 eggs from them by bleach treatment.[29] Hatch each pool of F_2 worms in M9 buffer in the absence of food. Under this condition, all worms would be synchronized at the early L1 stage. Transfer the synchronized F_2 worms to NGM/OP50 plates. When the F_2 population consists predominantly of young adult worms, wash the animals off each plate and transfer them, as separate pools, to SKM plates containing *P. aeruginosa* and incubate the plates at 25°. It is necessary to synchronize the F_2 population prior to exposure to pathogen because the time course of *C. elegans* killing by *P. aeruginosa* is stage dependent; young adults die about 18 hr faster than L4 larvae.[11] When wild-type young adult worms are exposed to *P. aeruginosa* grown under the above conditions, the worms begin to die after 30 hr.[11] To isolate mutants that are more susceptible to *P. aeruginosa*-mediated killing, pick dead worms after 16–24 hr of exposure and transfer them individually to NGM/OP50 plates in order to recover their progeny. For this step, use an *E. coli* OP50 strain that is resistant to 300 μg/ml streptomycin to select against *P. aeruginosa.* Mutants can be recovered because eggs are not infected; thus the surviving progeny from the dead worm will form the mutant strain. This feature obviates the need to perform clonal screens.

Although ENU mutagenesis produces a larger diversity of single nucleotide changes, it has not been the mutagen of choice because it is more toxic to *C. elegans* than EMS. De Stasio and S. Dorman have described a method that essentially is the same as the EMS mutagenesis protocol described above, except for the following modifications, which minimize the ENU toxicity while maintaining the mutagenic efficacy.[30] Prepare ENU stock solutions in fresh 100% ethanol instead of M9 buffer. This stock solution may be kept at $-20°$ and used within 2 weeks with little increase

---

[29] T. Stiernagle, *in* "*C. elegans:* A Practical Approach" (I. A. Hope, ed.), p. 51. Oxford University Press, New York, 1999.

[30] E. A. De Stasio and S. Dorman, *Mutat. Res.* **495,** 81 (2001).

in toxicity and change in mutagenicity. Mutagenize worms in 4 ml of 0.3–1.0 mM ENU in M9 buffer for 4 hr with agitation. The ENU solution should be inactivated after use in a final concentration of 10% $Na_2S_2O_3$ and 1% NaOH left at room temperature for 1 hr.

Mapping of Chemical-Induced Mutants. Chemical-induced mutants identified from the screens can be genetically mapped by means of molecular or visible markers. The use of molecular markers is preferred over morphological markers, such as *dpy, sma,* and *lon,* because some of the *sma* and *lon* mutants are not wild type with regard to susceptibility to pathogens (M.-W. Tan, unpublished, 2000). Wicks and colleagues have described a rapid method to map mutations by using the bulked segregant analysis (BSA) of molecular markers snip-SNPs.[31] Snip-SNPs are single-nucleotide polymorphisms (SNPs) that contain modification of a restriction enzyme recognition site and can be visualized as restriction fragment length polymorphisms (RELPs) after digestion of a PCR product with the appropriate restriction enzyme.[31] This technique takes advantage of the high density of "snip-SNP" polymorphisms between the CB4856 strain and N2 (the Bristol strain background in which the mutants were isolated). In order for this method to work, it is important to ensure that the CB4856 strain phenocopies the N2 strain for susceptibility to the pathogen used in the screen; CB4856 phenocopies N2's susceptibility to PA14-mediated slow killing.

Bulked segregant analysis involves comparing two pooled lysates of individuals from a segregating F_2 population originating from the same cross. To map a recessive mutation that showed enhanced susceptibility to pathogens (*esp*), for example, perform the following.

1. Cross L4 *esp* hermaphrodites with 5- to 6-fold more CB4856 males. This is to ensure complete outcrossing, which would result in cross-progeny that are heterozygous for the recessive *esp* mutation and for all the snip-SNP markers present in CB4856. The male frequency of about 50% in the F_1 generation is indicative of successful mating. Pick 6–8 F_1 cross-progeny to individual plates.

2. An F_1 animal that is heterozygous for the *esp* mutation should also be heterozygous for any snip-SNP markers tested. Thus to ensure that the animal is a heterozygous cross-progeny, perform single-worm PCR for an arbitrary snip-SNP after the worm has been allowed to produce progeny for 36 hr. Transfer a single worm into 2.5 μl lysis buffer [50 mM KCl, 10 mM Tris-HCI pH 8.3, 2.5 mM $MgCl_2$, 0.45% noxidet P-40 (NP-40), 0.45% Tween 20, 0.01% gelatin, freshly added 60 mg/ml proteinase K] in a thin-wall PCR tube and freeze the tube at $-80°$ for at least 1 hr. Lyse the worm to release genomic DNA by heating the tube at $60°$ for 60 min, followed by $95°$ for 15 min to inactivate proteinase K.

[31] S. R. Wicks, R. T. Yeh, W. R. Gish, R. H. Waterston, and R. H. A. P. Plasterk, *Nat. Genet.* **28,** 160 (2001).

3. Perform PCR on the lysate by adding 22.5 μl of a PCR mix [2.5 μl 10 ×
PCR buffer (100 mM Tris-HCl pH 8.3, 500 mM KCl, 15 mM MgCl$_2$, 200 μg/ml
gelatin), 100 mM each dNTPs, 0.5 μM each primer, 2 μl 30% sucrose, 1 μl
0.1% cresol red, 0.5–1.0 U *Taq* polymerase, and water to 22.5 μl] and run a PCR
reaction for 30–35 cycles. Digest the amplified product in the PCR mix by adding a
restriction cocktail (3 μl restriction enzyme buffer, 1 μl 30% sucrose, 0.5 μl 0.1%
cresol red, restriction enzyme, and water, with bovine serum albumin if required,
to 10 μl) directly to the PCR reaction tube. Incubate this mix for at least 2 hr at
the appropriate temperature. Because the buffer is not ideal, use at least 5 U of
restriction enzyme in each digest. With cresol red and sucrose in the reaction mix,
no further handling is required prior to loading the 2% agarose gel. The worm
should be heterozygous for any snip-SNP markers tested.

4. Once the identity of the F_1 cross-progeny is confirmed, allow the heterozygous hermaphrodites to self. It should produce F_2 progeny with a Mendelian proportion of three phenotypically wild type and 1 mutant (Esp).

5. The F_2 worms can only be accurately phenotyped by testing the susceptibility of the F_3 population to the pathogen. Therefore, each F_2 worm has to be singled
to an individual plate and allowed to produce F_3 progeny. After 36 hr, transfer each
F_2 parent into 2.5 μl lysis buffer and perform single worm lysis as described in
step 2.

6. Once the phenotype of the F_2 has been determined by performing killing
assays on each of the F_3 populations, pool 0.5 μl of each F_2 Esp mutant worm
lysate into an Eppendorf tube. Similarly, pool 0.5 μl of each wild-type worm lysate
into a *separate* Eppendorf tube. Each phenotype should consist of about 75 worms.

7. Dilute the remaining single-worm lysates to a total of 15 μl and freeze for
later use.

8. To establish linkage to a chromosome arm, use the 18 recommended snip-
SNPs (from the two arms and the center of each of the six chromosomes in
C. elegans[31]). For each snip-SNP, amplify DNA from 1 μl of each bulked lysate
using the corresponding primer pairs, digest the PCR products, load the products
side by side, and run them in parallel on 2% agarose gels.

9. For each snip-SNP, calculate R_{WT} (the ratio of the density of a CB4856-
specific band and a N2-specific band) in the lane from the wild-type lysate.
Similarly, calculate R_{Esp} (the ratio of the density of a CB4856-specific band and
an N2-specific band in the lane) from the Esp lysate. Determine the densities of
bands using EagleSight software (Stratagene, La Jolla, CA) or NIH Image with a
gel quantification module. Calculate the Map Ratio by taking the ratio of R_{Esp}/R_{WT}.
This method of determining Map Ratio provides an internal control for alteration
in the efficiency of elongation of the PCR product produced by the polymorphism
and for incomplete restriction.[31] The Map Ratio measures how linked the bial-
lelic marker is to the *esp* mutation. If a snip-SNP marker is linked to the *esp*
mutation, the CB4856-specific marker will be absent or present at low frequency

(because of rare recombination events) from the Esp-derived lysate and enriched in the wild-type-derived lysate. Thus, linked markers will have a Map Ratio that is significantly less than 1.

Once clear linkage to a chromosome has been established, further refinement of the map position can be accomplished by using additional 10–20 snip-SNPs flanking the *esp* mutation on the target chromosome. Additional confirmed snip-SNPs can be found at genomics.niob.knaw.nl/protocols/snipSNP. Using this approach, mutations can be mapped to 1-5cM resolutions using a single cross between a mutant and CB4856 and about 50–80 PCR reactions. For subsequent fine-scale snip-SNP mapping, select recombinants within the interval defined by the BSA. Once the mutation is localized to several cosmids, generate transgenic animals expressing those cosmid(s)[32] and assay these animals for rescue of the mutant phenotype.

B. Genome-Wide Gene Expression

Infection of a host by a pathogen triggers complex cascades of events that ultimately determine the outcome of the interaction. Transcription control of genes has been shown to play a key role in host–pathogen interaction.[33,34] The genome of *C. elegans* was completely sequenced in 1998 and predicted to contain 19,099 genes.[2] This has paved the way for the use of high-density arrays, such as oligonucleotide arrays containing 18,791 predicted worm genes and DNA microarray containing 17,871 predicted genes, to monitor genome-wide gene expressions during development.[35,36] As the microarray technology becomes more readily available, the use of this quantitative method for genome-wide and simultaneous analysis of expression profiles can be used to monitor gene expression in *C. elegans* when it is interacting with bacterial pathogens. Synchronized populations of *C. elegans* can be exposed to pathogen and RNA can be harvested from these worms after a specified time of exposure to the pathogen. The true power of microarray analysis comes not from analysis of single experiments, but rather from the analysis of many hybridizations to reveal common patterns of gene expression. This genome-wide transcriptional profile can be sorted and clustered into groups using a self-organizing map[37] and a hierarchical clustering algorithm[38] to

[32] C. C. Mello, J. M. Kramer, D. Stinchcomb, and V. Ambros, *EMBO J.* **10**, 3959 (1991).

[33] P. A. Cotter and J. F. Miller, *Curr. Opin. Microbiol.* **1**, 17 (1998).

[34] C. Svanborg, G. Godaly, and M. Hedlund, *Curr. Opin. Microbiol.* **2**, 99 (1999).

[35] A. A. Hill, C. P. Hunter, B. T. Tsung, G. Tucker-Kellogg, and E. L. Brown, *Science* **290**, 809 (2000).

[36] M. Jiang, J. Ryu, M. Kiraly, K. Duke, V. Reinke, and S. K. Kim, *Proc. Natl. Acad. Sci. U.S.A.* **98**, 218 (2001).

[37] P. Tamayo, D. Slonim, J. Mesirov, Q. Zhu, S. Kitareewan, E. Dmitrovsky, E. S. Lander, and T. R. Golub, *Proc. Natl. Acad. Sci. U.S.A.* **96**, 2907 (1999).

[38] M. B. Eisen, P. T. Spellman, P. O. Brown, and D. Botstein, *Proc. Natl. Acad. Sci. U.S.A.* **95**, 14863 (1998).

identify genes that show similar patterns of expression. The availability of on-line databases, such as WormBase,[39] WormPD,[40] and Intronerator,[41] in which every *C. elegans* gene is extensively annotated with sequence, RNA splicing pattern, data from published literature, and genetic information, would facilitate the analyses the extensive data generated from the microarray experiments.

C. Gene Inactivation by Double-Stranded RNA-Mediated Interference (RNAi)

The microarray technology is an efficient tool to reveal the identity of genes that have altered expression during infection. Typically hundreds or even thousands of genes will show altered expression in microarray experiments. To determine the function of these genes for defense against pathogen, an efficient, economical, and logistically feasible technology to inactivate each gene is essential. Mutants generated by this reverse genetics approach can then be tested in a relevant pathogenesis assay; this targeted gene inactivation approach therefore directly connects mutant phenotypes with known genes.

A powerful targeted-gene inactivation method called double-stranded (ds) RNA-mediated interference (RNAi) has been established in *C. elegans*. Exposing *C. elegans* to dsRNA causes a reduction in the level of mRNA for the corresponding endogenous gene, and thus suppresses the expression of the gene and, in most cases, phenocopies the loss-of-function phenotype due to chromosomal mutation for that gene.[42]

RNAi can be achieved by exposing worms to dsRNAs in any of the following ways: (1) injection of dsRNA, (2) ingestion of bacterially expressed dsRNAs, and (3) soaking in dsRNA. For the injection strategy, dsRNAs are injected into the syncytial gonad of an adult hermaphrodite. dsRNA can also be effectively delivered simply by soaking worms in a solution containing dsRNA and spermidine.[43,44] In the following, I describe the ingestion method, which we found to be the most convenient way to produce a large number of RNAi-induced mutant worms for pathogenesis assays. First discovered by Timmons and Fire,[45] this method has subsequently been modified to increase the effectiveness of RNAi.[46,47] The ingestion method is also favored over injection because it has been shown to be more effective

[39] L. Stein, P. Sternberg, R. Durbin, J. Thierry-Mieg, and J. Spieth, *Nucleic Acids Res.* **29**, 82 (2001).
[40] M. C. Costanzo, M. E. Crawford, J. E. Hirschman, J. E. Kranz, P. Olsen, L. S. Robertson, M. S. Skrzypek, B. R. Braun, K. L. Hopkins, P. Kondu, C. Lengieza, J. E. Lew-Smith, M. Tillberg, and J. I. Garrels, *Nucleic Acids Res.* **29**, 75 (2001).
[41] W. J. Kent and A. M. Zahler, *Nucleic Acids Res.* **28**, 91 (2000).
[42] A. Fire, *Trends Genet.* **15**, 358 (1999).
[43] H. Tabara, A. Grishok, and C. C. Mello, *Science* **282**, 430 (1998).
[44] I. Maeda, Y. Kohara, M. Yamamoto, and A. Sugimoto, *Curr. Biol.* **11**, 171 (2001).
[45] L. Timmons and A. Fire, *Nature* **395**, 854 (1998).
[46] R. S. Kamath, M. Martinez-Campos, P. Zipperlen, A. G. Fraser, and J. Ahringer, *Genome Biol.* **2**, 2.1 (2001).
[47] L. Timmons, D. L. Court, and A. Fire, *Gene* **263**, 103 (2001).

in older larvae and adults,[47] the worm stages typically used in pathogenesis assays. The following protocol is adapted from Kamath and colleagues.[46]

1. Obtain fragments designated for RNAi by PCR from genomic DNA or cDNA clones. If genomic DNA is used, choose a fragment that corresponds to several large exons in the target gene. Alternatively, for each of the predicted genes in the *C. elegans* genome, primer pair sequences, and conditions for PCR are available from WormBase, www.wormbase.org, and can be used to generate the desired PCR fragment. Clone the PCR fragment into a feeding vector L4440 (pPD129.36[45]). This vector contains two T7 polymerase promoters in opposite orientation separated by a multicloning site and the β-lactamase gene.

2. Transform RNAi plasmid into *E. coli* HT115(DE3).[47] Select transformants on LB plates containing 50 μg/ml ampicillin. HT115(DE3) is an RNase III (dsRNA-specific endonuclease)-deficient strain that contains a λDE3 lysogen, which contains an isopropylthiogalactoside (IPTG)-inducible T7 polymerase gene. The RNase III gene is interrupted by a Tn*10* transposon that confers tetracycline resistance to this strain. The *E. coli* HT115(DE3) strain can be obtained from CGC.

3. To induce the expression of dsRNA, grow bacterial strain in LB broth containing 50 μg/ml ampicillin for 6–18 hr. (Do not grow longer than 18 hr; cultures grown for shorter times sometimes give better results.) Seed bacteria on NGM plates that contain 1 mM IPTG, 50 μg/ml ampicillin. Do not include tetracycline in the NGM plates as this has been shown to significantly decrease the RNAi effects of several genes tested. Allow seeded plates to dry and induction to continue overnight at room temperature.

4. Transfer 10 gravid adult hermaphrodite N2 worms to seeded plates for egg-laying for 4–6 hr at 20°. To minimize the amount of OP50 bacteria transferred, wash worms in M9 buffer before transfer. Remove the adults after the egg-laying period and allow the progeny to develop at 15° and 22°; both temperatures are recommended because some genes give different phenotypes at 15° and 22° (J. Ahringer, personal communication, 2001). This will provide sufficient worms at the L4 or young adult stages to be tested in the killing assays described above. As a control, compare RNAi-induced mutant worms to N2 worms treated in the same manner except that they are grown on the bacteria strain that carries only the vector without insert.

There are weaknesses associated with RNAi. RNAi does not always phenocopy chromosomal mutation; only 50% of genes with known phenotypes are detectable by RNAi.[48] This is in part because certain tissues, such as neuronal cells, are refractory to RNAi.[49] Therefore failure to detect a phenotype does not indicate

[48] C. I. Bargmann, *Genome Biol.* **2** (2001).
[49] N. Tavernarakis, S. L. Wang, M. Dorovkov, A. Ryazanov, and M. Driscoll, *Nat. Genet.* **24**, 180 (2000).

that the target gene does not play a role in the phenotype in question. Because of these limitations, RNAi methods can augment, but not replace, the standard loss of function analysis due to physical genetic lesion. Nonetheless, these methods are very useful in identifying a smaller subset of genes identified by the genome-wide gene expression approach for further detailed studies.

An added advantage of using *C. elegans* is that it is relatively easy to genetically manipulate the host to test various hypotheses regarding the mechanism of actions of a particular virulence factor. Furthermore, many *C. elegans* mutants generated from the same N2 background are already available and can be obtained from the CGC, which currently maintains more than 3000 *C. elegans* genetic stocks. Where mutation of a specific host gene of interest is not available, one can generate with relative ease gene-specific loss-of-function or hypomorphic mutants using the RNAi technique described above. These mutants can then be tested for susceptibility or resistance against various bacterial mutants to further define the interactions between host- and pathogen-derived factors. For example, a set of existing *C. elegans* mutants with altered susceptibility to PA14 fast killing was used to show that phenazines, including the redox-active compound pyocyanin secreted by *P. aeruginosa,* mediate their toxic effects through generation of reactive oxygen species. The worm mutants *mev-1* and *rad-8* that are sensitive to oxidative stress are sensitive to *P. aeruginosa.* Conversely, the *C. elegans age-1* mutant that is more resistant to oxidative stress is also more resistant to *P. aeruginosa.*[12]

Hodgkin and colleagues have also taken advantage of the availability of a large number of known *C. elegans* mutants to identify host factors involved in infection by a new species of gram-positive bacterium, *Microbacterium nematophilum*. This coryneform bacterium infects the postanal and rectal cuticle of *C. elegans* and causes swelling of the underlying hypodermal tissue at the site of infection.[50] Although the infection does not result in worm mortality, it delays the growth of the infected worms by 20% or more. They screened 200 characterized mutants of *C. elegans* for enhanced susceptibility or resistance to infection. From this screen, worms with mutation in *srf-2, srf-3,* and *srf-5* did not display a swelling phenotype and grew at the normal rate when exposed to *M. nematophilum,* indicating that they are resistant to infection. *srf* mutants display altered surface antigenicity,[51] leading Hodgkin and colleagues to suggest that the change in cuticle surface properties of these mutants may alter the adherence or recognition by the pathogen, thereby preventing infection.

The *C. elegans*–pathogen model opens up the possibilities of exploiting the power of genetics and functional genomics to examine the interplay that occurs between a pathogen and its host by manipulating host factors involved in disease

[50] J. Hodgkin, P. E. Kuwabara, and B. Corneliussen, *Curr. Biol.* **10,** 1615 (2000).

[51] S. M. Politz, M. Philipp, M. Estevez, P. J. O'Brien, and K. J. Chin, *Proc. Natl. Acad. Sci. U.S.A.* **87,** 2901 (1990).

resistance, as well as to identify and study bacterial virulence factors. The conservation of many signaling pathways, particularly in innate immunity, suggests that many insights gained from the use of this invertebrate model will have general significance for host–pathogen interactions in vertebrate organisms, including humans.

Supplementary Materials

Additional Reading

The following two volumes contain detailed protocols for standard techniques used in *C. elegans* research. The first volume is a good introductory guide and very useful for researchers currently working outside the field.

1. I. A. Hope (ed.), "*C. elegans:* A Practical Approach." Oxford University Press, New York, 1999.
2. H. F. Epstein and D. C. Shakes, "*Caenorhabditis elegans:* Modern Biological Analysis of an Organism." Academic Press, San Diego, 1995.

The next two volumes contain vital background information about most aspects of *C. elegans* biology.

1. W. Wood (ed.), "The Nematode *Caenorhabditis elegans*." Cold Spring Harbor Laboratory Press, Cold Spring Harbor, N.Y., 1988.
2. D. L. Riddle, T. Blumenthal, B. J. Meyer, and J. R. Priess (eds.), "*C. elegans* II." Cold Spring Harbor Laboratory Press, Cold Spring Harbor, NY, 1977.

Web Resources

1. The *C. elegans* WWW server, elegans.swmed.edu. This comprehensive site has many useful features, including *C. elegans* specific literature search, listings of nematode labs, and links to the *C. elegans* Genetics Center (CGC) that curates and distributes *C. elegans* strains and to common *C. elegans* protocols.
2. WormBase,[39] www.wormbase.org, is the most comprehensive web-based resource on *C. elegans* genome and biology. It is updated every 2 weeks and contains the most current mapping, sequencing, and phenotypic information on *C. elegans*.
3. WormPD,[40] www.proteome.com/databases.

Acknowledgment

Supported by the Donald E. and Delia B. Baxter Foundation Scholar Award.

[3] Goldfish as an Animal Model System for Mycobacterial Infection

By KRISTIN M. RULEY, RENATE REIMSCHUESSEL, and MICHELE TRUCKSIS

Introduction

Tuberculosis (TB) has plagued the human race for centuries, earning such names as "the Consumption" and "White Death." Currently, one-third of the world's population is afflicted with this disease, and 3 million people die annually of TB.[1]

Mycobacterium tuberculosis, the causative agent of tuberculosis, was first isolated in 1882 by Robert Koch. Several factors have enabled this pathogen to survive in the human population for centuries. Tuberculosis is spread from an infected person to a new host through infected droplets made airborne by coughing and talking. In addition, persons infected with the bacteria can remain asymptomatic, harboring the pathogen in a latent stage, which can be reactivated later in life. Finally, *M. tuberculosis* is an intracellular pathogen that can survive inside the host's first line of defense, macrophages.

Human Disease

The most common form of tuberculosis in humans is pulmonary tuberculosis. The infection begins with the inhalation of a droplet containing the bacilli, *M. tuberculosis.* Approximately 6% of the inhaled bacilli reach the alveoli and produce tubercles; the remainder settle in the upper respiratory tract and are expelled.[2] In most bacterial infections, T cells secrete cytokines [including interferon (IFN)-γ] that activate the macrophages. The macrophage engulfs the bacteria and contains them in phagosomes. The infected phagosomes then fuse with a lysosome, which contains degradative enzymes that destroy the pathogen. Mycobacteria act much differently. Although some mycobacteria are engulfed and destroyed by activated macrophages, many that enter nonactivated macrophages survive and proliferate. The mycobacteria evade destruction by altering maturation of the macrophage. Although the exact mechanism is unknown, phagosomes containing *M. tuberculosis* and *M. marinum* resist acidification and fail to undergo phagosome–lysosome fusion. The infected macrophages are spread through the bloodstream resulting in a systemic infection. Approximately 2–6 weeks following infection, a cell-mediated

---

[1] S. Mundayoor and T. M. Shinnick, *Ann. N.Y. Acad. Sci.* **730**, 26 (1994).

[2] R. L. Riley, C. C. Mills, W. Nyka, N. Weinstock, P. B. Storey, L. U. Sultan, M. C. Riley, and W. F. Wells, *Am. J. Hyg.* **70**, 185 (1995).

immune response occurs. The host responds with development of the granuloma which contain the bacilli. The early hematogenous spread during primary infection, however, leads to dispersal of the tubercle bacilli and establishes a latent infection.

Models of Tuberculosis Infection

Humans are the primary natural host of *M. tuberculosis* infection; however, several animal models have been developed to study infection with this pathogen. The primary animal models used are the guinea pig, mouse, and rabbit models. Each of the commonly used mycobacterium models has its advantages and disadvantages. In addition to these small animal models, a primate model has been developed which most closely parallels human disease. It is uncommonly used because of its expense and the limited availability of primates.

Mouse Model

The mouse model is the most commonly used animal model to study *M. tuberculosis* infection. Its greatest utility is in studying the host immune response. In the model, mice are inoculated intravenenously or via aerosol. Soon after infection, the mouse immune system responds with macrophages, which engulf the bacteria and present mycobacterial protein antigens on their surface. CD4 T cells then release cytokines, including IFN-γ and tumor necrosis factor (TNF)-α. In addition to a CD4 cellular response, CD8 cells also play a role in controlling infection. When infected macrophages lyse and release mycobacterial antigens into the extracellular space, class I major histocompatability cells (MHCs) and CD8 cells respond. Thus, this model has identified that T cells, including CD4, CD8, and $\gamma\delta$, are involved in the acquired immune response to *M. tuberculosis* infection in mice, an observation also noted in human infection.[3] Despite the strong cellular response that mimics human infection, there are disadvantages to this model. First, mice are more resistant to *M. tuberculosis* infection than are humans. More importantly, they fail to develop caseous granulomas or a giant cell response, the pathologic hallmark of human tuberculosis infection.

Guinea Pig Model

The guinea pig has been used as an animal model for tuberculosis since the development of Koch's postulate in 1882. The inhalation of *M. tuberculosis* results in primary lesions in the lungs. Guinea pigs have a similar pulmonary physiology

---

[3] I. M. Orme, *Trends Microbiol.* **1,** 77 (1993).

and inflammatory response to that seen in humans.[4] In fact, they mount a delayed hypersensitivity reaction to purified protein derivative (PPD), a measurement of tuberculosis exposure in humans, and respond to antibiotics used to treat human tuberculosis. Early after infection, *M. tuberculosis* can be isolated from the tracheobronchial lymph nodes, and as soon as 21 days postinoculation, a systemic infection is seen, with isolation of bacteria in extrapulmonary organs such as the spleen.[4]

Although the guinea pig develops granulomas, a host response that contains the infection in humans, this animal is much more susceptible to *M. tuberculosis* infection than are humans.[2,4] Unlike humans, these animals inevitably succumb to the disease. Therefore, this animal model is not useful for studies examining latent infection or mechanisms of reactivation of disease.

Rabbit Model

Most rabbit studies are performed using *Mycobacterium bovis,* as the rabbit is relatively resistant to infection with *M. tuberculosis*. Rabbits can be inoculated via aerosol, as are humans. Upon infection with *M. bovis,* rabbits develop tubercules, which can lead to cavities developing in the lungs.[5] The cavitary lung lesions and the rabbit's cough reflex increases the biohazard associated with this model.

Primate Model

Finally, primates are also used as models to study *M. tuberculosis* infections. Primates are fairly susceptible to *M. tuberculosis* infection, often succumbing to an infection of 10^4 or 10^5 colony-forming units (cfu).[6] Monkeys can be infected by the aerosol route and clinically develop a disease like that seen in humans. In response to infection, primates mount a human-like immune reaction, including a T-cell response.[4] Primates can develop an acute and chronic state of disease[6] and are moderately protected by vaccination with Bacillus Calmette-Guérin (BCG).[4] Two major disadvantages of using primates as a model are the expense of the animals, and the physical and biohazard danger they pose toward human handlers.

All of these models use *M. tuberculosis,* which because of the airborne transmission of disease requires the use of a Biosafety Level 3 (BSL-3) facility. In addition, the long generation time of *M. tuberculosis* (20 hr) and the expenses involved in a BSL-3 laboratory have created the need for a new mycobacterial animal model.

[4] D. McMurray, *in* "Tuberculosis: Pathogenesis, Protection, and Control" (B. Bloom, ed.), p. 35. American Society for Microbiology, Washington, D.C., 1994.

[5] A. Dannenberg, *in* "Tuberculosis: Pathogenesis, Protection, and Control" (B. Bloom, ed.), p. 149. American Society for Microbiology, Washington, D.C., 1994.

[6] G. P. Walsh, E. V. Tan, E. M. de la Cruz, R. M. Abalos, L. G. Villahermosa, L. Young, R. V. Cellona, J. B. Nazareno, and M. A. Horwitz, *Nat. Med.* **2,** 430 (1996).

Goldfish and *Mycobacterium marinum:* A Model to Study Mycobacterial Infection

Our laboratory has established a novel model system to study mycobacterial pathogenesis, using the goldfish *Carassius auratus* and *Mycobacterium marinum.* *M. marinum,* the causative agent of fish tuberculosis, was first isolated in 1926 from a saltwater aquarium.[7] Although *M. marinum* infects more than 150 species of fish, it is an uncommon zoonotic human pathogen. Transmission to humans occurs through direct skin inoculation; it is not airborne.

M. marinum can be used to study *M. tuberculosis* infection because of their close genetic relationship. *M. marinum* is one of the two most closely related mycobacterium (outside the *M. tuberculosis* complex) to *M. tuberculosis,* based on 16S rRNA homology (99.4%) and DNA–DNA hybridization.[8] In addition to its close genetic relationship, another feature of *M. marinum* infection in goldfish is that it mimics *M. tuberculosis* human infection.

M. marinum, like *M. tuberculosis,* is an intracellular pathogen, surviving and replicating inside the macrophage. When *M. marinum* is injected into the peritoneal cavity of the fish, it disseminates, causing a systemic infection just as in human tuberculosis. The immune response of fish infected with *M. marinum* is to contain the infection by development of a granuloma, the hallmark of the human immune response to tuberculosis.

There are several advantages to using *M. marinum* and the goldfish to study mycobacterial infection. One of the most important advantages of this model is the use of standard laboratory precautions (Biosafety Level 2), instead of the BSL 3 facility required for *M. tuberculosis* research. In addition, *M. marinum* has a much shorter generation time, compared to *M. tuberculosis* (4 versus 20 hr). The goldfish model can be adapted as a chronic or acute disease model. This quality parallels the human infection, in that humans can harbor the bacterium for years and show little or no sign of disease (chronic), or can rapidly progress to active disease and succumb to the infection (acute disease in the newborn or immunocompromised host). Finally, *M. marinum* is a natural pathogen of the host, *C. auratus,* thus the fish model represents a natural infection model.

Two other animal models of *M. marinum* infection have been developed. One is the mouse footpad model used to simulate *Mycobacterium leprae* infection. In this model, mice were injected subcutaneously (footpad) or intravenously in an attempt to establish a systemic infection, but the model failed as the organism did not disseminate.[9] The second model is the leopard frog, *Rana pipiens.* This model is also a natural infection model. *M. marinum* infection in the frog causes a chronic

---

[7] J. D. Aronson, *J. Infect. Dis.* **39,** 315 (1922).

[8] T. Rogall, J. Wolters, T. Florh, and E. C. Bottger, *Int. J. Syst. Bacteriol.* **40,** 323 (1990).

[9] F. M. Collins, V. Montalbine, and N. E. Morrison, *Infect. Immun.* **11,** 1079 (1975).

granulomatous reaction, which appears 6 weeks postinoculation. Only on steroid (hydrocortisone) injection, 5 days a week, did the frog succumb to infection.[10]

Procedure

The *M. marinum* strain ATCC 927 is obtained from the American Type Culture Collection (Manassus, VA). The bacteria are grown with shaking at 30° as a dispersed culture in 7H9 (Difco, Detroit, MI) broth with 10% oleic acid–albumin–dextrose–catalase enrichment (OADC) until late log phase growth (OD_{600} 1.6–1.8). [OADC enrichment is prepared by combining 4.05 g NaCl, 25.0 g albumin fraction V, 10.0 g glucose, 20.0 mg catalase, and 15 ml oleic acid solution (120 ml distilled water, 2.4 ml 6 N NaOH, 2.4 ml oleic acid) with 460 ml distilled water. Adjust the pH to 7.0, stir for 1 hr to completely solubilize the albumin, and filter sterilize.] The bacteria are pelleted by centrifugation (6000 rpm, 10 min., 25°) and then resuspended in sterile phosphate-buffered saline (PBS) to produce a $10\times$ concentrated suspension. These inocula stocks are disaggregated by sonication for 3 min at power level 3, using a cup horn accessory attached to a cell disruptor (model W-220 F; Heat Systems-Ultrasonics, Inc., Farmingdale, NY).[11] The inocula are aliquoted into 4-ml cryogenic tubes and stored at $-80°$. The concentration of the organism in the inocula [colony-forming units (cfu) per ml] is determined by plating serial dilutions on 7H10 agar supplemented with OADC and incubating at 30° until colonies appear (approximately 10 days).

Goldfish, *C. auratus,* weighing 20 to 30 g, are obtained from a local fishery (Hunting Creek Fisheries, Thurmont, MD). The fish are acclimated to 20-gallon flow-through aquaria with a water temperature of 18–22°, and a photoperiod of 16 hr light and 8 hr dark. After 2 weeks of acclimation, the fish are moved to a negative-air-pressure room, where the mycobacteria experiments are performed.[11]

Prior to inoculation into fish, the frozen inocula stock of *M. marinum* is thawed, adjusted to the appropriate dose, and sonicated for 3 min to disaggregate the bacteria. Fish are inoculated intraperitoneally through the lateral abdominal musculature with 0.5 ml of the *M. marinum* inocula using a 25-gauge needle and tuberculin syringe (Fig. 1).[11] Simultaneously, control fish are inoculated with sterile PBS. Fish injected with *M. marinum* and control fish are housed in separate tanks. The dose of *M. marinum* inoculated into the fish can be adjusted to cause either acute or chronic disease. A dose of 10^8 or 10^9 cfu /ml of the wild-type strain ATCC 927 causes an acute infection in the fish, with a median survival time of 4 to 10 days. However, fish infected with a sublethal dose of 10^2 to 10^7 cfu develop a chronic inflammatory response, characterized by granuloma formation by 4 to 8 weeks postinoculation.

---

[10] L. Ramakrishnan, R. H. Valdivia, J. H. McKerrow, and S. Falkow, *Infect. Immun.* **65,** 767 (1997).
[11] A. M. Talaat, R. Reimschuessel, S. Wasserman, and M. Trucksis, *Infect. Immun.* **66,** 2398 (1998).

FIG. 1. Fish are inoculated intraperitoneally through the lateral abdominal musculature (arrow) using a 25-gauge needle and tuberculin syringe.

Assessment of Infection

Behavioral Observations

The fish are observed daily for changes in behavior and physical findings. Fish infected with a lethal dose of *M. marinum* ATCC 927 (10^8 CFU or higher) suffer from lethargy, depressed dorsal fins, loss of appetite, and loss of equilibrium which occurs about 5–7 days postinoculation. The fish are often observed settling motionless at the bottom of the tank. Occasionally, fish develop gross skin lesions, peritoneal distention, and ascites (Fig. 2).

Bacterial Load

To measure the persistence of *M. marinum* in the goldfish during infection, the bacteriological burden, expressed as CFU of bacteria in liver, spleen, and

FIG. 2. Skin lesions of fish infected with 10^7 cfu of *M. marinum* ATCC 927 characterized by hemorrhage (A) or scale loss (B).

kidneys, is determined. On sacrifice, fish are euthanized either by rapidly severing the spine, or anesthetizing with 150 mg/liter tricaine methanesulfonate (MS-222) (500 mg/liter). These methods are approved by both the American Veterinary Medicine Association Panel on Euthanasia and the Canadian Council on Animal Care. The liver, spleen, and kidneys are aseptically removed and placed in separate sterile tubes to which 500 μl of sterile PBS is added. Each organ is homogenized for 2–3 min using a PowerGen 700 homogenizer (Fisher Scientific, Newark, DE) set at power level 4. A separate sterile tip is used for each organ. Serial dilutions of the homogenate are plated on 7H10 supplemented with OADC and incubated at 30° until colonies appear (approximately 10–14 days).

Histology

On sacrifice of the fish, samples of all body organs are fixed in 10% neutral buffered formalin.[12] Once fixed in paraffin, 5-μm sections of tissue are prepared using a rotary microtome (American Optical, Buffalo, NY). The sections are de-waxed and stained with hematoxylin and eosin stain.[13]

A scoring system is used to measure and classify the extent of pathology seen in the goldfish organs. The severity scale is described as a score 0, normal; 1, minimal; 2, mild; 3, moderate; 4, marked; and 5, severe. Lesions are graded by considering the number of lesions present and the extent of organ affected. Lesions classified as minimal are scarce and have little effect on the organ. In contrast, severe lesions are usually numerous and may distort the architecture of the organ. However, a single, excessively large lesion would also be graded as severe, especially if its size would compromise organ function. Scores of mild, moderate, and marked represent varying degrees between these extremes.[13] Fish organs are assessed using this scoring system, to determine the granuloma score (GS) of each organ analyzed. The cumulative GS (CGS) is the total of the GS's of the liver, kidneys, spleen, heart, and peritoneum of each fish.[11]

Following injection of *M. marinum,* there is an initial peritonitis with a cellular infiltrate composed of macrophages and lymphocytes. In the following weeks, granulomata are found in the peritoneum and surrounding organs, including liver, spleen, intestinal serosa, and gonads. In addition, granulomata appear in extraperitoneal organs such as heart and kidney. Occasionally granulomata are found in muscles, eyes, gills, and brain.

The granulomata develop from foci of epithelioid macrophages. These foci develop small necrotic centers and thin one to three cell-layer fibrous capsules. Over time the granulomata enlarge, developing prominent fibrous capsules, compressing and replacing surrounding parenchyma. It is not uncommon to have a third of the spleen replaced with granulomata in fish that survive longer than 8 weeks.

[12] E. B. Prophet, B. Mills, J. B. Arrington, and L. H. Sobin, "Laboratory Methods in Histotechnology," p. 29. American Registry of Pathology, Washington, D.C., 1992.
[13] R. Reimschuessel, R. Bennett, and M. Lipinsky, *J. Aquat. Anim. Health.* **4,** 135 (1992).

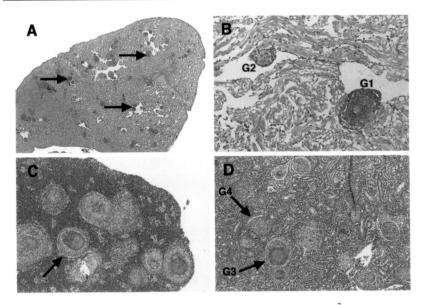

FIG. 3. Representative histopathology of fish infected 8 weeks earlier with 10^7 cfu of *M. marinum* ATCC 927. All histopathology sections were stained with hematoxylin and eosin stain. (A) Heart showing widespread granulomatous lesions (arrows). Original magnification: ×20. (B) Heart containing two typical granulomatous lesions. One granuloma (G1) is well formed with a necrotic center surrounded by epithelioid macrophages and fibroblasts. Lesion G2 is characteristic of granulomata found early after infection. It has a small eosinophilic core, surrounded by 2–3 cell layers of epithelioid macrophages, and 1–2 cell layers of fibroblasts. Original magnification: ×20. (C) Spleen showing diffuse granulomata. The granulomata have a necrotic core, surrounded by epithelioid macrophages, and an outer ring of fibroblasts and lymphocytes. Several of the lesions in this section have not yet been well encapsulated and are described as sheets of epithelioid macrophages. One granuloma has surrounded a macrophage aggregate (arrow). Original magnification: ×10. (D) Kidney with diffuse distribution of granulomata scattered throughout the section. This lesion would be graded as 3 or moderate. Many of the granulomata contain a necrotic bright eosinophilic core as well as pigmented macrophages. Original magnification: ×10.

Lesions in the heart and kidney are usually multifocal with a generally uniform distribution throughout the organ.

Figure 3 (A–D) illustrates the pathology seen with systemic *M. marinum* infection in the goldfish. The cardiac endothelium in fish is part of the reticuloendothelial system and is actively phagocytic. In Fig. 3A, many of these lesions in the heart are located in the endocardium or the myocardium near the luminal surface. Because of the number of granulomata in this section, the lesion would be graded 4, "marked." The endocardial lesions seen in Fig. 3B are probably the result of endothelial cell phagocytosis of *M. marinum* during periods of bacteremia. This section of heart contains two typical granulomatous lesions. They are located in the myocardium extending into the endocardium. Frequently the lesions bulge into the lumen of the ventricle. Here, one granuloma (G1) is well formed with a necrotic center surrounded by epithelioid macrophages and fibroblasts. Lesion G2 is characteristic of granulomata found early after infection: it is located primarily in the endocardium, and it has a small eosinophilic core, surrounded by only 2–3 cell layers of epithelioid macrophages and 1–2 cell layers of fibroblasts (Fig. 3B).

Figure 3C shows one-third of the spleen replaced with diffuse granulomata. Such a lesion would be graded as 4 or "marked." Some of the lesions bulge from the surface of the spleen and can, if large, appear as whitish nodules on gross examination. The spleen is a pigmented organ in the viscera of many fish species, including goldfish.

The fish trunk kidney is retroperitoneal and covered ventrally by the swim-bladder. It is a preferred site to culture when a systemic infection is suspected (Fig. 3D). The granulomata seen throughout the renal parenchyma are the result of vascular spread of the organisms by the renal portal system. They vary in size and the amount of encapsulation, from a sheet of epithelioid macrophages to granulo-mata with a prominent fibrous capsule. The goldfish trunk kidney contains normal hematopoietic tissue surrounding the renal tubules and glomeruli. These cells are not an inflammatory infiltrate.

Lethal Dose Experiment

The 50% lethal dose (LD_{50}) is the dose at which half of the fish infected with *M. marinum* die by 1-week postinoculation. To determine the LD_{50} dose, fish are inoculated with doses spanning one log lower and one log higher than the expected LD_{50} dose of *M. marinum*, in an attempt to bracket the actual inoculum that kills 50% of the fish in 1 week. The data is graphed on semilogarithmic paper with log of inoculum on the X axis and percent survival on the Y axis. A regression line connects the points. The inoculum dose corresponding to the 50% survival point can be interpolated from the regression line. This is a simple graphical analysis method based on the Reed and Meunch method of calculation.[14] Using this method, the LD_{50} dose of the wild-type *M. marinum* strain ATCC 927 in our model was 4.5×10^8 cfu, 1-week postinoculation.[11] The same technique can be applied to determine the LD_{50} of mutant strains of *M. marinum* in the fish model.

Applications of Goldfish Model of Mycobacterial Pathogenesis

Identification of Virulence Mutants

Rational design of a live *M. tuberculosis* vaccine strain centers on the iden-tification of virulence genes to target for attenuation. We have applied signature-tagged mutagenesis (STM) to the goldfish model of mycobacterial pathogenesis to

[14] L. J. Reed and H. Muench, *Am. J. Hyg.* **27**, 493 (1938).

identify virulence genes in *M. marinum*. STM is a transposon-based mutagenesis system that enables a large pool of mutants to be screened in the animal at once, eliminating the need to test each individually.[15] Each mutant has been "tagged" with a unique DNA sequence so that it can be identified in the pool. Those mutants that are represented in the input pool of mutants, but are not recovered from the animal on sacrifice (missing in the output pool), are potential virulence mutants. In these mutants, the tagged transposon has disrupted a gene necessary for the pathogen's survival in the host.

Assessment of Virulence Mutants

We have determined the LD_{50} of the wild-type *M. marinum* ATCC 927 strain in the goldfish model to be 4.5×10^8 cfu at 1 week postinoculation.[11] We have used this standard dose to rapidly screen potential *M. marinum* virulence mutants to assess whether the mutants are attenuated as compared to the wild-type strain. For these analyses, fish are inoculated with the LD_{50} dose (4.5×10^8 cfu fish) with mutant strains of *M. marinum* and survival over a 35-day course is determined.

Competition Assay

Another method to assess virulence is to determine the competitive index (CI) value of *M. marinum* mutant to *M. marinum* wild-type strain. In these assays, the *M. marinum* mutant strain and wild-type ATCC 927 strain, each at a dose of 10^7 cfu, are combined to form a mixed inoculum (each strain in equal proportions). This mixed inoculum is used to inoculate a group of six fish. The fish are sacrificed 1 week postinoculation, and the liver is removed and homogenized. Serial dilutions of the mixed inoculum (input) and the homogenized organs (output) are plated on 7H10 medium and 7H10 medium with kanamycin (50 μg/ml), and incubated at 30° until colonies appear. Both wild-type and mutant strains of *M. marinum* will grow on the 7H10 medium, however, only the mutant strains will grow on medium supplemented with antibiotic. The colonies are counted, and the number of wild-type (WT) and mutant colonies is determined. The Competitive Index is then calculated using the equation:

$$CI = [(\text{Mutant output/WT output})/(\text{Mutant input/WT input})]$$

Typically, the *CI* value is less than or equal to 0.6, when the mutant is attenuated.[16]

---

[15] M. Hensel, J. E. Shea, C. Gleeson, M. D. Jones, E. Dalton, and D. W. Holden, *Science* **269**, 400 (1995).

[16] A. J. Darwin and V. L. Miller, *Mol. Microbiol.* **32**, 51 (1999).

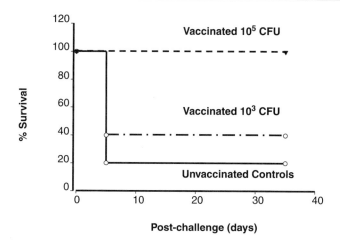

FIG. 4. Survival curves of 3 groups of fish challenged 4 weeks after vaccination. Groups 1 and 2 fish vaccinated with 10^3 cfu (dash–dot line) or 10^5 cfu (dashed line) of wild-type *M. marinum* ATCC 927, respectively. Group 3 fish vaccinated with PBS (placebo control) (solid line). Challenge dose was 4.5×10^8 cfu/fish of *M. marinum* ATCC 927 (LD$_{50}$).

Goldfish Model as a Vaccine Challenge Model

In addition to a mycobacterial pathogenesis model, our animal model can be used as a vaccine challenge model. A challenge model is used to demonstrate whether a vaccine strain can elicit protective acquired immunity and therefore protect against a challenge with a fully virulent wild-type strain. This application was demonstrated in an experiment with two groups of fish inoculated intraperitoneally with *M. marinum* ATCC 927 organisms at a dose of 10^3 or 10^5 cfu. A third group (the placebo group) was inoculated with PBS (Fig. 4). Four weeks postinoculation (Fig. 4, time 0) the fish were challenged intraperitoneally with an LD$_{50}$ dose of the wild-type *M. marinum* strain. Eighty percent of the fish in the placebo group died by 35 days postchallenge. Animals immunized with 10^3 cfu of *M. marinum* displayed partial protection, with 40% of the fish surviving to 35 days (the end of experiment). However, animals immunized with 10^5 cfu of *M. marinum* were completely protected, with 100% of the fish surviving until the end of the experiment. This established that protective immunity develops following an *M. marinum* infection. This model can thus be used to evaluate the protective efficacy of candidate vaccine strains.

Section II

Virulence and Essential Gene Identification

[4] Green Fluorescent Protein as a Marker for Conditional Gene Expression in Bacterial Cells

By Roy J. M. Bongaerts, Isabelle Hautefort, Julie M. Sidebotham, and Jay C. D. Hinton

Introduction

Bacterial pathogenesis results from a complex adaptation of the pathogen to its host. The necessity to resist the mammalian immune response has led to the selection of bacteria that have developed sophisticated virulence determinants. Expression of these determinants occurs in response to various environmental signals and is tightly regulated by a complex regulatory cascade.[1] Until recently, most virulence genes were identified from studies based on *in vitro* systems, which had little relevance to the true *in vivo* situation of bacterial infection. This has been remedied by the development of several powerful techniques to identify *in vivo* induced (*ivi*) genes such as *in vivo* expression technology (IVET)[2] and signature tagged mutagenesis (STM).[3] However, these two techniques do not yield detailed information about the expression levels of *ivi* genes during infection. To enable the study of spatial and temporal expression of *ivi* genes in the host, analysis at the mRNA or protein level is required. Traditional reporter systems have been used for many years to study bacterial gene expression, but we now need to develop new accurate reporter systems that allow the monitoring of gene expression at the individual bacterial cell level. Since most virulence genes respond to environmental signals, *ivi* gene expression is likely to vary with the stage of infection, the spatial localization within the host, and the particular cell type within that host tissue. Overall, the monitoring of virulence gene expression requires sensitive reporter systems to show up- and down-regulation and transient and low levels of virulence gene expression.

Green Fluorescent Protein as Reporter System to Monitor *in Vitro* and *in Vivo* Gene Expression

The green fluorescent protein (GFP) of the marine invertebrate *Aequorea victoria* is a single autofluorescent, acidic, compact, globular polypeptide with a molecular mass of 26 kDa. Assembly of the GFP fluorophore requires a series

---

[1] B. B. Finlay and S. Falkow, *Microbiol. Mol. Biol. Rev.* **61**, 136 (1997).

[2] M. J. Mahan, J. M. Slauch, and J. J. Mekalanos, *Science* **259**, 686 (1993).

[3] M. Hensel, J. E. Shea, C. Gleeson, M. D. Jones, E. Dalton, and D. W. Holden, *Science* **269**, 400 (1995).

of posttranslational intramolecular reactions, involving cyclization and autoxidation of amino acids Ser^{65}-Tyr^{66}-Gly^{67}. Mature GFP emits green light (508 nm) when excited with ultraviolet light (395 nm). Because no exogenous substrates or cofactors are required for its activity, it is a unique tool for monitoring gene expression, protein localization, and protein dynamics in both prokaryotic and eukaryotic living cells.[4,5]

Advantages and Drawbacks of Green Fluorescent Protein

GFP has been intensely studied in recent years and different types of GFP variants with altered characteristics have been developed. Wild-type GFP has a neutral excitation peak of 395 nm with a minor peak at 475 nm, and an emission peak at 508 nm. Classes of spectral variants include GFPs with shifted emission and excitation wavelengths, and higher and lower intensities of fluorescence compared to wild-type GFP.[6] A significant proportion of wild-type GFP molecules fail to fold and cyclize properly when synthesized at 37° and another class of GFP derivatives correct this problem. These thermostable variants fold correctly and are therefore significantly brighter. Combinatorial incorporation of multiple mutations of this class of variants has lead to a substantial increase in fluorescence,[7–9] exemplified by a mutagenesis study in *Escherichia coli* which resulted in the isolation of three distinct classes of GFP variants, all having red-shifted excitation maxima and folding more efficiently that the wild-type protein. The GFPs from these three classes contained the amino acid substitutions F64L and S65T for GFPmut1; S65A, V68L, and S72A for GFPmut2; and S65G and S72A for GFPmut3.[7] These variants have proved to be particularly useful for *in vivo* studies[10] because they are ideal for microscopic and flow cytometric analyses. Another class of GFP variants possesses enhanced levels of intracellular fluorescence without changes in the amino acid sequence through optimization of codon bias. These variants produce mRNA species that are more efficiently translated than mRNA produced from the wild-type coding sequence resulting in increased GFP production. Combining the mutations from these different classes has produced distinct GFP reporters with greatly enhanced levels of fluorescence. Because only one single fluorophore is formed from each GFP molecule synthesized, such increases in GFP fluorescence

[4] M. Chalfie, Y. Tu, G. Euskirchen, W. W. Ward, and D. C. Prasher, *Science* **263,** 802 (1994).

[5] M. B. Elowitz, M. G. Surette, P. E. Wolf, J. Stock, and S. Leibler, *Curr. Biol.* **7,** 809 (1997).

[6] R. Heim, D. C. Prasher, and R. Y. Tsien, *Proc. Natl. Acad. Sci. U.S.A.* **91,** 12501 (1994).

[7] B. P. Cormack, R. H. Valdivia, and S. Falkow, *Gene* **173,** 33 (1996).

[8] M. Chalfie, "Green Fluorescent Protein: Properties, Applications, and Protocols." John Wiley & Sons, New York, 1998.

[9] A. Crameri, E. A. Whitehorn, E. Tate, and W. P. Stemmer, *Nat. Biotechnol.* **14,** 315 (1996).

[10] R. H. Valdivia and S. Falkow, *Science* **277,** 2007 (1997).

TABLE I

Name	Excitation maximum[a] (nm)	Emission maximum (nm)	Estimated fluorescence intensity relative to wild type	Mutation	Reference[b]
GFP	395(475)[#]	508(503)	1		(1,2)
Wild type S65T	489	511	6	S65T	(2)
GFPmut1 (= EGFP)	488	507	35	F64L, S65T	(3)
GFPmut2	481	507	19	S65L, V68L, S72A	(3)
GFPmut3	501	511	21	S65G, S72A	(3)
GFPuv	396(476)	508	42	F100S, M154T, V164A	(4)
GFP5	(396)476[c]	508	111	V163A, I167T, S175G	(5)
GFP+	491	512	130	F64L, S65T, F99S, M153T, V163A	(6)
BFP	382	448	0.6	Y66H	(7)
	384	448	3	Y66H, V163A, S175G	(5)
CFP	433	475	3	Y66H, N146I, M153T, V163A, N212K	(8)
	432	480	1.5	Y66W, I123V, Y145H, H148R, M153T, V163A, N212K	(8)
YFP	513	527	6	S65G, V68L, S72A, T203Y	(9)
DsRED (= drFP583)	558	583	n.a.[d]	n.a.[d]	(10)

[a] The value in parentheses is a minor peak.

[b] Key to references: (1) M. Chalfie, Y. Tu, G. Euskirchen, W. W. Ward, and D. C. Prasher, *Science* **263**, 802 (1994); (2) R. Heim, D. C. Prasher, and R. Y. Tsien, *Proc. Natl. Acad. Sci. U.S.A.* **91**, 12501 (1994); (3) B. P. Cormack, R. H. Valdivia, and S. Falkow, *Gene* **173**, 33 (1996); (4) A. Crameri, E. A. Whitehorn, E. Tate, and W. P. Stemmer, *Nat. Biotechnol.* **14**, 315 (1996); (5) K. R. Siemering, R. Golbik, R. Sever, and J. Haseloff, *Curr. Biol.* **6**, 1653 (1996); (6) O. Scholz, A. Thiel, W. Hillen, and M. Niederweis, *Eur. J. Biochem.* **267**, 1565 (2000); (7) R. Heim, A. B. Cubitt, and R. Y. Tsien, *Nature* **373**, 663 (1995); (8) R. Heim and R. Y. Tsien, *Curr. Biol.* **6**, 178 (1996); (9) M. Ormo, A. B. Cubitt, K. Kallio, L. A. Gross, R. Y. Tsien, and S. J. Remington, *Science* **273**, 1392 (1996); (10) M. A. Wall, M. Socolich, and R. Ranganathan, *Nat. Struct. Biol.* **7**, 1133 (2000).

[c] Excitation intensity is similar at both wavelengths.

[d] n.a., not applicable.

are very important for improving sensitivity. An overview of GFP variants is shown in Table I.

Conventional reporter proteins are enzymes [e.g., β-galactosidase (LacZ), chloramphenicol acetyltransferase (Cat), and luciferase (Lux)] whose signal amplification is derived from multiple substrate cleavage by one molecule of reporter protein. However, activity of these reporters is also dependent on substrate levels and/or energy reserves within cells, whereas GFP needs no exogenous substrates or cofactors.[4,11] Moreover, the use of luciferase is complicated by high background in bacterial populations, making it unsuitable for reporting expression in individual bacterial cells.[12] The value of fluorogenic β-galactosidase substrates is reduced by the need to load these substrates into organisms by cell permeabilization or osmotic shock.[13] Several interesting reviews describe these types of reporter systems.[14–16]

The fluorescence signal of posttranslationally modified GFP depends largely on the rate of biosynthesis of functional protein and the rate of dilution as the cell divides.[12] Wild-type GFP is very stable with a half-life of at least 24 hr in *E. coli*.[17] Importantly, GFP fluorescence is stable under stress conditions, such as starvation,[18] allowing determination of gene expression *in vivo*. However, care must be taken when performing acid stress experiments since GFP fluorescence is reduced at low pH.[19,20] The availability of unstable GFP variants with half-lives of 40, 60, and 110 min enables the measurement of fast changes in expression patterns.[17] There is an excellent book on the use of GFP which is essential reading.[8] The development of new GFP variants is ongoing, leading to fluorescent proteins that are brighter, more stable, or have other improved characteristics.[21] An interesting phenomenon that might be exploited in the future is photoactivation of GFP with blue light in a low oxygen environment, resulting in red-emitting GFP.[5] In addition, the use of autofluorescent proteins such as red fluorescent drFP583 [commercially available as dsRED (Clontech, Palo Alto, CA)] and derivatives attracts growing attention.[22–24]

[11] C. E. Nwoguh, C. R. Harwood, and M. R. Barer, *Mol. Microbiol.* **17**, 545 (1995).

[12] R. Tombolini and J. K. Jansson, *Methods Mol. Biol.* **102**, 285 (1998).

[13] R. H. Valdivia and S. Falkow, *Curr. Opin. Microbiol.* **1**, 359 (1998).

[14] B. B. Christensen, C. Sternberg, J. B. Andersen, R. J. Palmer, Jr., A. T. Nielsen, M. Givskov, and S. Molin, *Methods Enzymol.* **310**, 20 (1999).

[15] C. Prigent-Combaret and P. Lejeune, *Methods Enzymol.* **310**, 56 (1999).

[16] I. Hautefort and J. C. D. Hinton, *Phil. Trans. R. Soc. Lond. B Biol. Sci.* **355**, 601 (2000).

[17] J. B. Andersen, C. Sternberg, L. K. Poulsen, S. P. Bjorn, M. Givskov, S. Molin, L. Molina, C. Ramos, M. C. Ronchel, and J. L. Ramos, *Appl. Environ. Microbiol.* **64**, 2240 (1998).

[18] R. Tombolini, A. Unge, M. E. Davey, F. J. deBruijn, and J. K. Jansson, *FEMS Microbiol. Ecol.* **22**, 17 (1997).

[19] G. H. Patterson, S. M. Knobel, W. D. Sharif, S. R. Kain, and D. W. Piston, *Biophys. J.* **73**, 2782 (1997).

[20] M. Kneen, J. Farinas, Y. Li, and A. S. Verkman, *Biophys. J.* **74**, 1591 (1998).

[21] O. Scholz, A. Thiel, W. Hillen, and M. Niederweis, *Eur. J. Biochem.* **267**, 1565 (2000).

[22] M. V. Matz, A. F. Fradkov, Y. A. Labas, A. P. Savitsky, A. G. Zaraisky, M. L. Markelov, and S. A. Lukyanov, *Nat. Biotechnol.* **17**, 969 (1999).

[23] M. A. Wall, M. Socolich, and R. Ranganathan, *Nat. Struct. Biol.* **7**, 1133 (2000).

[24] A. Terskikh, A. Fradkov, G. Ermakova, A. Zaraisky, P. Tan, A. V. Kajava, X. Zhao, S. Lukyanov, M. Matz, S. Kim, I. Weissman, and P. Siebert, *Science* **290**, 1585 (2000).

In summary, GFP variants are ideal for the study of development, cell biology, and bacterial pathogenesis; conventional reporter genes are limited by their inability to measure single-cell expression accurately in living bacterial cells.

Practical Considerations

The utility of GFP for a particular application depends very much on the type of bacteria and the precise research requirements. GFP can be used to localize bacteria in a particular environment[25,18] or to visualize several bacterial populations at the same time, by using combinations of different GFP variants.[26,27] Specific GFP-protein fusions can be localized in bacterial cells,[28] and GFP can also be used as an effective reporter of bacterial gene expression.[29] For each application it is important to decide whether to study live bacteria, or whether they should be fixed. Fixation allows permeabilization of cells (which provides a better access for antibodies), prevents antigen leakage, maintains cell structure, and stops all biological functions of cells, giving an instant image of what was occurring in the cell at the moment of fixation. A clear description of all types of fixatives and their advantages and disadvantages is available.[30] Fixation not only is critical from a safety point of view when working with pathogens, but also assists the study of temporal gene expression patterns through the study of sequential samples. Before choosing a fixative it is important to decide whether a sample will be analyzed directly, or whether additional staining will be necessary, which could affect GFP fluorescence.

Many studies have been performed using plasmid-borne GFP transcriptional fusions, as they provide a bright fluorescence signal and are simple to construct. However, care must be taken when studying *in vivo* expression with high copy number systems in animal models, as accurate measurements will rely on plasmid stability. We have found that selective pressure due to increased toxic GFP levels results in plasmid loss during *in vivo* studies (I. Hautefort, unpublished data, 2000). To avoid problems with plasmid stability in animal models and cultivated cell lines, we favor the use of single-copy stable integrated chromosomal GFP fusions that allow accurate quantification of GFP expression in individual cells. However, visualization of GFP expressed from a single copy requires a bright autofluorescent reporter molecule and very sensitive detection equipment.

[25] C. R. Beuzon, S. Meresse, K. E. Unsworth, J. Ruiz-Albert, S. Garvis, S. R. Waterman, T. A. Ryder, E. Boucrot, and D. W. Holden, *EMBO J.* **19,** 3235 (2000).

[26] G. V. Bloemberg, A. H. Wijfjes, G. E. Lamers, N. Stuurman, and B. J. Lugtenberg, *Mol. Plant Microbe Interact.* **13,** 1170 (2000).

[27] N. Stuurman, C. P. Bras, H. R. Schlaman, A. H. Wijfjes, G. Bloemberg, and H. P. Spaink, *Mol. Plant Microbe Interact.* **13,** 1163 (2000).

[28] A. Feucht and P. J. Lewis, *Gene* **264,** 289 (2001).

[29] R. H. Valdivia and S. Falkow, *Mol. Microbiol.* **22,** 367 (1996).

[30] E. Harlow and D. Lane, "Using Antibodies: A Laboratory Manual." Cold Spring Harbor Laboratory Press, Cold Spring Harbor, New York, 1999.

Methods to Detect or Measure Green Fluorescent Protein Expression

Several methods have been presented for assessing GFP fluorescence.[4] In this article we describe how to visualize individual GFP-expressing bacteria by epifluorescence microscopy, fluorometry, flow cytometric analysis, and cell sorting.

Epifluorescence Microscopy and Image Analysis of Bacterial Cells

A number of studies have involved the visualization of GFP, either to localize GFP–protein fusions within bacterial cells or to monitor GFP-expressing bacteria in their environment.[25,31] To capture images of infected tissues, cultivated cell lines or GFP-expressing biofilms, a very sensitive camera and image-grabbing system is required. Fluorescence and confocal microscopes are generally fitted with highly sensitive cooled charge-coupled device (CCD) cameras controlled by powerful analysis software. For quantitative imaging of fluorescence with a microscope, equipment requirements are even more specific, and accurate measurements of GFP fluorescence intensity *in vivo* is very difficult.[32]

Fading of fluorescent molecules is a common problem in microscopy. Each fluorochrome has a limited capacity for excitation and emission and the emitted light tends to decline over time. One can limit fading by minimizing the exposure time to the source of excitation light, and by including an antifading agent in the medium used to mount the sample. Different antifading reagents are available, most of which act by scavenging free radicals liberated by excitation of the fluorochromes. The free radicals attack unexcited fluorochromes and damage them, thus producing exponential fading. The most commonly used antifade compound is triethylenediamine (Dabco) because of its solubility and chemical stability, although *p*-phenylenediamine is a useful alternative.[33,34]

Protocols for Microscopy

We describe below a number of protocols that can be used to visualize GFP expression. They can also be used as a starting point for the development of specific protocols, as each application can require different conditions to obtain an optimal GFP signal.

MATERIALS

The equipment used in our laboratory to visualize low-level GFP expression is shown in parentheses.

[31] V. Sourjik and H. C. Berg, *Mol. Microbiol.* **37,** 740 (2000).
[32] D. W. Piston, G. H. Patterson, and S. M. Knobel, *Methods Cell Biol.* **58,** 31 (1999).
[33] G. D. Johnson and G. M. Nogueira Araujo, *J. Immunol. Methods* **43,** 349 (1981).
[34] G. D. Johnson, R. S. Davidson, K. C. McNamee, G. Russell, G. Goodwin, and E. J. Holborow, *J. Immunol. Methods* **55,** 231 (1982).

Epifluorescence microscope (e.g., Olympus BX51)

GFP (FITC) filter set (e.g., Excitation BP470–490nm, Emission DM500 dichroic beam splitter + BA515 barrier filters, Chroma)

Cooled CCD Camera [e.g., F-view (Norfolk Analytical, Hilgay, UK)]

Image analysis software [e.g., Analysis (SIS, Münster, Germany)]

Microscope slides and coverslips

Coating for slides [e.g., 1% (v/v) polyethyleneimine (PEI; Sigma St. Louis, MO)] or 0.01% (w/v) poly-(L-lysine)(Sigma)

Mounting medium [Mowiol 9% (w/v), Calbiochem, La Jolla, CA].

Note: For confocal microscopy Mowiol mounting medium is not appropriate since its refractive index is not well established.

Antifading agent, such as triethylenediamine (Dabco) or phenylenediamine (1 mg/ml, Sigma)

Cryostat (e.g., Reichert Jung Cryocut E, Leica)

Bunsen burner

Immersion oil

Phosphate-buffered saline (PBS) pH 7.4

Saponin (Calbiochem)

2-Methylbutane (isopentane, BDH, Poole, UK)

Liquid nitrogen

Freezing medium (OCT compound, Leica)

Supporting cork pieces to mount sample

IMAGING BACTERIA IN SUSPENSION

Method 1

1. Place uncoated slide on heating block at 45°.

2. Rapidly mix equal volumes of bacterial sample and low melting agarose in PBS (pH 7.4) at 45°, place quickly on slide, and apply coverslip.

3. Take slide directly from the block and cool at room temperature for 5 min in the dark.

4. Use phase contrast on an epifluorescence microscope to focus on bacteria, and a GFP filter set to detect GFP-expressing cells.

5. Capture images with a CCD camera using image analysis software and use for further applications.

Method 2

This method is only suitable for imaging bacteria that express high levels of GFP.

1. Coat slides for 2 min in 1% polyethyleneimine or for 1 hr at room temperature in 0.01% poly (L-lysine).

2. Wash slides for 30 min in deionized water or PBS and dry by centrifugation at 600g for 5 min at room temperature.

3. Apply 10 μl of bacterial sample on the slide and air dry.

4. Fix bacterial cells by passing quickly through flame of Bunsen burner. *Note:* Drying samples can reduce the amount of GFP fluorescence dramatically.

5. Wash 30 min in deionized water and air dry.

6. Add mounting medium on the sample and place coverslip on top.

7. Examine as described above in Method 1, steps 4 and 5.

IMAGING OF INFECTED ANIMAL TISSUE SECTIONS OR OF CULTIVATED CELL LINES GROWN ON COVERSLIPS

An increasing number of experiments are designed for localizing bacterial gene expression in particular infection models. The next two methods concern preparation of slides for sections of host infected tissue and observing bacterially infected mammalian cells.

Method 3: Tissue sample preparation, sectioning, and imaging

1. Fix tissue samples immediately after dissection of the bacterially infected animal host by incubating in glass or polypropylene tubes for 1 hr at room temperature in freshly prepared 4% paraformaldehyde (pH 7.4).

2. Wash thoroughly in PBS.

3. Incubate fixed tissue in 20% sucrose solution (in PBS) overnight to protect the tissue against alteration during cryopreservation.
Note: Add 0.1% sodium azide if the incubation period at 4° extends 16 hr to prevent microbial growth.

4. Place an aluminum beaker containing 2-methylbutane into liquid nitrogen to reduce its temperature rapidly to −40° to −60°.

5. Stick the tissue sample to a piece of cork with freezing medium (OCT compound), and snap freeze in cooled 2-methylbutane.

6. Once frozen, store samples at −80° until further processing. *Note:* Storage at −80° for several months does not cause significant loss of GFP fluorescence.

7. Cut frozen sample with cryostat, and collect thin sections (4–6 μm) directly onto freshly coated microscope slides [0.01% poly (L-lysine) (Sigma); see method 2]. Alternatively, thick cryosections (>20 μm) are collected by floating in PBS.

8. If necessary, stain sections or cells with antibodies using 0.03% (w/v) saponin as permeabilizing agent. Staining of thick sections floating in PBS allows better access of antibody to target epitopes.

9. Mount coverslips on slides in medium containing an antifading agent (1 mg/ml phenylenediamine)

10. Examine as described above in method 1, steps 4 and 5.

Method 4: Imaging of bacteria in mammalian cells grown on coverslips

1. Grow mammalian cells in tissue culture on 0.01% poly(L-lysine)-coated coverslips and infect with the pathogen of interest.

2. If necessary, stain sections or cells with antibodies using 0.03% (w/v) saponin as permeabilizing agent and proceed as described in method 3, steps 9 and 10.

Fluorometric Analysis

Fluorometry is an easy, fast, and commonly available technique for detecting and measuring fluorescence intensity. Initially used in biochemistry for determining the luminescence properties of tryptophan in proteins or as a monitor for protein conformation,[35,36] it has evolved into a useful tool for measuring levels of GFP expression and other fluorescent reporter systems in bacterial populations.

Fluorometry generally uses a quartz halogen lamp as light source and appropriate excitation (ex) and emission (em) filters, e.g., 485_{ex} nm/538_{em} nm for most GFP variants. Sample fluorescence should be measured in nonfluorescent cuvettes or microtiter plates using several dilutions of the samples, and compared with a recombinant GFP standard. Linear regression of the values obtained for defined concentrations of recombinant GFP allows the accurate quantification of fluorescing protein produced by a bacterial population at a particular time. Temperature-control and/or shaking options present on modern equipment allow the measurement of the kinetics of gene induction in living bacteria. Fluorometry can therefore be a fast and easy tool to screen for bacterial gene induction under various environments such as acidic pH or high osmolarity, on fixed samples or in real time, and lends itself to high throughput analysis. The detection limit of intracellular GFP in bacteria is ~10^3 molecules per bacterial cell, as described in the following example of a fluorometric analysis to investigate the variable effects of specific fixatives on GFP.

Protocols for Fluorometric Analysis

An increasing number of applications require the monitoring of temporal bacterial gene expression. Commonly, this is achieved by stopping the expression of *gfp* fusions at appropriate times by fixation. Similarly, localization of bacterial gene expression by microscopic observations of infected tissue sections or invaded cell lines usually requires fixation of the samples. Various fixatives are commonly used in microscopy or flow cytometry but can reduce GFP fluorescence. We compared the effect of several commonly used fixatives on GFPmut1 fluorescence in *Salmonella enterica* serovar *typhimurium*.

[35] B. R. Pattnaik, S. Ghosh, and M. R. Rajeswari, *Biochem. Mol. Biol. Int.* **42,** 173 (1997).
[36] E. G. Strambini and G. B. Strambini, *Biosens. Bioelectron.* **15,** 483 (2000).

TABLE II
FIXATION/PERMEABILIZATION PROTOCOLS[a]

Fixative/permeabilizing agents	Incubation time (min)	Incubation temperature
100% Acetone	10	Room temperature
100% Methanol	15	4°
50% Acetone / 50% methanol	1	Room temperature
48% Acetone / 48% methanol / 4% formalin	1.5	Room temperature
Buffered formol acetone (pH 6.6)	0.5	Room temperature
4 ml 12.4 mM Na$_2$HPO$_4$,		
1.22 mM KH$_2$PO$_4$		
16.6 ml 4% (v/v) formalin		
30 ml 100% acetone		
20 ml H$_2$O		
2.5% (v/v) Glutaraldehyde	60	Room temperature
3% (w/v) Paraformaldehyde	30	Room temperature
4% (v/v) Formalin	1	Room temperature

[a] Fixatives should be freshly prepared. See Fig. 1.

MATERIALS

Fluorometer with appropriate GFP filterset (e.g., excitation filter, 485 nm; HBW, 14 ± 2 nm; and emission filter, 538 nm HBW 25 ± 3 nm; Molecular Devices, Sunnyvale, CA).

Nonfluorescent flat bottom 96-well UV plate (Costar, Corning, NY).

GFP protein standard [e.g., purified recombinant enhanced green fluorescent protein (EGFP) (Clontech)].

PBS

0.22-μm membrane filter

METHOD TO ASSESS THE EFFECT OF FIXATIVES ON GFP FLUORESCENCE

1. Bacterial strains [*S. typhimurium* NCTC 12023 wild type [(identical to ATCC 14028s[37]] and JH2031, a 12023 derivative that contains a single chromosomal transcriptional *ssrA::gfpmut1* fusion) are grown in Luria–Bertani (LB) broth (10 g tryptone, 5 g yeast extract, 5 g NaCl per liter distilled, deionized water, pH 7.0) overnight at 37° in a shaking incubator at 250 rpm.

2. One ml of bacterial culture is harvested by centrifugation at room temperature in a microfuge at 6,000g for 5 min. *Salmonella* cells are then washed twice with 0.22 μm filtered PBS to remove the autofluorescent LB media. The number of bacteria is determined by optical density measurement (OD$_{600\,nm}$), adjusted to the same number for both strains and centrifuged again.

[37] J. Deiwick, T. Nikolaus, J. E. Shea, C. Gleeson, D. W. Holden, and M. Hensel, *J. Bacteriol.* **180**, 4775 (1998).

3. Bacterial pellets are resuspended in 1 ml of the fixative agent and treated as described in Table II.

4. Samples are subsequently washed three times, resuspended in PBS, and analyzed as follows: PBS is used as blank. Samples are twofold serially diluted in PBS. Duplicates of (diluted) samples are placed in individual wells of a 96-well microtiter plate. The highest concentration of bacteria was 1.2×10^8 cells per well. Duplicates of a series of recombinant EGFP dilutions (156 pg/ml to 5 ng/ml) are used to establish a standard curve.

5. The fluorescence of each well is determined using the GFP filter set on a fluorometer.

6. Quantitation of GFP fluorescence in terms of micrograms of recombinant EGFP protein is obtained by converting the relative fluorescence units (RFU) from the linear regression standard curve. The results are presented in Fig. 1. The amount of GFP fluorescence in the untreated JH2031 sample, corresponding to the fluorescence of 0.3 μg of EGFP protein, is set as the 100% reference.

Figure 1 shows the effect of seven fixation procedures on GFP fluorescence. Acetone fixation results in a good preservation of GFP fluorescence, but as it permeabilizes the membrane, GFP will leak out of cells and complicate flow cytometric (see Fig. 4 on p. 61) and microscopic analysis. Methanol-based fixation has a dramatic effect on GFPmut1 fluorescence and is an unsuitable fixative for GFP. Glutaraldehyde increases the autofluorescence of the wild-type *Salmonella* strain considerably (220% brighter) and is therefore not appropriate for GFP-based studies. However, GFPmut1 fluorescence is largely unaffected by fixatives such as paraformaldehyde, formalin, and buffered formol acetone. The requirement of subsequent antibody labeling procedures will determine which of these three fixatives is most suitable.

Fluorometry provides a quick and useful way of measuring GFP fluorescence of a bacterial population, but has several limitations:

1. First, most of the media in which bacteria are grown is autofluorescent. Complex media such as LB often contain flavonoids, which emit light in the same wavelength range as GFP. To prevent this problem, bacterial samples need to be washed and resuspended in a nonfluorescent solution such as PBS. To perform kinetic measurements, in real time, the choice of a minimal growth medium with a lower level of autofluorescence is an option. Unfortunately, this limits the number of environmental signals that can be tested and restricts the detectable fluorescence intensity range, preventing the measurement of low levels of fluorescence.

2. A minimal number of bacteria expressing GFP are required to produce measurable fluorescence. GFP fluorescence detection occurs as soon as its intensity reaches at least 3 times or more the level of the negative control (here non-GFP expressing bacteria). For example on a Molecular Devices fMax fluorometer, 5×10^5

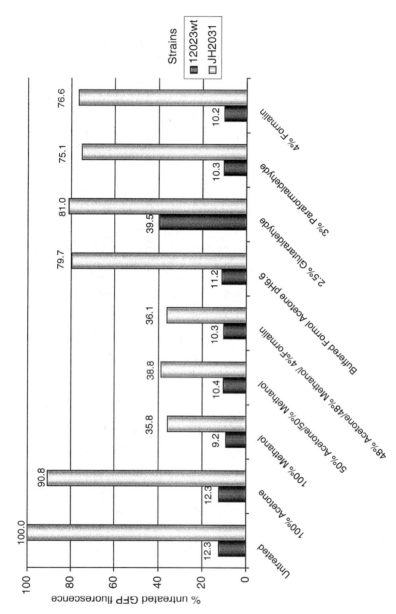

FIG. 1. Effect of various fixatives on GFP fluorescence. *Salmonella enterica* sv. *typhimurium* 12023 derivative (JH2031, light bars) containing a single copy chromosomal *ssrA::gfpmut1* fusion was treated with various fixatives and compared with the wild-type 12023 strain (dark bars) as described in the text and Table II. Bacterial cells were subsequently washed three times in filtered PBS and green fluorescence was measured in microtiter plates in an fMax fluorometer (Molecular Devices). A purified recombinant EGFP protein standard (Clontech) was used.

Salmonella cells containing a chromosomal *ssrA::gfpmut1* fusion (Fig. 1) is the detection limit. This corresponds to $\sim 10^3$ molecules of EGFP per bacterial cell.

3. Fluorometry measures the total sum of fluorescence intensities of all bacterial cells in a population, which is not a good indicator of single-cell gene expression, as it cannot discriminate between cell-to-cell variations in gene expression. In contrast, new fluorescence-based technologies and improvements of existing techniques such as flow cytometry allow monitoring fluorescence of a more limited number of bacterial cells, even down to the level of single bacterial cells. However, this requires expensive equipment and experience to optimize and interpret parameters and data (see below).

Overall, fluorometry is a relatively inexpensive and rapid technique for measuring and comparing GFP fluorescence which does not require specific skills or experience. It can be very useful for high-throughput screening experiments using particular environmental inducing or repressing conditions. Kinetics of gene expression can also be measured with equipment offering temperature-control and/or orbital shaking options. However, the relatively low sensitivity of most systems has led researchers to explore technologies that allow monitoring GFP gene fusion expression at the single bacterial cell level, such as flow cytometry.

Flow Cytometry of Microorganisms

Flow cytometry is a powerful technique for studying individual mammalian or bacterial cells within a population. It has been used extensively for eukaryotic studies, such as immune cell maturation and cytokine production.[38] However, the relatively small size of microorganisms is close to the detection limit of flow cytometers, but the use of stains, autofluorescent proteins, and suitable antibodies make it possible to discriminate them from background noise. Flow cytometry can now be used for rapid analysis of individual bacterial cells, permitting the quantitative analysis of microbial heterogeneity.[39]

Flow cytometry is a laser-based system capable of detecting many types of cells. Samples containing the cell population to be analyzed and/or sorted are carried in a fluidic stream to a specialized chamber where they are hit with a laser beam. Forward-scattered and 90° side-scattered light (respectively, FSC and SSC) is detected, and its intensity quantified and stored for each particle. Forward-scattered light gives information about the size of mammalian cells, but it is not accurate for small particles such as bacterial cells. Side-scattered light is a good indicator of mammalian cell granulometry and is useful for discriminating bacterial populations from other particles and electronic noise. When fluorescent stains are

[38] H. M. Shapiro, "Practical Flow Cytometry." Wiley-Liss, New York, 1995.
[39] G. Nebe-von-Caron, P. J. Stephens, C. J. Hewitt, J. R. Powell, and R. A. Badley, *J. Microbiol. Methods* **42,** 97 (2000).

used, the emitted fluorescence can also be detected by flow cytometry, amplified, and stored for each particle that passes through the laser beam.[8] This approach allows information concerning DNA content, the amount of protein, cell cycle, and membrane integrity to be collected for each bacterial cell analyzed within an entire population.[39,40]

The use of GFP in flow cytometry has facilitated major developments in eukaryotic and prokaryotic studies. In mammalian cells, GFP expression is often used as a selective marker for transfection experiments. In bacteria, the use of GFP has been based on the development of more soluble, brighter, blue- or red-shifted GFP mutants which were specifically selected for flow cytometric analysis.[6,7,9,41] Because bacterial cells are at the lower detection limit of most commercially available flow cytometers, it is very difficult to discriminate between bacterial populations and electronic noise or dust particles in the carrier fluidic stream. Optimization procedures such as 0.1 μm filtration of the carrier fluid and fluorescent labeling of bacteria using dyes, fluorophore tagged antibodies or the use of bright GFP as a reporter system can largely overcome this. The ability of flow cytometry to distinguish between different levels of fluorescence intensity allows the monitoring of gene expression in individual bacteria, and the sorting of cells displaying distinct levels of GFP expression.[42] The multiparameter information that is collected and stored for each bacterium or particle can be used to effectively display information on bacterial populations and on individual cells via histograms, dot plots, density plots, contour plots, and three-dimensional plots, depending on the type of information to be highlighted. Histograms depict one parameter with its intensity, whereas dot plots, density plots, and contour plots illustrate the distribution of particles within a population for two different parameters displayed (see Fig. 3). Three-dimensional plots allow separation of particle populations for more than two parameters at the same time. For each parameter detected, statistical analysis is possible at the level of each particle, allowing determination of gene induction in individual bacterial cells.

Protocol for GFP Analysis with a Benchtop Flow Cytometer

GFP detection in benchtop flow cytometers (e.g., Becton Dickinson FACScan or FACScalibur; Partec PASIII) can be performed with a standard argon ion laser, tuned at 488 nm for GFP excitation, with bandpass filters centered around 510–515 nm (e.g., 515/40 or 530/30). In the bigger and more complex flow cytometers (Coulter Epics or Altra, Becton Dickinson FACSvantage), tuneable lasers allow the efficient use of different GFP variants that possess excitation wavelengths distinct from 488 nm. The laser power must remain below 100 mW to avoid high noise levels on forward-scattered light detection.

[40] J. Vives-Rego, P. Lebaron, and G. Nebe-von-Caron, *FEMS Microbiol. Rev.* **24**, 429 (2000).
[41] K. R. Siemering, R. Golbik, R. Sever, and J. Haseloff, *Curr. Biol.* **6**, 1653 (1996).
[42] R. H. Valdivia and L. Ramakrishnan, *Methods Enzymol.* **326**, 47 (2000).

MATERIALS

Benchtop flow cytometer (e.g., FACScalibur, Becton Dickinson)
Suitable software (e.g., Cell Quest, Becton Dickinson)
PBS, pH 7.4, 0.22 μM filtered
Vortex
Sonicating water bath
Vacuum system
FACS (fluorescence activated cell sorter) tubes
Dulbecco's modified Eagle's medium (DMEM; GIBCO)
Fetal bovine serum (FBS; GIBCO)
24-Well tissue culture plates
1% (v/v) Triton X-100 (Sigma)
4% Formalin (Sigma)

SETUP

1. Switch on flow cytometer and computer according to manufacturers' description and acquire data from the bacterial sample to set up parameters for the amplification voltages. As detection of bacteria requires the collection of information with logarithmic amplifiers for both light scattering and fluorescence, the dynamic range of detection is increased at the expense of higher background noise. *Note:* Using 0.1 μM filtered PBS and sheath fluid can reduce background noise.

2. The voltages of the FSC photodiode and SSC photomultiplier (PMT) detectors are gradually increased until the bacterial population is detected and clearly visible.

3. The voltage of the fluorescence PMT detector is gradually increased to set the detection borders of the lowest and highest intensity. This requires the use of fluorescent and nonfluorescent bacterial cells. The voltage is usually increased until the nonfluorescing population remains within the first log decade of the plot scale. If several fluorescent labels are used, compensation is set up to adjust spectral overlap. Unlabeled bacteria, bacteria labeled with each fluorescent stain separately, and bacteria labeled with both stains are required to set up compensation. It is important that compensation for each color be set using the brightest stained population. *Note:* Once set up, compensation and PMT voltage should remain unmodified throughout acquisition.

4. The size of the sample will vary depending on the type of experiment. For eukaryotic samples a concentration of around 10^6 cells/ml is used. If the aim of the work is to look at the predominant bacterial population, fluorescence analysis of 5×10^3 individual bacterial cells (around 10^6 cells/ml using the lowest flow rate, 12 μl $\pm$ 3 μl/min) usually gives reliable quantification of gene expression. A greater number of bacteria is necessary to look at a specific subpopulation. This

number can easily be increased if the subpopulation of interest represents less than 1% of the total population.[8]

Optimization of Bacterial Flow Cytometry. Aggregation and clumping of bacterial cells often occurs during flow cytometric analysis of certain bacterial species, such as mycobacteria. To overcome this problem, detergents can be used and samples sonicated prior to analysis.[13] However, it is important to avoid reagents that affect GFP fluorescence.

The generation of aerosols during flow cytometric analysis of bacteria gives rise to safety concerns. The risk is particularly significant with droplet-forming flow cytometers (e.g., Coulter Epics or Altra, Becton Dickinson FACS Vantage). On these machines, droplets of carrier fluid can be formed which only contain one particle. Working with pathogens on droplet-forming flow cytometers requires the machine to be located in an appropriate containment facility and an effective decontamination procedure to be used. A long rinse followed by 10% (v/v) bleach solution is usually sufficient, or the use of 0.5% sodium dodecyl sulfate (SDS) followed by 95% (v/v) ethanol. However, the use of ethanol risks DNA precipitation in the tubing of the fluidic system. An attractive alternative is to avoid the use of live bacterial cells by treating with fixative prior to analysis, which is appropriate for many applications.

Detection of GFP fluorescence as a reporter of gene expression requires the gene promoter of interest to be transcribed at a reasonable level. To ensure signal detection, most studies have use plasmid-borne transcriptional GFP fusions as reporter of gene activity. Such multicopy GFP expression allows optimal separation of green fluorescing bacteria from background noise. Figure 2 illustrates the expression of an *ssrA::gfpmut1* transcriptional fusion in *S. typhimurium,* either carried on a plasmid (pJSG110) or as a single copy integrated on the chromosome (strain JH2031). We favor a merodiploid approach, where the *gfp* gene fusion of interest is integrated at a specific site on the chromosome which does not affect virulence.

GFP expression of the plasmid-borne *ssrA::gfpmut1* fusion is easily detectable above the autofluorescence from the wild-type strain, as the presence of multiple *gfp* copies results in increased GFP fluorescence. However, such high levels of GFP protein can be toxic for the cells, resulting in partial or complete loss of the plasmid in animal models or cultivated cell lines (I. Hautefort, unpublished, 2000). Such toxicity could give rise to variable plasmid copy number between cells, preventing the accurate measurement of gene expression. However, *Salmonella* cells expressing the chromosomal *ssrA::gfpmut1* fusion only produce fivefold more GFP fluorescence than the autofluorescence of the negative control, illustrating the challenge of measuring low levels of gene expression. Importantly, chromosomal integrated fusions have the advantage of genetic stability and are present at the same copy number as the wild-type gene of interest, making them more reliable than plasmid-borne fusions.

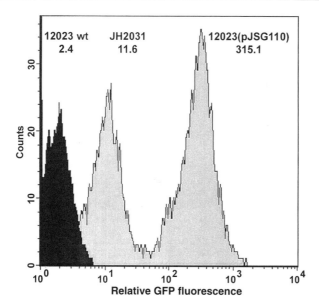

FIG. 2. Flow cytometric analysis of *S. typhimurium* 12023-derived strains expressing a *ssrA::gfpmut1* transcriptional fusion either from a plasmid (12023(pJSG110)), or as a single copy integrated on the chromosome (JH2031) compared with wild type 12023 (in black) that does not harbor any *gfp* gene. Strains were grown overnight in LB broth, fixed in 4% formalin solution for 1 min, and washed in PBS prior to their analysis by flow cytometry in a FACScalibur (Becton Dickinson). Relative GFP fluorescence of each strain is indicated.

Specific fluorescent antibodies that recognize the pathogen under study offer a useful method for discriminating bacteria from background noise. This approach allows the detector of the flow cytometer to be triggered first on the fluorescent antibody signal [for example, phycoerythrin (PE) that is measurable by a different detector than is used for GFP]. This PE-labeled population can be selected by drawing a region around it, a procedure called gating (see Fig. 5B). Only the GFP fluorescence of the gated population is subsequently measured, avoiding problems with background fluorescence. Figure 3 shows the flow cytometric separation of a mixture of GFP-expressing *Salmonella* and the corresponding wild-type strain.

Antibody labeling allows the discrimination between the bacterial population of interest and background noise, to generate robust experimental data (Fig. 3). However, suitable antibodies are not always available and the stability of some surface epitopes can be modified by fixation, preventing the use of antibodies. Intensity of the chosen fluorophore and its sensitivity to photobleaching also need to be considered. The use of live bacteria limits antibody use, and the stability of the antibody–antigen interaction can be lost on sorting. Although the use of antibodies allows a better distinction between the bacterial population and noise,

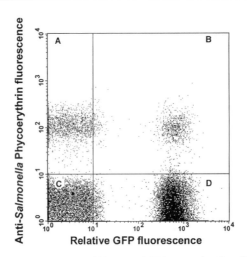

FIG. 3. Flow cytometric separation of wild-type and GFP-expressing *S. typhimurium* populations, either unlabeled or labeled with a phycoerythrin-conjugated anti-*Salmonella* antibody. Quadrants A and B show phycoerythrin-labeled SL1344 wild type and SL1344 expressing a plasmid-borne *rpsM::gfpmut3* fusion pFPV25.1 (29), respectively. Quadrants C and D show the same populations without label.

optimizing specific labeling procedures for the bacterial species of interest can be required.

The visualization of *gfp* fusions that are expressed at low levels requires a substantial increase in GFP brightness. This enables discrimination of the GFP-expressing bacteria from non-expressing cells and background noise. A variant of GFP, called GFP$^+$, has shown 130-fold brighter fluorescence than the wild-type protein (21). GFP$^+$ contains the GFP$_{uv}$ and GFPmut1 mutations, which improve the folding and enhance brightness of the protein, respectively (Table I). GFP$^+$ exhibits great promise for monitoring gene expression in bacterial cells. We tested the effect of acetone fixation on GFP$^+$ fluorescence (Table II) by flow cytometry. As shown in Figs. 4A and 4B, acetone changes the light scattering properties of the GFP-expressing *E. coli* cells that carry a plasmid-borne *rpsM::gfp$^+$* fusion. First, acetone treatment reduces the intensities of side- and forward-scattered light, suggesting modification of the bacterial cell surface. Second, the bacterial population is tightly clustered before treatment (Fig. 4A) while appearance of cell debris is visible in Fig. 4B. An important advantage of flow cytometry compared with fluorometry is shown in Fig. 4C. Although the 10% reduction in overall GFP$^+$ fluorescence is for both methods the same (Figs. 1 and 4C), flow cytometry discriminates two distinct populations after acetone treatment. GFP$^+$ fluorescence of the minor population, 25% of the bacteria, is reduced by 98%, likely due to GFP leakage. Therefore, acetone fixation cannot be trusted for flow cytometric analysis of GFP$^+$ fluorescence in bacteria.

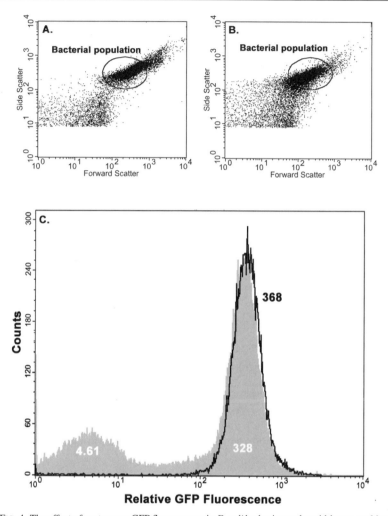

FIG. 4. The effect of acetone on GFP fluorescence in *E. coli* harboring a plasmid-borne *rpsM::gfp⁺* fusion measured by flow cytometry. (A) and (B) shows the light scattering properties of the bacterial cells before (A) and after (B) acetone treatment. GFP⁺ fluorescence was determined in the selected bacterial gated population. (C) compares GFP⁺ fluorescence before (solid line) and after (gray shaded line) acetone treatment. Numbers correspond to the relative GFP fluorescence intensities.

Fluorescence Activating Cell Sorting (FACS) of Bacteria. One of the most powerful features of flow cytometers is the ability to sort cells. Sorting of particles can be done on flow cytometers equipped with a droplet-forming flow-in-air sorting system (e.g., Coulter Epics or Altra, Becton Dickinson FACSVantage) or electromechanically on a benchtop flow cytometer (e.g., Becton Dickinson FACScalibur; Partec PASIII). Cell sorting has proved immensely important for eukaryotic research and is gaining popularity in microbiology. The use of stains, autofluorescent proteins, and suitable antibodies makes it possible to discriminate bacteria from background noise and facilitates microbial cell sorting. Although commercial droplet-forming flow cytometers are able to sort single cells rapidly, they are costly, mechanically complex, and require highly trained personnel for operation and maintenance. Cell sorting is suited to a wide range of applications and produces valuable multiparameter information at a single bacterial cell level. Publications have demonstrated the possibilities of this technique.[39,40]

Cell sorting is also possible on benchtop flow cytometers, although the mechanism of sorting does not enable the fast, accurate single-cell sorting in small volumes that can be achieved with flow-in-air machines. Benchtop flow cytometers lack the ability to sort bacteria and deposit individual cells onto agar plates, microscope slides, or microtiter plates. For further reading an excellent article describing protocols and application of GFP in bacteria and FACS in a benchtop flow cytometer is recommended.[42] An interesting development is the production of a disposable microfabricated fluorescence-activated cell sorter, which sorts by means of electroosmotic flow and has been used for enrichment of living GFP-expressing *E. coli* cells.[43]

Conditional Gene Expression during Invasion

Cultivated cell lines offer a simple model for the study of bacterial virulence gene expression. For pathogens such as *Salmonella typhi*, which lack an animal model, cultivated cell lines play an important role. Intracellular bacterial gene expression has been studied within various types of eukaryotic cells. Inoculum doses and incubation time necessary for the invasion of mammalian cells by the bacterial strain vary depending on the pathogen of interest. For example, *S. typhimurium* invades cultivated cells mainly within the first 45 min after infection.[44]

Protocol for Bacterial Invasion of Mammalian Cells

The protocol presented here is an example of how to measure and compare gene expression levels of a *gfp* transcriptional fusion in intracellular and extracellular bacteria, to determine the level of induction of gene expression during invasion of the host cell.

1. Mammalian cells used for bacterial invasion are grown to confluent monolayers in 24-well plates in tissue culture medium such as Dulbecco's modified Eagle's medium (DMEM) containing 10% fetal bovine serum. In parallel, the pathogen containing the *gfp* gene fusion to be tested is grown in appropriate medium for inducing invasion gene expression. For instance, *S. typhimurium*

---

[43] A. Y. Fu, C. Spence, A. Scherer, F. H. Arnold, and S. R. Quake, *Nat. Biotechnol.* **17,** 1109 (1999).
[44] B. D. Jones, C. A. Lee, and S. Falkow, *Infect. Immun.* **60,** 2475 (1992).

invasion genes are induced when the bacteria are grown overnight under high osmolarity and low oxygen.[45,46]

2. Infection is generally preceded by at least two washing steps of the cell monolayer in prewarmed DMEM. The multiplicity of infection (MOI) with bacteria can vary from 5 : 1 to 100 : 1 for the same pathogen depending on the study. Bacteria can be gently centrifuged at room temperature (5 min, 2000g) onto the monolayer prior to incubation in order to maximize the contact between pathogen and eukaryotic cells. Infected cells are incubated at 37° in 5% (v/v) CO_2 atmosphere, or in normal atmosphere when the tissue culture medium contains sodium bicarbonate.

3. At the end of the incubation the monolayer is washed at least twice with prewarmed DMEM. To compare bacterial gene expression inside and outside mammalian cells after incubation, the traditional gentamicin protection approach is not required. Prewarmed DMEM is added and incubation is carried on for as long as appropriate for the study (usually 2 to 5 hr). At the end of the experiment, the supernatant containing extracellular bacteria is removed, and bacteria are fixed and analyzed by flow cytometry.

4. The monolayer is washed three times with ice-cold PBS. Subsequently, 200 μl of 1% (v/v) Triton X-100 is added per well and incubated for 10 min on ice, while carefully pipetting the solution up and down to avoid foaming. Then 800 μl PBS is added per well and lysed cells are scraped from the bottom of the well with a pipette tip. Intracellular and extracellular bacteria are fixed in 4% formalin for 1 min at room temperature, washed three times in PBS, and analyzed by flow cytometry.

Figure 5C shows induction of *ssaG::gfpmut3* plasmid-borne fusion (*mig10*) in *S. typhimurium*, 6 hr after invasion of epithelial HEp-2 cells. We observed 400-fold *ssaG* induction, as previously observed during invasion of macrophages, HEp-2, or dendritic cells.[10] Figure 5A illustrates the use of a specific anti-*Salmonella* antibody to separate bacterial cells from HEp-2 cell debris. Without such labeling, it is very difficult to detect a small GFP-expressing bacterial population, because of the large amount of epithelial cell debris that causes the majority of acquired events.

Differential Fluorescence Induction (DFI): A Powerful Technique to Identify ivi Genes

Differential fluorescence induction (DFI) was developed to identify *ivi* genes and to measure *in vivo* gene expression levels.[10] It is an alternative and complementary technique to the IVET promoter-trap approach or to the STM negative

---

[45] J. E. Galan, K. Nakayama, and R. D. Curtiss, *Gene* **94,** 29 (1990).

[46] V. Bajaj, R. L. Lucas, C. Hwang, and C. A. Lee, *Mol. Microbiol.* **22,** 703 (1996).

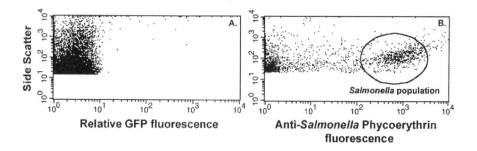

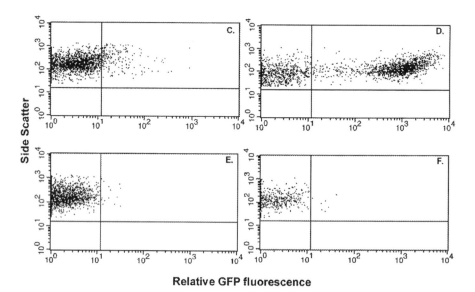

FIG. 5. Induction of *Salmonella ssaG* gene expression during invasion of epithelial HEp-2 cells. *S. typhimurium* SL1344 strain harbouring a plasmid-borne *ssaG::gfpmut3* fusion 10 (C,D) was used to infect HEp-2 cell monolayers in comparison with the wild-type SL1344 strain (E,F). Six hours after infection detergent was used to release intracellular bacteria which were analyzed by flow cytometry. To facilitate identification of the bacterial population among the green fluorescent HEp-2 cell debris (A), released bacteria were stained with a specific PE-conjugated anti-*Salmonella* antibody (B). Intracellular bacterial fluorescence was measured on the selected Salmonella gated population (D,F) in comparison with extracellular bacteria that remained outside the eukaryotic cells (C,E).

selection. DFI is based on the insertion of random chromosomal DNA fragments of the pathogen of interest into a plasmid vector upstream of a promoterless *gfp* gene and is described in Fig. 6. Bacteria containing *gfp* gene fusions are pooled and either used to infect cultured mammalian cells or exposed to various environmental stimuli *in vitro,* such as low pH. Bacteria with induced *gfp* expression are analyzed by

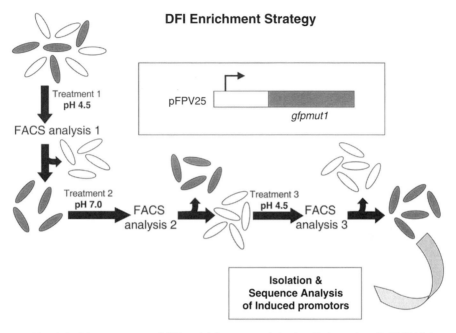

FIG. 6. Enrichment strategy of differential fluorescence induction. Redrawn from R. H. Valdivia and S. Falkow, *Mol. Microbiol.* **22**, 367 (1996).

flow cytometry and sorted on the basis of increased GFP fluorescence. The sorted bacterial population is subsequently grown in normal laboratory culture conditions and reanalyzed by flow cytometry. Only the bacteria which do not express GFP in these conditions are sorted, to eliminate constitutively expressed genes. A second selective round is performed and followed by flow cytometric analysis and sorting to ensure selection of genes that are induced only either during invasion or after exposure to a specific environmental signal. DNA fragments inserted upstream of the promoterless *gfp* gene that were responsible for GFP expression under the selective conditions are subsequently sequenced and identified.

DFI has identified genes of *S. typhimurium* that were induced on invasion of macrophages or dendritic or epithelial cells,[10] or under acidic conditions.[29] DFI has also been applied to other bacterial pathogens such as mycobacteria[47–49] and appears to be a promising approach for studying virulence gene expression during

[47] L. Kremer, A. Baulard, J. Estaquier, O. Poulain-Godefroy, and C. Locht, *Mol. Microbiol.* **17**, 913 (1995).

[48] S. Dhandayuthapani, L. E. Via, C. A. Thomas, P. M. Horowitz, D. Deretic, and V. Deretic, *Mol. Microbiol.* **17**, 901 (1995).

[49] L. Ramakrishnan, N. A. Federspiel, and S. Falkow, *Science* **288**, 1436 (2000).

infection by several pathogens. The potential of DFI has been enhanced by use of optical trapping, resulting in the identification of environmentally induced genes in single bacterial cells.[50]

However, DFI possesses several limitations; like IVET, it will not identify genes that already exhibit GFP expression *in vitro,* but are essential for successful infection of a host by the pathogen. More importantly, DFI will only identify genes that are switched from OFF to ON *in vivo;* it does not give information about the role of *ivi* genes during infection.[51]

Summary

To date, the majority of studies of bacterial gene expression have been carried out on large communities, as techniques for analysis of expression in individual cells have not been available. Recent developments now allow us to use reporter genes to monitor gene expression in individual bacterial cells. Conventional reporters are not suitable for studies of living single cells. However, variants of GFP have proved to be ideal for the study of development, cell biology, and pathogenesis and are now the reporters of choice for microbial studies. In combination with techniques such as DFI and IVET and the use of flow cytometry and advanced fluorescence microscopy, the latest generation of GFP reporters allows the investigation of gene expression in individual bacterial cells within particular environments. These studies promise to bring a new level of understanding to the fields of bacterial pathogenesis and environmental microbiology.

Acknowledgments

We are grateful to Reginald Boone, Kamal Ivory, Margaret Jones, Douglas Kell, Gerhard Nebe-von-Caron, Jonathan Porter, Howard Shapiro, and Raphael Valdivia for sharing expertise, and we thank Maria José Proenca for technical assistance. I.H. is supported by a Training and Mobility of Researchers fellowship from the European Union (contract number ERBFMRXCT9). R.J.M.B. and J.C.D.H. are supported by the BBSRC.

[50] D. Allaway, N. A. Schofield, M. E. Leonard, L. Gilardoni, T. M. Finan, and P. S. Poole, *Environ. Microbiol.* **3,** 397 (2001).
[51] I. Hautefort and J. C. D. Hinton, *Methods Microbiol.* **31,** 55 (2002).

[5] Genetic Methods for Deciphering Virulence Determinants of *Mycobacterium tuberculosis*

By MIRIAM BRAUNSTEIN, STOYAN S. BARDAROV,
and WILLIAM R. JACOBS, JR.

Introduction

More than 100 years have passed since Robert Koch made his momentous identification of *Mycobacterium tuberculosis* as the causative agent of tuberculosis. More than 50 years have passed since the introduction of effective chemotherapy for tuberculosis. Yet, currently one-third of the world's population is infected with the tubercle bacillus and more people are dying from tuberculosis than at any other time in history (www.who.int/inf-fs/en/fact104.html). This tragic situation is only worsening from the impact of HIV/AIDS on tuberculosis and the emergence of multidrug resistant *M. tuberculosis* strains.

Fortunately, recent advances in tuberculosis research are enabling efforts to understand the virulence of *M. tuberculosis*. The *M. tuberculosis* genome has been fully sequenced[1] and is now available through two public domains: genolist.pasteur.fr/TubercuList/ and www.tigr.org. In addition, powerful genetic tools for mycobacteria have been developed over the last 10 years. Most notably, allelic exchange (gene replacement) and transposon mutagenesis can now be performed in an efficient manner in *M. tuberculosis*.[2-4] Consequently, tuberculosis researchers can now inactivate single genes and create defined mutations in the *M. tuberculosis* genome. The phenotypic analyses of the resulting mutants can reveal the role individual gene products play in *M. tuberculosis* physiology and/or pathogenesis. This is an exciting time in tuberculosis research. This genetic power has already enlarged the list of genes known to participate in *M. tuber-culosis* virulence,[5] brought to our attention previously unknown components of

[1] S. T. Cole, R. Brosch, J. Parkhill, T. Garnier, C. Churcher, D. Harris, S. V. Gordon, K. Eiglmeier, S. Gas, C. E. Barry III, F. Tekaia, K. Badcock, D. Basham, D. Brown, T. Chillingworth, R. Connor, R. Davies, K. Devlin, T. Feltwell, S. Gentles, N. Hamlin, S. Holroyd, T. Hornsby, K. Jagels, A. Krogh, J. McLean, S. Moule, L. Murphy, K. Oliver, J. Osborne, M. A. Quail, M.-A. Rajandream, J. Rogers, S. Rutter, K. Seeger, T. Skelton, R. Squares, S. Squares, J. E. Sulston, K. Taylor, S. Whitehead, and B. G. Barrell, *Nature* **393,** 537 (1998).

[2] W. R. J. Jacobs, *in* "Molecular Genetics of Mycobacteria" (J. G. F. Hatfull and W. R. Jacobs, eds.), p. 1. ASM Press, Washington, D.C., 2000.

[3] W. R. Jacobs, Jr., G. V. Kalpana, J. D. Cirillo, L. Pascopella, S. B. Snapper, R. A. Udani, W. Jones, R. G. Barletta, and B. R. Bloom, *Methods Enzymol.* **204,** 537 (1991).

[4] V. Pelicic, J. M. Reyrat, and B. Gicquel, *Mol. Microbiol.* **28,** 413 (1998).

[5] J. A. Triccas and B. Gicquel, *Immunol. Cell Biol.* **78,** 311 (2000).

tuberculosis pathogenesis,[6,7] and increased our understanding of how the bacillus survives *in vivo*.[8]

This article will describe the methodologies for directed allelic exchange and transposon mutagenesis, employed by our laboratory to engineer mutant strains of *M. tuberculosis*. Allelic exchange protocols based on plasmid transformation or a recently developed mycobacteriophage delivery system will be presented. We will also describe the use of the mycobacteriophage delivery system for transposon mutagenesis of *M. tuberculosis*.

Genetic Methods in Mycobacteria

The ability to apply molecular genetics to the study of *M. tuberculosis* only recently became a reality. Efficient methods for DNA transfer, utilizing transformation by electroporation and phage transduction, are now available for the production of defined mutations in *M. tuberculosis*. The development of these techniques was impeded by the intrinsic difficulties of working with *M. tuberculosis,* which include slow growth (a generation time of 18–24 hr), biosafety considerations, and a tendency to grow in clumps (making isolation of individual clones problematic).

Transformation

Transformation is a phenotypic change resulting from exogenous DNA uptake by a bacterium. By means of electroporation, both fast- and slow-growing mycobacteria can be transformed with exogenous DNA, such as plasmids.[9] However, slow-growing mycobacteria are not highly competent for electroporation, which results in low transformation efficiencies. The transformation efficiency of individual *M. tuberculosis* strains varies between 10^2 and 10^5 transformants/μg DNA of pYUB412, a mycobacterial vector that integrates into the chromosome.[10,11] In our experience, the *M. tuberculosis* strain H37Rv exhibits higher transformation efficiency than either the Erdman or CDC1551 strains.

Genetic material can be electroporated into *M. tuberculosis* using three major types of plasmids: replicative, integrating, and "suicide" (Table I). The most commonly used replicating vectors contain oriM, a mycobacterial origin of replication

[6] L. R. Camacho, D. Enserguelx, E. Perez, B. Gicquel, and C. Guilhot, *Mol. Microbiol.* **34,** 257 (1999).

[7] J. S. Cox, B. Chen, M. McNeil, and W. R. Jacobs, Jr., *Nature* **402,** 79 (1999).

[8] J. D. McKinney, K. Honer zu Bentrup, E. J. Munoz-Elias, A. Miczak, B. Chen, W. T. Chan, D. Swenson, J. C. Sacchettini, W. R. Jacobs, Jr., and D. G. Russell, *Nature* **406,** 735 (2000).

[9] S. B. Snapper, L. Lugosi, A. Jekkel, R. E. Melton, T. Kieser, B. R. Bloom, and W. R. Jacobs, Jr., *Proc. Natl. Acad. Sci. U.S.A.* **85,** 6987 (1988).

[10] M. Braunstein, unpublished results (1999).

[11] M. S. Pavelka, Jr., and W. R. Jacobs, Jr., *J. Bacteriol.* **181,** 4780 (1999).

TABLE I

CLONING VECTORS USED IN GENETIC ANALYSIS OF MYCOBACTERIA

Name	Genotype	Description	Ref.
Plasmids			
pMV261	$colE1$, $oriM$, aph, $P_{hsp60}{'}$	Multicopy shuttle plasmid able to replicate in *E. coli* (*colEI*) and in mycobacteria (*oriM*). The *hsp60* promoter can be used to control expression of cloned genes	*a*
pMV361	$colE1$, $att\text{-}int^{L5}$, aph, $P_{hsp60}{'}$	Integrating vector. The *hsp60* promoter can be used to control expression of cloned genes	*a*
pYUB631	$colE1$, $oriM$, aph, $P_{hsp60}{'}\text{-}sacB$	Also known as pMP7. Replicative (oriM based) plasmid vector. The presence of *sacB* in pYUB631 confers an intrinsic instability to the replicating plasmid and provides a convenient negative selection for the plasmid loss	*b*
pYUB657	$colE1$, bla, hyg, $P_{hsp60}{'}\text{-}sacB$	Nonreplicative ("suicide") plasmid vector. The presence of sacB gene enables selection of cells that have undergone allelic exchange by an intramolecular recombination event	*b*
Cosmids			
pYUB328	$colE1$, bla, cos^{λ}	Cosmid vector containing *NotI-EcoRI-PacI* polilinker and two lambda *cos* sites. Useful vector for generation of mycobacterial cosmid libraries using lambda *in vitro* packaging system	*c*
pYUB572	$colE1$, bla, cos^{λ}	Smaller molecular size derivative of pYUB328. The *bla* gene flanked by *Bsp*HI sites allows for easy cloning of a DNA fragment containing a different antibiotic selection marker by blunt end ligation. The minimal lambda *cos* site (127 bp) allows cloning into phasmids using the lambda *in vitro* packaging system	*d*
pYUB854	$colE1$, $res\text{-}hyg\text{-}res$, cos^{λ}	Derivative of pYUB572. *bla* gene was replaced with a specialized *hyg* cassette flanked by $\gamma\text{-}\delta$ *res* sites recognized by the $\gamma\text{-}\delta$ resolvase. Convenient multiple cloning sites flanking the cassette allow easy directional cloning of flanking DNA fragments for constructing homologous recombination substrates	*d*
Phasmids			
phAE87	TM4*ts*::pYUB328	Conditionally replicating shuttle phasmid derivative of TM4 mycobacteriophage. pYUB328 cosmid inserted in a 354 bp deleted region of the phage	*e*
phAE159	TM4*ts*::pYUB328	Derivative of phAE87 with pYUB328 inserted in 5.6 kb deleted region of the phage	*f*
phAE94	TM4*ts*::Tn5367(*kan*).	Derivative of phAE87 containing Tn*5367* transposon	*g*
phAE175	TM4*ts*::Tn5370*mini*(hyg).	Derivative of phAE87 containing Tn*5370 mini* transposon	*h*

[a] C. K. Stover, V. F. de la Cruz, T. R. Fuerst, J. E. Burlein, L. A. Benson, L. T. Bennett, G. P. Bansal, J. F. Young, M. H. Lee, G. F. Hatfull, S. B. Snapper, R. G. Barletta, W. R. Jacobs, Jr., and B. R. Bloom, *Nature* **351**, 456 (1991).

[b] M. S. Pavelka, Jr., and W. R. Jacobs, Jr., *J. Bacteriol.* **181**, 4780 (1999).

[c] V. Balasubramanian, M. S. Pavelka, Jr., S. S. Bardarov, J. Martin, T. R. Weisbrod, R. A. McAdam, B. R. Bloom, and W. R. Jacobs, Jr., *J. Bacteriol.* **178**, 273 (1996).

[d] S. Bardarov and W. R. Jacobs, Jr., unpublished results (1998).

[e] C. Carriere, P. F. Riska, O. Zimhony, S. Kriakov, S. Bardarov, J. Burns, J. Chan, and W. R. Jacobs, Jr., *J. Clin. Microbiol.* **35**, 3232 (1997).

[f] J. Kriakov, unpublished results (2000).

[g] S. Bardarov, J. Kriakov, C. Carriere, S. Yu, C. Vaamonde, R. A. McAdam, B. R. Bloom, G. F. Hatfull, and W. R. Jacobs, Jr., *Proc. Natl. Acad. Sci. U.S.A.* **94**, 10961 (1997).

[h] J. S. Cox, B. Chen, M. McNeil, and W. R. Jacobs, Jr., *Nature* **402**, 79 (1999).

originally isolated from the *M. fortuitum* plasmid pAL5000.[12,13] The integrating vectors contain the L5 mycobacteriophage attachment/integration system (attP/int). The attP/int system promotes site-specific integration of these vectors at the attB site in the chromosome of mycobacteria.[13,14] The "suicide" vectors contain neither a mycobacterial origin of replication nor an integration (attP/int) system. Once introduced into a mycobacterial cell these plasmids do not partition into the daughter progeny cells and will be lost from the population. For a comprehensive discussion of vectors used in mycobacterial genetics the reader is referred to an excellent review by Pashley and Stoker.[15]

Transduction

Transduction is the natural transfer of genetic material to a bacterium via bacteriophage. There are two types of transductional DNA transfer—generalized and specialized. Generalized transducing phages exclusively contain bacterial DNA from the donor cell, and they transfer it to a recipient cell. In contrast, specialized transducing phages contain only a small piece of bacterial DNA along with phage DNA. Specialized transductional DNA transfer is characteristic of the lysogenic phages, which as part of their normal physiology integrate at specific sites (*attB*) in the bacterial chromosome. Specialized transducing phages are formed when a small piece of adjacent chromosomal DNA is packaged along with phage DNA into the phage head. With the development of molecular biology techniques for manipulating DNA, the definition of specialized transducing phages has been expanded to include phage "lines" that transduce any specific fragment of DNA.[16] Transductional DNA transfer is highly efficient. In proper conditions, nearly 100% of phage-sensitive cells can be infected by a transducing phage.

The relatively low and inconsistent transformation efficiency of *M. tuberculosis* has been a major limitation to methods that require the introduction of exogenous DNA, including allelic exchange and transposition. For this reason, our laboratory has developed specialized transducing mycobacteriophages to circumvent this problem. The phages used are derivatives of a temperature-sensitive mutant (PH101) of the broad host range mycobacteriophage TM4.[17] PH101 is able to

[12] A. Labidi, H. L. David, and D. Roulland-Dussoix, *Ann. Inst. Pasteur Microbiol.* **136B,** 209 (1985).
[13] C. K. Stover, V. F. de la Cruz, T. R. Fuerst, J. E. Burlein, L. A. Benson, L. T. Bennett, G. P. Bansal, J. F. Young, M. H. Lee, G. F. Hatfull, S. B. Snapper, R. G. Barletta, W. R. Jacobs, Jr., and B. R. Bloom, *Nature* **351,** 456 (1991).
[14] M. H. Lee, L. Pascopella, W. R. Jacobs, Jr., and G. F. Hatfull, *Proc. Natl. Acad. Sci. U.S.A.* **88,** 3111 (1991).
[15] C. Pashley and N. G. Stoker, *in* "Molecular Genetics of Mycobacteria" (J. G. F. Hatfull and W. R. Jacobs, eds.), p. 55. ASM Press, Washington, D.C., 2000.
[16] R. A. Weisberg, *in* "*Escherichia coli* and *Salmonella:* Cellular and Molecular Biology" (F. C. Neidhardt, ed.), p. 2442. ASM Press, New York, 1996.
[17] S. Bardarov, J. Kriakov, C. Carriere, S. Yu, C. Vaamonde, R. A. McAdam, B. R. Bloom, G. F. Hatfull, and W. R. Jacobs, Jr., *Proc. Natl. Acad. Sci. U.S.A.* **94,** 10961 (1997).

replicate and lyse mycobacteria at 30° but not at 37°. Importantly, PH101 still infects mycobacteria at both temperatures. When PH101 infects mycobacteria at 37° it only serves to introduce its genomic DNA into the cell. We use this phage as a delivery system by engineering it, *in vitro*, into a specialized transducing phage that carries either an allelic exchange substrate or transposon. The great advantage of using this phage delivery system is that essentially every cell in the population will be infected and receive the cloned allelic exchange substrate or transposon, which increases the probability of obtaining mutants.

Allelic Exchange or Gene Replacement

As soon as transformation protocols for slow-growing mycobacteria were developed, various allelic exchange experiments were attempted.[18,19] However, until recently, the engineering of specific mutations in *M. tuberculosis* was notoriously difficult. Initial attempts involved electroporation into *M. tuberculosis* of a linear DNA fragment containing a disruption/deletion of a gene marked with an antibiotic resistance cassette. It was hoped that a double-crossover homologous recombination event would occur between the marked allelic exchange construct and the wild-type allele in the chromosome and that antibiotic resistant transformants would represent such recombinants. Although success was eventually achieved with this method, once long (40 kb) linear substrates were used, it was rather inefficient.[20] By directly selecting for mutants, this approach is dependent on the probability of having many low frequency events occur at one—DNA uptake and two homologous recombination events. As detailed above, the transformation efficiency of *M. tuberculosis* is low. In addition, the frequency of a single homologous recombination event in *M. tuberculosis* is 10^{-4} to 10^{-5}.[11] Furthermore, it is known that slow-growing mycobacteria, such as *M. bovis* BCG and *M. tuberculosis,* integrate exogenous DNA into their chromosome by illegitimate as well as homologous recombination.[18,21] Although there have been few direct experimental comparisons between homologus and illegitimate recombination in *M. tuberculosis,* it appears that illegitimate recombination is more frequent with linear rather than circular DNA substrates.[22,23]

More efficient methods for allelic exchange were subsequently sought. Pretreatment of the DNA substrate with alkali and, most notably, with ultraviolet (UV) light has been reported to increase the frequency of homologous

[18] A. Aldovini, R. N. Husson, and R. A. Young, *J. Bacteriol.* **175,** 7282 (1993).

[19] B. J. Wards and D. M. Collins, *FEMS Microbiol. Lett.* **145,** 101 (1996).

[20] V. Balasubramanian, M. S. Pavelka, Jr., S. S. Bardarov, J. Martin, T. R. Weisbrod, R. A. McAdam, B. R. Bloom, and W. R. Jacobs, Jr., *J. Bacteriol.* **178,** 273 (1996).

[21] G. V. Kalpana, B. R. Bloom, and W. R. Jacobs, Jr., *Proc. Natl. Acad. Sci. U.S.A.* **88,** 5433 (1991).

[22] M. J. Colston, E. O. Davis, and K. G. Papavinasasundaram, *in* "Molecular Genetics of Mycobacteria" (J. G. F. Hatfull and W. R. Jacobs, eds.), p. 85. ASM Press, Washington, D.C., 2000.

[23] M. S. Pavelka, Jr., unpublished results (1998).

recombination.[24–26] It is important to note that with UV treatment comes the risk that unintended secondary mutations might arise in the allelic exchange substrate. Currently, the most successful transformation-based allelic exchange strategies are based on the premise that individual events be achieved in separate steps.[11,25–29] First, transformants containing a vector carrying an allelic exchange construct are obtained. A single transformant is selected and time is provided for homologous recombination. These strategies benefit greatly from the presence of counterselectable markers (described below). The transformation-based allelic exchange methods employed by our laboratory are detailed in the protocol section of this article.

Allelic exchange can also be achieved using the *in vitro* generated specialized transducing mycobacteriophages as a delivery system. Because this method serves to introduce the allelic exchange construct to nearly all mycobacterial cells in the population, a marked disruption/deletion mutant can be directly selected in a single step. This method is also described in detail in the protocol section of this article.

Transposition

Large libraries of random mutants are a prerequisite for a comprehensive genetic screen. An efficient way of generating such mutant collections is by transposon mutagenesis, which has been successfully used in diverse genera of bacteria.[30–32] Transposable elements are mobile DNA fragments that integrate into the chromosome by a mechanism that is independent of the general homologous recombination system of the host. Transposable elements are the ideal tools for genetics in mycobacteria. First, the insertion of a transposable element into a gene often leads to its inactivation. Second, the reversion frequency of the resulting "null" mutation is low since precise excision of the transposon and restoration of

[24] T. Parish, B. G. Gordhan, R. A. McAdam, K. Duncan, V. Mizrahi, and N. G. Stoker, *Microbiology* **145,** 3497 (1999).
[25] J. Hinds, E. Mahenthiralingam, K. E. Kempsell, K. Duncan, R. W. Stokes, T. Parish, and N. G. Stoker, *Microbiology* **145,** 519 (1999).
[26] T. Parish and N. G. Stoker, *Microbiology* **146,** 1969 (2000).
[27] M. K. Hondalus, S. Bardarov, R. Russell, J. Chan, W. R. Jacobs, Jr., and B. R. Bloom, *Infect. Immun.* **68,** 2888 (2000).
[28] F. X. Berthet, M. Lagranderie, P. Gounon, C. Laurent-Winter, D. Enserqueix, P. Chavarot, F. Thouron, E. Maranghi, V. Pelicic, D. Portnoi, G. Marchal, and B. Gicquel, *Science* **282,** 759 (1998).
[29] V. Pelicic, M. Jackson, J. M. Reyrat, W. R. Jacobs, Jr., B. Gicquel, and C. Guilhot, *Proc. Natl. Acad. Sci. U.S.A.* **94,** 10955 (1997).
[30] C. M. Berg and D. E. Berg, *in* "*Escherichia coli* and *Salmonella:* Cellular and Molecular Biology" (F. C. Neidhardt, ed.), p. 2588. ASM Press, New York, 1996.
[31] N. Kleckner, J. Roth, and D. Botstein, *J. Mol. Biol.* **116,** 125 (1977).
[32] N. Kleckner, J. Bender, and S. Gottesman, *Methods Enzymol.* **204,** 139 (1991).

gene function is rare or nonexistent. Most importantly, on integration transposons can introduce genetic markers, such as antibiotic resistance genes, which make the identification and clonal purification of the mutants easier. This is particularly relevant to mycobacteria because of the strong tendency of the cells to clump. The screening of transposon mutant libraries has successfully led to the identification of virulence determinants in other pathogenic bacteria.[33–40]

Numerous mobile genetic elements have been identified and characterized in mycobacteria.[41–46] Our laboratory has exploited one of these elements, IS*1096*, for transposon mutagenesis. IS*1096* is an insertion element isolated from *M. smegmatis*.[47] IS*1096* contains two open reading frames, one of which, *tnpA*, encodes the transposase responsible for its movement. The other open reading frame designated *tnpR* has a low homology to the resolvases of Tn*1000* and Tn*552*, but its role in transposition is not clear. IS*1096* was modified into an artificial transposon, Tn*5367(aph)*, which carries *aph* (encoding kanamycin resistance).[48]

Tn*5367* transposes relatively randomly by a nonreplicative ("cut-and-paste") mechanism with a frequency of 10^{-5}.[17,48] Using Tn*5367*, transposon mutagenesis has been applied to both fast- and slow-growing mycobacteria.[17,48,49] Since Tn*5367* carries the entire machinery (*tnpA* and *tnpR*) for excision and reintegration in the

[33] B. M. Ahmer, J. van Reeuwijk, C. D. Timmers, P. J. Valentine, and F. Heffron, *J. Bacteriol.* **180**, 1185 (1998).

[34] P. I. Fields, R. V. Swanson, C. G. Haidaris, and F. Heffron, *Proc. Natl. Acad. Sci. U.S.A.* **83**, 5189 (1986).

[35] J. E. Shea, J. D. Santangelo, and R. G. Feldman, *Curr. Opin. Microbiol.* **3**, 451 (2000).

[36] K. E. Unsworth and D. W. Holden, *Phil. Trans. R. Soc. Lond. B Biol. Sci.* **355**, 613 (2000).

[37] S. L. Chiang, J. J. Mekalanos, and D. W. Holden, *Ann. Rev. Microbiol.* **53**, 129 (1999).

[38] M. Hensel, J. E. Shea, C. Gleeson, M. D. Jones, E. Dalton, and D. W. Holden, *Science* **269**, 400 (1995).

[39] J. M. Mei, F. Nourbakhsh, C. W. Ford, and D. W. Holden, *Mol. Microbiol.* **26**, 399 (1997).

[40] F. Cockerill III, G. Beebakhee, R. Soni, and P. Sherman, *Infect. Immun.* **64**, 3196 (1996).

[41] R. A. McAdam, C. Guilhot, and B. Gicquel, "Transposition in Mycobacteria." American Society for Microbiology, Washington, D.C., 1994.

[42] J. W. Dale, *Eur. Respir. J. Suppl.* **20**, 633s (1995).

[43] D. Thierry, A. Brisson-Noel, V. Vincent-Levy-Frebault, S. Nguyen, J. L. Guesdon, and B. Gicquel, *J. Clin. Microbiol.* **28**, 2668 (1990).

[44] D. Thierry, M. D. Cave, K. D. Eisenach, J. T. Crawford, J. H. Bates, B. Gicquel, and J. L. Guesdon, *Nucleic Acids Res.* **18**, 188 (1990).

[45] R. A. McAdam, P. W. Hermans, D. van Soolingen, Z. F. Zainuddin, D. Catty, J. D. van Embden, and J. W. Dale, *Mol. Microbiol.* **4**, 1607 (1990).

[46] E. P. Green, M. L. Tizard, M. T. Moss, J. Thompson, D. J. Winterbourne, J. J. McFadden, and J. Hermon-Taylor, *Nucleic Acids Res.* **17**, 9063 (1989).

[47] J. D. Cirillo, R. G. Barletta, B. R. Bloom, and W. R. Jacobs, Jr., *J. Bacteriol.* **173**, 7772 (1991).

[48] R. A. McAdam, T. R. Weisbrod, J. Martin, J. D. Scuderi, A. M. Brown, J. D. Cirillo, B. R. Bloom, and W. R. Jacobs, Jr., *Infect. Immun.* **63**, 1004 (1995).

[49] C. Guilhot, B. Gicquel, and C. Martin, *FEMS Microbiol. Lett.* **77**, 181 (1992).

genome, it remains a possibility that precise excision of the transposon might occur. For this reason a derivative of Tn*5367* was constructed in which *tnpA* is relocated to a position outside of the transposon.[50] Once this transposon, *mini-Tn5370(hyg)*, which carries *hyg* (encoding hygromycin resistance), transposes into the chromosome, it will never move again because of the absence of *tnpA*.

Different approaches have been used to introduce transposons into *M. tuberculosis*. "Suicide" vectors (discussed above) carrying transposons have been used.[48] With these vectors, only in the event of transposition will stable antibiotic resistant transformants be obtained. Although successful in fast-growing mycobacteria, this method is of limited use in slow-growing mycobacteria because of the low transformation frequency. In order to generate the large libraries of mutants we desired, our laboratory turned to conditionally replicating specialized transducing mycobacteriophages to deliver transposons. Using this method we have succeeded in generating large libraries of mutants in *M. tuberculosis*.[7,17] Pelicic and colleagues[29] have developed an alternative method for large-scale transposition in which the transposon is electroporated into the cell on a replicating vector that has a temperature sensitive origin of replication (*oriM^{ts}*). First, transformants are obtained at the permissive temperature for plasmid replication. A period of growth follows allowing for transposition to occur. The final step is the selection for cells in which the transposon has inserted into the genome and the delivery plasmid has been lost; this is achieved by selecting antibiotic resistant mutants at the nonpermissive temperature for plasmid replication. This method is also successful in generating large mutant libraries; however, the time required to generate a mutant collection using *oriM^{ts}* vector delivery is significantly longer than that required with the phage delivery system.

Application of New Mycobacterial Genetic Tools

Efficient allelic exchange and transposon mutagenesis are having a tremendous impact on tuberculosis research. The genome sequence revealed many promising candidates for factors involved in virulence. The role played by such factors in *M. tuberculosis* can now be determined through the construction of defined mutants. If a given gene plays a nonredundant role in pathogenesis, then elimination of the corresponding gene product should alter the virulence property of the strain. By testing defined mutants in virulence assays, our understanding of how *M. tuberculosis* establishes an infection, interacts with the host, causes disease, and persists in a latent state will greatly increase.

The importance of allelic exchange methodology to tuberculosis research is exemplified by the dramatic increase in the number of *M. tuberculosis* mutants

[50] J. S. Cox, unpublished results (1998).

reported over the past few years.[8,11,20,24,25,27,51–60] The ability to directly mutate a specific gene by means of gene replacement has already increased our awareness of genes involved in tuberculosis pathogenesis. *M. tuberculosis* mutants in *icl, pcaA, hma*, and *erp* genes were constructed by allelic exchange; all of these mutants are attenuated in the mouse model of tuberculosis.[8,28,51,52] With this knowledge, research is now underway to identify the precise role played by the corresponding proteins in pathogenesis.

The mycobacterial transposition systems has enabled two independent research groups to apply signature-tagged transposon mutagenesis (STM) to *M. tuberculosis*.[6,7] The STM method, which has been previously used with other bacterial pathogens, enables *in vivo* selection of attenuated mutants from pools of transposon-containing mutants.[61–65] The method requires large numbers of transposon mutants with individual mutants containing distinct transposons marked with unique DNA sequence tags. The attenuated mutants are identified as distinctly marked bacteria present among a pool of mutants used in an initial inoculum to infect the host but absent or underrepresented in the host after a defined period of infection. Between these two groups, 13 different genes were identified that, when disrupted by a transposon, lead to attenuation for growth in the lungs of mice. Notably, many of these genes cluster in a chromosomal region associated with the synthesis and transport of the complex mycobacterial lipid phthioceroldimycocerosate (PDIM).

[51] E. Dubnau, J. Chan, C. Raynaud, V. P. Mohan, M. A. Laneelle, K. Yu, A. Quemard, I. Smith, and M. Daffe, *Mol. Microbiol.* **36,** 630 (2000).

[52] M. S. Glickman, J. S. Cox, and W. R. Jacobs, Jr., *Mol. Cell* **5,** 717 (2000).

[53] T. P. Primm, S. J. Andersen, V. Mizrahi, D. Avarbock, H. Rubin, and C. E. Barry III, *J. Bacteriol.* **182,** 4889 (2000).

[54] M. Jackson, C. Raynaud, M. A. Laneelle, C. Guilhot, C. Laurent-Winter, D. Ensergueix, B. Gicquel, and M. Daffe, *Mol. Microbiol.* **31,** 1573 (1999).

[55] L. Y. Armitige, C. Jagannath, A. R. Wanger, and S. J. Norris, *Infect. Immun.* **68,** 767 (2000).

[56] M. S. Glickman, S. M. Cahill, and W. R. Jacobs, Jr., *J. Biol. Chem.* **276,** 2228 (2001).

[57] P. Chen, R. E. Ruiz, Q. Li, R. F. Silver, and W. R. Bishai, *Infect. Immun.* **68,** 5575 (2000).

[58] T. Parish and N. G. Stoker, *J. Bacteriol.* **182,** 5715 (2000).

[59] G. R. Stewart, S. Ehrt, L. W. Riley, J. W. Dale, and J. McFadden, *Tuber. Lung Dis.* **80,** 237 (2000).

[60] O. Dussurget, G. Stewart, O. Neyrolles, P. Pescher, D. Young, and G. Marchal, *Infect. Immun.* **69,** 529 (2001).

[61] J. E. Shea, M. Hensel, C. Gleeson, and D. W. Holden, *Proc. Natl. Acad. Sci. U.S.A.* **93,** 2593 (1996).

[62] N. Autret, I. Dubail, P. Trieu-Cuot, P. Berche, and A. Charbit, *Infect. Immun.* **69,** 2054 (2001).

[63] P. Lestrate, R. M. Delrue, I. Danese, C. Didembourg, B. Taminiau, P. Mertens, X. De Bolle, A. Tibor, C. M. Tang, and J. J. Letesson, *Mol. Microbiol.* **38,** 543 (2000).

[64] A. L. Jones, K. M. Knoll, and C. E. Rubens, *Mol. Microbiol.* **37,** 1444 (2000).

[65] T. E. Fuller, S. Martin, J. F. Teel, G. R. Alaniz, M. J. Kennedy, and D. E. Lowery, *Microb. Pathog.* **29,** 39 (2000).

General Materials and Methods for Genetic Manipulations of *M. tuberculosis*

Mycobacterial Strains

1. *M. tuberculosis* strains: Erdman, H37Rv, and CDC1551 (all virulent strains)
2. *Mycobacterium smegmatis* mc^{2}155 (*ept*), a high frequency transformation derivative of *M. smegmatis* mc^{2}6.[66]

Plasmids, Cosmids, and Shuttle Phasmids

Vectors routinely used in our laboratory for genetic studies of mycobacteria are listed in Table I. All DNA manipulations are performed in *E. coli* strains DH5α or HB101 using standard methods.[67] Some vectors, such as pMV261, contain the highly promiscuous mycobacterial *hsp60* promoter that can be used in transcriptional or translational fusions to control gene expression in mycobacteria.

Growth of Mycobacterial Strains[68]

Growth and Media for M. tuberculosis. M. tuberculosis must be grown in a Biosafety level 3 (BSL3) facility using appropriate protocols. Liquid growth of large volumes (50 to 100 ml) of *M. tuberculosis* is carried out in 1 liter plastic roller bottles on a roller apparatus at 5 rpm at 37°. Smaller volumes can be grown in small 25-ml plastic bottles or 15-ml plastic test tubes on a slow (100–150 rpm) shaker.

Middlebrook 7H9 broth for *M. tuberculosis* (7H9-ADS-Tw):
  4.7 g of Middlebrook 7H9 broth base (Difco, Detroit, MI)
  100 ml ADS (see below)
  10 ml 50% (v/v) Glycerol
  2.5 ml 20% (v/v) Tween 80
  Appropriate supplements/antibiotics (see below)
  900 ml Distilled water
Mix together with stirring. Filter sterilize through a 0.22-μm pore membrane.

Middlebrook 7H10 agar for *M. tuberculosis* (7H10-ADS agar):
  19 g Middlebrook 7H10 agar (Difco)
  900 ml Distilled water

[66] S. B. Snapper, R. E. Melton, S. Mustafa, T. Kieser, and W. R. Jacobs, Jr., *Mol. Microbiol.* **4,** 1911 (1990).

[67] J. Sambrook, E. F. Fritsch, and T. Maniatis, Molecular Cloning: A Laboratory Manual. Cold Spring Harbor Laboratory Press, Plainview, NY, 1989.

[68] M. H. Larsen, *in* "Molecular Genetics of Mycobacteria" (J. G. F. Hatfull and W. R. Jacobs, eds.), p. 313. ASM Press, Washington, D.C., 2000.

Autoclave for 20 min. Let cool on bench, with stirring, for approximately 30 min. Add the following components:

100 ml ADS (see below)

10 ml 50% (v/v) Glycerol

Appropriate supplements/antibiotics (see below)

2.5 ml 20% Tween 80 (optional). The incorporation of Tween 80 into agar plates will produce softer colonies that are easier to pick

Mix by stirring and pour 30 ml per plate.

Supplements for M. tuberculosis.

Albumin–dextrose–saline (ADS) enrichment:

8.1 g NaCl

50 g Bovine serum albumin fraction V (BSA; Boehringer Mannheim)

20 g Dextrose

950 ml Distilled water

Dissolve NaCl and BSA in distilled water, add dextrose, and raise final volume to 1000 ml. Centrifuge the ADS at 4500g at 4° for 10 min to pellet insoluble material. Filter sterilize through a 0.22-μm pore membrane. An alternative to ADS enrichment is oleic acid-albumin-dextrose-catalase (OADC) (Middlebrook OADC Enrichment, Becton Dickinson Microbiology; recipe also found in Ref. 68.) We have found that *M. tuberculosis* grows at a faster rate when OADC is used as supplementation.

20% Tween 80:

20 ml Tween 80 (Fisher Scientific)

80 ml Distilled water

Dissolve completely; may require heat at 56°. Sterilize by filtration through a 0.22-μm pore membrane. Store at room temperature.

Sucrose (Sigma) use at a final concentration of 3%.

Sucrose can be added, using a 50% sucrose stock, to 7H10-ADS plates after autoclaving.

Antibiotics for *M. tuberculosis:*

Hygromycin B (Roche Diagnostics/Boerhinger Mannheim): use at a final concentration of 50–75 μg/ml

Kanamycin monosulfate (Sigma): use at a final concentration of 20 μg/ml

Amino acid and/or vitamin supplements may be added at standard concentrations[69] if trying to generate amino acid or vitamin auxotrophs.

[69] D. L. Provence and R. Curtiss III, *in* "Methods in General and Molecular Bacteriology" (P. Gerhardt, ed.), p. 317. American Society of Microbiology, Washington, D.C., 1994.

Media and Growth of M. smegmatis. M. smegmatis is grown and manipulated at Biosafety level 2 (BSL2). Liquid growth of *M. smegmatis* can be carried out in glass or plastic flasks on a slow shaker (100–150 rpm) at 37°.

Middlebrook 7H9 broth for *M. smegmatis* (7H9-Tw):
    4.7 g of Middlebrook broth base (Difco)
    10 ml 20% Dextrose
    10 ml 50% Glycerol
    2.5 ml 20% Tween 80
    980 ml Distilled water
Mix together with stirring. Filter sterilize through a 0.22-μm pore membrane. Enrichment with ADS/OADC is not necessary for fast-growing mycobacteria.

Luria Broth for *M. smegmatis* (LB-Tw):
    10 g Bacto-tryptone
    5 g Bacto-yeast extract
    5 g NaCl
    Distilled water up to 1 liter
Autoclave 20 min. Add Tween 80 to a final concentration of anywhere between 0.05% and 0.5%.

Middlebrook 7H10 agar for *M. smegmatis* (7H10 agar):
    19 g Middlebrook 7H10 agar (Difco)
    980 ml Distilled water
Autoclave for 20 min. Let cool on bench, with stirring, for approximately 30 min and then add the following components:
    10 ml 20% Dextrose
    10 ml 50% Glycerol
Mix by stirring and pour plates. It is important that Tween 80 *not* be incorporated into 7H10 plates that are used as bottom agar for the preparation of phage lysates in *M. smegmatis* (see below).

Protocol 1: Transformation of M. tuberculosis

1. In a plastic 1 liter roller bottle, inoculate 100 ml of 7H9-ADS-Tw medium with 1 ml of a frozen stock of saturated (OD $A_{600} > 1.0$) *M. tuberculosis*.

2. Incubate at 37° until the culture reaches mid-log phase of growth (OD A_{600}0.5–1.0). Depending on the starter stock this takes between 6 and 12 days.

3. Harvest 45 ml of culture in a 50-ml screw cap tube by centrifugation for 10 min at 1500g at room temperature (approximately 25°). Discard the supernatant.

4. Resuspend the pellet well in 4.5 ml of 10% glycerol. Then raise the final volume up to 45 ml with additional 10% glycerol. Centrifuge the cells as done previously and again discard the supernatant.

5. Repeat the wash described in step 4 one additional time.

6. Following the second wash, resuspend the final pellet in 4.5 ml of 10% glycerol.

7. Place appropriate volume of DNA (100 ng–1 μg) into a microcentrifuge tube. Add 400 μl of the prepared cells to the DNA, mix well by pipetting up and down, and transfer the mixture into an electroporation cuvette (0.2-cm electrode gap width). Make sure to eliminate any air bubbles.

8. Set the electroporation apparatus (Gene-Pulser II, Bio-Rad, Richmond, CA) to 2.5 kV and 25 μF and set the pulse controller (Bio-Rad) to 1000 Ω. Place the cuvette into the holder and expose to one pulse (time constants are generally between 15.0 and 25.0 ms).

9. Transfer the electroporated cells into 1.0 ml of 7H9-ADS-Tw medium in a 15-ml tube.

10. Place the tubes in a shaking incubator at 37° for 12–24 hr to allow for cell recovery and phenotypic expression of antibiotic resistance.

11. Plate the cells on selective medium and incubate at 37°.

12. After 3–4 weeks inspect plates for transformants.

Protocol 2. Transformation of M. smegmatis. The same protocol can be used as that presented for *M. tuberculosis* with the following modifications:

1. *M. smegmatis* culture is prechilled to 4° and all subsequent steps in the preparation of the electrocompetent cells are carried out at 4°.

2. Following electroporation, 1 ml of 7H9-Tw is added to the cells in a 15 ml tube.

3. Place the tube in a shaking incubator at 37° for 3–4 hr.

4. Plate the cells on selective medium and incubate at 37°.

5. After 2–3 days inspect the plates for transformants.

Note: It has been observed that a higher transformation efficiency is achieved when *M. smegmatis* is grown in LB-Tw (0.5% final) prior to preparation of electrocompetent cells.[23]

Mycobacteriophage Methods

Protocol 3: Phage Infection of M. smegmatis and Titering of Mycobacteriophage. *M. smegmatis* is used as the host for production and titering of mycobacteriophage.

1. In 5 to 10 ml of 7H9-Tw broth, inoculate a well-isolated single colony of *M. smegmatis* mc^2 155 that is taken from a 7H10 agar plate.

2. Incubate the culture at 37° with shaking until late-log phase is reached (approximately OD A$_{600}$ 1.5). Tween 80 inhibits mycobacteriophage infections; however, *M. smegmatis* produces a Tween hydrolase that will destroy most of the Tween in the media during this growth period (2–3 days).

3. Prepare 10-fold serial dilutions (10^{-1} to 10^{-7}) of a phage lysate in 1-ml volumes of MP phage buffer (recipe below).

4. Carefully add 100 μl of each phage dilution to 200 μl of the *M. smegmatis* mc^{2}155 cells in a sterile disposable glass test tube.

5. Incubate the phage–cells mix at 30° for 30 min to allow for adsorption of the phage to the mycobacterial cells.

6. To each phage–cells mix add 3.5 ml molten top agar (56°), gently mix, and quickly pour onto 7H10 bottom agar plates. Spread the top agar evenly by tilting the plate with a circular motion until the entire surface of the plate is covered.

7. Allow the top agar to solidify and incubate the plates at 30° in an inverted position to avoid condensation on the plates.

8. Phage plaques will appear in the lawn of *M. smegmatis* at 2 to 3 days. On plates containing well-separated phage plaques, count the total number of plaques.

9. The titer of the lysate is expressed as:
Plaque-forming units (pfu)/ml $= 10 \times$ (number of the plaques) $\times$ (dilution factor)

Recipes. MP phage buffer:
   50 mM Tris/HCl pH7.5
   150 mM NaCl
   10 mM MgSO$_4$·7H$_2$O,
   2 mM CaCl$_2$

Top agar:
   0.6% Agar Noble (Difco) in distilled water

7H10 bottom agar plates:
   Follow the recipe for Middlebrook 7H10 agar for *M. smegmatis* (7H10 agar).
   It is critical that Tween *not* be incorporated into these plates (see above).

Protocol 4: Confirming Mycobacteriophage Temperature-Sensitive Phenotype

1. Aliquot 200 μl of late-log *M. smegmatis* mc^2 155 cells (see Protocol 3) into disposable glass tubes.

2. Add 3.5 ml of top agar (56°) to the tubes, gently mix, and quickly pour onto 7H10 bottom agar plates.

3. Transfer individual phage plaques (using a sterile plastic pipette tip) onto duplicate plates containing the *M. smegmatis* mc^{2}155 (see steps 1 and 2). Pick plaques from a plate that was grown at the permissive temperature (30°) and has well-separated plaques. Patch phage plaques in an orderly fashion using 50-square grids. Test 20 to 50 phage plaques.

4. Incubate one of the duplicate plates at the permissive temperature (30°) and the other at the nonpermissive temperature (37°).

5. Following confirmation of the *ts* phenotype, pick a phage plaque for future work. Isolate a plaque from the plate grown at the permissive temperature that corresponds to a plaque that fails to grow (no clearance in the cell lawn) at the nonpermissive temperature. Use a sterile Pasteur pipette or precut plastic pipette tip (P-1000). Soak the agar plug in 0.5 ml of MP phage buffer overnight. The phage will elute out from the plug during this incubation.

Protocol 5: Preparation of High Titer Mycobacteriophage Stocks from Plate Lysates

1. Grow up *M. smegmatis* mc^{2}155 in 7H9-Tw to late-log phase of growth (see Protocol 3).

2. Prepare 10-fold serial dilutions from a phage lysate or phage plug in MP buffer (see Protocol 4) using 2 ml volumes and titer the phage (see Protocol 3). Save the remainder of the phage dilutions at 4° for later use.

3. Incubate the plates from each phage dilution at 30° until well-formed individual plaques appear on high dilution plates. It is important to incubate the plates until the bacterial cell lawn reaches dense growth.

4. Choose the plates giving confluent lysis of the bacterial lawn. The highest phage titers are obtained from plates in which plaques are touching each other such that they resemble lace.

5. Using the appropriate saved dilutions (from step 2), prepare 10 to 15 additional "lacy" phage plates.

6. Once confluent lysis of the cells is reached, add 4 ml MP phage buffer to each plate.

There are two alternative methods for the subsequent processing of the lysate. Both methods will be presented below.

Recovering lysate from top agar

7a. Incubate the plates for 30 min at room temperature.

8a. Using a glass spreader, scrape the top agar off each plate and transfer it along with recovered phage buffer into a 50-ml Oak Ridge centrifuge tube.

9a. Centrifuge the agar and lysate for 30 min at 23,000g at 4° in an appropriate centrifuge.

10a. Filter the supernatant through a 0.45-μm filter and store the lysate at 4°.
11a. Check for contamination of the lysate, with viable *M. smegmatis* cells, by streaking some of the lysate on 7H10 agar.
12a. Determine the titer of the phage lysate (see Protocol 3).

Recovering lysate from phage buffer

7b. Incubate the plates for 2 hr at 4°. Transfer the plates to room temperature and shake gently for 1 hr. The cold temperature causes the agar to constrict and higher temperature aids in releasing the phage from the gel matrix.
8b. Using a plastic pipette, recover the phage buffer from the top of the plate.
9b. Filter the lysate through a 5.0-μm filter and then through a 0.45-μm filter.
10b. Check for contamination with viable *M. smegmatis* cells by streaking some of the lysate on 7H10 agar.
11b. Determine the titer of the phage lysate (see Protocol 3).
A good high titer lysate is 10^{10}–10^{11} pfu/ml.

Protocol 6: Storage of Phage. High titer phage lysate stocks can be conveniently stored at 4° for at least 4 months without any significant loss in titer. For long-term storage, it is recommended that the lysates be frozen at −70°.

1. In a 2.0-ml cryovial tube combine equal volume of phage lysate and 50% glycerol.
2. Quick-freeze the sample in liquid nitrogen and store at −70°.
3. To isolate phage from the frozen stock, scratch the surface of the frozen lysate with a plastic pipette and drop onto a lawn of *M. smegmatis* (see Protocol 4).
4. Incubate the plate at 30° until a macroplaque appears. Use a sterile Pasteur pipette or precut plastic pipette tip (P-1000) to isolate the macroplaque. Soak the plug in 0.5 ml of MP phage buffer overnight (see Protocol 4).
5. Expand the phage to a high titer (see Protocol 5).

Protocol 7: Isolation of M. tuberculosis Genomic DNA. An optional step in this protocol is to add glycine to a final concentration of 1% 12–24 hr before harvesting the culture. The presence of glycine will act to weaken the cell wall and help to increase the DNA yield.

1. Grow *M. tuberculosis* in 7H9-ADS-Tw with appropriate antibiotics/supplements.
2. Harvest 10 ml of *M. tuberculosis* culture (OD $A_{600} > 0.4$) in a 15-ml screw cap tube by centrifugation for 10 min at 1500g. Discard the supernatant.

3. Resuspend the pellet in 1 ml of GTE solution (recipe below). Transfer to a 2-ml microcentrifuge tube. Centrifuge for 10 min at 12,000 rpm in a microcentrifuge. Discard the supernatant.

4. Resuspend the pellet in 450 μl GTE. Add 50 μl of lysozyme (10 mg/ml). Mix gently and incubate overnight at 37°.

5. Add 100 μl of 10% sodium dodecyl sulfate (SDS) and mix gently. Add 50 μl of proteinase K (10 mg/ml) and mix gently. Incubate at 55° for 20 to 40 min.

6. Add 200 μl 5 M sodium chloride (NaCl) and mix gently (NaCl blocks the binding of DNA to cetrimide in step 7).

7. Preheat cetrimide saline solution (CTAB) to 65° (recipe below). Add 160 μl CTAB and mix gently. Incubate at 65° for 10 min.

8. Add an equal volume of chloroform : isoamyl alcohol (24 : 1), shake gently to mix, and spin in microcentrifuge for 5 min at 12,000 rpm.

9. Transfer aqueous layer to fresh microcentrifuge tube and repeat extraction with equal volume of chloroform : isoamyl alcohol (24 : 1). These extractions remove CTAB-protein/polysaccharide complexes.

10. Transfer 800 μl of the aqueous layer to fresh microcentrifuge tube. Wipe the tube with Vesphene II to disinfect outer surface and finish protocol in a BSL2 laboratory.

11. Add 560 μl of 2-propanol; mix gently by inversion until the DNA has precipitated out of solution. Spin in microcentrifuge for 10 min at 12,000 rpm.

12. Remove supernatant and wash pellet with 70% (v/v) ethanol. Spin in microcentrifuge for 5 min at 12,000 rpm.

13. Remove supernatant and air-dry the DNA pellet. Do not overdry. Add 50–100 μl of Tris–EDTA buffer to the pellet and store at 4° overnight to allow the pellet to dissolve.

Recipes

GTE solution:
50 mM Glucose
25 mM Tris-HCl (pH 8.0), 10 mM EDTA

Cetrimide solution (CTAB): Dissolve 4.1 g of NaCl in 90 ml distilled water. Add 10 g of cetrimide (Sigma). Cetrimide is hexadecyltrimethylammonium bromide. Heat the solution to 65° while stirring to get cetrimide to go into solution.

Protocol 8: Isolation of Mycobacteriophage DNA

1. Take 0.5 ml of high titer phage lysate (see Protocol 5) and add 2.5 μl DNase (1 mg/ml) and 5 μl RNase (10 mg/ml).

2. Incubate the reaction mix at 37° for 30 min.

3. Add 25 μl STE lysis buffer (recipe below).

4. Add 10 μl proteinase K (10 mg/ml).

5. Incubate at 56° for 30 min.

6. Perform two phenol : chlorophorm : isoamyl alcohol (25 : 24 : 1) extractions.

7. Perform two chlorophorm : isoamyl alcohol (24 : 1) extractions.

8. To the aqueous phase containing the phage DNA add 0.1 volume 3 M sodium acetate.

9. Carefully overlay 2.5 volumes of 95% ethanol and mix slowly by gentle inversion.

10. Depending on the amount of the DNA, either spool the phage DNA using a glass rod or pellet the DNA by centrifugation in a microcentrifuge at 12,000 rpm for 20 min.

11. Wash the DNA with 70% ethanol and dry at room temperature. Do not overdry.

12. Resuspend the DNA in 50–100 μl 10 mM Tris–HCl pH 8.0.

This protocol can be scaled up when larger amounts of phage DNA are desired.

Recipes

STE lysis buffer:
1% SDS
50 mM Tris–HCl pH 8.0
400 mM EDTA
Prepare fresh. Heat up EDTA to 50° prior to adding SDS to avoid precipitation.

Specific Methods for Obtaining Mutants in *M. tuberculosis*

Transformation as Means of Introducing Allelic Exchange Constructs. Transformation is one way of introducing allelic exchange constructs into *M. tuberculosis*. There are primarily two transformation-based gene replacement protocols currently being used with success. These methods both share the feature that first transformation is carried out to introduce the allelic exchange substrate and then homologous recombination is completed. The methods differ in the types of vectors being used. To date, a systematic analysis of the amount of flanking DNA that is needed on each side of the deletion has not been conducted. In our experience 0.7–1.0 kb of flanking DNA on each side of the deletion works well to achieve homologous recombination. Of course, the use of larger stretches of flanking DNA will only serve to increase the efficiency of the process.

Allelic Exchange Utilizing Replicating Plasmids. This method involves transformation with a replicating mycobacterial vector that contains three important features: (1) a deletion allele of the gene of interest (your favorite gene—*yfg*) with a selectable marker, such as *hyg* (encoding hygromycin resistance), marking the deletion site, (2) a second marker on the vector backbone, such as *aph* (encoding kanamycin resistance), which selects for the presence of the plasmid in the host cell, and (3) a counterselectable marker *sacB*. Expression of *sacB* in mycobacteria is lethal in the presence of sucrose.[70] Thus, transformants expressing *sacB* are unable to grow in the presence of sucrose [i.e., such cells are sucrose sensitive (Suc^S)]. This enables counterselection, later on, of cells in the population that have lost the replicating vector (Suc^R).

The first step is to obtain a single hygromycin-resistant (Hyg^R) transformant with a vector of the type described above. The second step is to provide the opportunity for homologous recombination and loss of the replicating plasmid. This is done by growing up a Hyg^R transformant, without selection for the plasmid, and plating the culture on agar containing hygromycin and sucrose. The resulting Hyg^R Suc^R colonies are ideally mutants in which the *hyg*-marked deletion allele has replaced the wild-type gene in the chromosome and the replicating delivery vector has been lost. To ensure this is the case Hyg^R Suc^R colonies are screened for the presence of the vector; in the case outlined here *aph* marks the vector and kanamycin resistance is screened. This is a very important step because in addition to arising from plasmid loss sucrose resistance can also result from inactivation of the *sacB* gene on the vector. Inactivation of *sacB* occurs at a frequency of 10^{-4} to 10^{-5}.[29] Colonies that are Hyg^R Suc^R Kan^S should represent cells that have undergone allelic exchange. Southern bolt analysis is used to confirm the gene replacement. A variation of this method, which involves a thermosensitive (*oriM^{ts}*) counterselectable replicating vector, has been developed.[29] This vector is maintained in mycobacteria at 32°, but it is unstable and lost at 39°. This temperature sensitivity of the vector facilitates the plasmid loss. However, our group has found that even without the presence of a thermosensitive mycobacterial origin of replication, *sacB*-containing derivatives of the commonly used episomal mycobacterial vector pMV261[13] are lost at a frequency that enables a successful second step.[8,27] Although the plasmid loss may not be as efficient as that obtained with an *oriM^{ts}-sacB* vector, with these vectors periods of growth at 32° are not required, which serves to speed up the process. For this method our laboratory has used the vector pYUB631 (Fig. 1), which is also referred to as pMP7,[8,11,27] to construct an *icl* and a *leuD* mutant of *M. tuberculosis*. In our experience we have found it preferable to clone the allelic exchange substrate between the *sacB* and *aph* genes.[71] By separating *sacB* and *aph* the possibility of simultaneously deleting both of these genes, to generate undesired Hyg^R Suc^R Kan^S clones that have not undergone allelic exchange, is significantly reduced. Using this strategy it

[70] V. Pelicic, J. M. Reyrat, and B. Gicquel, *J. Bacteriol.* **178**, 1197 (1996).
[71] J. D. McKinney, unpublished results (1998).

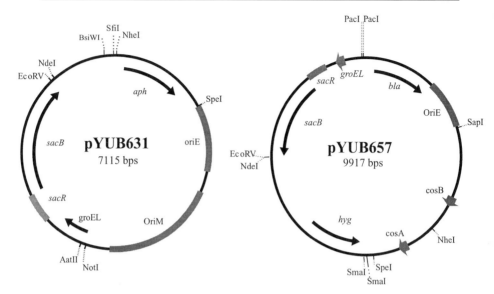

Fɪɢ. 1. Circular genetic map of the plasmids pYUB631 and pYUB657 used as replicating (oriM-based) (pYUB631) and nonreplicating ("suicide") (pYUB657) vectors for the delivery of homologous DNA substrates for allelic exchange in *M. tuberculosis* (see text for details about their usage). The presence of *sacB* in pYUB631 confers an intrinsic instability to the replicating plasmid and provides a convenient negative selection for the plasmid loss. The presence of *sacB* on pYUB657 enables selection of cells that have undergone allelic exchange—by an intramolecular recombination event. Shown are convenient unique restriction enzyme sites for cloning a deletion allele of *yfg* (your favorite gene). Adapted with permission from M. S. Pavelka, Jr., and W. R. Jacobs, Jr., *J. Bacteriol.* **181,** 4780 (1999).

has been found that 30–70% of Hyg^R Suc^R Kan^S clones have undergone a correct allelic exchange event and lost the replicating plasmid.[71,72]

Protocol 9: Allelic Exchange Using Replicating Vector

1. Construct the appropriate counterselectable mycobacterial replicating vector carrying a *hyg* marked deletion allele of the gene of interest (throughout this article we will refer to this gene as "your favorite gene" or *yfg*).

2. Transform *M. tuberculosis* with the recombinant replicating vector and select for hygromycin resistant transformants on 7H10-ADS plates containing hygromycin (50 μg/ml) (see Protocol 1). The transformants should have a Hyg^R Suc^S Kan^R phenotype.

3. Select a well-isolated transformant and inoculate it into 5–10 ml of 7H9-ADS-Tw medium containing hygromycin (50 μg/ml).

---

[72] M. S. Glickman, unpublished results (1999).

4. Incubate in a shaking incubator at 37° for approximately 7–10 days so that the culture is close to saturated.

5. Plate serial 10-fold dilutions onto 7H10-ADS agar supplemented with hygromycin (50 μg/ml) and 3% sucrose.

6. Screen Hyg^R Suc^R colonies for kanamycin sensitivity. Individual colonies (about 100) are picked and patched onto 7H10-ADS agar plates containing hygromycin (50 μg/ml) with and without kanamycin (20 μg/ml). Incubate plates at 37°.

7. Prepare genomic DNA (see Protocol 7) from a subset of colonies that are Hyg^R Suc^R Kan^S and perform Southern analysis to confirm the anticipated allelic exchange event.

In the case of a gene where phenotype is expected on the deletion, the phenotype can also be monitored during the mutant construction process.

Allelic Exchange Utilizing "Suicide" Counterselectable Vector. A schematic outline of the method is shown in Fig. 2. This method involves transformation with a "suicide" vector that contains three important features: (1) a deletion of the gene of interest −*yfg* (preferably unmarked), (2) a selectable marker, such as *hyg*, on the vector backbone, and (3) the counterselectable marker *sacB*. The vector we use to carry out this allelic exchange method is pYUB657[11] (Fig. 1). Upon transformation, this suicide vector is unable to replicate in mycobacteria and stable Hyg^R transformants arise by means of a single-crossover homologous recombination event between the mutated allele on the vector and the wild-type allele on the chromosome. These Hyg^R recombinants are referred to as single-crossover clones. They contain the suicide vector integrated into the chromosome, which results in a tandem duplication of the gene of interest (both a mutated and a wild-type allele are present). The next step in this process is to select for recombinants that have undergone a second homologous recombination event (this is an intramolecular recombination event) between the two alleles of the gene of interest. This recombination event is selected by virtue of the fact that it eliminates ("loops out") all the intervening vector sequences including *sacB*. Suc^R colonies are selected on sucrose-containing media and then screened for the loss of the vector (in this case *hyg*), to distinguish the homologous recombinants (Hyg^S Suc^R) from the spontaneous sucrose-resistant clones (Hyg^R Suc^R). Depending on the site of this intrachromosomal recombination event it results in either a mutant or wild-type allele being left behind in the chromosome. The Hyg^S Suc^R clones are subjected to Southern analysis to determine which clones contain the desired allelic exchange event. This method has been successfully used by our laboratory to generate a *lysA* and a *secA2* mutant of *M. tuberculosis*.[11,73]

[73] M. Braunstein and W. R. Jacobs, Jr., unpublished results (2001).

First Step: Integration of the suicide plasmid
into the chromosome by homologous recombination

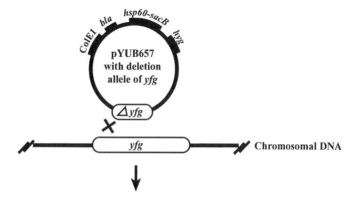

Single Cross-Over Strain

Second Step: A second recombination event generates
a wild-type or deletion allele in the chromosome

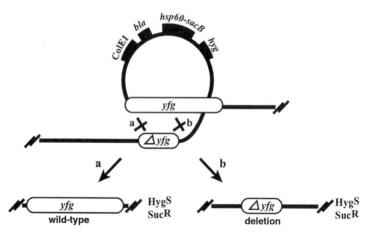

FIG. 2. A schematic of transformation-based allelic exchange via a "suicide" counterselectable plasmid. The first step depicts a single homologous recombination event between *yfg* (your favorite gene) on the chromosome and the deleted allele of *yfg* present on the plasmid. This event results in the production of a single-crossover strain. The second step depicts the subsequent intrachromosomal recombination event between the wild-type *yfg* allele and the deleted allele of *yfg*. When the site of this recombination event is **a**, then the intervening vector sequences are lost and the wild-type allele is left behind in the chromosome. When the site of recombination is **b**, the deletion allele of *yfg* remains in the chromosome and the wild-type allele along with the vector sequences are lost. If the homologous DNA flanking each side of the deletion is of equivalent length then recombination at sites **a** and **b** should occur with roughly equal frequency. If only wild-type alleles are obtained it strongly suggests that an *yfg* deletion is inviable.

There are several important advantages of this method over the others described in this article. First, it allows construction of unmarked in-frame deletions. Strains with such mutations are antibiotic sensitive, which allows greater flexibility in future experiments, as only a limited number of antibiotic markers are available for *M. tuberculosis*. In addition, mutant strains that are vaccine candidates should not contain any antibiotic resistance determinants. Furthermore, in-frame deletion mutants will not have polar effects on other genes in an operon. Thus, the phenotypes observed with an in-frame deletion mutant are most likely a reflection of the gene of interest. Second, this method can be used to determine whether a gene is essential. Because of the methodology involved, the second homologous recombination step can produce either wild-type or mutant alleles. The exclusive recovery of wild-type alleles strongly suggests that a deletion of the gene of interest is lethal. In contrast, with other methods the inability to obtain a deletion mutant may reflect the essential nature of that particular gene, the essential nature of a gene located downstream of the gene but in the same operon, or simple failure in the methodology. Subsequently a definitive experiment can be carried out to prove with certainty that the gene of interest is essential. The same experiment is performed in a merodiploid strain (a strain that contains an additional copy of the gene) and if the inability to delete the gene of interest in the chromosome is now overcome, it demonstrates the absolute requirement of that gene for viability. In other words, the inability to delete the gene in a haploid strain and the ability to delete the gene in a merodiploid strain demonstrates the essential nature of the gene. Using such a method, *glnE* of *M. tuberculosis* has been shown to be an essential gene.[58]

Protocol 10: Allelic Exchange Using "Suicide" Counterselectable Vector

1. Construct the appropriate recombinant "suicide" counterselectable vector carrying an unmarked deletion in the gene of interest −*yfg*.

2. Transform *M. tuberculosis* with the "suicide" counterselectable vector and select for hygromycin resistant transformants. These transformants should have a Hyg^R Suc^S phenotype.

3. Inoculate individual transformants into 5–10 ml of 7H9-ADS-Tw medium containing hygromycin (50 μg/ml).

4. Make genomic DNA (see Protocol 7) from individual transformants and perform Southern analysis to confirm that a single-crossover homologous recombination event has occurred.

5. Take a single-crossover strain and inoculate into 5–10 ml of 7H9-ADS-Tw medium containing hygromycin (50 μg/ml). Grow until near saturated.

6. Take 0.5 ml of saturated culture and add to 9 ml of 7H9 medium without antibiotic.

7. Incubate at 37° for approximately 7–10 days so that the culture is close to saturated.

8. Plate serial dilutions onto 7H10-ADS agar supplemented with sucrose 3%.

9. Suc^R colonies are screened for hygromycin sensitivity. About 100 colonies are picked and patched onto 7H10-ADS agar plates with and without hygromycin (50 μg/ml). Incubate plates at 37°.

10. Make genomic DNA (see Protocol 7) from a subset of colonies that are Hyg^S Suc^R and evaluate by Southern analysis to determine the allele (mutant or wild type) present in individual recombinants.

This approach is best taken with the H37Rv strain of *M. tuberculosis* because of the higher transformation efficiency. Expect a low number of Hyg^R transformants (between 1 to 4) that represent single-crossover strains from an individual transformation with H37Rv. It is recommended that several electroporations be performed to ensure that a homologous recombinant will be obtained.

Phage Delivery as Means of Introducing Allelic Exchange Constructs

Specialized transducing mycobacteriophages provide an alternative method for delivering exogenous DNA for allelic exchange in *M. tuberculosis*. As we mentioned above, transductional DNA transfer is highly efficient and the experimental conditions of a phage infection can be adjusted in such a way that nearly every bacterial cell in the population will receive phage DNA. The high transduction frequency enables rare recombination events such as double-crossover homologous recombinants to be selected in a single step. Using this method, mutants are obtained 3–4 weeks following a phage infection. This contrasts with transformation based methods that take 8–12 weeks before mutants are generated.

Shuttle Phasmids

The construction of recombinant specialized transducing mycobacteriophages involves shuttle phasmids. These shuttle phasmids are mycobacteriophage vectors that are able to replicate in *E. coli* as a plasmid and replicate in mycobacteria as a phage.[2,3,74] These chimeric DNA molecules consist of an *E. coli* cosmid (λ *cos* sites, a selectable marker, and a ColE1 origin of replication) integrated into a nonessential region of the mycobacteriophage genome. Phasmids can be propagated in a variety of ways. They can be transformed into *E. coli* where they replicate as plasmids. They can be packaged *in vitro* into λ phage particles and transduced into *E. coli* where they replicate as high molecular weight (40–50 kb) cosmids. Finally, they can be transfected by electroporation into mycobacteria where they replicate as lytic mycobacteriophages.

[74] W. R. Jacobs, Jr., M. Tuckman, and B. R. Bloom, *Nature* **327,** 532 (1987).

For phage delivery a conditional mycobacteriophage was needed, such that the phage would introduce DNA into the bacteria but would not lyse the recipient cell. Such a conditional mycobacteriophage was obtained by screening for a temperature sensitive (*ts*) mutant of the TM4 mycobacteriophage that lyses *M. smegmatis* at 30° but not at 37°.[17,75] This mutant phage, clone PH101, has a very low frequency of reversion to the wild-type lytic phenotype ($>10^{-9}$ pfu/ml). Two shuttle phasmid vectors—phAE87[75] and phAE159[76]—have been constructed from PH101. These two shuttle phasmids differ only in the size of deletions in nonessential regions of the phage genome (300 bp for phAE87 and 5.6 kb for phAE159). A cosmid is located at the deletion site in each of these phasmids. It is the cosmid features of the shuttle phasmid that allow DNA manipulations to be undertaken in *E. coli*. Exogenous DNA can be cloned into both of these shuttle phasmids.

The construction of conditional phages carrying allelic exchange substrates is shown in Fig. 3. The final product is a specialized transducing phage containing a marked mutated gene integrated into a nonessential region of the TM4^{ts} phage genome.

Construction of Recombinant Shuttle Phasmid

Construction of Recombinant Cosmid Containing Allelic Exchange Substrate. With the phage delivery system there is an upper limit to the size of the allelic exchange construct that can be cloned into the mycobacteriophage genome in shuttle phasmids. This limit is set by the size of DNA that can be packaged into λ phage heads for the *in vitro* packaging and *E. coli* transduction step and by the size of DNA that can be packaged into TM4 phage heads for the mycobacteriophage lytic step. The size of the wild-type TM4 genome is 52,797 bp.[77] We have shown experimentally that phAE87, which contains a 300-bp deletion, is able to accommodate a maximum of 6.0 kb of exogenous DNA into the phage heads. This defines the upper limit of the head packaging capacity as approximately 58 kb. Any larger phage molecules are unstable and will result in the production of undesired large deletions in the cloned exogenous DNA. Although it has not been directly tested, the larger deletion in phAE159 (5.6 kb) should theoretically increase the upper size limit of this recombinant phasmid to approximately 10 kb. The allelic exchange substrate should contain 0.7–1.0 kb of flanking DNA on each side of the gene deletion. The deletion site should be marked with a selectable marker, such as *hyg*.

[75] C. Carriere, P. F. Riska, O. Zimhony, J. Kriakov, S. Bardarov, J. Burns, J. Chan, and W. R. Jacobs, Jr., *J. Clin. Microbiol.* **35**, 3232 (1997).

[76] J. Kriakov, unpublished results (2000).

[77] M. E. Ford, C. Stenstrom, R. W. Hendrix, and G. F. Hatfull, *Tuberc. Lung Dis.* **79**, 63 (1998).

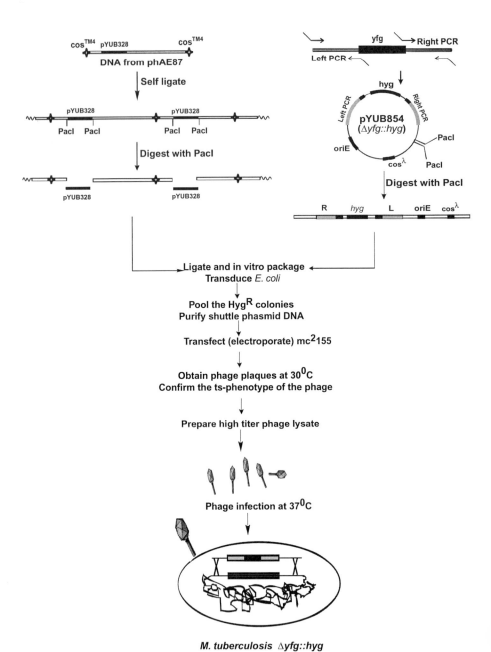

FIG. 3. Schematic representation of the steps involved in the construction of a specialized transducing mycobacteriophage and the method used for one-step allelic exchange in *M. tuberculosis*. *Upper left:* Steps involved in the preparation of the phasmid for cloning of the recombinant cosmid containing the deletion allele of *yfg*. *Upper right:* One of the many strategies for the construction of the recombinant DNA substrate and its cloning into the cosmid. In this presentation cosmid pYUB854 is used.

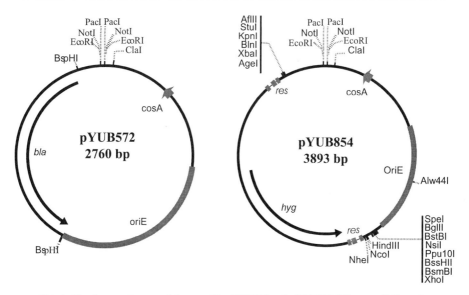

FIG. 4. Circular genetic map of the cosmids pYUB572 and pYUB854. When an allelic exchange substrate is first constructed in any other plasmid vector, cosmid pYUB572 is the cosmid of choice. A DNA fragment can be subcloned into the *Bsp*HI sites. This subcloning will replace the *bla* gene in the cosmid and can be selected for as Hyg^R transformants. The cosmid pYUB854 is used when the two flanking arms of the recombinant DNA substrate are being cloned as PCR products. Convenient multiple cloning sites allow for their directional cloning of each of the PCR fragments on both sides of the *hyg* cassette.

Protocol 11: Construction of Cosmids Containing Allelic Exchange Substrate. pYUB572 and pYUB854 (Fig. 4) are cosmid vectors that can be used in the construction of phasmids carrying allelic exchange substrates.

For pYUB572

1. A marked allelic exchange substrate is constructed in a cloning vector of choice.
2. The allelic exchange substrate is subcloned, by blunt-end ligation, into *Bsp*HI cut pYUB572.

For pYUB854

1. Design two pairs of PCR primers to amplify the LEFT and RIGHT flanking DNA arms. These primers should have 5′ and 3′ extensions containing unique restriction endonuclease sites to facilitate cloning into pYUB854. The location of the LEFT 3′and the RIGHT 5′ primer will determine the size of the deletion being generated.

2. Purify the PCR-generated products and sequentially clone into the respective unique restriction sites in pYUB854. Once this cloning is complete the LEFT and RIGHT flanking DNA arms will be in their original orientation being separated by the *res-hyg-res* cassette.

The specialized *res-hyg-res* cassette which marks the deletion in pYUB854 contains the *hyg* gene flanked by DNA binding sites for the site-specific recombinase: γ,δ-resolvase.[78] The presence of these flanking *res* sites will allow for the precise excision and unmarking of the deletion when resolvase is expressed in the host strain. A controlled resolvase expression system is currently being developed for this purpose.[79] It is important to note that the presence of *res* sites limits the range of *E. coli* strains that may be used during the construction of recombinant phages. Strains that exhibit resolvase activity will lead to unstable constructs. For this reason, avoid using any *E. coli* strain that has an F factor. We routinely employ the *E. coli* HB101 strain.

Protocol 12: Cloning Allelic Exchange Construct into Shuttle Phasmid. The cosmids pYUB572 and pYUB854 posses the required elements of a λ *cos* site and a unique *Pac*I cloning site which enables the cloning of these vectors as cosmids into the shuttle phasmid. The shuttle phasmids phAE87 and phAE159 contain a resident cosmid, pYUB328, which carries *bla* (encoding ampicillin resistance) and two unique *Pac*I sites. *Pac*I cloning is used to replace this resident cosmid with the allelic exchange carrying cosmid.

1. Prepare shuttle phasmid DNA as phage DNA from the appropriate phage lysate (see Protocol 8).
2. Purified phasmid DNA is first self-ligated through the TM4 *cos* sites so that long phage head-to-tail concatamers are formed. Self-ligate approximately 7.5 μg of phage DNA in a standard 50-μl ligation reaction with T4 DNA ligase (150 μg/ml final DNA concentration in the ligation). Because of the large amount of phage DNA a gradual increase in the viscosity of the ligation mix may be observed.
3. Heat inactivate the ligase by incubating the reaction at 65° for 30 min.
4. Digest the oligomerized phage DNA with *Pac*I. On digestion, a fragment containing pYUB328 (3.7 kb in size) and a large fragment representing the phage genome (approximately 50 kb) can be observed by agarose gel electrophoresis.
5. Linearize the cosmid carrying the allelic exchange substrate (pYUB572 or pYUB854 derivatives—see Protocol 11) by digestion with *Pac*I.

---

[78] M. R. Sanderson, P. S. Freemont, P. A. Rice, A. Goldman, G. F. Hatfull, N. D. Grindley, and T. A. Steitz, *Cell* **63**, 1323 (1990).
[79] S. Bardarov and W. R. Jacobs, Jr., unpublished results (2002).

6. Heat inactivate *Pac*I in both phasmid and cosmid digestions at 65° for 30 min.

7. Using a 10 : 1 concentration ratio, ligate *Pac*I-digested phasmid with the *Pac*I linearized recombinant cosmid (i.e., in a standard 20-μl reaction ligate approximately 1.0 μg of phage DNA with 100 ng of cosmid DNA).

8. Using commercially available λ packaging extracts (Giga Pack, Stratagene), *in vitro* package the ligation mix into λ phage heads by virtue of the specific λ *cos* site on the cosmid vector. Proper-sized phage molecules are size-selected by the highly specific packaging constraints of the λ phage heads.

9. Take 50 μl of phage packaging–ligation mix and use to transduce 200 μl of *E. coli* HB101. Follow the transduction protocol provided by the manufacturer of the λ packaging extract (Stratagene). *E. coli* competent for transduction with phage lambda (λ) are prepared by standard protocols.[67] Because of the presence of the *res-hyg-res* cassette in the cosmids in all transduction experiments *E. coli* HB101 is used (see Protocol 11).

10. Select transductants containing the recombinant shuttle phasmid by selection on LB agar containing hygromycin (150 μg/ml).

Protocol 13: Converting Recombinant Shuttle Phasmid into Phage

1. Pick individual Hyg^R clones (see Protocol 12) and isolate phasmid DNA by a standard alkaline lysis method.[67] Alternatively, all the Hyg^R *E. coli* transductants obtained can be pooled by scraping them off a plate to which 3.0 ml TES buffer has been added. Pellet the cells by centrifugation at 1500g for 10 min. Use an appropriate portion of the washed cells for phasmid DNA purification. The phasmids isolated at this step represent recombinant phages carrying the allelic exchange substrate.

2. Electroporate the purified phasmid DNA into *M. smegmatis* mc^2 155 (see Protocol 2). When a pool of phasmid DNAs, rather than a single phasmid clone, is used in this electroporation there is a higher probability that functional mycobacteriophages will be obtained.

3. Following electroporation, add 1.0 ml of 7H9 broth without Tween and incubate for 30 min. at 30° to allow for phage infection.

4. Mix 100 μl of electroporated cells with 200 μl of late-log *M. smegmatis* mc^2 155 cells. Add 3.5 ml of top agar and pour onto 7H10 bottom agar plate (see Protocol 3).

5. Take 800 μl of the remainder of the electroporated cells, mix with 3.5 ml of top agar, pour onto 7H10 bottom agar plate. (In steps 4 and 5 cells infected with phage are being plated among a lawn of *M. smegmatis* mc^2 155. For step 5 there are enough untransformed cells around to serve as a lawn so additional *M. smegmatis* mc^2 155 does not need to be added.)

6. Incubate plates at 30° in an inverted position. Two to 3 days later plaques should become visible on the lawn of *M. smegmatis*.

7. Pick individual plaques and confirm the ts-phenotype (see Protocol 4). Choose a single ts-plaque and use it to produce a high titer lysate (see Protocol 5).

Note: Recombinant mycobacteriophages (phasmids) should be stored as phages, rather than as plasmids in *E. coli* (see Protocol 6). This is due to the instability exhibited by these large molecules in *E. coli*.

Recipe

TES buffer:
50 m*M* Tris-HCl pH 7.6
150 m*M* NaCl
10 m*M* EDTA–Na$_2$

Protocol 14: Allelic Exchange via Mycobacteriophage Delivery

1. In a plastic 1-liter roller bottle, inoculate 100 ml of 7H9-ADS-Tw medium with 1.0 ml of a frozen stock of *M. tuberculosis*. Grow the culture to OD A_{600} of 1.0, which correlates to approximately 2.5×10^8 cfu/ml.

2. Harvest 10 ml of culture for each phage infection to be performed. Using a 50-ml screw cap tube centrifuge the required amount of culture for 10 min at room temperature at 1500*g*. Discard the supernatant.

3. Resuspend the pellet well in an equal volume of MP phage buffer. Centrifuge the cells as done previously and again discard the supernatant.

4. Resuspend the cell pellet carefully in 1/10 of the original volume in MP phage buffer.

5. Prewarm an appropriate amount of high-titer phage lysate to 37°–39°.

6. In a 2-ml screw cap microcentrifuge tube gently mix 1 ml of cells (approximately 2×10^9 cfu/ml) with 1 ml of high titer phage stock (10^{10} pfu/ml). The desired multiplicity of infection (MOI) is 10.

7. Incubate the mix in a heating block for 4–6 hr at 37°–39°.

8. Centrifuge the cell–phage mix in a microcentrifuge for 10 min at 10,000 rpm. Use a P1000 and remove the supernatant from the pellet.

9. Resuspend the pellet in 1 to 2 ml of 7H9-ADS-Tw. Plate 0.2 ml of cells on a selective plate with hygromycin (75 μg/ml) and any appropriate supplements. Plate out the entire sample over the appropriate number of plates.

10. Incubate the plates at 37° for 3–4 weeks.

11. Make genomic DNA (see Protocol 7) from a subset of Hyg^R colonies and follow up with Southern analysis to confirm allelic exchange.

It is important that a "no-phage" control be included in the experiment. To 1 ml of cells add 1 ml of MP phage buffer. Process this control in parallel with the other samples. Take 100 μl of the "no-phage" control, make dilutions of 10^{-5} to 10^{-7}, and plate onto 7H10-ADS plates with no antibiotic in order to determine the number of viable CFU in the phage infection. Take the remaining 900 μl of the "no-phage" control and plate it out over the appropriate number of antibiotic-containing 7H10-ADS plates to determine the frequency of spontaneous antibiotic resistance.

MP phage buffer: See Protocol 3.

Transposition. As described above, our laboratory has found that the mycobacteriophage delivery system is an efficient way to deliver transposable elements to *M. tuberculosis*. We commonly use the transposons Tn*5367(aph)* and *mini-*Tn*5370(hyg)*, which are present in the shuttle phasmids phAE94 and phAE175, respectively. These shuttle phasmids were constructed by *Pac*I cloning of transposon-containing cosmids into phAE87 in the manner presented above for the construction of the specialized transducing mycobacteriophages.[17,50]

From a single infection with these phages, it is possible to generate a large representative transposon library containing more than 10^4 independent mutants in *M. phlei, M. bovis* BCG, and *M. tuberculosis*.

Protocol 15: Transposition via Mycobacteriophage Delivery. The method presented in Protocol 14 can be adapted for transposon delivery by simply substituting phAE94 or phAE175 as the phage used in the infection. An alternative, which was followed in Bardarov and Jacobs,[79] is presented below. This more involved protocol may serve to increase the efficiency of the phage infection. Protocol 15 can also be applied to allelic exchange via mycobacteriophage delivery, if desired.

1. In a plastic 1-liter roller bottle, inoculate 100 ml of 7H9-ADS-Tw medium with 1 ml of a frozen stock of *M. tuberculosis*. Grow the culture to an OD A_{600} of approximately 1.0.

2. Harvest 10 ml of culture for each phage infection to be performed. Using a 50-ml screw cap tube, centrifuge the required amount of culture for 10 min at room temperature at 1500g. Discard the supernatant.

3. Resuspend the pellet well in equal volume of Phage Adsorption buffer (recipe below). Centrifuge the cells as done previously and again discard the supernatant.

4. Repeat the wash described in step 3 one additional time.

5. Carefully resuspend the cell pellet in an equal volume of Phage Adsorption buffer and, for each infection, transfer 10 ml of the cell suspension to a separate 25-ml plastic media bottle. Incubate the cell suspension standing at 37° for at least 24 hr. This step serves to remove any residual Tween 80 that could interfere with the phage infection.

6. Transfer the 10 ml cell suspension to a 15-ml screw cap tube. Centrifuge the cells at 1500g at room temperature, resuspend carefully in 1 ml Phage Adsorption buffer, and transfer to a 2-ml screw cap tube.

7. Add the appropriate volume of high titer phage lysate phAE94 or phAE175 to achieve an MOI of 10 (see Protocol 14 for details).

8. Incubate the tube in a heating block at 37° for 4 to 6 hr to allow for adsorption of the phage.

9. Transfer the cell–phage suspension to a prepared 1-liter roller bottle containing 50 ml 7H9-ADS-Tw (with any necessary supplements as required, but no antibiotics) and incubate on the roller for 18 to 24 hr at 37°.

10. In a 50-ml centrifuge tube, harvest the cells at room temperature by centrifugation for 10 min at 1500g. Discard the supernatant.

11. Resuspend the pellet in 2 ml of 7H9-ADS-Tw and plate 0.2 ml per 7H10-ADS agar plate containing the appropriate antibiotics and supplements. Plate out the entire reaction.

12. Incubate the plates at 37° for 3–4 weeks.

13. Make genomic DNA (see Protocol 7) from a subset of antibiotic resistant colonies and follow up with Southern analysis to confirm random transposition.

Once again, it is important to include a no-phage control, which is used to derive the number of cfu/ml present in the phage infection and the frequency of spontaneous antibiotic resistance in the experiment. The transposition frequency is calculated as the number of antibiotic resistant cells minus the number of spontaneous antibiotic resistant cells/total number of cells in the phage infection.

Phage Adsorption buffer:
7H9-ADS
2% Glycerol (do not add Tween 80)

Signature Tagged Mutagenesis

As mentioned earlier in this article, this transposon mutagenesis system has been utilized for signature tagged mutagenesis (STM) in *M. tuberculosis*.[7] STM is a powerful method that enables the identification of transposon mutants that are defective for growth in an animal host. Attenuated mutants are identified as distinctly marked bacteria present among a pool of mutants used in an initial inoculum to infect the host but absent or underrepresented in the host after a defined period of infection. The hallmark of STM is that the transposons in individual mutants contain distinct DNA tags. For this purpose, random DNA tags were ligated into *mini*-Tn*5370(hyg)* to generate 48 distinct transposons. Each of these transposons was then cloned into the *Pac*I site of phAE87 to generate 48 distinct shuttle phasmids. In order to generate the STM library of mutants, 48 separate phage infections were performed and pools of 48 mutants were generated by picking a mutant from each of the 48 infections into a 48-well plate. The different pools of mutants were then used in mouse infection experiments. STM has also been used in combination with a *ts*-vector delivery system.[6] For additional details of these STM experiments the reader is referred to Cox *et al.*[7] and Camacho *et al.*[6]

Complementation Analysis of M. tuberculosis Mutants

The true power of being able to manipulate individual genes lies in the characterization of the resulting mutant phenotypes. It is from these phenotypes that the function/activity of the corresponding gene product can be deduced. Therefore, it is extremely important that the mutants being generated be characterized in a careful manner and that complementation experiments be undertaken. In order to be assured that the gene disrupted/deleted is responsible for the observed phenotype, a wild-type allele of the gene of interest should be introduced into the mutant strain and the mutant phenotype reevaluated. If the mutant phenotype is complemented (reversed/eliminated) by the presence of the wild-type gene, it proves that the phenotype is due to the mutation in the gene of interest. This is a particularly important control when dealing with mutants that are marked by a selectable marker or transposon. This is because of the possibility that the insertion is having a polar effect on neighboring genes in an operon. An advantage of making mutants with the "suicide" vector approach is that in-frame and unmarked mutations can be produced, which eliminates concerns about polar effects on other genes. However, even with nonpolar mutants it is still advisable that a complementation experiment be performed to ensure that the observed phenotype is a result of the gene deletion rather than an unintended second mutation in the strain.

Complementation experiments require the construction of a mycobacterial vector that carries a wild-type allele of the gene of interest and the introduction of this construct into the mutant strain by transformation. A complementation construct should have the gene under the control of a functional promoter. It is ideal if the gene is being introduced under the control of its own promoter; this will ensure the proper timing and degree of expression. However, if this is not possible an endogenous promoter such as *hsp60* can be used.[27] In addition, it is preferable if the construct is one that integrates into the chromosome in single copy (such as pYUB412, pMV306, pMV361 derivatives). This will also help to produce an appropriate level of gene expression, and integrating constructs should be more stably maintained throughout future experiments.

Acknowledgments

We gratefully acknowledge M. Glickman, J. Kriakov, J. McKinney, and M. Pavelka for permission to discuss their unpublished results, S. Daugelat and M. Pavelka for the critical reading of this article, and S. S. Bardarov, Jr., for the graphic art assistance.

[6] Analysis of Gene Function in Bacterial Pathogens by GAMBIT

By BRIAN J. AKERLEY and DAVID J. LAMPE

Introduction

The GAMBIT approach ("genome analysis and mapping by *in vitro* transposition") is a versatile system for creation and phenotypic analysis of pools of transposon mutants that combines *in vitro* transposon mutagenesis and genetic footprinting. It was originally developed for functional genomic studies of two naturally transformable bacterial pathogens, *Haemophilus influenzae* and *Streptococcus pneumoniae,* but is applicable to any haploid organism that can be efficiently transformed with linear DNA including members of such diverse bacterial genera as *Bacillus, Neisseria, Campylobacter,* and *Helicobacter.* GAMBIT owes its versatility to its simplicity. The method creates a bank of transposon mutations spanning the entire genome or within user-defined subgenomic regions and uses an analytical PCR-based approach, genetic footprinting,[1] to locate the positions of genes essential for growth or viability under any specific experimental condition. Transposition is conducted using purified components *in vitro* and therefore does not require complex genetic arrangements such as conditional replicons, organism-specific transposase expression systems, or "suicide" DNA delivery systems. The only requirement for conducting this procedure in any naturally transformable organism is the presence of a gene within the transposon conferring antibiotic resistance or other selectable traits.

This article is intended as a practical guide to the use of the GAMBIT approach for analysis of genetic function on a genomic scale or to study individual genomic regions of interest. As part of this discussion, versatile components of this method will also be described including transposase purification, *in vitro* transposon mutagenesis with *mariner*-derived transposons, and genetic footprinting. This article will focus on studies in naturally transformable bacteria; however, elements of the GAMBIT system have been adapted to genetic analysis of organisms in which DNA is introduced by other means and descriptions of some of these approaches can be found elsewhere.[2,3]

---

[1] I. R. Singh, R. A. Crowley, and P. O. Brown, *Proc. Natl. Acad. Sci. U.S.A.* **94,** 1304 (1997).

[2] E. J. Rubin, B. J. Akerley, V. N. Novik, D. J. Lampe, R. N. Husson, and J. J. Mekalanos, *Proc. Natl. Acad. Sci. U.S.A.* **96,** 1645 (1999).

[3] S. M. Wong and J. J. Mekalanos, *Proc. Natl. Acad. Sci. U.S.A.* **97,** 10191 (2000).

Mariner Transposition: General Features and Parameters Affecting Efficiency

GAMBIT was developed using minitransposons derived from the *mariner*-family transposon *Himar1*. *Himar1* and the related *Mos1* element, which was the earliest active *mariner* to be characterized,[4] are small DNA-mediated (Class II) transposons, originally ~1.3 kb in length, encoding a single transposase protein flanked by 28-bp inverted repeats. The repeats are recognized by the transposase, a single protein which acts *in trans,* without additional host factors. The only known site specificity for insertion is a TA dinucleotide in the target sequence. Even this specificity can be inhibited by adding $MnCl_2$ to the reaction causing approximately 50% of insertions to occur at sites with no recognizable consensus sequence, providing a highly random mutagenesis system.[5] Although the precise mechanism of transposition by *mariner* elements is under investigation, current data strongly support a "cut-and-paste" mechanism similar to that mediated by the Tn*10* transposase and also proposed for Tc*3*,[5–7] and transposition is independent of DNA replication.

As the terminal repeats are the only *cis*-acting sequences within the transposon that are essential for transposition it was possible to reengineer elements by removing the transposase gene and introducing restriction sites to allow convenient cloning of antibiotic cassettes as markers for transposition. We constructed *magellan1*,[8] a minitransposon that contains the inverted repeats that are the only *cis*-acting elements required for transposition. This construct carries the Tn*903* kanamycin resistance gene flanked by *Mlu*I restriction sites internal to the inverted repeat sequences.

Conditions and factors affecting *in vitro* transposition reactions include time, temperature, divalent cation concentration, transposase concentration, and transposon size.[9] Reactions conducted at 30° containing 10 mM $MgCl_2$ produce optimum yields of transposition events. Slight variations ($\pm 3°$ or ± 5 mM $MgCl_2$) do not drastically affect the yield. Transposition events accumulate linearly with time over a 6 hr period and show a steep dependence on transposase concentration centered on an optimal value of 10 nM with equimolar target and donor DNA concentrations (10 fM). The distance between the inverted repeats (transposon size) also influences the transposition reaction. Efficiency is maximal with elements of approximately 1 kb or less and declines exponentially with increasing transposon size. However,

---

[4] J. W. Jacobson, M. M. Medhora, and D. L. Hartl, *Proc. Natl. Acad. Sci. U.S.A.* **83**, 8684 (1986).

[5] D. J. Lampe, M. E. Churchill, and H. M. Robertson, *EMBO J.* **15**, 5470 (1996).

[6] H. W. Benjamin and N. Kleckner, *Proc. Natl. Acad. Sci. U.S.A.* **89**, 4648 (1992).

[7] H. G. van Luenen, S. D. Colloms, and R. H. Plasterk, *Cell* **79**, 293 (1994).

[8] B. J. Akerley, E. J. Rubin, A. Camilli, D. J. Lampe, H. M. Robertson, and J. J. Mekalanos, *Proc. Natl. Acad. Sci. U.S.A.* **95**, 8927 (1998).

[9] D. J. Lampe, T. E. Grant, and H. M. Robertson, *Genetics* **149**, 179 (1998).

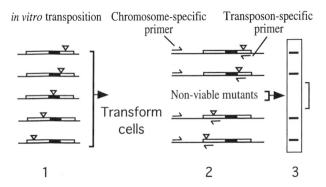

FIG. 1. The GAMBIT procedure (adapted from Ref. 8. Copyright © 1998 National Academy of Sciences, U.S.A.). Random *mariner* insertions (triangles) in a linear DNA target are generated by *in vitro* transposition (1). Mutant pools are created by transformation into naturally competent bacteria and selection for an antibiotic resistance gene carried by the transposon followed by PCR of the complex pool (2) and gel electrophoresis to map insertion sites (3). Regions that do not sustain insertions *in vivo* (black box) are putatively essential for growth or viability.

it is possible to mobilize *mariner* transposons of at least 10 kb in length indicating that, although efficiency is decreased, the upper size limit is quite large. Each of these factors has been incorporated into the protocol given in this article.

Many forms of regulation have been identified for transposons. Presumably, such regulatory mechanisms have evolved as a result of selection in nature against detrimental effects on host fitness. It is possible that mutational events modulate the activity of *mariner* transposases during the course of their evolution. If so, it should be possible to increase transposase activity by mutation. A genetic screen for hyperactive mutants in the transposase expressed in *Escherichia coli* resulted in the isolation of two mutant enzymes termed A7 and C9 exhibiting 5- or 7-fold more activity *in vitro,* respectively. Use of the C9 enzyme increased the net yield of transposon mutants isolated via the *in vitro* transposition procedure in *H. influenzae* by approximately 3.4-fold.[10]

The GAMBIT Procedure

Overview

The GAMBIT approach consists of a series of steps that create a bank of mutants having transposon insertions within a discrete region of the chromosome followed by localization of the insertions by the PCR-based mapping approach, genetic footprinting (Fig. 1). Target DNA mutagenized *in vitro* via the *mariner* system is introduced into bacteria by transformation and homologous recombination.

[10] D. J. Lampe, B. J. Akerley, E. J. Rubin, J. J. Mekalanos, and H. M. Robertson, *Proc. Natl. Acad. Sci. U.S.A.* **96,** 11428 (1999).

Recombinants are selected for drug resistance encoded by the transposon. Insertions in genes essential for growth or viability are lost from the pool during growth. PCR with primers that hybridize to the transposon, reading outward from the inverted repeats in both directions together with primers to specific chromosomal sites in the mutagenized region, yields a product corresponding to each mutation in the pool. Chromosomal regions in which mutations abrogate or greatly reduce viability contain no insertions as assessed by gel electrophoresis of the PCR products.

Despite the connotation of the acronym, there is very little gambling involved in using GAMBIT as the odds clearly favor the investigator. For example, based on a typical density of 500 insertions in a 10-kb target DNA region, the theoretical probability of failing to recover an insertion in an average 1-kb gene as a result of a stochastic effect is approximately 10^{-23}. Therefore, lack of insertions in a given region can be considered definitive evidence of the loss of mutants having insertions in that region except in rare instances or as a result of several caveats that will be discussed in the context of the procedure. The stepwise procedure used to conduct GAMBIT in *H. influenzae* and *S. pneumoniae* is given below.

I. Transposase Purification

Currently *Himar1* transposase is not commercially available. *Himar1* transposase is poorly soluble and purification has been complicated by the formation of inclusion bodies.[5] We have discovered that the transposase retains all of its activity as a maltose binding protein (MBP) fusion that appears to be freely soluble, and the MBP tag can be used for affinity purification on amylose resin. The expression plasmid pMALC9 contains the hyperactive form of *Himar1* cloned into the MBP fusion vector, pMAL-cri from New England Biolabs (Beverly, MA). Thus, transposase purification is now relatively straightforward. A stepwise procedure to purify and store the MBP-*Himar1* transposase protein is outlined below:

*Buffers Used**

Column Buffer (CB): 20 mM Tris-HCl (pH 7.4), 200 mM NaCl, 1 mM EDTA.
Transposase Wash Buffer (TWB): 20 mM Tris-HCl (pH 7.4), 200 mM NaCl, 1 mM EDTA, 2 mM dithiothreitol (DTT), 10% v/v glycerol.
Transposase Elution Buffer (TEB): 20 mM Tris-HCl (pH 7.4), 200 mM NaCl, 1 mM EDTA, 2 mM DTT, 10% glycerol, 10 mM maltose.

Protocol

1. Transform pMALC9 (available from D. Lampe) into *E. coli* TB1 cells (the latter available free from NEB). Select with ampicillin (Am, 100 μg/ml) on LB agar plates at 37° overnight.

*All of the above solutions are made 1× with respect to the complete, mini, EDTA-free protease inhibitor cocktail from Roche (Indianapolis, IN).

2. Grow a 5 ml overnight culture in LB broth with Am.

3. Inoculate 80 ml of LB-Am with 800 μl of the overnight culture. Grow with shaking at 37° until the OD_{600} of the cell culture is ~0.5.

4. Induce protein production by the addition of isopropylthiogalactoside (IPTG) to a final concentration of 0.3 mM. Continue shaking for 2 hr.

5. Harvest the cells by centrifugation in Oak Ridge tubes at 4000g for 10 min at 4°. Discard the supernatant and resuspend the cell pellets in a total of 10 ml of cold CB. Freeze the cell suspension overnight at −20°.

6. Disrupt the cells by any reliable method such as repeated sonication, multiple freeze–thaw cycles, or a French press. We have had the best success using a single pass through a French press. Failure to recover protein or poor yields often result from a failure to disrupt the cells. For an overview of cell disruption methods, see Cull and McHenry (elsewhere in this series).[11]

7. Spin out the insoluble material at maximum rpm in an Oak Ridge tube at 4°. Alternatively, split the lysate into several 1.5-ml tubes and spin at 13,000 rpm in a microcentrifuge tube cooled to 4°.

8. Discard the pellets and pool the supernatant into a sterile, capped 15-ml tube on ice.

9. Prepare the amylose resin (New England Biolabs) that will bind the MBP-*Himar1* transposase by thoroughly resuspending the resin stock and aliquoting 500 μl into a 1.5-ml tube. Pellet the resin at 13,000 rpm in a microfuge and discard the supernatant. Resuspend the pellet in 1.0 ml of TWB. Repeat the pelleting and resuspension procedure a total of 3 times.

10. Transfer the resuspended amylose resin to the cell lysate in the 15-ml tube on ice. Incubate the resin and lysate together at 4° on a rocking platform for 1 hr.

11. Pellet the resin in the 15-ml tube in a clinical centrifuge at 4° at high speed. Carefully remove the supernatant and discard it. Resuspend the pellet in 2 ml of TWB and transfer the resuspended resin into two 1.5-ml centrifuge tubes.

12. Repeatedly pellet and wash the resin as in step 9, using TWB. We recommend a total of ca. 4 wash steps. After the final wash, combine the resin from both tubes into a single tube.

13. Elute the bound protein by adding 400 μl of TEB. Mix well and leave on ice. Continue to occasionally mix the tube over the course of ca. 5 min.

14. Pellet the resin at 13,000 rpm for 2 min at 4°. Carefully remove the supernatant and transfer it to a clean 1.5-ml tube. This is the purified transposase. Aliquot this solution into ca. 10 μl volumes and freeze at −80°. Examine the purity and concentration of the preparation by SDS–PAGE and Bradford analysis.[12]

For quality control of the procedure, we recommend analysis of the purified transposase in addition to the soluble and insoluble fractions of the cell lysates

[11] M. Cull and C. S. McHenry, *Methods Enzymol.* **182,** 147 (1990).
[12] M. M. Bradford, *Anal. Biochem.* **72,** 248 (1976).

from cultures before and after IPTG induction on a 4–20% gradient SDS–PAGE gel. If low yields are obtained, failure to lyse the cells is typically at fault. The typical concentration of MBP-transposase after using the procedure outlined above is approximately 100 μg/ml. The molecular weight of the MBP-transposase fusion protein is 83,000.

Updates to this protocol and an annotated sequence of pMALC9 can be found on the Internet (www.home.duq.edu/~lampe/transposase).

II. In Vitro Transposon Mutagenesis with Mariner-Based Minitransposons

Protocol

1. Combine target DNA (75 fmol), transposon donor DNA (50 fmol), and re-action buffer to produce the following final concentrations: 10% glycerol (v/v), 25 mM HEPES pH 7.9, 250 μg/ml acetylated bovine serum albumin (BSA), 2 mM DTT (dithiothreitol), 100 mM NaCl, 5 mM MgCl$_2$. Acetylated BSA can be ob-tained from NEB (often supplied with restriction enzymes).

2. Start the transposition reaction by adding *Himar1* C9 transposase to 10 nM final concentration, mix, and incubate at 30° for 6 hr.

3. Purify DNA from the reaction mixture by phenol/chloroform extraction, chloroform extraction, ethanol precipitation, and washing with 70% ethanol.

4. Resuspend DNA in water and incubate for 20 min at 37° in a reaction containing 25 μM dNTPs, 50 mM NaCl, 10 mM Tris-HCl (pH 7.9 at 25°), 1 mM DTT, and 1.5 units T4 DNA polymerase (NEB). Terminate the reaction by heating at 75° for 10 min.

5. Adjust the reaction mixture to 50 mM Tris-HCl (pH 7.5 at 25°), 10 mM MgCl$_2$, 10 mM DTT, 1 mM ATP. Incubate with 40 units T4 DNA ligase (NEB) 4 hr or overnight at 16°. Terminate the reaction by heating at 75° for 10 min.

6. Desalt the reaction by ethanol precipitation followed by a 70% ethanol wash. Transform target cells according to the appropriate transformation protocol for that organism and grow under conditions that select for the expression of the selectable marker present on the *mariner* minitransposon.

III. Detection of Insertion Sites by Genetic Footprinting

Protocol

1. Pool cells after transformation and selection for mutants containing trans-poson insertions. Resuspend to an approximate OD$_{600}$ of 0.1 in 10 mM Tris pH 8.0. Use 1 μl of this suspension as the template for PCR.

2. Combine template, a primer specific to the inverted repeats (50 pmol), a chromosome-specific primer (50 pmol), *Taq* polymerase combined with *Pfu* poly-merase at a 10 : 1 unit ratio, and reaction buffer as directed by the *Taq* polymerase manufacturer.

3. To detect insertions up to 10 kb from the chromosomal primer position, conduct PCR using 30 cycles of "two-step" amplification under the following conditions: 95° for 30 sec, 68° for 5 min with an additional extension time of 15 sec added per cycle. For templates derived from *H. influenzae*, we use primers with 62° theoretical melting temperatures designed using the MacVector program.

4. Resolve the resulting PCR products by gel elecrophoresis.

Critical Parameters of Procedure

Success with this procedure is influenced by several primary factors. The method relies on highly efficient *in vitro* transposition. The transposase is very sensitive to freeze–thaw cycles. In practice, thawing the enzyme more than four times decreases its activity to an unacceptable level. For naturally competent organisms, transformation involves a single-stranded intermediate. Therefore, the reactions that extend and join the gaps left by transposase are critical. We have not conducted a comprehensive study of optimal conditions for the gap repair reactions and evaluating parameters affecting these steps may further improve efficiency. Finally, the transformation efficiency must reach a certain minimal level and it is advisable to include positive controls for transformation in each experiment. Under conditions where we obtain 10^6 transformants per 1 μg of a control DNA we obtain $\sim$800 transposon mutants in *H. influenzae* using the above procedure and a 10-kb target PCR product.

Variations

There are many possible variations that can be incorporated into this technique. For example, several other *in vitro* transposition systems (such as Tn5 and Tn7) have been developed[13–15] that could substitute for the *mariner* system. Several options are available for analysis as well. To determine the precise positions of transposon insertions that are within 2 kb of the chromosomal primer, one of the PCR primers can be labeled fluorescently or radioactively and the reaction products visualized on polyacrylamide gels. We have used a fluorescently labeled primer and an Applied Biosystems DNA sequencing apparatus to map insertions.[2] To examine larger regions of DNA (up to $\sim$10 kb) at lower resolution, fragments can be seperated by agarose gel electrophoresis and visualized by staining with ethidium bromide.[8]

[13] I. Y. Goryshin and W. S. Reznikoff, *J. Biol. Chem.* **273,** 7367 (1998).
[14] M. L. Gwinn, A. E. Stellwagen, N. L. Craig, J. F. Tomb, and H. O. Smith, *J. Bacteriol.* **179,** 7315 (1997).
[15] R. J. Bainton, K. M. Kubo, J. N. Feng, and N. L. Craig, *Cell* **72,** 931 (1993).

Considerations in Interpreting Results

In general, several issues must typically be considered in interpretation of mutant phenotypes generated by insertional mutagenesis. Not all mutations in a given gene confer the same phenotype. For example, the orientation of the antibiotic resistance cassette in the transposon with respect to the disrupted gene can result in transcriptional polarity, reducing transcription of the downstream genes. However, by examining a pool containing multiple mutations in each gene, GAMBIT can avoid this caveat. The antibiotic cassette in the transposon carries an outward reading promoter that can, in some cases, substitute for native chromosomal promoters. Transposon mutants in the pool that contain nonpolar insertions within operons needed for growth or survival will be favored while mutants containing mutations deleterious because of polarity will be lost from the pool. Another consideration is that PCR can generate spurious products. Though expensive, use of a labeled chromosome-specific primer is an effective means of avoiding some forms of PCR artifacts. As a standard procedure, we conduct independent reactions with at least two chromosome-specific primers to interrogate a given mutagenized region. A shift in the banding pattern of the resulting PCR products on gels corresponding to the distance between the primers verifies that the products seen are specific.

Application of GAMBIT to Bacterial Pathogenesis Studies

The *in vitro mariner* mutagenesis procedure itself has immediately apparent uses for classical genetics and generalized *mariner* mutagenesis by *in vitro* transposition has been applied to numerous pathogens.[8,16,17] However, the GAMBIT procedure also provides an important counterpart to genomic investigations by providing a versatile means to interrogate a discrete segment of the chromosome. Genome sequencing has revealed regions of heterogeneity between bacterial strains including putative pathogenicity islands. In addition, global expression profiling via DNA microarrays or proteomic approaches can identify candidate virulence genes based on coexpression patterns. A rapid means of mutagenizing these loci (short of designing deletion constructs on a gene-by-gene basis) is desirable for integrating the large databases of candidates produced by genome-scale approaches with the next step in functional characterization. Mutagenizing only the region of interest means that a more cumbersome large-scale screen is not needed. The footprinting procedure outlined in this article can then be applied to pools before and after selection for growth or survival in an animal model to locate regions essential for colonization or persistence.

---

[16] D. Hendrixon, B. Akerley, and V. DiRita, *Mol. Microbiol.* **40**, 214 (2001).

[17] V. Pelicic, S. Morelle, D. Lampe, and X. Nassif, *J. Bacteriol.* **182**, 5391 (2000).

Acknowledgments

We thank Eric Rubin for helpful comments on the manuscript and Sangita Chakraborty and Matthew Butler for troubleshooting and refining the transposase purification protocol. Application of the GAMBIT approach to *H. influenzae* and other pathogens was supported by a Horace H. Rackham Faculty Research Grant and the Cancer Research Fund of the Damon Runyon–Walter Winchell Foundation Fellowship DRG-1371 to B.J.A. Work on *Himar1* transposase was supported by grants to D.J.L. from NIH–National Institute of Allergy and Infectious Diseases and NSF–Microbial Genetics.

[7] Microbial Gene Expression Elucidated by Selective Capture of Transcribed Sequences (SCOTS)

By France Daigle, Joan Y. Hou, and Josephine E. Clark-Curtiss

A. Introduction

Selective capture of transcribed sequences (SCOTS) is a technique that can be used to evaluate gene expression by microorganisms in a diversity of environments, *in vitro* or *in vivo*.[1] This approach does not require well-characterized genetic tools to manipulate the organism of interest, or extensive knowledge of the genome of the organism. SCOTS is applicable to any microorganism from which nucleic acids can be isolated.

The potential applications of SCOTS are numerous. We have used SCOTS to study gene expression of mycobacteria and salmonella in a specific environment (the macrophage). SCOTS could be used to analyze gene expression of a pathogen present at different sites (i.e., different organs or tissues) within the host. SCOTS could also be used to compare gene expression between mutant and wild-type strains grown under a specific condition. Bacterial gene expression during exposure to or growth in a specific environment (temperature, medium composition, etc.) could also be analyzed by SCOTS, and enrichment could be used to compare gene expression between two *in vitro* conditions. Genes from a pathogenic strain that are expressed under a specific condition and that are absent from a nonpathogenic strain could be identified using SCOTS. It is also possible that cDNA mixtures obtained by SCOTS could be used as probes in microarray technology, although one would be unable to quantitate the level of expression of any given gene with such probes. SCOTS could be adapted to study expression of host genes in response to the presence of a specific pathogen.

Pathogenic mycobacteria and salmonella are able to survive and multiply within macrophages. Thus, our impetus in developing the SCOTS technique was

---

[1] J. E. Graham and J. E. Clark-Curtiss, *Proc. Natl. Acad. Sci. U.S.A.* **96**, 11554 (1999).

to provide a means to study gene expression by these pathogens while they were growing intracellularly.

Understanding infectious diseases requires elucidating interactions between the infected host and the pathogen. Pathogenic bacteria have developed myriad approaches to colonize and persist within the host by targeting specific cells and organs that provide favorable environments for survival and replication of the bacteria. Hosts, however, are not always willing victims and have developed ways to eliminate or wall off these foreign invaders. The consequences of the initial interactions between hosts and pathogens depend on which adversary produces the more superior armamentarium at the time of infection.

During the past two decades, many investigators have studied how various bacterial pathogens adhere to host cells, resulting in an extensive body of knowledge about the roles that specific pili or fimbriae, other adhesins, and bacterial surface proteins play in affording bacteria modes of entry into their hosts (reviewed in Ref. 2). Similarly, efforts of numerous investigators have been directed toward defining ways that various pathogens invade their hosts after attachment. The result has been an impressive amount of detailed understanding of the invasion properties of a number of pathogens and the genes and gene products involved in cell invasion and intracellular lifestyle. Intracellular pathogens must possess additional genes that facilitate entry into and promote survival and replication of the bacteria within host cells.

During the past decade, a number of approaches have been developed to study *in vivo* gene expression in different bacterial species. Two-dimensional gel electrophoresis analyses have been used to identify proteins synthesized intracellularly.[3–6] A fluorescence-based technique, differential fluorescence induction (DFI), is a positive selection method in which promoter sequences fused to the promoterless green fluorescent protein (GFP) gene have been used to detect in *vivo* gene expression.[7] Signature-tagged mutagenesis (STM) is a negative selection approach in which tagged mutants that cannot survive *in vivo* are identified.[8] The *in vivo* expression technology (IVET) is a positive selection for promoters that are turned on *in vivo* but not *in vitro*.[9] Subtractive hybridization of cDNA was also successfully used for identification of a *Mycobacterium avium* gene expressed following

[2] B. B. Finlay and S. Falkow, *Microbiol. Mol. Biol. Rev.* **61,** 136 (1997).
[3] K. Z. Abshire and F. C. Neidhardt, *J. Bacteriol.* **175,** 3734 (1993).
[4] N. A. Buchmeier and F. Heffron, *Science* **248,** 730 (1990).
[5] L. Burns-Keliher, C. A. Nickerson, B. J. Morrow, and R. Curtiss III, *Infect. Immun.* **66,** 856 (1998).
[6] B. Y. Lee and M. A. Horwitz, *J. Clin. Invest.* **96,** 245 (1995).
[7] R. H. Valdivia and S. Falkow, *Science* **277,** 2007 (1997).
[8] M. Hensel, J. E. Shea, C. Gleason, M. D. Jones, E. Dalton, and D. W. Holden, *Science* **269,** 400 (1995).
[9] M. J. Mahan, J. M. Slauch, and J. J. Mekalanos, *Science* **259,** 686 (1993).

phagocytosis by human macrophages.[10] Many other approaches are being developed and will be discussed in different articles in this book. However, differential gene expression in bacteria within infected host cells or tissues has been limited because of low numbers of bacteria in these systems and instability of bacterial mRNA. There are also significant difficulties involved in separating bacterial mRNA from ribosomal RNA (rRNA) and host RNA. SCOTS has been used to identify bacterial genes that are expressed during growth of the bacteria within macrophages.[1,11,12,13] SCOTS allows the selective capture of bacterial cDNAs from total cDNA prepared from infected cells or tissue by hybridization to biotinylated bacterial genomic DNA. cDNA mixtures obtained are then enriched for sequences that are preferentially transcribed during growth in the host by additional hybridizations to bacterial genomic DNA in the presence of cDNA similarly prepared from bacteria grown *in vitro*.[1,12]

Each of these approaches has advantages and disadvantages. Many of the approaches used for identification of *in vivo* gene expression require a considerable quantity of starting material, which can be difficult to obtain in some cases and often may not represent bacterial numbers found in tissues during natural infections. In methods using nucleic acids or proteins, it is sometimes difficult to differentiate between host and pathogen products. Many of these approaches also require well-developed genetic manipulation systems of the pathogen including ability to mutagenize the pathogen, genetic transfer mechanisms, ability to construct libraries, and cloning vectors that replicate in the pathogen of interest. Also, with some approaches that use insertional mutagenesis, some of the genes identified could be the result of a polar mutation. Moreover, the need for good animal models is also essential to study the host–pathogen interactions regardless of the approaches used for identification of *in vivo* expressed genes. All of these methods should contribute to an understanding of relevant physiological attributes displayed by bacterial pathogens *in vivo*. It should be mentioned that no single approach can provide all of the information necessary for understanding microbial pathogenesis. Each approach, sometimes in combination with one or more other approaches, will provide superior results in some cases but not others.

As with other methods used for identification of *in vivo* gene expression, SCOTS can provide insights about host–pathogen interactions. Moreover, SCOTS uses wild-type strains without any genetic modifications. SCOTS recovers relevant genes, essential or not, that are expressed by a microbe *in vivo* or in a particular environment. SCOTS can also detect expression from small numbers of microbial cells, which should be applicable to samples obtained from living tissues in natural disease states, including human biopsies.

---

[10] G. Plum and J. E. Clark-Curtiss, *Infect. Immun.* **62**, 476 (1994).

[11] B. J. Morrow, J. E. Graham, and R. Curtiss III, *Infect. Immun.* **67**, 5106 (1999).

[12] J. Y. Hou, J. E. Graham, and J. E. Clark-Curtiss, *Infect. Immun.* **70**, 3714 (2002).

[13] F. Daigle, J. E. Graham, and R. Curtiss III, *Molec. Microbiol.* **41**, 1211 (2001).

SCOTS has been used successfully to study gene expression of the following bacteria growing in macrophages in culture: *Mycobacterium tuberculosis,*[1] *M. avium,*[12] *Salmonella typhi,*[13] and *S. typhimurium.*[11] In addition, SCOTS has been used to identify genes expressed by pathogens from infected animal tissues such as chickens infected with avian pathogenic *Escherichia coli* (APEC)[14] or with *S. typhimurium,*[15] or mice infected with *S. typhimurium.*[16]

In this article, we will describe SCOTS, a technique that combines total RNA isolation, followed by cDNA synthesis, hybridization with biotinylated genomic DNA, PCR amplification, cloning, and Southern blot hybridization, all of which are commonly used techniques in molecular biology.

B. Growth Conditions

First, the investigator should determine the appropriate growth condition in which microbial gene expression will be analyzed. In the methods described below and previously, we wanted to study gene expression by mycobacteria and salmonella during growth under conditions we consider to be relevant to the pathogenesis of these bacteria: *in vivo* growth within human macrophages.

1. Macrophage Infections

a. Mycobacteria. Peripheral blood is collected from healthy, tuberculin skin-test-negative human volunteers and mononuclear cells are isolated by passage through Ficoll gradients (Amersham Pharmacia Biotech, Piscataway, NJ), according to the manufacturer's instructions. Monocytes are cultivated in Teflon wells in supplemented RPMI 1640 medium with 20% autologous serum at 37°, 5% (v/v) CO_2 to allow differentiation of the monocytes into adherent macrophages.[17] Supplemented RPMI 1640 medium is RPMI 1640 (Life Technologies, Rockville, MD) supplemented with 4 mM L-glutamine, 25 mM HEPES (N-2-hydroxyethylpiperazine-N'-2-ethanesulfonic acid, pH 7) (Sigma, St. Louis, MO), and 1% MEM (modified Eagle's medium) nonessential amino acids (Sigma). Five days later, nonadherent monocytes are removed from the cultures by washing, fresh supplemented RPMI 1640 medium with 20% autologous serum is added, and the cultures are incubated an additional 7 days to allow further maturation of the macrophages.[18] Mature, adherent macrophages are infected at a multiplicity of 0.5–1 bacterium : 1 macrophage with a predominantly single-cell suspension of *M. tuberculosis* or *M. avium,*[19] which has been grown, as described below, in

[14] C. M. Dozois, personal communication (2001).

[15] J. O. Hassan and R. Curtiss III, in preparation (2002).

[16] C. A. Nickerson, personal communication (2001).

[17] L. S. Schlesinger and M. A. Horwitz, *J. Immunol.* **147,** 1983 (1991).

[18] T. M. Kaufman and L. S. Schlesinger, personal communication (1997).

[19] D. L. Clemens and M. A. Horwitz, *J. Exp. Med.* **181,** 257 (1995).

Middlebrook 7H9 broth + supplements to mid-logarithmic phase.[1,12] Prior to infection, the mycobacteria are suspended in supplemented RPMI 1640 medium with 10% autologous serum and are then added to the macrophage monolayers (which are now also in supplemented RPMI 1640 with 10% autologous serum) and incubated for 2 hr at 37°, 5% (v/v) CO_2. Infected monolayers are washed with RPMI 1640 and are then maintained at 37° in supplemented RPMI 1640 medium with 1% autologous serum, and medium is changed 16 hr postinfection. The monolayers are washed 3 times with supplemented RPMI medium before lysis.[1,12] The generation times of *M. tuberculosis* and *M. avium* growing within the macrophages are approximately 18–20 hr, as determined by colony-forming units (cfu) and microscopic enumeration of acid-fast bacilli.[1,12]

Infected macrophage monolayers are lysed at 18, 48, and 110 hr after infection with *M. tuberculosis*[1] and at 48 and 110 hr after infection with *M. avium*.[12] Forty-eight hours was initially selected as a relatively early time point after infection to study intracellular mycobacterial gene expression. By 48 hr, the mycobacteria have adapted to the macrophage environment and are actively multiplying. Later, macrophages infected with *M. tuberculosis* are also lysed 18 hr after infection to produce a cDNA mixture representing genes expressed at an earlier time after infection. By approximately 120 hr after infection, the infected macrophages are beginning to detach from the tissue culture flask surfaces and are no longer intact. At 110 hr, the infected macrophages are still healthy and adherent and there are no apparent extracellular mycobacteria, so this time has been chosen as a late time point in this infection model.

For mycobacterial infections, only one human donor provides blood for macrophages for a given infection. Thus, the volume of blood that can be drawn is limited to one unit (400 ml) per bleed. Macrophages obtained from each unit of blood are divided into separate tissue culture flasks that are then used for the individual infection times. These restrictions have been applied because humans respond differently to infection by mycobacteria and it is unknown whether mycobacteria respond identically to macrophages from different donors.

b. Identification of S. typhi Genes Expressed in THP-1 Cells. The goal of this study is to identify genes expressed during the early stage interactions of *S. typhi* in macrophages. Under these conditions, bacterial genes should encode products involved in the adaptive responses that allow survival of *S. typhi* following phagocytosis. The human monocyte cell line, THP-1 (ATCC, TIB-202, Manassas, VA), is maintained in RPMI 1640 containing 10% fetal calf serum (FCS) (Atlanta Biologicals, Atlanta, GA), 25 mM HEPES, 2 mM L-glutamine, and 1% MEM nonessential amino acids (RPMI+). A stock culture of these cells is maintained as monocyte-like, nonadherent cells at 37° in an atmosphere containing 5% (v/v) CO_2. THP-1 cells are differentiated by addition of $10^{-7} M$ phorbol 12-myristate 13-acetate (PMA) for 24 to 48 hr before infection. Wells are seeded with 2×10^6 cells in 6-well tissue culture dishes and are infected for 2 hr with *S. typhi* ISP1820

(χ 3744). Bacteria are grown in Luria–Bertani (LB) broth to mid-logarithmic phase and are washed in phosphate-buffered saline (PBS); the optical density of the suspension is adjusted in PBS, and then 10 μl is added to the cell monolayer at a multiplicity of infection of 10 bacteria : 1 macrophage and centrifuged for 5 min at 800g to synchronize phagocytosis. After incubation for 20 min at 37° (T_0), the infected cells are washed three times with prewarmed Hanks' buffered salt solution (HBSS) and are incubated with RPMI+ that contains a high concentration of gentamycin (100 μg ml^{-1}) to kill extracellular bacteria. At 2 hr and 24 hr after infection, the monolayers are gently scraped and RNA is extracted, as described below.

Comments. These infection protocols and parameters are dependent on the type of pathogen studied and on the host model available and should be adjusted by the investigator, depending on the objective of the studies.

2. In vitro Growth Conditions

The *in vitro* growth conditions also need to be carefully chosen by the investigator, since bacterial transcripts from *in vitro* grown cells will be used to prepare cDNA mixtures to enrich for capture of bacterial cDNA molecules representing genes that are more specifically expressed *in vivo*. For example, the mycobacterial strains are grown in Middlebrook 7H9 broth, as described below. Middlebrook 7H9 medium is a partially defined, fairly minimal medium and has been chosen because (1) the mycobacterial strains are grown in this medium prior to infection and (2) we believe that the nutrients available to the mycobacteria might be more similar to those available within the macrophage phagosomes. In our hands, *M. tuberculosis* and *M. avium* are unable to multiply in supplemented RPMI 1640 medium with human serum at the concentrations used during macrophage infections.

a. Mycobacteria. Mycobacterial strains are grown in Middlebrook 7H9 broth (Difco Laboratories, Detroit, MI), supplemented with oleic acid–albumin–dextrose–catalase (OADC; Microbios, Phoenix, AZ or Remel, Inc., Lenexa, KS), or ADC, and 0.05% Tween 80 (Sigma). Freezer stock cultures (1 ml) are diluted 1 : 10 into Middlebrook 7H9 broth + OADC or ADC and Tween 80 and are grown in flasks at 37° in a New Brunswick Gyrotory Shaker with moderate aeration (100 rpm). When larger cultures are needed, the seed cultures are diluted 1 : 10 into fresh medium. Cultures are never diluted more than three times after removal from the freezer. Cultures are harvested for RNA isolation at early and mid-log phase and at late log–early stationary phases of growth.

b. Salmonella. *S. typhi* is grown standing overnight (18 hr) in tissue culture medium RPMI+ at 37° in an atmosphere containing 5% (v/v) CO_2. *S. typhi* invasion and survival within macrophages following growth in RPMI+ is similar to *S. typhi* invasion and survival within macrophages following growth in standing overnight Luria broth. This condition has been chosen to eliminate any genes induced by the cell culture conditions used for macrophage infection.

3. Isolation of Nucleic Acids

a. Chromosomal DNA. For *Salmonella,* total bacterial chromosomal DNA is prepared using a small-scale preparation method as described.[20] For mycobacteria, chromosomal DNA is isolated from Middlebrook broth-grown *M. tuberculosis* and *M. avium* as described previously,[21] except that the bacilli are disrupted in tubes containing silica–zirconium beads by agitation in a Mini-Bead Beater (Bio Spec Products, Inc., Bartlesville, OK). In addition, carbohydrates associated with mycobacterial DNA are removed by treatment with cetyltrimethylammonium bromide (CTAB), followed by organic extraction and ethanol precipitation,[22] rather than by dissociation by centrifugation of the DNA through ethidium bromide–cesium chloride density gradients.[21]

b. RNA. Total RNA is isolated from 10^6 infected macrophages or from 10^9 broth-grown bacteria. Total RNA is obtained from mycobacteria by mechanical disruption in the Mini-Bead Beater as described above, followed by organic extraction with hot (80°) guanidinium thiocyanate–phenol–chloroform, as described by Chomczynski and Sacchi.[23] For *S. typhi,* RNA is extracted using TRIzol reagent (Life Technologies), according to the manufacturer's instructions. All RNA samples are treated with RNase-free DNase I (Ambion Inc., Austin, TX) for 20–30 min at 37° and the enzyme is then heat-inactivated at 75° for 5 min. RNA concentrations and integrity are determined by spectrophotometer readings and agarose gel electrophoresis, respectively.

Comments. The number of broth-grown bacteria used for isolation of total RNA has been chosen empirically: 10^9 bacteria give good yields of high quality RNA. No experiments have been done to determine the smallest number of broth-grown bacteria that would give adequate yields. We hypothesize that the eukaryotic RNA in the infected macrophages serves as "carrier RNA" to facilitate recovery of total RNA from the smaller number of bacteria present in the macrophages. However, no experiments have been done to test this hypothesis. In our experience, the use of commercial kits for RNA isolation is not recommended because most of the short transcripts (<200 nucleotides) are selectively excluded.

4. cDNA Synthesis

A 5-μg sample of total RNA isolated from infected macrophages or other growth conditions is converted to first-strand cDNA by random priming with

[20] F. M. Ausubel, R. Brent, R. E. Kingston, D. M. Moore, J. G. Seidman, J. A. Smith, and K. Struhl, "Current Protocols in Molecular Biology." Wiley Interscience, New York, 1991.
[21] J. E. Clark-Curtiss, W. R. Jacobs, M. A. Docherty, L. R. Ritchie, and R. Curtiss III, *J. Bacteriol.* **161**, 1093 (1985).
[22] E. B. Hill, L. G. Wayne, and W. M. Gross, *J. Bacteriol.* **112**, 1033 (1972).
[23] P. Chomczynski and N. Sacchi, *Anal. Biochem.* **162**, 156 (1987).

Moloney murine leukemia virus (MMLV) reverse transcriptase, Superscript II (Life Technologies), according to the manufacturer's instructions. In a number of independent experiments, we have obtained superior results using the MMLV reverse transcriptase compared to using AMV (avian myeloblastosis virus) reverse transcriptase. Primers with a defined sequence at the 5' end and random nonamers at the 3' end (PCR primer-dN9) are used for both first- and second-strand cDNA synthesis.[24,25] Different terminal sequences are added to cDNA mixtures prepared from each growth condition. This allows comparative analysis of cDNA clones from different growth conditions. Each cDNA mixture is amplified by the polymerase chain reaction (PCR) using the defined primers for each set of cDNA mixtures for 25 cycles of amplification using parameters suitable for each given primer.

Comments. The defined terminal sequence primers used for each growth condition should be derived from different linkers or adaptors that will not hybridize with the genome of the bacterium being studied. The terminal sequences should be designed to include a restriction site to facilitate cloning of individual cDNA molecules. In our experiments, the lower limit of bacteria present in infected macrophages from which complex cDNA mixtures have been prepared is 8×10^5 to 1×10^6. We think that SCOTS can be used to detect bacterial RNA transcripts from a lower number of bacteria in infected host cells, but the limit is not yet established and should be tested.

5. Selective Capture of Transcribed Sequences (SCOTS)

a. Ribosomal RNA Genes. Ribosomal RNA (rRNA) is the major class of RNA molecules present in total RNA preparations (>82%).[26] It is necessary to significantly decrease the amount of these abundant sequences in order to capture the cDNA molecules representing bacterial mRNA transcripts, which constitute only a small portion of total prokaryotic RNA (≈4%).[26] A plasmid containing a ribosomal operon of the bacterium being studied should be constructed. Alternatively, PCR amplification of the ribosomal operon could also be done if sequences are available. In the SCOTS experiments that have been done in our laboratory, plasmids containing the rRNA operon have been used (see Fig. 1.).

b. Biotinylation. Photobiotinylation of genomic DNA is realized as follows: photoactivatable biotin (PAB) acetate (Clontech, Palo Alto, CA, or Sigma) is dissolved in sterile distilled water to a final concentration of 1 μg/μl. Ten μl of bacterial chromosomal DNA (at a concentration of 0.5–1.0 μg/μl in sterile, distilled water) is added to a microfuge tube and an equal volume of PAB acetate

[24] P. Froussard, *Nucleic Acids Res.* **20,** 2900 (1992).

[25] D. Grothues, C. R. Cantor, and C. L. Smith, *Nucleic Acids Res.* **21,** 1321 (1993).

[26] F. C. Neidhardt and H. E. Umbarger, "*Escherichia coli* and *Salmonella typhimurium:* Cellular and Molecular Biology," p. 13. ASM Press, Washington, D.C., 1996.

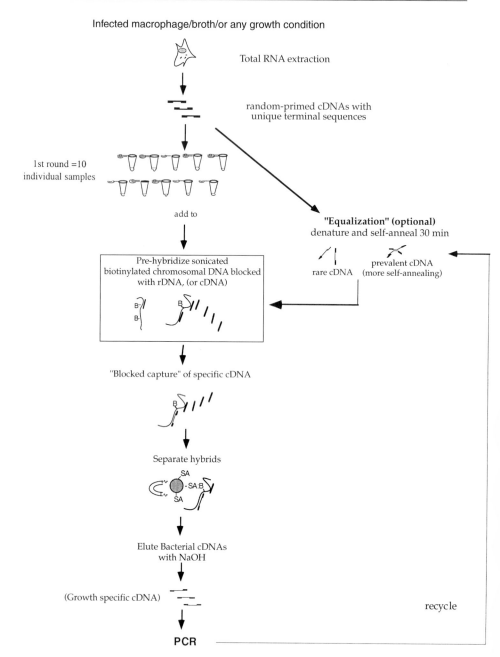

FIG. 1. Flow chart showing selective capture of transcribed sequences (SCOTS).

solution is also added to the tube. After the solutions are mixed, the cap of the tube is removed and the tube is placed in crushed ice so that the top of the tube is 2–3 cm below a 300 watt incandescent lightbulb. The mixture is photoactivated for 30 min, while being kept on ice. A fresh volume of PAB acetate solution, equal to the first volume of PAB acetate, is added to the tube and an additional 30 min photoactivation is done to ensure even biotinylation of the chromosomal DNA. A 10-fold dilution of the biotinylated solution in TE pH 9.0 is extracted with iso-butanol (repeated 3–4 times) until the solution becomes clear, and the chromo-somal DNA is then precipitated. The ribosomal DNA plasmid is added at a ratio of 17 : 1 (5 μg of rDNA to 0.3 μg of chromosomal DNA). The mixture is then sonicated to obtain a smear of DNA with most fragments within a size range of 1 to 5 kb (determined by agarose gel electrophoresis). The sonicated, biotinylated chromosomal DNA and rDNA mixture is then precipitated with ethanol and re-suspended in 10 mM EPPS [N-(2-hydroxyethyl)piperazine-N'-(3-propanesulfonic acid)] and 1 mM EDTA (EPPS–EDTA buffer).

Comments. As 0.3 μg of chromosomal DNA is needed for each round of SCOTS (10 individual reactions for the first round) and each round of enrichment reaction, we suggest that a sufficient amount of biotinylated chromosomal DNA be prepared so that all successive rounds of SCOTS and enrichment can be done with the same preparation of biotinylated chromosomal DNA.

c. Blocking of Ribosomal Loci on Chromosomal DNA. The sonicated mixture of biotinylated chromosomal DNA fragments (0.3 μg) and plasmid DNA con-taining the ribosomal RNA genes of the bacterium (5 μg) is denatured by boiling under oil for 3 min in 4 μl of EPPS–EDTA buffer indicated above. One microliter of 1 M NaCl is added to the tube and the fragments are hybridized for 30 min at a temperature 20° below the T_m of the bacterial DNA.[1] This step allows the rDNA to hybridize to the rDNA sequences on the bacterial chromosomal DNA, thereby rendering these sites unavailable for hybridization of cDNA corresponding to the rRNA genes present in the cDNA mixtures.

d. Self-Annealing of cDNA (Equalization). In a separate reaction, total ampli-fied cDNA (3 μg) from infected macrophages is denatured as described earlier and allowed to reanneal for 30 min at the same hybridization temperature used above.[1] The self-annealing step has been incorporated into the protocol to reduce the number of cDNA molecules corresponding to abundant mRNA sequences and to increase the representation of rare cDNAs.[27,28] This equalization step may not be necessary, since self-annealing of abundant cDNA molecules undoubtedly occurs throughout the hybridization period. Moreover, during PCR amplification, it has been shown that rehybridization of PCR products may interfere with primer

[27] M. S. Ko, *Nucleic Acids Res.* **18,** 5705 (1990).
[28] W. E. Hahn, D. E. Pettijohn, and J. Van Ness, *Science* **197,** 582 (1977).

binding or extension, thus decreasing the rate of amplification.[29] This effect is greater on the abundant PCR products (i.e., abundant cDNA molecules); the net effect is to diminish the differences in product abundance and effectively equalize the representation of individual cDNA molecules in the cDNA mixtures.[29] Satis- factory cDNA mixtures have been prepared from *M. avium* without performing the self-hybridization step.[12]

e. Hybridization–Capture.[1] The cDNA is then added to the rDNA blocked- genomic DNA, and hybridization is allowed to proceed for 18 to 24 hr (final volume, 10 μl). After hybridization, DNA mixtures are diluted to 500 μl with water and bacterial cDNA–chromosomal DNA hybrids are removed from the hybridization mixture by binding to 60 μg of streptavidin-coated magnetic beads (Dynabeads M-280, Dynal, Lake Success, NY) resuspended in 500 μl of 10 mM Tris HCl pH 7.5 and 1 mM EDTA. After a 10 min incubation at 37°, the beads are separated using a magnet and washed 3 times at 65° with 20 mM NaCl and 0.1% SDS. Captured cDNA is then eluted with 100 μl of 0.4 N NaOH/0.1 M NaCl, precipitated with glycogen (10 μg, Roche Molecular Biochemicals, Indianapolis, IN) as a carrier, and amplified by PCR using primers corresponding to the defined terminal sequences added during preparation of the cDNA described above.

Comments. For *M. avium,* 60 μg of washed beads resuspended in 1 ml of 1 M NaCl, 5 mM EPPS, and 0.5 mM EDTA is added to the hybridization mixture. This solution is incubated at 43° for 30 min prior to magnetic separation.

f. First Round of SCOTS. For each growth condition, in the first round of SCOTS, 10 separate samples of the cDNA mixture are captured by hybridiza- tion to biotinylated rDNA–blocked chromosomal DNA in parallel reactions. This is done to enhance the likelihood of recovering cDNA molecules corresponding to a more complete diversity of transcripts present at the time of RNA prepara- tion for each growth condition. After the first round of SCOTS, the 10 amplified cDNA preparations are combined (for each growth condition), denatured, and again hybridized to fresh samples of rDNA-blocked, biotinylated chromosomal DNA for two successive rounds of SCOTS.

Comments. In some cases, only five individual samples are used for the first round of SCOTS. cDNA mixtures prepared this way demonstrate a level of com- plexity similar to that of cDNA mixtures prepared using 10 samples in the first round (assessed by hybridization to restriction endonuclease-digested chromoso- mal DNA as described below).

g. Successive Rounds of SCOTS. At least three rounds of SCOTS are done with cDNA mixtures from each condition. After obtaining cDNA mixtures by SCOTS, the cDNA mixtures are used as probe templates for analysis of cloned cDNA molecules and PCR-amplified control gene fragments, and for competitive

[29] F. Mathieu-Daude, J. Welsh, T. Vogt, and M. McClelland, *Nucleic Acids Res.* **24,** 2080 (1996).

hybridization enrichment as described below. The bacterial cDNA mixture obtained after three rounds of SCOTS should contain a complex population of cDNA molecules representing genes expressed in the growth condition tested. The increasingly detectable complexity of the bacterial cDNA molecules recovered after each round of SCOTS compared to the original cDNA preparation should be verified. This is done by removing a sample of the cDNA mixture recovered after each round of SCOTS, labeling the samples, and using them as individual probes in Southern hybridizations against the bacterial genomic DNA. With each round of SCOTS, the cDNA probes should hybridize to an increased number of fragments of restriction endonuclease-digested chromosomal DNA and to rDNA fragments with decreasing intensity (see Graham and Clark-Curtiss,[1] which illustrates these results).

Comments. In our experiments, three rounds of SCOTS have been chosen because we observed that the level of rDNA is low and the bacterial cDNA complexity is high (as tested above by hybridization to chromosomal DNA). However, depending on the experiment, two rounds of SCOTS might be sufficient; in other cases, four rounds or more might be necessary. The investigator can assess the cDNA preparation after each round of SCOTS as described above. Alternatively, the investigator can pick 10–20 random cDNA clones obtained after 2 or 3 rounds of SCOTS and hybridize these with a labeled ribosomal DNA probe. If the number of cDNA clones that hybridize with the rDNA is low, this SCOTS-purified cDNA preparation can be used for further analysis.

6. Enrichment for Bacterial cDNA Molecules from a Specific Growth Condition

This step is used to enrich the bacterial cDNA populations for sequences preferentially transcribed or up-regulated under a specific condition, such as growth in macrophages. This competitive hybridization enrichment strategy is similar to SCOTS, except that in addition to being blocked with rDNA, the chromosomal DNA is also blocked with cDNA obtained from bacteria grown in another condition that is being compared to selectively reduce cDNAs that are expressed in both conditions.[1] Briefly, denatured, biotinylated genomic DNA (0.3 μg) is first hybridized with rDNA and an excess of denatured cDNA (10-fold, 30 μg) from the other growth condition (e.g., broth), which is obtained after three rounds of SCOTS, as described above. SCOTS-purified bacterial cDNA (3 μg) from infected macrophages is self-annealed for 30 min before hybridization to the blocked genomic DNA at a temperature 20° below the T_m of the bacterial chromosomal DNA for 18 to 24 hr, as described above. The hybrids are removed from the hybridization solution by binding to streptavidin-coated magnetic beads, as described above. Bacterial cDNAs are eluted, precipitated, and PCR-amplified using terminal sequences specific for the cDNA molecules from the macrophage-grown bacteria.

Individual SCOTS-purified cDNA molecules can then be cloned into a plasmid vector to generate cDNA libraries, which can be used for further analysis (as described in Section 7).

7. Preparation of cDNA Libraries

Depending on the interests of the investigator, the cDNA mixture could be cloned after three rounds of SCOTS, which represents a complex mixture of the genes transcribed from a particular growth condition, or after three rounds of enrichment, if the interest is to identify genes expressed specifically in a particular growth condition.

a. Vector and Strain Used for cDNA Library Preparation. Using the restriction site included in each terminal sequence of the primer that has been used for the cDNA synthesis, individual cDNA molecules are cloned into pBluescript II SK(+).[1,12] Cloning can also be performed after the PCR amplification by using the Original TA Cloning kit (Invitrogen, Carlsbad, CA) according to the manufacturer's instructions.[13] Transformation of *Escherichia coli* K12 (DH5α) bacterial cells is performed by standard techniques.[30] This step can be done with any combination of strain and cloning vector available that has an unique restriction site corresponding to the restriction site in the terminal sequence of the primer at the discretion of the investigator.

b. Screening of Clones. Inserts are PCR-amplified using primers specific to the cloning vector (e.g., universal T7 and M13 primers) or to the defined terminal sequences used for cDNA synthesis. Individual inserts are screened by hybridization with probes made from cDNA mixtures obtained by three rounds of SCOTS either from broth-grown bacteria or from infected human macrophages. In our studies, cloned inserts that do not hybridize to a SCOTS-purified cDNA probe prepared from bacteria grown *in vitro* but do hybridize to the cDNA probe mixtures prepared from bacteria grown in macrophages are chosen for further characterization and analysis.

Comments. Two independent SCOTS procedures have been performed on cDNA prepared from *S. typhi* grown in macrophages, using a different terminal primer for cDNA synthesis. Clones obtained after three rounds of enrichment from each experiment have been found to hybridize with both cDNA probe mixtures, which confirms the reproducibility of SCOTS.[31] In another set of experiments to confirm the reproducibility of SCOTS, cDNA mixtures have been purified by SCOTS on three separate occasions (6–11 months apart) from a cDNA preparation from *M. avium* grown for 48 hr in macrophages. Each of these cDNA mixtures hybridized to a subset of individual cDNA clones isolated from one of the mixtures.[12]

[30] J. H. Miller, "A Short Course in Bacterial Genetics: A Laboratory Manual for *Escherichia coli* and Related Bacteria." Cold Spring Harbor Laboratory Press, Cold Spring Harbor, NY, 1992.
[31] F. Daigle, unpublished data (2000).

8. Analysis of cDNA Clones

a. Sequence Determination of Inserts. PCR-amplified inserts that hybridize with a cDNA probe from a specific growth condition are sequenced using one of the universal primers and the ABI Prism Big Dye primer cycle sequencing kit according to the manufacturer's instructions (PE Applied Biosystems). Other sequencing strategies could also be used.

b. Comparison of Sequences with Databases. Database searches and DNA and protein similarity comparisons can be carried out with the BLAST algorithm[32] available from the National Center for Biotechnology Information (NCBI) at the National Library of Medicine. If the genome sequence of the studied bacteria is available, it should also serve as a useful tool.

c. Analysis of Genes Predicted to Be Coexpressed with Sequences Identified by SCOTS. Some transcripts identified by SCOTS may correspond to a gene that belongs to an operon. This suggests that all genes in this operon should be expressed in the growth condition tested. In our cases, to determine whether the gene(s) are coexpressed intracellularly, PCR-amplified DNA fragments corresponding to other genes in the operon are hybridized to the SCOTS-derived cDNA probe mixture from infected macrophages. Positive hybridization of these putative coexpressed genes confirms the prediction. The same strategy could be used to predict whether or not an interactive gene product is also expressed with genes identified by SCOTS that belong to regulons or act in conjunction with one or more protein products (e.g., two-component systems).

d. Comparison of Gene Expression under Other Growth Conditions. The genes identified by SCOTS in our studies correspond to genes that are expressed in the specific condition of growth within macrophages. We can test whether the identified genes are also expressed when the bacteria are growing under other conditions or are specifically expressed following phagocytosis. Bacterial cDNA probes can be prepared by three rounds of SCOTS from different growth conditions, such as bacteria grown to different phases of growth in broth or grown under a variety of stresses that can mimic the macrophage environment (such as low pH or low iron) or from bacteria grown in different types of eukaryotic cells. These cDNA mixtures can then be used as probes against previously identified cDNA clones on Southern blots.

e. Construction of Mutants and Role in Virulence. Determining the potential role of specific genes first identified using SCOTS can be pursued by inactivation of these genes. Effects of such mutations on survival in macrophages or virulence in animal models (if available) can be determined by comparing strains carrying the mutated genes to their wild-type parent strains.

f. Expectations. We cannot definitively predict the number of different cDNA molecules that might be recovered in a cDNA mixture from any given growth

---

[32] S. F. Altschul, W. Gish, W. Miller, E. W. Myers, and D. J. Lipman, *J. Mol. Biol.* **215**, 403 (1990).

condition. One can hypothesize about the kinds of conditions an organism might encounter in a specific growth environment and the kinds of responses that an organism might make to those conditions. In an *in vitro* environment, the investigator might be able to design the conditions so that the organism's response to a single parameter (e.g., acidic pH) could be determined. However, in an *in vivo* environment such as a eukaryotic cell, many different variables are operative simultaneously. Even for the most well-characterized bacterial species, we do not have a complete understanding of all the physiological responses operating during growth in complex environments, although application of high-density DNA microarray technology (see articles 12–16 in this volume) promises to facilitate such understanding.

Moreover, SCOTS can be used to determine the expression of any given gene of interest by Southern hybridization of the PCR-amplified gene with the SCOTS-derived cDNA probe mixture from the desired growth condition.

Discussion/Conclusion

Understanding the basic mechanisms by which bacterial pathogens invade, survive, and multiply in host cells and tissues is likely to contribute to the development of methods to prevent and control infectious diseases. Many methods have been used for *in vivo* analysis of bacterial gene expression. Techniques such as signature-tagged mutagenesis (STM) and *in vivo* expression technology (IVET) seldom identify the same genes, even when the same infection model and pathogens are used. This suggests that the use of more than one approach will be necessary to obtain a comprehensive picture of bacterial gene expression associated with host interaction, including the production of specific virulence factors as well as the basic physiology and adaptive responses that allow bacterial colonization of the host.

Analysis of microbe–host interactions by SCOTS is a powerful tool for the global analysis of microbial gene expression inside the natural host. Moreover, any potential candidate antigen can be tested for its expression when bacteria are growing in the host cell. Understanding the metabolic pathways used by bacteria growing in host cells could suggest potential targets for development of new antibiotics. The information obtained by SCOTS can lead to identification of specific proteins produced at high levels within antigen-presenting cells, providing candidates for development of effective vaccines against bacterial infections.

Acknowledgments

The authors acknowledge R. Curtiss III and C. M. Dozois for critical review of the manuscript. Methods and observations from the authors' laboratories arose during research supported by the National Institutes of Health (AI35267, AI38672, AI46428) to J.E.C.-C. and (AI24533) to R.C. III and by a Canadian Natural Sciences and Engineering Research Council postdoctoral fellowship to F.D.

[8] Identification of Essential Genes in *Staphylococcus aureus* Using Inducible Antisense RNA

By Yinduo Ji, Gary Woodnutt, Martin Rosenberg, and Martin K. R. Burnham

Introduction

Conditional disruption of gene expression is an important approach for addressing information on genes essential for bacterial growth or pathogenesis. This information is particularly useful for validating molecular targets for antibiotic discovery and vaccine development. Antisense technology is an effective approach to down-regulate expression of specific genes. It has been used widely to interfere with eukaryotic gene expression through injection of synthetic oligonucleotides complementary to mRNA[1] and by the synthesis of antisense RNA from DNA cloned in an antisense orientation.[2] However, the antisense method has not been used routinely to inhibit gene expression in bacteria, even though there is evidence that antisense regulation occurs naturally in bacteria during plasmid, phage, and chromosomal replication.[3] Reports have demonstrated that antisense RNA can effectively down-regulate gene expression in various bacterial systems.[4,5]

Combining an antisense strategy with a regulated expression system is useful in identifying and characterizing essential genes critical to bacterial growth *in vitro* and *in vivo*. Moreover, such a strategy offers a unique approach to the study of bacterial pathogenesis and definition of virulence factors. Regulated antisense RNA can be used to decrease the expression of known genes during different stages of infection.

The Tn*10*-encoded *tet* repressor has been successfully used to regulate expression of specific genes in prokaryotic cells[6] and has also been employed in a *xyl/tet* chimerical promoter system which has been shown to be strongly inducible in *Bacillus subtilis* using subinhibitory concentrations of tetracycline.[7] We

[1] A. Fire, S. Xu, M. K. Montgomery, S. A. Kostas, S. E. Driver, and C. C. Mello, *Nature* **391**, 806 (1998).

[2] D. S. Kernodle, R. K. R. Voladri, B. E. Menzies, C. C. Hager, and K. M. Edwards, *Infect. Immun.* **65**, 179 (1997).

[3] E. G. H. Wagner and R. W. Simons, *Ann. Rev. Microbiol.* **48**, 717 (1994).

[4] L. Good and P. E. Nielsen, *Nat. Biotechnol.* **16**, 355 (1998).

[5] S. A. Walker and T. R. Klaenhammer, *Appl. Environ. Microbiol.* **66**, 310 (2000).

[6] M. Stieger, B. Wohlgesinger, M. Kamber, R. Lutz, and W. Keck, *Gene* **226**, 243 (1999).

[7] M. Geissendorfer and W. Hillen, *Appl. Microbiol. Biotechnol.* **33**, 657 (1990).

constructed a *tet* regulatory system in *S. aureus*[8,9] and initially cloned an antisense *hla* (α-hemolysin) fragment downstream of the inducible *xyl/tet* promoter–operator fusion to demonstrate that this *tet* regulatory system can function in *S. aureus* on induction of antisense to *hla*. It was found possible to down-regulate expression of the chromosomal *hla in vitro* and to eliminate virulence of *S. aureus* in an animal model of infection using anhydrotetracycline (ATc), a nonantibiotic analog of tetracycline, as inducer.[8] By creating a random library of inducible antisense clones we have been able to implement a strategy to find genes relevant to both viability and pathogenesis.

Construction of Staphylococcal Chromosomal DNA Random Library

In order to construct a more random antisense library the fragments of genomic DNA in the range of 200 op to 800 op are produced by mechanical shearing of chromosomal DNA of the clinically derived strain *Staphylococcus aureus* WCUH29 in a nebulizer. Forty μg chromosomal DNA in 1 ml of 50% (v/v) glycerol, 0.3 *M* sodium acetate (pH 7.5) solution is added to the chilled nebulizer chamber, the nitrogen pressure adjusted to 40 psi, and the DNA solution nebulized for 30–60 sec at 4° and aliquoted into two 1.5-ml Eppendorf tubes containing 0.9 ml of 100% ethanol, then mixed and chilled at −20° overnight. The nebulized DNA is precipitated by centrifugation at 14,000 rpm for 10 min and washed with 70% (v/v) ethanol, air-dried, and dissolved in 50 μl of 10 m*M* Tris-HCl (pH 8.5). Since mechanical shearing of DNA can result in broken ends or produce a mixture of 3′-OH, 5′-P termini and 3′-P, 5′-OH termini, the nebulized DNA is digested with *Bal*31 nuclease to convert all the ends to 3′-OH and 5′-P termini suitable for blunt-end ligation. Briefly, in the digestion solution 50 μg DNA fragments are incubated with 2.5 U *Bal*31 (New England Biolabs, Beverly, MA) in the reaction buffer at 30° for 5 min and 10 μl of 0.5 *M* EDTA is added into the reaction tube to stop further digestion. The *Bal*31-digested DNA fragments are electrophoresed in TAE buffer in 0.8% agarose gel containing 0.05 μg/ml of ethidium bromide. A slice of the gel covering 200 to 800 bp is excised and put into an Eppendorf tube. The DNA fragments are purified from the gel using QIAEXII gel extraction kit (Qiagen, Valencia, CA). Subsequently the sized DNA fragments are ligated into an inducible vector pYJ335, a plasmid carrying ampicillin and erythromycin resistance markers and able to replicate in *Escherichia coli* and *S. aureus* (Fig. 1A). The ligated DNA is electroporated into 25 μl of *E. coli* ElectroMax DH10B Cells (Gibco-BRL, Gaithersburg, MD) in a 0.1 cm

[8] Y. Ji, A. Marra, M. Rosenberg, and G. Woodnutt, *J. Bacteriol.* **181**, 6585 (1999).
[9] L. Zhang, F. Fan, L. M. Palmer, M. A. Lonetto, C. Petit, L. L. Voelker, A. S. John, B. Bonkosky, M. Rosenberg, and D. McDevitt, *Gene* **255**, 297 (2000).

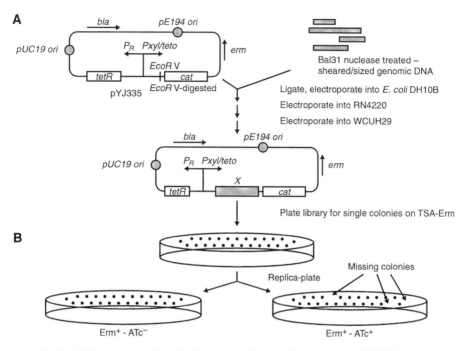

FIG. 1. (A) Construction of the inducible antisense *S. aureus* library. *S. aureus* WCUH29 chromosomal DNA was sheared, treated with *Bal*31, sized in the range of 200 bp to 1000 bp, and ligated into the *Eco*RV site of downstream *Pxyl/teto* in ATc inducible vector pYJ335 [for construction of this plasmid, see Y. Ji, A. Marra, M. Rosenberg, and G. Woodnutt, *J. Bacteriol.* **181**, 6585 (1999)]. The ligation of DNA was electroporated into *E. coli* DH10B using Amp as selection. Transformants were pooled and incubated in LB-Amp (100 μg/ml) for amplification and purification of library plasmid DNA, which was electroporated into *S. aureus* laboratory strain RN4220 and after passage into *S. aureus* clinical strain WCUH29. (B) Screening for essential genes. The electrotransformants, WCHU29 carrying library plasmid DNA, were selected on TSA-Erm (5 μg/ml) and duplicated onto TSA-Erm plates with ATc and without ATc. Growth-defective and conditional lethal colonies on TSA-Erm with ATc were selected after overnight incubation.

cuvette at 1.8 kV, 200 Ω, and 25 μF using the Bio-Rad (Hercules, CA) Gene Pulser unit. Transformed cells are incubated in 900 μl of S.O.C. medium (Gibco-BRL) at 37° for 45 min and plated (25 μl) onto LB-agar plates (ampicillin 100 μg/ml). Colonies are picked for PCR (polymerase chain reaction) analysis of diversity of insertion by mixing a colony in 50 μl PCR supermixture (Gibco-BRL) using two plasmid-specific primers, tetRfor1399 (5′CAATACAATGTAGGCTGC 3′) and catUrev (5′ AGTTCATTTGATATGCCTCC 3′). If the library appears representative, colonies are collected from 20 LB-agar-Amp plates containing 500 to 1000 cfu (colony-forming units)/plate and inoculated into 100 ml of LB-Amp medium and incubate overnight.

Screening Growth Defect and Lethal Colonies

S. aureus electrocompetent cells are prepared by the following procedure. The *S. aureus* strain is inoculated in 500 ml of TSB medium and incubated at 37° with shaking (190 rpm) till $OD_{600 nm}$ 0.4–0.5. The bacterial cells are harvested by centrifugation at 8000 rpm for 10 min at 4° and washed four times in ice-cold sterilized 0.5 M sucrose with 0.5, 0.25, 0.125, and 0.0625 times the original culture volume. Finally, the cells are resuspended with 2 ml of sterilized 10% glycerol and aliquoted into 1.5-ml Eppendorf tubes (50 μl/tube) and stored in a −80° freezer.

Plasmid library DNA prepared from the collection of *E. coli* clones is subsequently electroporated into 50 μl of *S. aureus* laboratory strain RN4220 competent cells at 1.8 kV, 100 Ω resistance, and 25 μF capacitance; then the electrotransformants are spread for single colonies on TSA-Erm (5 μg/ml) plates. The colonies are collected and inoculated into 50 ml TBS-Erm and incubated with shaking at 37° overnight. Plasmid library DNA is purified from this culture using a QIAprep Miniprep Kit (Qiagen) and electroporated into *S. aureus* WCUH29 since this strain cannot accept foreign DNA directly from *E. coli*.[10] To screen for lethal and growth defect events, the colonies are duplicated onto TSA-Erm plates either with inducer or without inducer (Fig. 1B). Colonies which grow normally on TSA-Erm without ATc, but which are missing or grow poorly on the replica TSA-Erm plates containing ATc after overnight incubation, will be evident. We have found that direct electroporation of the plasmid library isolated from RN4220 into WCUH29 could produce a more random library in WCUH29 (about 71% of the lethal and growth defective events represent different unique genes) in comparison with transduction using Φ11 (only 4.5% of lethal and deficient events represent different genes).

The colonies displaying defective growth upon induction are isolated from the TSA-Erm plates and retested by streaking part of each colony onto the TSA-Erm-ATc and onto TSA-Erm after incubation overnight. To further confirm the phenotype, plasmid DNA is purified from each strain and electroporated back into *S. aureus* WCUH29 competent cells and selected on TSA-Erm plates. After growth, colonies on the plate are replicated onto TSA-Erm containing different concentrations (100 to 1000 ng/ml) of ATc. In these experiments *S. aureus* WCUH29 carrying the parent vector pYJ335 (i.e., strain YJ335) grow normally on TSA-Erm at the different concentrations of ATc. In contrast, the *S. aureus* strains carrying the transformed plasmids do not grow or grow poorly with different levels of ATc induction.

DNA Sequencing and Bioinformatic Analysis

To determine the orientation and to identify the specific DNA fragments leading to lethal or growth defective events after induction, the DNA fragments are obtained

---

[10] R. P. Novick, "Molecular Biology of the Staphylococci." VCH Publishers, New York, 1990.

by PCR amplification and sequenced using plasmid-specific primers tetRfor1399 and catUrev. The DNA sequence data is analyzed by using BLAST Homology Search against all bacterial genomes database developed by GlaxoSmithKline Pharmaceuticals Research and Development. An antisense orientation to the cloned fragment in the recombinant plasmid is indicated if the direction of the query sequence using catUrev primer is the same as the direction of subject amino acid sequence in the protein identified. Only 33% of the *S. aureus* strains displaying growth deficient or lethal phenotype when induced with ATc carry DNA fragments in the antisense orientation in the repression vector. Of these, 33% have chimeric or rearranged fragments at the cloning site (see detailed results in Ref. 11

Quantitative Titration of Essential Genes *in Vitro*

The ability to titrate down a gene product *in vitro* provides a powerful approach for quantitative analysis of gene essentiality. To characterize the titration of essential genes *in vitro,* the growth of *S. aureus* strains containing different essential antisense RNA constructs is determined by incubation of bacteria with different doses of ATc. The density of cells is determined at $OD_{600\,nm}$ after overnight incubation at 37° or by using kinetic assay. Dose-dependent inhibition of growth is observed during induction with various concentrations of inducer for *S. aureus* antisense mutant strains carrying essential antisense fragments. In contrast, the control strain *S. aureus* carrying pYJ335 does not show any obvious difference of growth in the presence of inducer.[11]

Quantitative Titration of Essential Genes *in Vivo*

Because the tetracyclines are available in many body compartments after oral dosing, this regulated antisense system provides a unique tool for determining essentiality of a gene during infection and for studying pathogenesis in this organism. To titrate the expression level of expression of essential genes *in vivo,* we have chosen a murine model of hematogenous pyelonephritis as it results in a localized kidney infection from which bacteria are readily recovered.[8] Five mice per group are infected with about 10^7 cfu of bacteria via an intravenous injection of 0.2 ml of bacterial suspension into the tail vein using a tuberculin syringe. Different doses of ATc are given orally in 0.2-ml doses (containing 5 μg/gram weight of Erm) to infected mice on days 1, 2, and 3 after infection. The mice are sacrificed by carbon dioxide overdose 2 hr after the last dose of Tc induction. Kidneys are aseptically removed and homogenized in 1 ml of PBS for enumeration of viable bacteria. As a control, approximately 5 log cfu of YJ335 should be recovered from infected kidneys at day 3 either in the presence or absence of ATc induction. Similarly, about 5 log cfu of antisense mutants which are wild-type *S. aureus* carrying

[11] Y. Ji, B. Zhang, P. Warren, G. Woodnutt, M. Burnham, and M. Rosenberg, *Science* **293**, 2266 (2001).

essential gene antisense constructs should be recovered from infected kidneys in the absence of induction. In contrast, no bacteria or less than 1 log cfu bacteria should be recovered from infected kidneys following induction of antisense using $0.5 \mu g$/gram mouse of ATc.[11] Also, the effect of essential antisense RNA induction on the survival of bacteria should be ATc-dose dependent. On some occasions induction of antisense *in vivo* does not lead to bacterial clearance although *in vitro* induction gives rise to cessation of growth. This suggests to us that interference of expression of the gene concerned *in vivo* has consequences that are not bactericidal.

[9] Transposomes: A System for Identifying Genes Involved in Bacterial Pathogenesis

By Les M. Hoffman and Jerry J. Jendrisak

Transposome Formation from Tn5 Transposable Elements

Transposons are DNA elements that can move from one genetic location to another. Bacterial transposons of the *Tn* class are used extensively as research tools in molecular biology. They contain defined terminal inverted repeats and encode a transposase that excises the element from a donor site and rejoins it to DNA at a second location. The molecular details of Tn5 transposition have been well characterized owing to its single subunit transposase and to the development of an *in vitro* transposition system.[1] Tn5 normally has a very low rate of transposition in bacteria. A hyperactive triple mutation of the Tn5 transposase was created which facilitates *in vitro* transposition studies.[1]

When hyperactive Tn5 transposase was combined with a transposon with inverted repeat ends containing a mosaic of outer end (OE) and inner end (IE) sequences,[2] *in vitro* transposition efficiency was significantly improved. The mosaic ends (MEs) are 19 base pairs (bp) long and are the only sequences required for transposase recognition. Any sequence between ME repeats can be mobilized by transposition. During the excision of the transposable element from the donor DNA, a synaptic complex is formed in which the transposon ends are brought together by the dimerization of the transposase.

Transposition intermediates called "transposomes" are complexes between two transposase molecules and any DNA with ME inverted repeat ends.[3] Transposomes

[1] I. Y. Goryshin and W. S. Reznikoff, *J. Biol. Chem.* **273,** 7367 (1998).
[2] M. Zhou, A. Bhasin, and W. R. Reznikoff, *J. Mol. Biol.* **276,** 913 (1998).
[3] I. Y. Goryshin, J. Jendrisak, L. M. Hoffman, R. Meis, and W. S. Reznikoff, *Nat. Biotechnol.* **18,** 97 (2000).

may be considered as stable synaptic complexes. In the absence of magnesium ions, transposomes are catalytically inactive. In the intracellular milieu, for example, magnesium ions activate the transposase and allow insertion of the transposon into cellular DNA by a cut-and-paste mechanism.[3,4] The combination of a hyperactive Tn5 transposase[1] and transposons containing hyperactive (ME) inverted repeat ends produces an efficiency of insertion about three orders of magnitude greater than that of wild-type Tn5.

The transposome system eliminates the host barrier for *in vivo* transposition. If the transposome is electroporated into a bacterial cell, the transposon can be integrated directly into genomic (or episomal) DNA. The Tn5-based transposon inserts randomly[5] and can create knockouts in genes whose functions are not essential for growth under the given conditions. Genomic mutagenesis is thus possible without conjugations between bacterial strains and without using "suicide" vectors unable to replicate within the recipient strain. The absence of a transposon-borne transposase gene prevents later "hopping" of the transposon, which is effectively locked in place.

Transposons have been introduced into genomic DNA by several previous strategies. The most frequently used method is by transformation with suicide vectors encoding the transposable element and a transposase. Because the plasmid is unable to replicate within the host, the transposon selectable marker is functional only after integration into the chromosome. Phage infection may also be used to mutagenize bacterial chromosomes by transposition. Bacteria are infected with a phage lambda(λ) derivative that is unable to either replicate or form lysogens and carries a transposon. The transposon is maintained only if the transposable element has been incorporated into the chromosome or into a replicating episome. A disadvantage of phage delivery of transposons is that phages are species- and strain-specific: lambda infects only lambda-sensitive *Escherichia coli*.

Both of the above methods have the disadvantage of requiring transposons or vectors encoding a transposase, which may cause instability of the transposon within the chromosome. The system described herein eliminates the instability inherent with transposase-containing transposable elements, because the transposase is supplied as an *in vitro* reagent to the transposon DNA.

Tn5 transposomes may be introduced into the chromosomes of many bacterial species by electroporation.[3,4] The efficiency of insertion varies from organism to organism, but it is usually high enough to produce a library of knockout mutations. In addition, transposomes can be used to deliver genomic control elements, screenable markers, or reporter genes whose activities can be measured under different metabolic states.

[4] L. M. Hoffman, J. J. Jendrisak, R. J. Meis, I. Y. Goryshin, and W. S. Reznikoff, *Genetica* **108,** 19 (2000).

[5] R. J. Meis, *Epicentre Forum* **7,** 5 (2000).

Transposome Formation

The commercial EZ::TN system (Epicentre Technologies, Madison WI) includes preformed transposomes containing one of several antibiotic resistance markers and, in some cases, a conditional origin of replication for rescue of genomic sequences surrounding the transposon insertion. It is also possible to construct new transposons with the modular pMOD-2 <mcs> high copy vector (Epicentre Technologies). After cloning the desired DNA fragment(s) into the multiple cloning site, the transposon may be cut from the vector by restriction endonuclease digestion or it can be amplified by PCR. Transposomes are formed *in vitro* by a simple reaction between DNA and hyperactive transposase.

Transposome formation reaction:

     1 volume 100 ng/μl transposon DNA
     2 volumes 1 unit/μl EZ::Tn transposase
     1 volume 100% glycerol

Mix, incubate 30 min at 37°.

Transposon Insertion Site Localization Methods

There are several avenues to find the exact location of disruption of genomic DNA by transposons (Fig. 1). The first two rely on cloning the regions of the bacterial chromosome into which the transposon is inserted. The third method analyzes the transposon insertion in total genomic DNA by direct sequencing. All three methods require isolation of genomic DNA from the insertion clones to be analyzed.

Isolation of Genomic DNA from Transposition Clones

The MasterPure DNA Purification Kit is used according to the product literature (Epicentre, Madison WI, www.epicentre.com). Briefly, 1.5 ml of an *Escherichia coli* cultured overnight is pelleted in a low speed centrifuge at 3000 rpm for 15 min. Cells are resuspended in 300 μl of 1× Cell and Tissue Lysis solution into which is added 1 μl of 50 mg/ml proteinase K. The cells are incubated at 70° for 15 min with occasional agitation. One μl of RNase A (5 mg/ml) is added and the incubation is continued for 30 min at 37°. After briefly chilling the lysate on ice, 175 μl of MPC Protein Precipitation Reagent is added and the mixture is mixed by vortex and centrifuged at 10,000–12,000 rpm in a microcentrifuge. The supernatant is recovered by pipetting and is transferred to a new tube. DNA is precipitated by the addition of 0.5 ml of 2-propanol, followed by a 5 min centrifugation at 10,000–12,000 rpm. The DNA is rinsed by adding 200 μl of 70% (v/v) ethanol, mixing, and centrifuging as before. The pellet (usually visible)

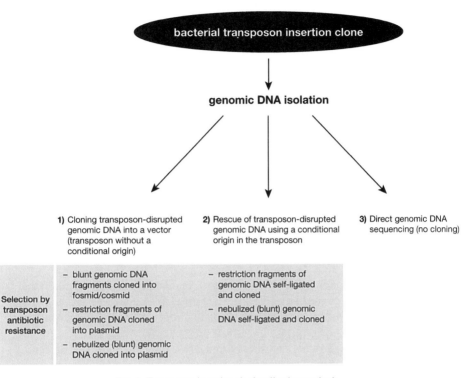

FIG. 1. Transposon insertion site localization methods.

is resupended in 35 μl of $T_{10}E_1$ buffer (10 mM Tris-HCl, pH 7.5, 1 mM EDTA). Optimal quantitation of DNA is by Hoechst 33258 fluorescent dye binding using the DyNA Quant fluorimeter (Hoefer, San Francisco, CA). The protocol yields from 10 to 15 μg of chromosomal DNA per 1.5 ml of *E. coli* culture.

1. The genomic DNA of each clone may be isolated, cleaved by one or more restriction endonucleases that do not cut within the transposon, or sheared by mechanical force, then cloned into a suitable plasmid or cosmid vector.

Cloning Transposon Insertions from Genomic DNA

Many transposons do not contain conditional origins of replication. In such cases, the genomic loci of transposon insertions can be restricted or randomly sheared from chromosomal DNA and cloned into cosmid, fosmid, or plasmid vectors. Fosmid or cosmid vectors are often chosen for rescue cloning because of the selectivity for large clone inserts of *in vitro* packaging. Fosmids have the additional advantage of being single copy episomes, preventing instability of clones with "toxic" genes. It is imperative to use a vector encoding an antibiotic resistance gene different from that of the transposon to be rescued.

A convenient fosmid lacking a kanamycin resistance gene is pCC1FOS, which accepts inserts with blunt ends. Bacterial DNA prepared with the MasterPure DNA Purification Kit (Epicentre Technologies) is largely of the correct size to be packaged in cosmid or fosmid clones. The ligation between genomic DNA and pCC1FOS is packaged into lambda phage coats and transfected into *E. coli,* producing a library of clones with 35–40 kb DNA inserts. The packaging reaction is infected into *E. coli* cells which are plated on media selective for the transposon.

The oriV enables the amplification of the fosmid when it is induced in certain *E. coli* hosts expressing the *trfA* gene under control of an arabinose-induced promoter. One such *E. coli* strain for fosmid amplification is EC300 (Epicentre Technologies). Addition of 0.01–0.1% arabinose to logarithmic phase EC300 fosmid clone cultures leads to a significant (6- to 20-fold) increase in the copy number of the fosmid.

An alternative way to prepare genomic fragments for cloning is by nebulization.

(a) Ten μg genomic DNA (400–600 ng/μl) is diluted to a final concentration of 10–15 μg/ml in 10 mM Tris-HCl, pH 7.5, 1 mM EDTA, and 20% glycerol. The DNA is sheared in a nebulizer (obtained from Invitrogen, Carlsbad, CA) for 30 sec at 10 psi of N_2.

(b) End-fill and phosphorylate the sheared DNA using the EndIt Kit (Epicentre Technologies) in a volume of 200 μl.

(c) Heat kill the reaction at 70° for 20 min, then precipitate with 1 volume of 2-propanol.

(d) Wash the DNA pellet with 70% ethanol and resuspend it in 20 μl of $T_{10}E_1$ buffer.

The nebulized DNA fragments will be mainly 2–7 kb in length and are suitable to be cloned into any blunt-ended plasmid cloning vector that does not encode the transposon antibiotic resistance marker.

2. If the transposon carries a conditional origin of replication, such as the plasmid R6K *gamma* origin, genomic DNA can be cleaved by the appropriate restriction endonucleases and ligated with itself. No vector is necessary in this situation, but care needs to be taken that the enzymes chosen do not have recognition sites within the transposable element.

Rescue of Transposons Using a Conditional Origin of Replication

One to five micrograms of genomic DNA are isolated with the MasterPure DNA Purification Kit as previously described. The DNA is digested to completion with a restriction endonuclease leaving staggered ends, using the conditions recommended by the manufacturer. One microgram of restricted genomic DNA is self-ligated in a volume of 200 μl (a final DNA concentration of 5 μg/ml) using the FastLink Ligation Kit (Epicentre Technologies). One microliter of the ligated DNA is transformed by electroporation into *E. coli* EC100D *pir*⁺ or *E. coli* EC100D *pir-116* (containing the pi protein, from Epicentre Technologies, Madison, WI). Selection of the rescue plasmid is accomplished using the antibiotic resistance marker of the transposon.

3. Without rescuing the transposon-containing region, the genomic DNA can be directly sequenced with primers corresponding to the transposon ends. The sensitivity of automated sequencing with fluorescent dye terminators has allowed genomic sequencing of bacteria with genomes larger than 6 Mb.[4] Two to three micrograms of genomic DNA is sufficient to obtain 400–450 nt of sequence, enough to unambiguously assign the position of the transposon insertion within the genome.

Direct Genomic Sequencing

Chromosomal DNA is subjected to direct sequencing using the BigDye Terminator cycle sequencing kit (PE Biosystems, Foster City, CA). Genomic DNA is sequenced in a 40 μl (2×) reaction containing 5–12 pmol primer, 2.5 μg genomic DNA, and 16 μl of Applied Biosystems BigDye terminator mix.[6]

Cycle sequencing is performed with the following program: 4 min at 95°, followed by 60 cycles of 30 sec at 95° and 4 min at 60°. Reaction products are purified using Edge Biosystems Centriflex Gel Filtration Cartridges (Edge Biosystems, Gaithersburg, MD), concentrated by ethanol precipitation, washed with 70% ethanol, and resuspended in 22 μl of Template Suppression Reagent (Applied Biosystems, Foster City, CA). After denaturing at 95° for 5 min, the samples are injected into an ABI 310 Genetic Analyzer (Applied Biosystems, Foster City, CA) and analyzed with ABI version 3.3 sequence analysis software. Genomic transposition sites are located using BLAST programs maintained at the NCBI Web site of the National Library of Medicine (www.ncbi.nlm.nih.gov/BLAST).

The cut-and-paste insertion mechanism of Tn5 causes the direct duplication of nine nucleotides at the ends of the transposon.[3] This duplication must be considered when the exact site of chromosomal interruption is to be located by sequencing from both ends of the transposon.

Techniques for Preparation of Electrocompetent Cells

1. Streak for single colonies from −70° glycerol stock onto a plate of the appropriate medium. Start a 50-ml culture in broth at 37°, shaking at 200 rpm

---

[6] C. R. Heiner, K. L. Hunkapiller, S. M. Chen, J. I. Glass, and E. Y. Chen, *Genome Res.* **8,** 557 (1998).

overnight. Inoculate 25 ml of overnight culture into 1 liter of broth (containing no NaCl and prewarmed to 37°), which is grown at 37° with shaking at 200 rpm to $A_{600} = 0.6$–0.75. Chill on ice. Spin culture 8000 rpm 10 min and resuspend in 200 ml of ice-cold 10% glycerol.

2. Spin 8000 rpm for 10 min and resuspend in 150 ml cold 10% glycerol.

3. Spin 8000 rpm for 10 min and resuspend in 100 ml cold 10% glycerol.

4. Spin 8000 rpm for 10 min and decant, removing most of the 10% glycerol.

5. To pellet add 1–2 ml 10% glycerol. Resuspend gently with a 1-ml pipettor. Dilute a 10-μl aliquot of the cells with 3 ml of 10% glycerol. Its A_{600} should be between 0.7 and 0.85, which indicates an A_{600} of the undiluted cells of 200–250. If the cell concentration is too low, they can be pelleted in a microcentrifuge at 10,000 rpm for 5 min and brought to the desired volume.

6. Aliquot 110 μl cells into prechilled 1.5-ml microcentrifuge tubes. Freeze at $-70°$.

Electroporation of Competent E. coli Cells

1. Thaw electrocompetent cells on ice and aliquot 50 μl per sample into 500-μl microcentrifuge tubes on ice.

2. Add DNA in 1 μl of a low salt solution such as $T_{10}E_1$.

3. Add sample to a sterile 2-mm gap electroporation cuvette. Electroporate at 2.5 kV for *E. coli* in an Eppendorf multiporator. Optimal settings for other instruments may vary.

4. Add approximately 0.3 ml of LB broth to cell and rinse cells from the cuvette with a 1-ml pipettor. Add cells to remainder of 1 ml of LB broth and shake at 370 for 30 min to 1 hr.

5. Plate 10–100 μl of the outgrowth on the appropriate selective plates.

Specialized Transposons for Mutagenesis of Genes for Exported or Membrane Peptides

Most virulence determinants are found on the surface of bacteria, or they are secreted into the medium where they may interact with the host. It is well documented that exported proteins play crucial roles in the process of infection.[7] We constructed a transposon for mutagenesis and identification of genes encoding exported, and often virulence related, proteins. Transposon tagged bacteria with "knockouts" of exported protein genes could then be tested for virulence in host organisms.

The transposon Tn*phoA,* delivered by conjugation with an *E. coli* strain harboring a suicide plasmid, has been used to produce translational fusions of alkaline

---

[7] B. B. Finlay and S. Falkow, *Microbiol. Mol. Biol. Rev.* **61,** 136 (1997).

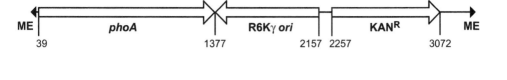

FIG. 2. ⟨phoA/R6Kγori/KAN-2⟩ transposon (3340 bp).

phosphatase (*phoA*) with the genes of many bacterial species.[8] *phoA* is only active when it has been translocated across the cytoplasmic membrane into the periplasm, where formation of disulfide bonds can occur. In the cytoplasm, disulfide bond formation cannot occur, and the *phoA* is inactive. The fusion of *phoA* with genes encoding exported or surface proteins can be used to identify transposon insertions.[9,10] Analyses with *phoA* fusions measure transcription as well as translation and protein export.

There are many obvious drawbacks to Tn*phoA*. Conjugative transfer of Tn*phoA* into many bacterial species is difficult because of antibiotic sensitivities and resistances, and it requires special *E. coli* donor strains. Most of the Tn*phoA* mutants isolated in one study had multiple insertions, and many had the entire suicide delivery vector cointegrated at the insertion site.[11]

To simplify in-frame, randomized fusions with open reading frames (ORFs) of exported or surface protein genes, we constructed an EZ::Tn transposon named ⟨*phoA*/R6K *gamma* ori/Kan⟩ (Fig. 2). A PCR product of the *E. coli phoA* gene, encoding amino acid residues 6–450 of mature *phoA,* was cloned in to the unique XmnI site of transposon ⟨R6K *gamma* ori⟩ by standard techniques. The final construct has an open reading frame (ORF) from the first triplet of the transposon ME through the truncated *phoA* gene.

Transposon ⟨*phoA*/R6K *gamma* ori/Kan⟩ features a promoterless, signal peptide-free *phoA* cassette (similar to Tn*phoA*), a conditional R6K *gamma* origin of replication, and the Tn903 kanamycin resistance gene (Fig. 2). The ⟨*phoA*/R6K *gamma* ori/Kan⟩ containing transposome was electroporated into *E. coli* DH10B by the protocols given in the section above. The outgrowth was plated on LB plates containing 25 μg/ml kanamycin and 40 μg/ml of 5-bromochloro-3-indoyl

[8] R. K. Taylor, C. Manoil, and J. J. Mekalanos, *J. Bacteriol.* **171,** 1870 (1989).

[9] C. Manoil and J. Beckwith, *Proc. Natl. Acad. Sci. U.S.A.* **82,** 8129 (1985).

[10] C. Manoil, J. J. Mekalanos, and J. Beckwith, *J. Bacteriol.* **172,** 515 (1990).

[11] A. Afsar, J. A. Johnson, A. A. Franco, D. J. Metzger, T. D. Connell, J. G. Morris, Jr., and S. Sozhamannan, *Infect. Immun.* **68,** 1967 (2000).

TABLE I
INSERTION SITES FOR ⟨phoA/R6K gamma ori/Kan⟩ TRANSPOSOME[a]

Insertion clone	Gene designation	Amino acid before insertion	In-frame insertion	Transposon orientation
1	yebF	108	Yes	Forward
2	ompT	116	Yes	Forward
3	(No annotation)	30	Yes	Forward
4	acrA	214	Yes	Forward
5	dacC	218	Yes	Forward
6	borW	10	Yes	Forward
1B	fadL	389	Yes	Forward
2B	phoT(pstA)[b]	NA	No	Forward
3B	rol(cld)	89	Yes	Forward
5B	phoR[b]	NA	Yes	Backward
6B	osmE	106	Yes	Forward

[a] In the genome of E. coli DH10B.
[b] Knockout of repressor of phoA.

phosphate (BCIP). Blue color indicating alkaline phosphatase activity developed within 24 hr and intensified during the next day or so.

Colonies with varying shades of blue were picked into LB broth with 25 μg/ml of kanamycin and grown overnight with shaking at 37°. Approximately 5 ml of each culture was centrifuged for 10 min at 3500 rpm in a tabletop centrifuge and the genomic DNA was isolated using the MasterPure DNA Isolation Kit (Epicentre).

Approximately 2.5 μg of genomic DNA was sequenced directly using primers specific for the ends of the phoA transposon. The positions of the transposons within the E. coli genome were determined by BLAST analysis. Using the GenBank annotated ORFs (open reading frames) of the transposon insertion sites, we found the orientation with respect to the mutated gene, reading frame, and position in the ORF of the transposon in each of 11 clones.

As the data in Table I show, the blue colonies indicated either that the insertion of the phoA transposon was in frame with an exported protein gene, or that the transposon had "knocked out" a gene repressing the expression of the endogenous E. coli phoA gene, rendering the expression of phoA constitutive (clones 2B and 5B). It is worth noting that the blue selection afforded by phoA fusions works only when the host cell background alkaline phosphatase activity is sufficiently low. One solution for high backgrounds is to construct or find a strain of the bacterium that has low alkaline phosphatase levels.

Clone 6 has an insertion of phoA into a segment with high similarity to the borW gene of phage lambda, a virulence factor for survival of E. coli in animal serum.[12]

[12] J. J. Barondess and J. Beckwith, J. Bacteriol. 177, 1247 (1995).

It is interesting to note that the genotype of DH10B *E. coli* is lambda$-$,[13] but other prophages are known to exist in the *E. coli* genome. Because there are sequence differences between the insertion site in *E. coli* DH10B and wild-type lambda, it is likely that the transposon in clone 6 inserted into a lambda-like prophage.

Transposome Mutagenesis in Organisms without Characterized Genetic Systems

Lack of gene transfer systems can hamper the application of standard transposition tools to many organisms. The transposome system requires only a transposon carrying a selectable marker functioning in the bacterium, and highly electrocompetent cells.

The bacterium *Xylella fastidiosa* is the causative agent of Pierce's disease in grapevines and it infects a variety of plant species. Although the complete genomic sequences of several strain of *X. fastidiosa* are available, the genetics of the bacterium are very rudimentary.[14,15] No shuttle plasmids with replicons that function in *X. fastidiosa* are available for gene transfer experiments.

We attempted to use the standard kanamycin resistance encoding transposome to randomly mutagenize the chromosome of *X. fastidiosa*. The Kan-2 transposome (Epicentre Technologies) was electroporated into electrocompetent *X. fastidiosa* cells. Electrocompetent *X. fastidiosa* were produced by a variation of the method given above for *E. coli*.[16]

A library of several hundred transposon insertion clones was generated and representative clones were shown by Southern hybridization to have incorporated transposon sequences into their chromosomes.[16] All insertion clones tested have stable transposons, as expected for transposome-mediated mutants. Sequencing from the transposon ends allowed precise assignment of insertion sites within the *Xylella* chromosome. The individual insertion clones have been inoculated into grapevines to identify mutants unable to survive and produce a disease state (M. R. Guilhabert, personal communication, 2000). Avirulent mutants may contain transposon disruptions of pathogenicity genes.

Transposome Insertions with Phenotypic Consequences

One valuable feature of transposome mutagenesis is its ability to cause readily detectable phenotypic changes that can be mapped to specific genes. We used an

[13] D. Lorrow and J. Jessee, *BRL Focus* **12,** 19 (1990).

[14] A. J. Simpson *et al., Nature* **406,** 151 (2000).

[15] P. B. Monteiro, D. C. Teixeira, R. R. Palma, M. Garnier, J. M. Bove, and J. Renaudin, *Appl. Env. Microbiol.* **67,** 2263 (2001).

[16] M. R. Guihabert, L. M. Hoffman, D. Mills, and B. C. Kirkpatrick, *Mol. Plant Microbe Interact.* **14,** 701 (2001).

easily screenable characteristic of *Proteus vulgaris* to indicate that a gene knockout had occurred. *Proteus vulgaris* is normally a motile organism that "swarms" on agar plates. Approximately 50 genes in *P. vulgaris* are involved in the swarming phenotype and present a large number of targets for transposon inactivation. We prepared electrocompetent *P. vulgaris* and electroporated cells with a kanamycin resistance transposome using the same techniques as for *E. coli*. Approximately 10^5 kanamycin resistant colonies per microgram of transposon DNA were generated. One *P. vulgaris* transposition clone was nonswarming, indicating a possible gene knockout.

Genomic DNA from four *P. vulgaris* transposition clones, including the non-swarming clone, was purified and directly sequenced as previously described. The transposon insertion in the nonswarming mutant clone was located within the fimbrial gene (pilus protein) homolog *atf,* one of at least 45 genes involved in the *Proteus* swarming phenotype. Thus, EZ::TN transposome disruption of the pilus protein gene led to the loss of swarming that is probably related to virulence. Our results demonstrate rapid production of a gene "knockout" and correlation of cell phenotype with the genomic locus by direct genomic DNA sequencing.

Signature-Tagged Mutagenesis Applied to Transposomes

In signature-tagged mutagenesis, a set of transposons is created containing unique oligonucleotide regions that act as tags for each transposon.[17] The trans-posons are mobilized into a bacterial pathogen and hundreds to thousands of transposon insertion clones are set into arrays and then pooled. The pools are used to inoculate into a host, or used to grow under specific conditions. After recovery of the bacteria from the host organism or from the selective growth situation, the oligonucleotide regions are amplified from the pooled transposon tags and used to probe an array of the original input clones. The inability of a transposon mutant to survive in the host or growth condition will lead to the absence of signal for that tag in the array hybridization.

The STM technique has been applied to at least 10 human pathogens.[18] It should be possible to apply STM technology to mutagenesis with the transposome system. Pooling many individual mutant bacterial clones for introduction into a single host leads to considerable savings in effort.

There are several methods for screening the recovered bacterial pools for the presence of individual clones. The original STM article[17] described a filter

---

[17] M. Hensel, J. Shea, C. Gleeson, M. Jones, E. Dalton, and D. Holden, *Science* **269,** 400 (1995).
[18] L. Hamer, T. M. De Zwaan, M. V. Montenegro-Chamorro, S. A. Frank, and J. E. Hamer, *Curr. Opin. Chem. Biol.* **5,** 67 (2001).

hybridization method to detect transposition clones which had not survived. A more recently described screening technique uses the PCR.[19] DNA from pooled clones is amplified with primers specific for each of a small number (12–36) of known, characterized, 21 nt oligonucleotides incorporated into the transposon. Each PCR contains a primer specific for one tag and a common primer that anneals to transposon sequences outside the tag. The tag-specific primers have a 14 nt invariant region and a variable 7 nt near their 3′ ends. PCR conditions can be selected in which amplification is primer specific, and in which "dropouts" of individual members of pooled insertion clones are easily identified.

Future Directions for Transposome-Based Studies of Virulence Genes

The gram-positive bacteria present special challenges for the introduction of episomal elements by electroporation. Gram-positive cells are surrounded by many layers of peptidoglycan, which may reduce the efficiency of plasmid transformation by electroporation. The same difficulty with electroporation may apply to transposomes, and the efficiency of *in vivo* transposon insertion is usually lower for gram-positives than for gram-negatives. However, there have been several recent reports of successful mutagenesis of gram-positives by the electroporation of transposomes (*Corynebacterium diphtheriae*, Diana Marra; *Mycobacterium smegmatis*, and *Mycobacterium bovis* BCG, Keith Derbyshire; *Mycobacterium tuberculosis*, Bonnie Plikaytis; all personal communications, 2001).

Mycobacteria are difficult to transform by electroporation, probably because of their complex lipid outer coats, and not because of a large number of peptidoglycan layers. Nevertheless, the coryneform bacterial species listed above appear to be amenable to transposome mutagenesis.

The ⟨*phoA*/R6K *gamma* ori/Kan⟩ transposon can also be inserted *in vitro* into plasmids, cosmids, fosmids, or BACs to help distinguish exported protein ORFs. Because *in vitro* transposon insertion is not limited to nonessential genes, more exported protein genes may be found than with *in vivo* transposition. Genes with unknown functions may also be characterized by *in vitro* fusion with *phoA*.

The *in vitro* fusion method does, however, require the expression of the cloned *phoA* fusion genes in *E. coli*. If knockouts in the original species are desired, the *in vitro* *phoA*-fused genes may be reintroduced by electroporation or by conjugation. Within the original host, recombination exchanges the mutant gene with the wild type, and antibiotic resistance is used to select for mutant recombinants containing the transposon.

---

[19] D. E. Lehoux, F. Sanschagrin, and R. C. Levesque, *BioTechniques* **26,** 473 (1999).

Conclusions

With new transposition tools called transposomes, many barriers to the insertion of transposons into genomic DNA are removed. No delivery vector is needed because the transposome DNA is not replicated within the target cell. Compatibility of the vector replicon with the target cells is therefore not a consideration. The contents of the transposon are unrestricted, and the only structural requirement is a 19-bp inverted repeat (ME) sequence at each end.

With conventional suicide vector delivery of transposons, there must be expression of the transposase polypeptide prior to the insertion reaction. With transposomes, the transposase is active and ready for cut-and-paste transpositions on arrival in the bacterial cytosol. Its close proximity to the transposon DNA in a complex allows the insertion reaction to occur rapidly and efficiently.

Electroporation of transposomes is an effective method for mutagenesis of many bacterial species, both gram-positive and gram-negative. The occurrence of multiple transposon insertions within the same cell is negligble.

With existing *in vivo* transposon insertion techniques, the transposon is introduced into the host cell on a plasmid. In many cases the entire plasmid, including the transposon, is integrated into the chromosome. Vectorless delivery of transposable elements by transposomes eliminates such Campbell-type recombination.

There are many possible variations in the "payload" part of the transposon, and several are discussed which may be useful for identifying virulence genes in bacteria. These include alkaline phosphatase translational fusion transposons and sets of transposons tagged with short, individually identifiable tags (STM).

Applications of transposomes for analysis and identification of pathogenesis genes are only beginning to be published. As more varieties of transposome are constructed and more species are targeted, there will be new discoveries of genes involved in host–pathogen interactions.

Acknowledgments

The expert advice of Dr. Bill Reznikoff and Dr. Igor Goryshin is appreciated. We thank Katie Loomis, Joanne Decker, and Darin Haskins for valued technical assistance. We thank Dr. Judy Meis for comments on the manuscript.

[10] Functional Screening of Bacterial Genome for Virulence Genes by Transposon Footprinting

By YOUNG MIN KWON, LEON F. KUBENA, DAVID J. NISBET, and STEVEN C. RICKE

Introduction*

Recent progress in microbial genome projects has accumulated genomic sequences of diverse microorganisms, including bacterial pathogens. However, the biological functions of many open reading frames (ORFs) identified by the genome sequences are yet to be determined. Comprehensive methods for functional screening of bacterial genomes are critical to facilitate the investigation of the biological functions of bacterial genes.[1] Such methods are essential for screening of bacterial genomes for virulence genes by infection of live host animals. The simplest approach to identify mutants that are attenuated during infection in live animals is to assess a large collection of mutants individually for attenuation during animal infection. This type of conventional approach has been applied to isolate transposon insertion mutants of *S. typhimurium* that are attenuated in the mouse and that have a colonization defect in the chicken intestine.[2,3] However, the very large number of mutants necessary to saturate the whole genome makes this approach unrealistic for comprehensive screening of a microbial genome for virulence genes.

Simultaneous Identification of Bacterial Virulence Genes for *in Vivo* Survival

Advances in genetic techniques have allowed the development of several comprehensive methods for functional screening of microbial genomes, including signature-tagged mutagenesis (STM), genetic footprinting, and deletion and parallel analysis.[4–6] Although these approaches use different genetic strategies, they

---

* Mention of a trademark, proprietary product, or specific equipment does not constitute a warranty by USDA and does not imply its approval to the exclusion of other products that may be suitable.

[1] E. J. Strauss and S. Falkow, *Science* **276**, 707 (1997).

[2] F. Bowe, C. J. Lipps, R. M. Tsolis, E. Groisman, F. Heffron, and J. G. Kusters, *Infect. Immun.* **66**, 3372 (1998).

[3] A. K. Turner, M. A. Lovell, S. D. Hulme, L. Zhang-Barber, and P. A. Barrow, *Infect. Immun.* **66**, 2099 (1998).

[4] M. Hensel, J. E. Shea, C. Gleeson, M. D. Jones, E. Dalton, and D. W. Holden, *Science* **269**, 400 (1995).

[5] V. Smith, D. Botstein, and P. O. Brown, *Proc. Natl. Acad. Sci. U.S.A.* **92**, 6479 (1995).

[6] E. A. Winzeler *et al., Science* **285**, 901 (1999).

share a common principle underlying the methods: to compare a pool of multiple mutants for their component mutants before and after selection in parallel to identify the mutants with competitive disadvantages in the given selective conditions. All of the methods have a great potential to be used for identification of virulence genes in animal infection. However, STM is the only one that has been applied to the identification of virulence genes of microbial pathogens in animal infection models.

STM combines the insertion of random sequence tags into transposons to differentiate the individual mutants within a complex pool of transposon mutants with a comparative hybridization strategy to identify the transposon mutants that are missed after recovery from host tissues.[4] Since STM allows screening of multiple transposon mutants simultaneously, this aproach has been used successfully with diverse microorganisms to screen the virulence genes in animal infection models.[7,8] The original protocol of STM has undergone several technical refinements to increase the sensitivity of the assay, to allow the use of nonradioactive detection methods, and to avoid the prescreening process for the tagged mutants that hybridize efficiently without cross-hybridization.[9] However, STM still requires a laborious step of tagging transposon mutants with sequence tags.

Principle of Transposon Footprinting

We devised an alternative experimental approach for efficient screening of bacterial genomes for virulence genes without prior information on DNA sequences. This approach, termed "transposon footprinting," is based on simultaneous amplification of multiple transposon-flanking sequences and subsequent analysis of the amplified DNA by gel electrophoresis.

Genomic DNA is isolated from a pool of random transposon mutants and digested with a 4 bp-recognition restriction enzyme. The digested fragments are then ligated to a linker with 5'-noncomplementary stretches (Y linker, Fig. 1) and used as a template in a PCR (polymerase chain reaction) amplification using a linker-specific primer (Y linker primer) and a transposon-specific primer as previously described.[10] The Y linker primer is made from the Y region and therefore cannot anneal to the Y linker itself. The fragments that do not have the transposon sequence will, therefore, not be amplified in the PCR reaction. However, the transposon primer anneals to the fragments containing transposon sequences during the first cycle of PCR and extends DNA synthesis into the Y region of the ligated Y linker, thus providing the annealing site for Y linker primer. Consequently,

[7] S. L. Chiang, J. J. Mekalanos, and D. W. Holden, *Ann. Rev. Microbiol.* **53**, 129 (1999).

[8] J. E. Shea, J. D. Santangelo, and R. G. Feldman, *Curr. Opin. Microbiol.* **3**, 451 (2000).

[9] J.-M. Mei, F. Nourbakhsh, C. W. Ford, and D. W. Holden, *Mol. Microbiol.* **26**, 399 (1997).

[10] Y. M. Kwon and S. C. Ricke, *J. Microbiol. Methods* **41**, 195 (2000).

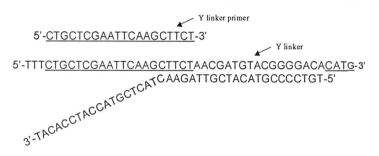

FIG. 1. Schematic diagram of Y linker and Y linker primer.

only the DNA fragments containing the transposon-specific sequences allow the amplification so that the lengths of the amplified transposon-flanking sequences usually are unique for each distinct transposon insertion site. The function of the noncomplementary region on the 5' end of the Y linker was to prevent the extension of DNA to the Y linker region by annealing with the complementary sequences. If the noncomplementary region on the 5' end of the Y linker was not present, the digested fragments that do not have transposon sequences also would provide an annealing site for Y linker primer and would be amplified with two Y linker primers annealed at both ends.

The PCR products amplified from a pool of transposon mutants generate "transposon footprints" when separated on a gel, where each DNA band serves as a marker for the corresponding transposon mutant within the pool. Therefore, comparing the transposon footprints of both inoculum (input pool) and the mutants recovered from host animal tissue after infection (output pool) allows simultaneous identification of the mutants attenuated during infection (Fig. 2).

Procedures of Transposon Footprinting

The protocols below are described for transposon footprinting as applied to Tn5 mutants of *Salmonella typhimurium* in a mouse infection model. The methods for transposon mutagenesis and animal infection experiments should be modified and optimized for the choice of microbial pathogen and animal model of infection.

Transposon Mutagenesis

Tn5 mutant strains of *S. typhimurium* ATCC 14028s are generated by conjugal transfer of a suicide plasmid pUT/Km.[11] To ensure that the mutants have unique insertions and do not originate from siblings, a large number of small-scale conjugation matings are conducted in parallel and one mutant is isolated from

---

[11] M. Herrero, V. de Lorenzo, and K. N. Timmis, *J. Bacteriol.* **172,** 6557 (1990).

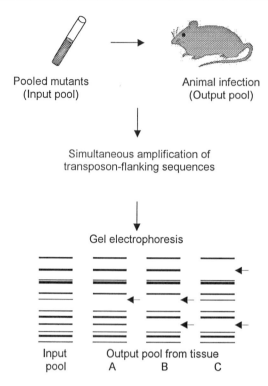

FIG. 2. Schematic diagram of transposon footprinting.

each mating as follows: the donor strain, *Escherichia coli* SM10 λpir carrying pUT/Km, is grown overnight with vigorous shaking (200 rpm) at 37° in 100 ml of LB medium containing 200 μg/ml ampicillin (Ap) and 100 μg/ml kanamycin (Km) to ensure maintenance of the delivery plasmid. The recipient strain, nalidixic acid-resistant strain of *Salmonella typhimurium* ATCC 14028s, is separately grown under the same conditions but without antibiotics. Each culture of donor and recipient cells is centrifuged and the pellets are resuspended in 10 ml PBS for 10× concentration. The concentrated cell suspensions of donor and recipient cells are mixed thoroughly in equal volume and kept on ice. Twenty μl of the mixture is spotted on a LB plate without antibiotics (15–20 spots/plate), and the LB plates are incubated overnight at 37°. The cells on each spot are collected and resuspended in a sterile microcentrifuge tube containing sterile phosphate-buffered saline (PBS) (100 μl/tube) by using a disposable inoculating loop. An appropriate amount of the cell suspension (50–100 μl) is then plated on small brilliant green agar (BGA) plate (15 × 60 mm) containing nalidixic acid (25 μg/ml) and kanamycin (150 μg/ml) and incubated overnight at 37°. We usually get several colonies of exconjugants on each BGA plate. To exclude exconjugants generated by integration of whole delivery plasmids, one colony is picked from each BGA plate and patched in replicate on both a LB plate containing amplicillin (200 μg/ml) and on LB plate containing nalidixic acid (25 μg/ml) and kanamycin (100 μg/ml). As the delivery plasmids carry a *bla* gene external to the minitransposon, the Ap[R] mutants are discarded and only the true Tn5 mutants generated by authentic transposition events (Ap[S]) are used for subsequent studies. The Ap[S] Tn5 mutants on the LB plate containing nalidixic acid (25 μg/ml) and kanamycin (100 μg/ml) are grown in LB medium and stored at $-70°$ as glycerol stocks.

Preparation of Input and Output Pools of Transposon Mutants

The Tn5 mutants of *S. typhimurium* are individually grown in the wells of microtiter plates with LB broth containing nalidixic acid (25 μg/ml) and kanamycin (100 μg/ml) by overnight incubation at 37°. To prepare the input pools, the cultures of Tn5 mutants (100 μl/mutant) are pooled together and diluted in sterile PBS to prepare inoculum of appropriate cell concentration. The pool size and cell concentration in inoculum need to be determined experimentally depending on the bacterial pathogen, animal model of infection, and route of infection of the experiments.[7] After mouse infection with input pools of *S. typhimurium* Tn5 mutants, the mutants in the tissues of infected mice are recovered as previously described.[4] Genomic DNA is then isolated from the input and output pools of mutants.

Design and Preparation of Y Linker

There are three factors to be considered for design of Y linker sequences for transposon footprinting. First, the Y linker should contain a 5′-noncomplementary region from which the linker-specific primer sequences are derived. Second, the linker sequences need to be designed such that the linker-specific primer does not have a nonspecific annealing site on the chromosome. Also, the Y linker sequences need to be designed to have a 3′ end that is compatible with the restriction endonuclease used for digestion of the genomic DNA. If a Y linker is designed to satisfy the above three requirements, the feasibility of the Y linker for transposon footprinting should be tested experimentally.

The oligonucleotide sequences that we have used to make the Y linker are as shown in Table I. First, linker B needs to be phosphorylated at the 5′ end using T4 polynucleotide kinase (PNK). Reaction mixture of 20 μl containing 9 μl of linker B (350 ng/μl), 6 μl of LoTE buffer (3 mM Tris-HCl, 0.2 mM EDTA, pH 7.5), 2 μl of 10× PNK buffer, 2 μl of 10 mM ATP, and 1 μl of PNK (10 unit/μl; New England BioLabs, Beverly, MA) is incubated for 1 hr at 37°. The PNK is then inactivated by heating for 10 min at 65°, and 9 μl of linker A is added to a final volume of 29 μl. The mixture is heated at 95° for 2 min using a heating block and

TABLE I

SEQUENCES OF OLIGONUCLEOTIDES

Oligonucleotides (bp)	Nucleotide sequences (5′→3′)
Linker B(44)	TTTCTGCTCGAATTCAAGCTTCTAACGATGTACGGGGACACATG
Linker A(40)	TGTCCCCGTACATCGTTAGAACTACTCGTACCATCCACAT
Y linker primer(20)	CTGCTCGAATTCAAGCTTCT
Tn5 primer(20)	GGCCAGATCTGATCAAGAGA

slowly cooled to room temperature by turning off the heat to allow annealing of the two partially complementary oligonucleotides. The concentration of the resulting Y linker is 200 ng/μl.

Preparation of Templates for PCR Amplification

Genomic DNA is isolated from transposon mutant pools by either the method using guanidium thiocyanate or QIAamp DNA MiniKit (Qiagen, Valencia, CA).[12] The genomic DNA is then completely digested with *Nla*III; the reaction mixture of 50 μl containing 18 μl of doubly distilled H_2O, 5 μl of 10× restriction enzyme buffer, 5 μl of 10× BSA, 20 μl of genomic DNA (40–100 ng/μl), and 2 μl of *Nla*III (New England BioLabs, Beverly, MA) is incubated for 3 hr at 37°. After digestion, the reaction volume is increased to 210 μl by adding 160 μl of LoTE buffer and extracted once with an equal volume of phenol/chloroform/isoamyl alcohol (25 : 24 : 1). Two hundred μl of extracted DNA is then concentrated by ethanol precipitation using glycogen as the carrier of DNA; 200 μl of extracted DNA, 3 μl of glycogen (Boehringer Mannheim, Indianapolis, IN), 100 μl of 10 M ammonium acetate, and 700 μl of 100% ethanol are added to make 1000 μl, mixed thoroughly, and centrifuged for 15 min at 13,000 rpm with a microcentrifuge. The pellet is then washed twice with 70% ethanol and resuspended in 30 μl of doubly distilled H_2O.

The next step is to ligate the *Nla*III-digested genomic DNA of transposon mutant pools to Y linker. A reaction mixture (20 μl) containing 5 μl of Y linker (200 ng/μl), 2 μl of 10× ligase buffer, 2 μl of doubly distilled H_2O, 10 μl of *Nla*III-digested DNA, and 1 μl of ligase (1 unit/μl) is prepared and incubated overnight at room temperature. The reaction is then diluted 10-fold by adding 180 μl of doubly distilled H_2O, and heated for 10 min at 65° to inactivate ligase activity. The ligation products are then used as template in the PCR amplification.

Simultaneous Amplification of Transposon-Flanking Sequences

For simultaneous amplification of transposon-flanking sequences, a reaction mixture containing 5 μl of PCR buffer [166 mM (NH$_4$)$_2$SO$_4$, 670 mM Tris (pH 8.8), 67 mM MgCl$_2$, 100 mM 2-mercaptoethanol], 1 μl of each primer (Tn5 primer

[12] D. G. Pitcher, N. A. Saunders, and R. J. Owens, *Lett. Appl. Microbiol.* **8,** 151 (1989).

and Y linker primer, 350 ng/μl), 3 μl of dNTPs mix (25 mM for each dNTP), 3 μl of dimethyl sulfoxide (DMSO), 10 μl template DNA, and 26 μl of doubly distilled H$_2$O in 49 μl is incubated at 95° for 2 min to denature template DNA in a PCR machine (PTC-200 DNA engine; MJ Research, Waltham, MA). As a next step, 1 μl *Taq* DNA polymerase (5 units/μl; Applied Biosystems, Foster City, CA) is added during a hot-start incubation at 80° to prevent nonspecific priming. The target sequences are then amplified through 30 cycles of 95° for 30 sec, 58° for 1 min, and 70° for 1 min, followed by a final extension cycle at 70° for 10 min.

Generation of Transposon Footprints by Gel Electrophoresis

The amplified fragments are separated by electrophoresis using either agarose gel of high concentration (2–3%) or polyacrylamide gel. As the number of PCR products with different lengths increases, it is more difficult to get distinct separation of the bands with agarose gel electrophoresis. We routinely analyze 10 μl of each PCR product on 6% precast polyacrylamide gel (TBE gel; Invitrogen, Carlsbad, CA) and the gels are stained by ethidium bromide to visualize transposon footprints.

Characterization of Significant DNA Fragments by Cloning and Sequencing

Comparison of the input and output pools allows identification of the PCR products that are present in the transposon footprint of the input pool but not in that of the output pool. Each of the PCR products thus identified represents transposon mutants that are attenuated during animal infection. The next step is to characterize the PCR products by cloning and sequencing, which will reveal the insertion sites of the transposon in corresponding mutants.

The PCR products of significance are isolated from the polyacrylamide gels by the method of Frost and Guggenheim[13] and concentrated by ethanol precipitation. DNA bands are excised from polyacrylamide gels with a sterile blade and then incubated at 94° for 90 min to elute DNA in 100 μl of 2× PCR buffer (20 mM Tris-HCl, pH 9.0 at 25°, 100 mM KCl, 0.2% Triton X-100; Promega, Madison, WI). After a brief centrifugation, the DNA-containing eluant (~95 μl) is removed to a fresh tube; the volume is increased to 210 μl by adding 160 μl of LoTE buffer and extracted once with equal volume of phenol/chloroform/isoamyl alcohol (25:24:1). Two hundred μl of extracted DNA is then concentrated by ethanol precipitation using glycogen as the carrier of DNA; 200 μl of extracted DNA, 3 μl of glycogen, 100 μl of 10 M ammonium acetate, and 700 μl of 100% ethanol are added to make 1000 μl, mixed thoroughly, and centrifuged for 15 min at 13,000 rpm with a microcentrifuge. The pellet is then washed twice with 70% ethanol and resuspended in 10 μl of doubly distilled H$_2$O.

[13] M. R. Frost and J. A. Guggenheim, *Nucleic Acids Res.* **27**, e6 (1999).

The concentrated PCR products are then cloned into the pCR 2.1 TOPO vector (Invitrogen, Carlsbad, CA) and sequenced. The obtained sequences are analyzed for homologs with DNA database searches.

Retrieval of Corresponding Mutants

Once the PCR products that are missing after selection are identified and characterized, it might be often necessary to further characterize the phenotypes of the corresponding mutants, including verification of their virulence phenotypes by individual animal infection (e.g., competitive index or LD_{50}). The retrieval of the target mutants can be achieved by using the individual or subset of mutants in an input pool as templates in PCR amplification with a gene-specific primer designed from the obtained transposon-flanking sequences and a transposon-specific primer. Only the mutant corresponding to a particular PCR product or subset of mutants containing it will give positive amplification of the PCR product of expected length. We found it is convenient to conduct PCR using intact cells as templates as follows: Ten microliters of culture of individual mutant or mixed culture of multiple mutants is incubated at 98° for 10 min to lyse cells. A reaction mix containing 5 μl of PCR buffer [166 mM $(NH_4)_2SO_4$, 670 mM Tris (pH 8.8), 67 mM $MgCl_2$, 100 mM 2-mercaptoethanol], 1 μl of each primer (Tn5 primer and gene-specific primer, 350 ng/μl), 3 μl of dNTPs mix (25 mM for each dNTP), 3 μl of DMSO, 1 μl of *Taq* DNA polymerase (5 units/μl), and 26 μl of doubly distilled H_2O in 40 μl is added to the lysed cells during a hot-start incubation of 80° to prevent false priming. The target sequences are amplified through 30 cycles of 95° for 30 sec, 58° for 1 min, and 70° for 1 min, followed by a final extension cycle at 70° for 10 min.

Identification of Virulence Genes of *Salmonella typhimurium* by Transposon Footprinting

The genomic DNA samples of *S. typhimurium* Tn5 mutant pools obtained from the mouse infection study of Tsolis *et al.*[14] were generously provided by Dr. R. M. Tsolis (Department of Medical Microbiology and Immunology, Texas A&M University System Health Science Center, College Station, TX) and were used to generate the transposon footprints in this study. Briefly, the bank of 260 Tn5 mutants of *S. typhimurium* IR715, a nalidixic acid-resistant derivative of ATCC 14028, is divided into 10 input pools of 24–30 mutants and is used for the intragastric inoculation of pairs of 6-week-old BALB/c mice at a dose of 10^9 cfu/animal. At day 5 postinfection, the cells of *S. typhimurium* Tn5 mutants are recovered from Peyer's patches and spleens.

---

[14] R. M. Tsolis, S. M. Townsend, E. A. Miao, S. I. Miller, T. A. Ficht, L. G. Adams, and A. J. Bäumler, *Infect. Immun.* **67,** 6385 (1999).

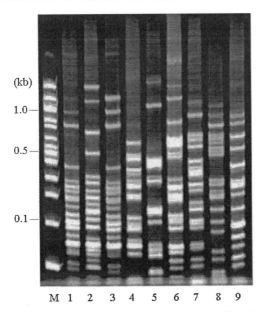

FIG. 3. Transposon footprints of input pools. A 100-bp DNA ladder (New England BioLabs) was used as a standard marker (M).

If a PCR product present in the transposon footprint of an input pool is missing from those of the same organs of both animals, the corresponding mutant is considered to have a colonization defect. Comparison of the transposon footprints of the input pools and corresponding output pools allows the identification of a total of 12 mutants with colonization defects (Figs. 3 and 4). The PCR products corresponding to the important Tn5 mutants are excised from the input pools, sequenced, and subjected to the BLASTX search algorithms (www.ncbi.nlm.nih.gov/BLAST).

Two of the mutations are in previously identified *S. typhimurium* virulence gene (*bcfC* and *spvB*), three are in homologs of known virulence genes of *S. typhimurium* and *E. coli* (*ompC, sirC,* and *orgA*), and seven are in homologs of known genes of *S. typhimurium* and *E. coli* (Table II).

Other Considerations

Restriction Enzymes for Preparation of PCR Templates

The transposon footprints in Fig. 3 show that the numbers of DNA bands in the transposon footprints are less than 20 for all of the input pools that consist of 25–30 mutants per pool. One possible explanation of the results is either that the transposon-flanking sequences were too long to be amplified in the PCR

TABLE II
CHARACTERIZATION OF IDENTIFIED GENES[a]

Clone	Transposon insertion site	Identity/similarity (%)[b]
1-2	Outer membrane porin (*ompC*) homolog (EC)	34/62(54)
1-3	Hypothetical 30.4 kDa protein in *dsrB-vsr* region homolog (EC)	60/87(68)
1-4	Phosphoserine phosphatase (*serB*) homolog (EC)	25/73(34)
2-3	Transcriptional regulator (*sirC*) homolog (ST)	29/66(43)
2-4	Glyoxylate-induced protein (*glxB2*) homolog (EC)	78/82(95)
2-6	Hypothetical 14.4 kDa protein in *murZ-rpoN* region homolog (EC)	65/92(70)
2-7	Bovine colonization factor (*bcfC;* ST)	56/57(98)
3-4	Phosphoadenosine phosphosulfate reductase (*cysH;* ST)	37/37(100)
4-2	Hypothetical 37.3 kDa protein in *leuS-gltL* region homolog (EC)	42/83(50)
4-3	Oxygen-regulated invasion protein (*orgA*) homolog (ST)	37/48(77)
5-1	Outer membrane esterase (*apeE;* ST)	30/30(100)
6-2	Virulence-associated protein (*spvB;* ST)	45/45(100)

[a] ST, *Salmonella typhimurium;* EC, *Escherichia coli.*
[b] Similarities ranging from 30 to 100% identity at the amino acid level were determined.

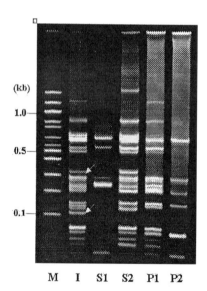

FIG. 4. Identification of virulence genes by transposon footprinting. The amplified Tn5-flanking sequences from inoculum (I) and the bacteria recovered from the spleen (S) and Peyer's patches (P) of mouse were separated by PAGE. The PCR products that are present in the input pool but not in the output pools are indicated by arrows. A 100-bp DNA ladder (New England BioLabs) was used as a standard marker (M).

amplification conditions used in this study for some of the mutants, or that multiple DNA fragments with similar length were not separated distinctively because of the limited resolution of the gel electrophoresis system used. If the first reason is true for some of the mutants, it should be possible to amplify the transposon-flanking sequences either by using a 4-bp recognition restriction enzyme other than *Nla*III to prepare a template for PCR or by using a protocol for long PCR.[15] It has been demonstrated that the protocol for amplification of transposon-flanking sequences using the Y linker was successfully used to amplify Tn*phoA*-flanking sequences of *Vibrio cholerae* Tn*phoA* mutants using restriction enzyme *Sau*3AI.[16]

Microorganisms and Mutagenesis Systems for Transposon Footprinting Analysis

The transposon footprinting approach can be combined with any insertional mutagenesis system, including insertion–duplication mutagenesis and restriction enzyme-mediated insertion as well as other transposon mutagenesis systems.[17,18] It has been suggested that a *mariner*-based transposon has the potential to be universal tool for transposon mutagenesis for a variety of microorganisms, because of its broad host range for *in vivo* transposition with little site specificity beyond the known requirement for the dinucleotide TA.[19] Many bacterial species that do not have a transposon mutagenesis system have been added to the list of the microorganisms that can be randomly mutagenized using the *mariner*-based transposon system, including *Haemophilus influenzae* and *Streptococcus pneumoniae*, *Mycobacterium smegmatis, Campylobacter jejuni, Methanosarcina acetivorans, Pseudomonas aeruginosa, Streptomyces coelicolor, Neisseria meningitides,* and *Helicobacter pylori*.[19–27] We have demonstrated that *mariner*-based transposon

[15] S. Cheng, C. Fockler, W. M. Barnes, and R. Higuchi, *Proc. Natl. Acad. Sci. U.S.A.* **91,** 5695 (1994).

[16] A. Ali, Z. H. Mahmud, J. G. Morris, Jr., S. Sozhamannan, and J. A. Johnson, *Infect. Immun.* **68,** 6857 (2000).

[17] A. Polissi, A. Pontiggia, G. Feger, M. Altieri, H. Mottl, L. Ferrari, and D. Simon, *Infect. Immun.* **66,** 5620 (1998).

[18] J. S. Brown and D. W. Holden, *Curr. Opin. Microbiol.* **1,** 390 (1998).

[19] E. J. Rubin, B. J. Akerley, V. N. Novik, D. J. Lampe, R. N. Husson, and J. J. Mekalanos, *Proc. Natl. Acad. Sci. U.S.A.* **96,** 1645 (1999).

[20] B. J. Akerley, E. J. Rubin, A. Camilli, D. J. Lampe, H. M. Robertson, and J. J. Mekalanos, *Proc. Natl. Acad. Sci. U.S.A.* **95,** 8927 (1998).

[21] N. J. Golden, A. Camilli, and D. W. K. Acheson, *Infect. Immun.* **68,** 5450 (2000).

[22] D. R. Hendrixson, B. J. Akerley, and V. J. DiRita, *Mol. Microbiol.* **40,** 214 (2001).

[23] J. K. Zhang, M. A. Pritchett, D. J. Lampe, H. M. Robertson, and W. W. Metcalf, *Proc. Natl. Acad. Sci. U.S.A.* **97,** 9665 (2000).

[24] S. M. Wong and J. J. Mekalanos, *Proc. Natl. Acad. Sci. U.S.A.* **97,** 10191 (2000).

[25] A. M. Gehring, J. R. Nodwell, S. M. Beverley, and R. Losick, *Proc. Natl. Acad. Sci. U.S.A.* **97,** 9642 (2000).

[26] V. Pelicic, S. Morelle, D. Lampe, and X. Nassif, *J. Bacteriol.* **182,** 5391 (2000).

[27] B. P. Guo and J. J. Mekalanos, *FEMS Immun. Med. Microbiol.* **30,** 87 (2001).

mutants of *Campylobacter jejuni* can be successfully used to generate transposon footprints from input pools and outputs pool recovered from the ceca of 1-day-old chickens after oral gavages of the input pools. We are continuing the screening with transposon footprinting to identify virulence genes of *C. jejuni* required for colonization in the ceca of 1-day-old chicks (unpublished results, 2001). We expect that transposon footprinting, in conjunction with *mariner*-based transposons, should find wide application for the functional analysis of the genome of diverse microorganisms in a variety of selective conditions, including live animal models of infection.

Conclusion

Transposon footprinting is a PCR-based approach for functional screening of bacterial genome. We have shown that this approach can be successfully used to screen bacterial genomes for virulence genes using an animal model of infection. The main advantage of transposon footprinting as compared to STM is that this approach is technically simpler because the sequence tagging on the transposon is not needed and a hybridization technique is not required. The only requirements for transposon footprinting are that the microorganism have an appropriate insertional mutagenesis system available and that it can be maintained as a haploid organism. Another advantage of transposon footprinting is that it is very feasible to analyze multiple samples simultaneously. This feature would facilitate the comparative functional analysis of bacterial genomes in different selective conditions, such as different animals, tissues, or infectious stages. It is expected that the use of transposon footprinting will facilitate functional screening of diverse bacterial genome to assign the open reading frames identified by genome sequencing their biological functions including the ones involved in virulence.

Acknowledgments

We thank Renée M. Tsolis (Department of Medical Microbiology and Immunology, Texas A&M University System Health Science Center) for the generous gift of DNA samples, and Deborah A. Siegele (Department of Biology, Texas A&M University) for guidance throughout this study.

[11] Use of LexA-Based System to Identify Protein–Protein Interactions *in Vivo*

By DAYLE A. DAINES, MICHÈLE GRANGER-SCHNARR, MARIA DIMITROVA, and RICHARD P. SILVER

Introduction

The characterization of protein–protein interactions can reveal much about the structure and function of individual proteins and protein complexes. Many methods designed to identify these interactions have been reported, including yeast two-hybrid and plasmid-based bacterial systems.[1–3] Most determine interaction *in vivo* using the expression of an essential gene, such as an antibiotic or bacteriophage resistance, or a colorimetric or fluorescent reporter protein as an indicator of a positive response.

In this article, we describe a system for analyzing both homo- and hetero-dimerization of full-length proteins and protein fragments in an *Escherichia coli* background.[4,5] As with other two-hybrid systems, this method takes advantage of the unique nature of a repressor protein pivotal in the SOS response, LexA.[6] This protein consists of two domains, one that recognizes and binds to a specific operator site on DNA, and one that functions as a dimerization domain. The LexA dimerization domain can be removed and replaced with another protein or protein fragment. Because the repressor is active only as a dimer, homodimerization of the fused moiety will allow the chimeric LexA to bind to its operator site and repress transcription of a gene carried on the chromosome of the reporter strain. Further, a LexA DNA binding domain (DBD) mutated such that it recognizes a different operator sequence can be used in conjunction with a wild-type LexA DBD in assays for heterodimerization of proteins and protein fragments. This is accomplished by engineering a hybrid operator site controlling reporter gene transcription which contains one half-site that is recognized by the mutant LexA DBD and one half-site that is recognized by the wild-type LexA DBD.[4] Using compatible plasmids, both the wild-type LexA fusion protein and the mutant LexA fusion protein are simultaneously expressed in the reporter strain. Only a heterodimeric protein consisting of one mutant LexA fusion subunit and one wild-type LexA

[1] S. Fields and O.-K. Song, *Nature* **340,** 245 (1989).

[2] F. J. Germino, Z. X. Wang, and S. M. Weissman, *Proc. Natl. Acad. Sci. U.S.A.* **90,** 933 (1993).

[3] G. Di Lallo, P. Ghelardini, and L. Paolozzi, *Microbiology* **145,** 1485 (1999).

[4] M. Dmitrova, G. Younès-Cauet, P. Oertel-Buchheit, D. Porte, M. Schnarr, and M. Granger-Schnarr, *Mol. Gen. Genet.* **257,** 205 (1998).

[5] D. A. Daines and R. P. Silver, *J. Bacteriol.* **182,** 5267 (2000).

[6] M. Schnarr and M. Granger-Schnarr, *Nucleic Acids Mol. Biol.* **7,** 170 (1993).

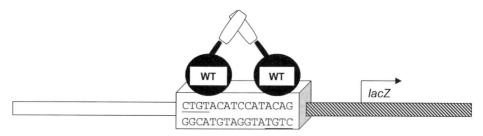

Homodimers are able to repress the **wild type operator** in strain SU101

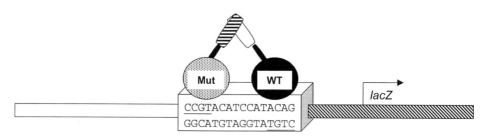

Heterodimers are able to repress the **hybrid operator** in strain SU202

FIG. 1. Overview of LexA system. Homodimers repress transcription of *lacZ* in strain SU101, and heterodimers repress transcription in strain SU202.

fusion subunit can recognize and bind to the hybrid operator on the chromosome, thereby allowing heterodimerization to be monitored by reporter gene repression (Fig. 1). This system has proven to be a rapid, economical, and reproducible tool for the initial identification of interactions between proteins and protein fragments in a bacterial background.[4,5,7]

Materials and Reagents

Bacteria and Media

Competent cells of SU101, SU202 (Table I) and a general cloning strain such as DH5α. Luria-Bertani (LB) broth (10 g Bacto-typtone, 5 g NaCl, and 5 g yeast extract per liter) and agar plates (15 g Bacto-agar per liter), MacConkey agar plates (with lactose).

[7] S. Enz, S. Mahren, U. H. Stroeher, and V. Braun, *J. Bacteriol.* **182,** 637 (2000).

TABLE I
PLASMIDS AND BACTERIAL STRAINS USED IN LexA SYSTEM

Plasmids	Origin of replication	MCS Frame	Resistance (μg/ml)
pSR658	ColE1	A	Tc(12)
pSR659	p15A	A	Ap(100)
pSR660	ColE1	B	Tc(12)
pSR661	p15A	B	Ap(100)
pSR662	ColE1	C	Tc(12)
pSR663	p15A	C	Ap(100)

Strains	Relevant genotype[a]	Resistance (μg/ml)[b]
SU101	lexA71::Tn5(Def)sulA211	Cm(20) Km(30)
	Δ(laclPOZYA)169/F'lacI^{q}lacZΔM15::Tn9	
SU202	lexA71::Tn5(Def)sulA211	Cm(20) Km(30)
	Δ(laclPOZYA)169/F'lacI^{q}lacZΔM15::Tn9	

[a] Strains differ in the LexA operator sequence controlling lacZ expression. SU101 (wt) operator sequence is CTGT (N)$_8$ ACAG. SU202 (mut) operator sequence is CCGT (N)$_8$ ACAG [M. Dmitrova, G. Younès-Cauet, P. Oertel-Buchheit, D. Porte, M. Schnarr, and M. Granger-Scharr, Mol. Gen. Genet. 257, 205 (1998)].
[b] Resistance encoded by transposable elements.

Chemicals and Antibiotics

Toluene, 2-mercaptoethanol, isopropyl-β-D-thiogalactopyranoside (IPTG). Z-buffer: 60 mM Na$_2$HPO$_4$, 40 mM NaH$_2$PO$_4$, 10 mM KCl, 1 mM MgSO$_4$, sterilize by filtering. Add 50 mM 2-mercaptoethanol just before use. o-Nitrophenyl-β-D-galactopyranoside (ONPG), 4 mg/ml in Z-buffer. Tetracycline (12 μg/ml), ampicillin (100 μg/ml).

Experimental Protocol

Procedure for Homodimerization

To determine homodimerization of a known protein, ligate the gene in frame with the LexA DBD in one of the ColE1 origin of replication plasmids (pSR658, pSR660, or pSR662; see Fig. 2 for reading frames). Use the ligation mix to transform competent cells of the homodimerization reporter strain, SU101 (Table I). If no convenient restriction sites exist in the gene of interest, design oligonucleotide primers that incorporate the desired sites in the proper frame and amplify the gene by polymerase chain reaction (PCR). The product can then be digested with the appropriate restriction enzymes and ligated to the chosen vector. Because transformation mixes grow very slowly if placed immediately on MacConkey agar,

pSR658 MCS-A Frame

GGT GAA CCG GTT CTG TAC GAC GAT GAC GAT AAG GAT CGA TGG GGA

 G   E   P   V   L   Y   D   D   D   D   K   D   R   W   G

      SacI XhoI BglII PstI   *KpnI*

TCC GAG CTC GAG ATC TGC AGC TGG TAC CAT ATG GGA ATT CGA AGC

 S   E   L   E   I   C   S   W   Y   H   M   G   I   R   S

TTG GCT GTT TTG GCG GAT GAG AGA AGA TTT TCA GCC TGA

 L   A   V   L   A   D   E   R   R   F   S   A   *

pSR660 MCS-B Frame   *SacI XhoI*

GGT GAA CCG GTT CTG TAC GAC GAT GAC GAT AAG GAT CCG AGC TCG

 G   E   P   V   L   Y   D   D   D   D   K   D   P   S   S

BglII PstI   *KpnI*

AGA TCT GCA GCT GGT ACC ATA TAT GGG AAT TCG AAG CTT GGC TGT

 R   S   A   A   G   T   I   Y   G   N   S   K   L   G   C

TTT GGC GGA TGA

 F   G   G   *

pSR662 MCS-C Frame

GGT GAA CCG GTT CTG TAC GAC GAT GAC GAT AAG CAT CGA TGG

 G   E   P   V   L   Y   D   D   D   D   K   H   R   W

          XhoI BglII PstI   *KpnI*

ATC CGA CCT CGA GAT CTG CAG CTG GTA CCA TAT GGG AAT TCG

 I   R   P   R   D   L   Q   L   V   P   Y   G   N   S

AAG CTT GGC TGT TTT GGC GGA TGA

 K   L   G   C   F   G   G   *

FIG. 2. Reading frames of the ColE1 origin of replication plasmid set, starting at amino acid 85 of the LexA protein. Unique restriction sites in each plasmid's MCS are noted.

use LB agar plates with appropriate antibiotics for the initial transformation. Pick a transformant and sequence the insert to determine that it is in frame with the chosen plasmid's LexA DBD. Perform a Western blot on the transformant using anti-LexA antisera (Invitrogen, Carlsbad, CA) to ensure that the fusion protein is being expressed and is of the expected size. If correct, streak the transformant onto MacConkey agar with 12 μg/ml Tc to determine if there is a pale colony phenotype relative to the color of a control colony of the chosen unfused plasmid in SU101. A colony that is expressing high levels of β-galactosidase will appear magenta on fresh MacConkey agar. Repressed colonies appear pale pink to white and display less precipitation in the surrounding medium than the control colonies. The repression required to distinguish a different color on MacConkey agar quantitates to usually $\geq 25\%$ more than the control colonies, which are considered to be fully expressed at 0% repression. If the transformant's phenotype indicates repression, isolate the plasmid and use to retransform competent cells of SU101. This proves that the phenotype is due to homodimerization of the fused moiety and is not an artifact, such as a rare *lacZ* mutation in the reporter strain. The level of repression of the transformant should then be quantitated.

Quantitation of Repression

Perform β-galactosidase assays using standard procedures[8] with the following changes. Because the reporter strains SU101 and SU202 constitutively transcribe *lacZ* from the strong *sulA* promoter, grow each putatively repressed colony overnight in LB broth with appropriate antibiotics and 1 mM isopropyl-β-D-thiogalactopyranoside (IPTG) to ensure that the plasmids expressing the LexA DBD fusions have been induced for a number of hours. This allows for the degradation of any preexisting β-galactosidase enzyme and results in a more accurate measure of reporter gene repression. The control cultures should be treated identically.

Procedure for Heterodimerization

Characterization of heterodimerizing proteins or protein fragments can be accomplished by expressing each from compatible plasmids (one from the ColE1 and one from the p15A origin of replication plasmid sets) in strain SU202 if both moieties are known, similar to the above homodimerization protocol. However, it may also be desirable to characterize interactions of a known protein with a number of unknown fragments. For this "bait and prey" approach to determine protein heterodimerization, ligate the "bait" gene in one of the p15A origin of replication plasmids (pSR659, pSR661, or pSR663; see Fig. 3 for reading frames) and use the

[8] J. H. Miller, "A Short Course in Bacterial Genetics," p. 72. Cold Spring Harbor Laboratory Press, Cold Spring Harbor, New York, 1992.

pSR659 MCS-A Frame

GGT GAA CCC TCG ATT GCG CGC GAT CTG TAC GAC GAT GAC GAT AAG GAT CGA

G　　E　　P　　S　　I　　A　　R　　D　　L　　Y　　D　　D　　D　　D　　K　　D　　R

　　　BamHI　　SacI XhoI BglII　　　　　　KpnI　　　　　　　　HindIII

TGG GGA TCC GAG CTC GAG ATC TGC AGC TGG TAC CAT ATG GGA ATT CGA AGC

W　　G　　S　　E　　L　　E　　I　　C　　S　　W　　Y　　H　　M　　G　　I　　R　　S

TTG GCT GTT TTG GCG GAT GAG AGA TTT TCA GCC TGA

L　　A　　V　　L　　A　　D　　E　　R　　R　　F　　S　　A　　*

pSR661 MCS-B Frame　　　　　　　　　　　　　　　　　　　　　　　　BamHI

GGT GAA CCC TCG ATT GCG CGC GAT CTG TAC GAC GAT GAC GAT AAG GAT CCG

G　　E　　P　　S　　I　　A　　R　　D　　L　　Y　　D　　D　　D　　D　　K　　D　　P

SacI XhoI BglII　　　　　　　KpnI　　　　　　　　　HindIII

AGC TCG AGA TCT GCA GCT GGT ACC ATA TGG GAA TTC GAA GCT TGG CTG TTT

S　　S　　R　　S　　A　　A　　G　　T　　I　　W　　E　　F　　E　　A　　W　　L　　F

TGG CGG ATG AGA GAA GAT TTT CAG CCT GAT ACA GAT TAA

W　　R　　M　　R　　E　　D　　F　　Q　　P　　D　　T　　D　　*

pSR663 MCS-C Frame

GGT GAA CCC TCG ATT GCG CGC GAT CTG TAC GAC GAT GAC GAT AAG GAT CGA

G　　E　　P　　S　　I　　A　　R　　D　　L　　Y　　D　　D　　D　　D　　K　　D　　R

　　　BamHI　　　　　　BglII　　　　　　KpnI　　　　　　　　　HindIII

TGG ATC CGA CCT CGA GAT CTG CAG ATG GTA CCA TAT GGG GAT TCG AAG CTT

W　　I　　R　　P　　R　　D　　L　　Q　　M　　V　　P　　Y　　G　　D　　S　　K　　L

GGC TGT TTT GGC GGA TGA

G　　C　　F　　G　　G　　*

Fɪɢ. 3. Reading frames of the p15A origin of replication plasmid set, starting at amino acid 85 of the LexA protein. Unique restriction sites in each plasmid's MCS are noted.

TABLE II
RESTRICTION ENZYMES COMPATIBLE WITH PLASMID SITES

Plasmid	Restriction sites[a]	Compatible overhangs
pSR658, pSR660, pSR662	*Bgl*II. *Kpn*I, *Pst*I, *Sac*I,[b] *Xho*I[c]	*Bgl*II: *Bam*HI, *Bst*YI, *Bcl*I, *Dpn*II *Kpn*I: None identified *Pst*I: *Bsi*HKAI, *Bsp*1286I, *Nsi*I, *Sbf*I *Sac*I[b]: *Ban*II, *Bsi*HKAI, *Bsp*1286I *Xho*I[c]: *Ava*I, *Sal*I
pSR659, pSR661, pSR663	*Bam*HI, *Bgl*II, *Hind*III, *Kpn*I, *Sac*I,[b] *Xho*I[c]	*Bam*HI: *Bcl*I, *Dpn*II, *Bgl*II, *Bst*YI *Bgl*II: *Bam*HI, *Bst*YI, *Bcl*I, *Dpn*II *Hind*III: None identified *Kpn*I: None identified *Sac*I[b]: *Ban*II, *Bsi*HKAI, *Bsp*1286I. *Xho*I[c]: *Ava*I, *Sal*I.

[a] Restriction sites that are unique. Other sites occur in the vectors.
[b] Note that there is no *Sac*I site in pSR662 or pSR663.
[c] Note that pSR663 lacks a *Xho*I site.

ligation mix to transform competent cells of the heterodimerization reporter strain SU202. Pick a colony and sequence the insert to determine that it is correct. Using anti-LexA antisera, perform a Western blot of the transformant to ensure that the fusion protein is expressed and is of the correct size. To construct the "prey" library, ligate a size-fractionated partial restriction digest of the chromosome (or the DNA of interest) into the ColE1 origin of replication set of plasmids (pSR658, pSR660, and pSR662; see Fig. 2 for reading frames). This allows all three reading frames of the restriction fragments to be analyzed. Select a number of different restriction enzymes to ensure complete coverage of the genome (for a list of compatible enzymes, see Table II). Use the ligation mixes to transform competent cells of a general cloning strain such as DH5α, to attain the highest transformation efficiency. Recover the resulting transformants and isolate plasmid pools using a kit such as Wizard Plasmid Minipreps (Promega, Madison, WI). Transform competent cells of the reporter strain SU202 that express the compatible bait plasmid with aliquots of these plasmid pools. Individually streak or replica-plate the SU202 transformants onto fresh MacConkey agar plates with appropriate antibiotics to identify those that express fragments which repress reporter gene transcription, relative to the control colony (strain SU202 carrying the bait plasmid and the chosen unfused compatible vector). Isolate the plasmids carried by all putatively heterodimerizing transformants and use to retransform competent cells of SU202 carrying the bait plasmid to ensure that the phenotype is due to the interaction of the two subunits. Repression is quantitated as above.

Although the bait protein can be initially expressed from the ColE1 origin of replication plasmid set, building a comprehensive library of chromosomal DNA fragments in the p15A origin of replication plasmid set may be more labor-intensive, since these plasmids are maintained at a lower copy number. Once fusions that heterodimerize with the bait protein have been identified, the interactions should be analyzed when expressed in the opposite vector pair, as strongly homodimerizing fusions can sometimes interfere with heterodimerization when expressed from the higher copy number ColE1 origin of replication plasmids.[5]

Discussion

This system has been successful in determining interactions between proteins involved in the biosynthesis of the polysialic acid capsule of *E. coli* K1. These interactions include those between full-length proteins as well as those between full-length proteins and protein fragments.[5] However, it is not necessary to restrict the system's use to studying only *Escherichia coli* protein interactions. A size-fractionated partial restriction digest of *Haemophilus influenzae* strain Rd KW20 chromosomal DNA ligated to the ColE1 plasmid set isolated a number of strongly homodimerizing fusions.[9] Two of these fragments mapped to *hsdM*, the gene encoding the dimeric methylase of a type I restriction modification complex. This confirms the system's utility in identifying interactions occurring between fusions originating from heterologous DNA.

Another feature of this system is that the detection of interaction between fusions is not disrupted by the tandem expression of unfused subunits in the host strain. Chloramphenicol acetyltransferase (CAT) fused to the wild-type LexA DBD efficiently repressed *lacZ* transcription in SU101, even though this strain carries a transposon that expresses subunits of the same type I CAT.[5] This data also illustrates that some multimeric proteins can be studied using this system, since CAT is a well-characterized homotrimer.

Once interacting fusions have been identified, a number of analyses can be performed. Fragments of each fusion can be ligated to the LexA DBD to determine which areas are interacting. Site-directed mutagenesis can be performed on the full-length protein or protein fragments to identify residues that are essential for interaction. The deduced prey fusion can be used as bait to screen a library of fragments to discover whether it may be interacting with more than one protein.

The limitations of this method are those that plague any LexA-based system.[10] There may be steric hindrance between fused fragments that prevents a functional dimer from binding to the operator site, or the fused moiety may not fold properly. Overexpression of some fusions may result in intracellular complex formation or toxicity. Some protein–protein interactions *in vivo* are significant but weak, are transient, or occur between proteins located in the periplasm or the outer

---

[9] D. A. Daines and A. L. Smith, unpublished results (2000).

[10] E. A. Golemis and R. Brent, *Mol. Cell. Biol.* **12,** 3006 (1992).

membrane. Others involve one subunit interacting with a higher-order multimer (i.e., pentamer) of another. It is doubtful that these interactions will be identified using this system.

Although the different copy numbers of the compatible plasmids in each set increase the system's range of use, they also increase its complexity. Expression of a strongly homodimerizing fragment in the ColE1 set of plasmids may overcome the low affinity of the wild-type LexA DBD for the hybrid operator in SU202. Therefore, each putative heterodimerizing pair should be studied when expressed from its reciprocal fusion. As presented, we consider this method to be a rapid and reproducible tool for initially determining whether proteins are interacting *in vivo*. Further characterization of any identified interactions should include biochemical techniques such as gel filtration and protein cross-linking.

Section III

Global Gene Expression: Microarrays and Proteomics

[12] *Borrelia burgdorferi* Gene Expression Profiling with Membrane-Based Arrays

By CAROLINE OJAIMI, CHAD BROOKS, DARRIN AKINS,
SHERWOOD CASJENS, PATRICIA ROSA, ABDALLAH ELIAS, ALAN BARBOUR,
ALGIS JASINSKAS, JORGE BENACH, LAURA KATONAH, JUSTIN RADOLF,
MELISSA CAIMANO, JON SKARE, KRISTEN SWINGLE, SIMON SIMS,
and IRA SCHWARTZ

Introduction

Lyme disease, a multisystem disorder with possible neurologic, cardiac, and arthritic manifestations, is caused by an infection with the spirochete *Borrelia burgdorferi*,[1–5] which is transmitted to humans during feeding of certain species of *Ixodes* ticks.[2–4] Lyme disease is the most prevalent arthropod-borne disease in the United States. Cases of Lyme disease are widely found in North America, Europe, and Asia, and are reported to be increasing in both number and geographic distribution.[5] Despite intensive study in recent years little is known regarding the pathogenesis of *B. burgdorferi* infection at the molecular level.

The *B. burgdorferi* B31 genome is unique among fully sequenced bacterial genomes in that it consists of a linear chromosome 910,725 bp in length and a collection of 9 circular and 12 linear plasmids.[6] Among the 1689 putative open reading frames (ORFs), 855 are chromosome encoded and 834 are plasmid encoded.[6,7] The mapping of genes encoding important structural proteins (e.g., OspA and OspC) and some metabolic enzymes (e.g., GuaA and GuaB) to plasmids suggests that these extrachromosomal elements are an integral part of the genome. There

---

[1] J. L. Benach, E. M. Bosler, J. P. Hanrahan, J. L. Coleman, G. S. Habicht, T. F. Bast, D. J. Cameron, J. L. Ziegler, A. G. Barbour, W. Burgdorfer, R. Edelman, and R. A. Kaslow, *N. Engl. J. Med.* **308,** 740 (1983).

[2] A. G. Barbour and D. Fish, *Science* **260,** 1610 (1993).

[3] D. H. Spach, W. C. Liles, G. L. Campbell, R. E. Quick, D. E. Anderson, Jr., and T. R. Fritsche, *N. Engl. J. Med.* **329,** 936 (1993).

[4] A. C. Steere, *Proc. Natl. Acad. Sci. U.S.A.* **91,** 2378 (1994).

[5] D. H. Walker, A. G. Barbour, J. H. Oliver, R. S. Lane, J. S. Dumler, D. T. Dennis, D. H. Persing, A. F. Azad, and E. McSweegan, *JAMA* **275,** 463 (1996).

[6] C. M. Fraser, S. Casjens, W. M. Huang, G. G. Sutton, R. Clayton, R. Lathigra, O. White, K. A. Ketchum, R. Dodson, E. K. Hickey, M. Gwinn, B. Dougherty, J. F. Tomb, R. D. Fleischmann, D. Richardson, J. Peterson, A. R. Kerlavage, I. Quackenbush, S. Salzberg, M. Hanson, R. van Vugt, N. Palmer, M. D. Adams, J. Gocayne, and J. C. Venter, *Nature* **390,** 580 (1997).

[7] S. Casjens, N. Palmer, R. van Vugt, W. M. Huang, B. Stevenson, P. Rosa, R. Lathigra, G. Sutton, J. Peterson, R. J. Dodson, D. Haft, F. Hickey, M. Gwinn, O. White, and C. M. Fraser, *Mol. Microbiol.* **35,** 490 (2000).

is considerable variation in plasmid content among isolates and some of these extrachromosomal elements are lost on serial propagation.[8] More than 90% of the *B. burgdorferi* plasmid ORFs are unrelated to any known bacterial sequences.[7] The novel genes found on the *B. burgdorferi* plasmids may, therefore, contribute to the ability of this pathogen to survive and maintain its complex life cycle.

The Lyme disease spirochete experiences considerable change in environment as it alternates between warm-blooded mammals and ticks. Variation in gene expression under different physiological conditions, such as temperature and pH, has been examined in *B. burgdorferi*. Several proven or putative outer surface lipoproteins of *B. burgdorferi* are differentially expressed at temperatures that correlate with infection of ticks or mammals.[9–13] Such studies confirm that *B. burgdorferi* modulates its gene expression in response to environmental changes. Differential expression of proteins in this manner may play a role in pathogenesis. Therefore, knowledge regarding the synthesis of *B. burgdorferi* proteins during its life cycle should provide a greater understanding of the pathogenesis of Lyme disease.

The advent of bacterial genomics provides a comprehensive approach for elucidation of potential virulence genes. The *B. burgdorferi* B31 genome sequence contains relatively few ORFs with recognizable homology to other characterized bacterial virulence genes.[6] Mining of *B. burgdorferi* genomic information for clues to virulence must, therefore, be as broadly based as possible. In the absence of high-throughput genetic screening tools for this organism, one of the most attractive approaches for finding virulence factors is probing of whole genome arrays. In practice, the approach involves the cultivation of bacteria under a defined set of growth conditions, isolation, and labeling of total RNA. The arrayed ORFs can then be "queried" by hybridization with the labeled total mRNA (or cDNA) "probe." Since the identities of all the spots (ORFs) on the array are known, the complete gene expression profile of an organism under a given set of conditions can be analyzed with a single array.

High-density membrane array technology is gaining increasing use in the study of bacterial gene expression. Whole genome membrane arrays of *Escherichia coli, Bacillus subtilis,* and several other bacteria are commercially available from Sigma-Genosys (The Woodlands, TX). These have been employed for global gene

[8] W. J. Simpson, C. E. Garon, and T. G. Schwan, *Microb. Pathog.* **8,** 109 (1990).

[9] T. G. Schwan, J. Piesman, W. T. Glode, M. C. Dolan, and P. A. Rosa, *Proc. Natl. Acad. Sci. U.S.A.* **92,** 2909 (1995).

[10] B. Stevenson, T. G. Schwan, and P. A. Rosa, *Infect. Immun.* **63,** 4535 (1995).

[11] D. R. Akins, S. F. Porcella, T. G. Popova, D. Shevchenko, S. I. Baker, M. Li, M. V. Norgard, and J. D. Radolf, *Mol. Microbiol.* **18,** 507 (1995).

[12] R. Ramamoorthy and M. T. Philipp, *Infect. Immun.* **66,** 5119 (1998).

[13] J. A. Carroll, N. El Hage, J. C. Miller, K. Babb, and B. Stevenson, *Infect. Immun.* **69,** 5286 (2001).

TABLE I
OVERVIEW OF BACTERIAL GENE EXPRESSION PROFILING USING MEMBRANE ARRAYS

1. Growth of bacterial cultures under various physiological conditions
2. Isolation of total RNA from each culture
3. Generation of radioactively labeled cDNA from all RNA samples for use as probes
4. Hybridization of labeled cDNAs to duplicate arrays
5. Data acquisition by phosphorimaging of arrays
6. Data analysis of expression patterns using image analysis and statistical software

expression analysis, demonstrating the potential benefits that genome array technology can provide for such studies.[14–16]

General Experimental Design

The development of gene array technology allows researchers to study the relative mRNA levels of hundreds to thousands of genes simultaneously. The overall approach for an expression profiling experiment is summarized in Table I.

The sample-to-sample variations in hybridization signals that are not due to the variation in the physiological conditions under study should be randomly distributed and decrease dramatically with the analysis of multiple samples from each treatment group. Therefore, each set of experiments should be repeated at least twice and in reciprocal manner (i.e., switching membranes) to facilitate more robust statistical analysis. Genome array data are semiquantitative and should be used as a screening tool. After identification of candidate genes by array analysis, differences in gene expression should be confirmed by a quantitative method, such as real-time RT-PCR (reverse transcription–polymerase chain reaction).

The *B. burgdorferi* membrane arrays described here contain PCR-amplified ORFs from *B. burgdorferi* strain B31 MI, the strain whose genome sequence has been elucidated.[6,7]

There are 1689 putative ORFs identified for strain B31 (see below) of which 1628 (96.4%) have been spotted in duplicate onto positively charged nylon membranes. Following spotting, each array is cross-linked with UV light. The arrays contain two printing fields and are 8 cm × 24 cm in size. Each field of the array consists of a primary grid composed of 16 rows by 24 columns of secondary grids. Each secondary grid contains six genes spotted as 3 rows and 2 columns. The

[14] H. Tao, C. Bausch, C. Richmond, F. R. Blattner, and T. Conway, *J. Bacteriol.* **181,** 6425 (1999).
[15] Y. Wei, J. Lee, C. Richmond, F. R. Blattner, J. A. Rafalski, and R. A. LaRossa, *J. Bacteriol.* **183,** 545 (2001).
[16] A. Petersohn, M. Brigulla, S. Haas, J. D. Hoseisel, U. Volker, and M. Hecker, *J. Bacteriol.* **183,** 5617 (2001).

four corners of the primary grid contain *B. burgdorferi* genomic DNA spotted in duplicate. These genomic DNA spots act as positive controls and facilitate array template alignment during image analysis.

Methods

ORF Selection

ORF selection for the *B. burgdorferi* genome is more complex than simply choosing a sequence from within each ORF identified by the computerized analysis of the genome sequence. The strain B31 MI chromosome contains a number of paralogous genes, but is a "typical" bacterial chromosome in that genes (ORFs) can be recognized quite unambiguously because of their close packing (the chromosome is 94% protein coding) and by similarity to previously characterized genes from other organisms. On the other hand, the 21 plasmids contain many patches of paralogous sequence (a majority of which are unique to *B. burgdorferi*). These paralogs are often extremely similar and could confuse any hybridization-based analysis of expression of individual members of paralogous gene sets. There are 175 paralogous gene families in the genome, and 107 of these are composed wholly or partly of plasmid genes; relationships within these families range continuously from identical genes which cannot be differentiated by hybridization to barely recognizably similar genes which are unlikely to cross-hybridize.[7] Automated ORF identification by a "normally" curated GLIMMER[17] analysis has found many examples of apparently truncated genes on the plasmids with at least 152 mutationally decaying pseudogenes present on the plasmids (decaying relative to a putatively intact paralog elsewhere on the plasmids; see Ref. 7 for details). In addition, the plasmids contain a number of small ORFs that overlap other longer ORFs or pseudogenes. A DNA array that uses double-stranded probes such as is described here will not discriminate between overlapping ORFs whether they are oriented in the same direction or not.

These problems have been dealt with in a manner that makes a useful but imperfect array whose limitations are known. A decision has thus been made to overtly ignore this "paralog problem," because solving it would require much experimental verification of any solution used. Thus, products are prepared from all ORFs without experimental determination as to whether the arrayed PCR products will cross-hybridize with other paralogous ORFs under our experimental conditions. Thus, any observed expression of more than one member of a paralogous family will have to be examined in more detail by other methods, if feasible. This approach represents a viable compromise between not including members of highly similar

---

[17] A. L. Delcher, D. Harmon, S. Kasif, O. White, and S. L. Salzberg, *Nucleic Acids Res.* **27,** 4636 (1999).

paralogous gene sets in the analysis and exerting the great time and effort required to devise ways of distinguishing such paralogous sequences in the analysis.

Given the apparently decaying state of the plasmid-borne pseudogenes, their expression seems unlikely to be important to the organism. However, since it is still of interest to know if these regions are transcribed, they are included in the array analysis. Pseudogenes and overlapping genes are managed by manually curating the published strain B31 MI ORF list. First, all pseudogenes are incorporated into the ORF list; this involves identifying all regions of nucleotide similarity among the plasmids and the plasmid-like rightmost 7.2 kbp of the chromosome.[7] Then, all overlapping ORFs (including the pseudogenes) are compared and among any overlapping ORF set, only the largest is retained. Finally, scattered on the linear plasmids are unique several Kilobase pair regions that appear to contain no ORFs. To query whether these regions are transcribed, primer sets are included for the largest such regions. In this way a list of 1697 "gene features" has been generated for *B. burgdorferi* B31 MI.

PCR Primer Design

A proprietary software program (Sigma-Genosys) is used to design PCR primers with optimal priming efficiency for the majority of the ORFs. The program was developed to find primers that match certain primer specifications, starting at both ends of the ORF sequences and moving inward until the appropriate priming sequences are found. The design specifications include: (a) a minimum length of 18 bases, (b) a %G + C content that is within the range of 30% to 70%, (c) a melting temperature of at least 55°, and (d) a G or a C present in one of the last three 3' base positions in order to help anchor the priming end of the oligonucleotide. In this fashion, suitable primers have been found that will amplify a significant portion of the ORF in the majority of the cases. Typically, yield issues are encountered when trying to amplify very long or very short PCR targets. For longer ORFs, sequences are truncated from the 5' end, such that most PCR products will not exceed 2000 bp. For ORFs less than 250 bp, primers are designed to amplify the entire coding sequence, from start to stop codons. Hence, these shorter ORFs will represent the maximal hybridizable domain for array studies. However, as primers are forced to prime at the extreme ends of the shorter ORFs, priming efficiency may be compromised as the above specifications may not always be met.

Melting temperatures of the primers are calculated using the nearest neighbor method,[18] assuming 50 mM salt and 50 μM DNA concentrations. The average melting temperature for all primers is 66.3° ($\pm$ 5.9°) and the average %G + C content for all primers is 35.9% ($\pm$ 8.06%).

[18] K. L. Breslauer, R. Frank, H. Blocker, and L. A. Marky, *Proc. Natl. Acad. Sci. U.S.A.* **83,** 3746 (1986).

PCR primers pairs are designed to amplify each of the 1697 "gene features" described above. The primers are synthesized and provided in 96-well plates (Sigma-Genosys) with both forward and reverse primers dried down together in each well. This facilitates the parallel processing of amplification reactions. Each well contains 2000 pmol of both forward and reverse primers. Primers are resuspended in 200 μl of water.

B. burgdorferi Growth

B. burgdorferi isolates are cultured at 35° in BSK-H medium (Sigma, St. Louis, MO) supplemented with 6% serum (complete BSK-H medium) in a CO_2 incubator and spirochete densities are measured by fluorescence microscopy as described.[19] For temperature-shift experiments, spirochetes are first grown at 35° to a density of 5×10^7 per ml in BSKII medium,[20] diluted 100-fold into fresh medium, and then incubated at 23° until a density of approximately 5×10^7 per ml. Identical aliquots are removed and diluted 100-fold into fresh 100 ml of BSKII medium, and the cultures are incubated at 23° and 35°, respectively. Cells are harvested by centrifugation at 12,000g after reaching mid-exponential phase (5×10^7 per ml).

Preparation of DNA from B. burgdorferi Cultures

Many commercial methods for preparation of bacterial DNA are available. Here, we describe isolation from 10 ml of *B. burgdorferi* cultures grown at 33° using the Isoquick nucleic acid extraction kit (Orca Research, Bothell, WA).

1. Harvest the cells by centrifugation (11,000g, 5 min). Discard the supernatant.
2. Resuspend pelleted cells in 100 μl of reagent A (sample buffer) and incubate for 5–15 min at room temperature.
3. Lyse the cells and stabilize the nucleic acid by mixing with an equal volume of Reagent 1 (Lysis Buffer).
4. Shake Reagent 2 (Extraction Matrix) vigorously and add 500 μl to the sample lysate.
5. Add 400 μl of Reagent 3 (Extraction Buffer) to the sample. Vortex 10 sec and incubate at 65° for 10 min. Vortex briefly after 5 min.
6. Centrifuge at 12,000g for 5 min at room temperature.
7. Transfer the aqueous phase to a microcentrifuge tube. Avoid transferring the colored organic phase or interface layer.
8. Shake Reagent 2 (Extraction Matrix) vigorously and add 500 μl to the aqueous phase sample. Vortex 10 sec to mix.

[19] G. P. Wormser, D. Liveris, J. Nowakowski, R. B. Nadelman, L. F. Cavaliere, D. McKenna, D. Holmgren, and I. Schwartz, *J. Infect. Dis.* **180,** 720 (1999).
[20] A. G. Barbour, *Yale J. Biol. Med.* **57,** 521 (1984).

9. Centrifuge at 12,000g for 5 min at room temperature.

10. Transfer the aqueous phase to a new microcentrifuge tube.

11. Add 0.1 volume of Reagent 4 sodium acetate and an equal amount of 2-propanol. Mix gently to precipitate the DNA.

12. Centrifuge at 12,000g for 5 min.

13. Discard the supernatant and allow the pellet to air-dry.

14. Isolated DNA is resuspended in 50 μl of RNase-free water.

DNA isolated in this manner can be used as the template for PCR amplification.

PCR Amplification

PCR is carried out in a 50 μl reaction mixture containing 10 mM Tris-HCl (pH 8.3), 1.5 mM MgCl$_2$, 50 mM KCl, 100 μM each of the four dNTP, 1.25 units of *Taq* DNA polymerase, and 20–30 pmol of each primer. Typically, amplification is carried out for 35 cycles at temperatures appropriate for the primers employed. The success of the PCR reactions is monitored by electrophoresis of a 10-μl sample of the reaction mixture through a 1 to 2% agarose gel in 1× TBE buffer. The amount of amplified product is determined by densitometric analysis of agarose gels using known quantities of molecular size markers. Gel analysis of representative PCR amplified ORFs is shown in Fig. 1. An amplification reaction is scored as successful if a single product is within 5% of the expected size of the ORF predicted from the genomic sequence. Unsuccessful PCR reactions are of three types: (i) no product observed by ethidium bromide staining; (ii) multiple bands observed; or (iii) reactions resulting in single products of unexpected size. Reactions that fail

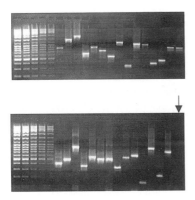
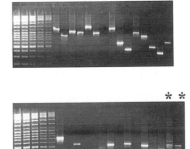

FIG. 1. Electrophoretic separation of the products from PCR amplification of *B. burgdorferi* genomic DNA. The first five lanes represent decreasing amounts of the GeneRuler 100-bp DNA ladder mix (MBI Fermentas). Arrow indicates a PCR failure due to the lack of any product and asterisks mark lanes scored as PCR failures due to multiple bands.

are repeated using different amplification conditions, for example, adjustment of annealing temperature. Those ORFs for which PCR fails after several attempts are eliminated and not represented on the arrays. Approximately 96% of the "gene features" (1628/1697) have been successfully amplified using this approach.

Synthesis of Labeled cDNA Hybridization Probe

RNA Preparation for Membrane Array Screening. The methodology for RNA extraction is a critical step for a successful expression profiling experiment. It is important to purify RNA free of contaminating genomic DNA which may contribute to competitive hybridization to all spots on the array. Bacteria of the genus *Borrelia* are fastidious organisms, growing slowly and requiring a complex growth medium with low oxygen tension.[20] In order to achieve enough high-quality RNA for several cDNA labeling reactions, large volume cultures (>50 ml) are needed. Many methods and several commercial kits are widely available for bacterial RNA isolation. On the basis of our experience, the RNAzol B extraction kit (TEL TEST) and the Ultraspec-II RNA isolation system (BIOTECX, Houston, TX) give the most reproducible yields of high quality RNA from *B. burgdorferi* cultures. The RNAzol procedure described below has been successfully used for 10–100 ml of *B. burgdorferi* cultures.

1. Divide the cultures into 15-ml polypropylene tubes.
2. Pellet the cells by centrifugation at $10,000g$ for 10 min at $4°$.
3. Transfer the pellet along with 0.5 ml of medium to a smaller 1.5-ml microcentrifuge tube and centrifuge for 5 min at $4°$ in a microcentrifuge at $14,000g$. Remove the medium completely from the pelleted cells.
4. Resuspend the pellets by the addition of 0.2 ml of RNAzol B per 10^6 cells. Lyse the cells by passing the suspension through a pipette several times or by use of an 18-gauge needle and syringe.
5. Add 0.2 ml chloroform per 2 ml of suspension, cover the samples tightly, shake vigorously for 15 sec (do not vortex), and incubate on ice for 5 min.
6. Centrifuge the suspension at $14,000g$ for 10 min at $4°$.
7. Transfer the aqueous phase to a fresh tube. Avoid taking the interphase or the lower phase. Add an equal volume of isopropanol and mix thoroughly by inversion. Incubate at $-20°$ for at least 1 hr to precipitate the RNA.
8. Centrifuge RNA precipitates for 30 min at $4°$ at $14,000g$.
9. Remove the supernatant and wash the RNA pellet once with 0.8 ml of 75% ethanol, mix thoroughly by inversion, and centrifuge at $8000g$ at $4°$.
10. Remove the supernatant and air-dry the RNA pellet at room temperature for 15 min.
11. Resuspend the RNA in 100 μl of RNase-free water. Heat the sample at $65°$ for 10 min and vortex to dissolve the RNA pellet.

12. Perform DNase I treatment using the DNA-free kit (Ambion, Austin, TX) as follows. To the 100 μl RNA sample, add 0.1 volume of 10$\times$ DNase I buffer and 3 μl of DNase I (10 U/μl). Mix gently and incubate at 37° for 45 min.

13. Resuspend the DNase inactivation reagent by flicking or vortexing the tube. Add 0.1 volume of the slurry to the sample, flick the tube to disperse the reagent in the RNA preparation, and incubate the tube for 2 min at room temperature.

14. Centrifuge the tube for 1 min to pellet the DNase inactivation reagent and remove the supernatant solution containing the RNA to a fresh tube.

15. Add 0.1 volume of 0.2 M NaCl and an equal volume of 2-propanol and incubate at −20° for at least 1 hr.

16. Centrifuge at 14,000g at 4° for 15 min, followed by a wash with 75% ethanol and air-drying of the pellet at room temperature.

17. Dissolve the RNA pellets in RNase-free H_2O, add 1–2 μl of RNasin, and store the samples at −70° for later use.

The concentration of the RNA is estimated spectrophotometrically using the equation $A_{260} = 1/44 \times C$, where C is the concentration in μg/μl.

Preparation of Labeled cDNA

One approach for generating cDNA from microbial RNA for hybridization with membrane arrays is random hexamer-primed synthesis using reverse transcriptase. A major drawback of this method is that the majority of the label is incorporated into rRNA as opposed to mRNA. Since there is no effective method for purifying mRNAs from total RNA in bacteria, the sensitivity of mRNA transcript detection is reduced. Alternatively, primers specific for the 3′ end of each ORF may be used and this approach generally results in labeled cDNAs with a substantially higher specific activity compared to random hexamer-primed cDNA. It is recommended to use ^{33}P rather than ^{32}P for expression analysis with membrane arrays because signal detection with ^{33}P yields a sharper image with well-defined spots on the arrays that allow more accurate quantitation. Because of the high AT content of the *B. burgdorferi* genome, [^{33}P]dATP is employed during probe synthesis.

Probe Synthesis

Hybridization probes are generated by standard cDNA synthesis using a mixture containing 1697 ORF-specific 3′-end oligonucleotides as primers.

1. 5 μg of RNA is mixed with 1 μl (1.0 pmol of each primer) of the ORF-specific primer mixture.

2. The primers are annealed to the RNA by heating at 65° for 10 min.

3. The sample is cooled to 42° and dTTP, dGTP, and dCTP (final concentration, 0.4 mM each), 200 U of Superscript II (Life Technologies, Carlsbad, CA),

10 U of RNase inhibitor, and 20 μCi of [α-^{33}P]dATP (Amersham, Piscataway, NJ; 2500 Ci/mmol) are mixed in a final volume of 20 μl of first-strand buffer.

4. The cDNA synthesis reaction mixture is incubated at 42° for 2 hr.

5. The labeling efficiency is determined by TCA precipitation. Under the above conditions, specific activities of 2–5 × 10^6 cpm/μg of input RNA are achieved.

Hybridization

The hybridization steps are performed in roller bottles using a Hybaid hybridization oven.

1. Before hybridization, the arrays are rinsed in 2× SSC for 10 min and then prehybridized for at least 30 min at 50° in 10 ml of 1× Perfect Hyb Plus hybridization buffer (Sigma).

2. Labeled cDNA probe, generated from 5 μg of RNA as described above, is denatured at 95° for 5 min and added to a fresh 10 ml of 1× Perfect Hyb Plus hybridization buffer, and arrays are hybridized overnight at 50°.

3. After hybridization, the membranes are washed twice with 2× SSC, 0.05% sodium dodecyl sulfate (SDS) for 30 min at room temperature and then twice with 0.1× SSC, 0.1% SDS at 65° for 30 min.

4. Arrays are wrapped in clear plastic wrap and exposed to a PhosphorImager screen (Molecular Dynamics, Sunnyvale, CA) for 48 hr. For comparison of different isolates or a given isolate grown at different growth conditions, the same membranes are probed reciprocally (i.e., switching the membranes) after stripping in alkaline stripping solution.

Data Acquisition

The exposed phosphoroimager screen is scanned with a pixel size of 50 or 100 μm on a Storm 840 PhosphorImager (Molecular Dynamics). The image files are analyzed using ArrayVision software (Imaging Research, St. Catharines, Ontario, Canada). A template which contains the spot layout of the array is overlaid on the phosphorimage and the pixel intensity of each spot on the array is determined. ArrayVision offers several methods for background correction including sampling of regions surrounding the spotted ORFs and user-defined regions. The latter method allows one to define specific areas of the membrane (e.g., regions spotted with buffer lacking DNA) to be used as background. Multiple custom background areas are designated and an average background value is calculated from all the background pixels sampled. This single value is subtracted from every spot in the array. Alternatively, user-defined regions can be locally subtracted from the nearest spotted ORFs in which different spotted ORFs will have varying degrees of background subtracted based on the local background within that region of the membrane.

Validation of Array Quality

Hybridization of labeled genomic DNA to the membrane arrays allows assessment of signal variation between spots due to factors such as size, amount of DNA spotted, and cross-hybridization to members of gene families. To ensure that DNA samples are successfully deposited on the membranes and to assess differential hybridization to target genes, ^{33}P-labeled DNA probes derived from randomly sheared B31 genomic DNA by random hexamer-directed synthesis are hybridized to the arrays. Hybridization signals are quantified with ArrayVision software after direct import of the phosphorimager files. After background subtraction, the overall spot normalization function of ArrayVision is used to calculate the normalized intensity values of individual spots. This procedure involves two steps: (1) calculation of the intensity of an average spot by dividing the sum of the intensities of all the PCR product specific signals on the array by the total number of spots and (2) dividing the intensity of the individual spot by the intensity of this average spot. An example of a *B. burgdorferi* array probed with labeled genomic DNA is shown in Fig. 2.

Statistical Analysis

The intensities for each spot are exported from ArrayVision into ArrayStat (Imaging Research, St. Catharines, Ontario, Canada) for statistical analysis. ArrayStat uses novel, rigorously validated algorithms to generate sensitive estimates of measurement error. With the error estimates in hand, the software applies classical dependent and independent statistical tests to confirm differences in gene expression between two samples (e.g., *B. burgdorferi* grown at two different temperatures). ArrayStat performs a $\log_{10}$ transformation of the raw pixel data, determines the common error, and removes outliers. The values are normalized by the mean across conditions by an iterative process and a false positive error correction

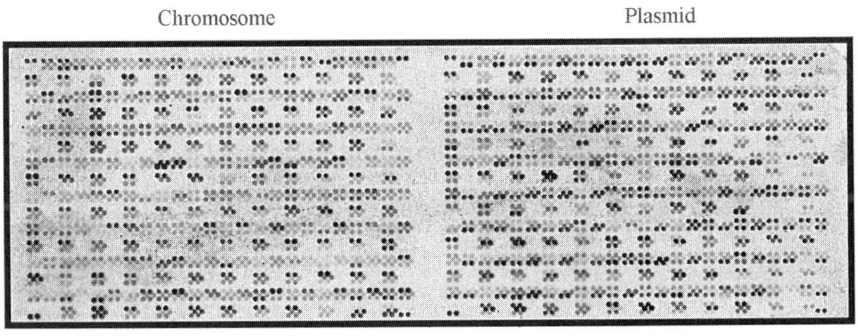

Chromosome   Plasmid

FIG. 2. *B. burgdorferi* gene array. Random hexamers were used for primer-directed synthesis of ^{33}P-labeled DNA probes derived from randomly sheared B31 genomic DNA.

is achieved by application of the false discovery rate method.[21] For our studies, the z-test for two independent conditions using the false discovery rate method with a nominal alpha of $p \leq 0.01$ was used as the basis for computing confidence intervals.

An alternative statistical approach to Arraystat also was performed using macro functions found within the Microsoft Excel spreadsheet program. Briefly, after ArrayVision was used to enumerate a probed membrane array, the raw pixel data was exported into an Excel spreadsheet. First, the raw pixel intensity data was converted into percent intensities, by taking each pixel intensity value per spot and dividing it by the sum of the combined pixel intensity values for all spots and multiplying by 100. After the normalization step, a filter was used to identify and delete all spots that were not found to be statistically different from background. It is our experience that any spot with a pixel intensity less than two standard deviations away from the average background intensity does not have sufficient information to warrant further analysis and is subsequently removed from the data population. The spots remaining after this step were then subjected to traditional descriptive statistics (e.g., mean spot intensity, standard deviation of spot intensity between replicates, expression ratio between conditions tested, log transformation of the expression ratios). To identify spots (i.e., genes) that were differentially regulated between the conditions analyzed, a two-tailed, unpaired Student t-test was performed with significance being placed on spots with probabilities ≤ 0.01.

Differential Gene Expression in B. burgdorferi

Gene expression analysis of *B. burgdorferi* B31 grown at 23° and 35° was carried out. Differences in expression over 3.5 orders of magnitude could be readily discerned and quantitated. At least minimal expression from 83% of the arrayed ORFs could be detected at 35°. A total of 215 ORFs were differentially expressed at the two temperatures; 133 were expressed at significantly greater levels at 35° and 82 were more significantly expressed at 23°. Of these 215 ORFs, 126 are characterized as hypothetical. Thus, a significant portion of the expressed genome of *B. burgdorferi* is represented by these proteins of unknown function. Plasmids encode 137 (64%) of the differentially expressed genes. Of particular interest is linear plasmid ip54 which contains 76 annotated putative genes; 31 of these are differentially expressed at the two temperatures. These findings demonstrate the utility of the arrays in exploring global gene regulation in *B. burgdorferi* and underscore the important role plasmid-encoded genes may play in adjustment of *B. burgdorferi* to growth in diverse environmental conditions.

---

[21] Y. Benjamini and Y. Hochberg, *J. R. Stat. Soc. Ser. B—Methodological* **57**, 289 (1995).

Acknowledgments

We thank Jianmin Zhong for PCR amplification and helpful discussions. James D. Frost III authored the primer design program. This work was supported by Grants AI45801 (I.S.), RR15564 (D.A.), AI37248 (A.B.), AI27044 (I.B.), AI29735 (I.R.), and AI42345 (I.S.) from the National Institutes of Health, from the American Heart Association (003013N to D.A.), and from the G. Harold & Leila Y. Mathers Charitable Foundation (I.S.), and by a fellowship from the Arthritis Foundation, New York Chapter (C.O.).

[13] Transcript Profiling of *Escherichia coli* Using High-Density DNA Microarrays

By STEPHEN K. PICATAGGIO, LORI J. TEMPLETON, DANA R. SMULSKI, and ROBERT A. LAROSSA

Introduction

High-density DNA microarrays represent a powerful tool with which to simultaneously measure the expression of every gene in a cell. By identifying the genes that are expressed differentially in response to either specific mutations or modified environmental conditions, we are provided with a view of global responsiveness that improves our understanding of microbial physiology and homeostasis on a genomic scale. This technology complements other methods that measure the cellular content of polypeptides and small molecules. As a consequence of the power of comprehensive transcript profiling, the focus of research is shifting away from the characterization of individual genes and reactions in a metabolic pathway to the complex interactions among metabolic pathways and the regulatory networks that determine the behavior of the entire cell.

The *Escherichia coli* genome is composed of more than 4.5 megabase pairs of DNA and is predicted to encode 4290 proteins. Even though this organism has been studied extensively for over 50 years and its genome is completely sequenced, the function of about 40% of these genes remains unknown. Furthermore, metabolism in a cell that grows, adapts to changing environmental conditions, and replicates in less than 30 min is necessarily both complex and dynamic. Regulatory control exercised by feedback inhibition, covalent modification of polypeptides, and the frequency by which individual genes are transcribed and translated takes place in short time frames ranging from milliseconds to minutes. However, our ability to study global gene expression in prokaryotes originally lagged behind those in eukaryotes because of the absence of $3'$-mRNA polyadenylated tails and the rapid turnover of bacterial transcripts. With the advent of suitable techniques for rapid RNA isolation and labeling. *E. coli* gene arrays have been used to study the transition between exponential and stationary phase, growth in rich versus minimal media, and responses to carbon source, protein overexpression, chemical exposure, starvation, heat shock, osmotic stress, and various mutations.[1-8] These experiments have demonstrated that global gene expression profiles can be used in a meaningful way to study microbial physiology. A good correlation has been shown to exist between DNA microarray-derived mRNA levels and protein abundance in *E. coli*.[3,9]

There are three types of DNA hybridization arrays that are used commonly to generate gene expression profiles: macroarrays, microarrays, and high-density oligonucleotide arrays. Macroarrays have DNA spotted at relatively low density (approximately 50 spots/cm^2) on a nylon membrane support and employ ^{33}P-labeled cDNA probes. Microarrays have DNA attached at higher density (approximately 250–1000 spots/cm^2) onto a glass surface and employ fluorescently labeled cDNA probes in a competitive hybridization scheme that permits analysis of both control and experimental samples on the same microarray. High-density oligonucleotide arrays (Affymetrix, Santa Clara, CA) provide the highest density available (60,000 spots/cm^2) by using a photolithography-based combinatorial method to synthesize 25-mers directly on a silicon substrate. The fundamentals of these different DNA hybridization arrays have been reviewed elsewhere.[10]

The microarray method is based on the following steps: (1) PCR (polymerase chain reaction) fragments representing each protein-encoding gene in the *E. coli* genome are amplified and mechanically spotted onto treated glass slides at high density. (2) Total RNA is isolated from control and experimental cultures. (3) Labeled cDNA is prepared from each RNA sample using either of two fluorescent dyes with distinct excitation and emission spectra. (4) The differentially labeled cDNAs containing equal amounts of incorporated dye are mixed. (5) The mixed probe is dispensed onto the slide, which is then incubated under conditions that permit competitive hybridization of the labeled probes to their corresponding genes on the high-density microarray. (6) After incubation, the unbound label is

[1] S. E. Chuang, D. L. Daniels, and F. R. Blattner, *J. Bacteriol.* **175,** 2026 (1993).
[2] C. S. Richmond, J. D. Glasner, R. Mau, H. Jin, and F. R. Blattner, *Nucleic Acids Res.* **27,** 3821 (1999).
[3] S. M. Arfin, A. D. Long, E. T. Ito, L. Tolleri, M. M. Riehle, E. S. Paegle, and G.-W. Hatfield, *J. Biol. Chem.* **275,** 29672 (2000).
[4] H. Tao, C. Bausch, C. Richmond, F. R. Blattner, and T. Conway, *J. Bacteriol.* **181,** 6425 (1999).
[5] M.-K. Oh and J. C. Liao, *Biotechnol. Prog.* **16,** 278 (2000).
[6] Y. Wei, J.-M. Lee, C. Richmond, F. R. Blattner, J. A. Rafalski, and R. A. LaRossa, *J. Bacteriol.* **183,** 545 (2001).
[7] A. B. Khodursky, B. J. Peter, N. R. Cozzarelli, D. Botstein, P. O. Brown, and C. Yanofsky, *Proc. Natl. Acad. Sci. U.S.A.* **97,** 12170 (2000).
[8] D. P. Zimmer, E. Soupene, H. L. Lee, V. F. Wendisch, A. B. Khodursky, B. J. Peter, R. A. Bender, and S. Kustu, *Proc. Natl. Acad. Sci. U.S.A.* **97,** 14674 (1998).
[9] Y. Wei, J.-M. Lee, D. R. Smulski, and R. A. LaRossa, *J. Bacteriol.* **183,** 2265 (2001).
[10] W. M. Freeman, D. J. Robertson, and K. E. Vrana, *BioTechniques* **29,** 1042 (2000).

washed away and the microarray is scanned using a confocal laser scanner. Because the DNA bound on the microarray is in vast molar excess compared to the fluorescent probe, the hybridization signals represent the relative level of each gene's expression. By comparing the control and experimental hybridization signals one can identify differentially expressed genes.

As with any new technology, it is important to standardize the methods to generate reproducible and reliable results. We describe here the techniques that we have used successfully for comprehensive transcript profiling of *E. coli* using high-density DNA microarrays.

Array Preparation

Amplification of E. coli Genes for DNA Microarrays

Each of the 4290 predicted open-reading frames (ORFs) making up the *E. coli* genome is amplified twice to prevent contamination of the high-density microarrays with the genomic DNA template used to generate the initial PCR product. In the first round, full-length ORFs are amplified using genomic DNA (30 ng) as the template and gene-specific primer pairs at 0.5 μM (Sigma Genosys, The Woodlands, TX) in a 50-μl PCR containing 2 units of ExTaq polymerase (Panvera, Madison, WI) and dNTPs (Amersham Pharmacia Biotech, Arlington Heights, IL) at 0.2 mM, as described.[6] The PCRs are conducted in 96-well microtiter plates using a thermocycler for 25 cycles of denaturation at 95° for 15 sec, annealing at 64° for 15 sec, and extension at 72° for 1 min. The primary PCR products are then diluted 500-fold and 2 μl of the diluted products is used as the template in a second round of amplifications to eliminate contamination of the amplified ORFs with genomic DNA. Duplicate 50 μl PCRs are performed in the reamplification as described above. The resulting PCR products are analyzed by gel electrophoresis in 1.0% (w/v) agarose (Sigma, St. Louis, MO). We found that greater than 95% of these second-round PCR products were of the predicted size. The reamplified PCR products are purified using a 96-well PCR purification kit (Qiagen, Inc., Valencia, CA). The PCR products are dried using a vacuum centrifuge and resuspended in 20 μl of 5 M Na$_2$SCN to a concentration of $\geq$0.1 ng/nl. The *E. coli* gene library is then transferred into twelve 384-well microtiter plates using a Biomek FX Robot (Beckman Coulter Inc., Fullerton, CA) and stored at $-80°$.

Transfer of PCR Products to High-Density DNA Microarrays

To construct the high-density microarray, we use a Generation III robotic DNA spotter (Molecular Dynamics, Sunnyvale, CA) to transfer the amplified ORFs onto treated glass slides (Molecular Dynamics) at about 55% relative humidity and 23–25°. The Generation III spotter uses a 12-tip pen set to transfer 0.5-nl aliquots of up to 4500 PCR products in duplicate from a set of 384-well microtiter plates onto as many as 36 replicate slides. Consequently, the entire set of *E. coli* ORFs is represented in duplicate on a single slide. After baking at 80° for 2 hr, the microarrays are stored in a microscope slide box in a vacuum desiccator at room temperature.

Cell Sampling Methods

Harvest

Many laboratory strains of *E. coli* have life cycles of approximately 20–30 min. With half-lives ranging from 40 sec to 20 min,[11] the stability of each mRNA plays an important role in the regulation of gene expression beyond transcriptional or translational initiation. Furthermore, mRNAs within a single transcriptional unit can have different functional half-lives.[11] In many *E. coli* strains, the mRNA degradation curves appear biphasic. Consequently, it is essential to recover intact mRNA as soon as possible after the cells are sampled. Many commercial kits for RNA isolation first require bacterial cells to be harvested by centrifugation. Clearly, mRNA degradation can occur during this interval unless potent RNase inhibitors are also included. Otherwise, the resulting measurements may be subject to systematic errors due to differential mRNA stability. To minimize the effect of differential mRNA stability on the outcome of microarray-based measurements, we rapidly freeze cell cultures in liquid nitrogen prior to RNA isolation. Duplicate 5-ml culture samples are collected with a sterile 10-ml syringe and immediately dispensed into a 50-ml conical polypropylene centrifuge tube (Corning, Corning, NY) containing approximately 40 ml of liquid nitrogen in an ice bucket filled with dry ice. The sample should be quickly dispensed as a steady stream while minimizing boiling of the liquid nitrogen as well as the amount of time before the sample is completely frozen. The liquid nitrogen is allowed to boil off before the cap is sealed and then the tube is immediately stored at −80°. A properly collected sample should appear as a frozen "stalactite." Alternatively, for experiments run at low cell density shaved ice is added to 50-ml culture samples and the cells are harvested immediately by centrifugation at 10,000g for 2 min at 4° prior to RNA isolation.[6]

RNA Isolation from Frozen Cultures

Tubes containing frozen samples are placed in an ice bucket filled with dry ice. A small amount of frozen sample (approximately 1 g) is removed with a single-edged razor blade and placed into a prechilled coffee grinder containing two equal-sized pieces of dry ice. The sample is ground for approximately 30 sec to a fine powder. Additional dry ice is added as necessary to keep the sample frozen. The ground sample is immediately transferred to a 50-ml conical centrifuge tube containing 15 ml of RNA extraction buffer [0.2 M Tris, 0.25 M NaCl, 50 mM

---

[11] S. Petersen, S. Rech, and J. D. Friesen, *Mol. Gen. Genet.* **166,** 329 (1978).

EGTA, 0.3 M p-aminosalicylic acid (Sigma), and 20 mM triisopropylnaphthalene-sulfonic acid (Acros Organics, NJ)], mixed briefly, and then poured into another 50-ml conical centrifuge tube containing 15 ml phenol : chloroform : isoamyl alcohol (25 : 24 : 1, v/v). The sample is vortexed vigorously for 1 min and placed on ice. Samples are centrifuged at 2000g in a swinging bucket centrifuge for 10 min at 4°. The aqueous phase is collected and similarly extracted twice more. The interface containing cell debris and protein should be avoided. RNA is precipitated overnight at $-20°$ after the addition of 0.1 volume 3 M sodium acetate (pH 4.8) and 2.5 volumes ethanol. The RNA is treated with 10 units of RNase-free DNase I (Gibco-BRL, Gaithersburg, MD) in a 100 μl reaction mixture containing 20 mM Tris-HCl (pH 8.3), 2 mM MgCl$_2$, 50 mM KCl for 1 hr at 37°. DNase I is inactivated by heating at 80° for 10 min. The RNA is extracted twice with equal volumes of phenol : chloroform : isoamyl alcohol (25 : 24 : 1, v/v), precipitated with 2.5 volumes of ethanol, and purified using a RNA Easy Kit (Qiagen, Inc.) according to the manufacturer's instructions. Typically, 30–75 μg of purified total RNA is recovered from each sample.

RNA Isolation from Cells Collected by Centrifugation

The cell pellet is resuspended in an ice-cold mixture containing 100 μl of Tris-HCl (10 mM, pH 8.0) and 350 μl of 2-mercaptoethanol supplemented RLT buffer (Qiagen, Inc., RNeasy Mini Kit). The cell suspension is immediately transferred to a chilled 2-ml microcentrifuge tube containing 100 μl 0.1 mm zirconia/silica beads (Biospec Product Inc., Bartlesville, OK). The cells are broken by vigorous agitation at room temperature for 25 sec with a Mini-Beadbeater (Biospec Products Inc.). Debris is removed by centrifugation at 16,000g for 3 min at 4° and the resultant supernatant is mixed with 250 μl of ethanol. This mixture is loaded onto RNeasy columns (Qiagen, Inc.) and RNA is isolated according to the manufacturer's instructions, with a single exception. Residual genomic DNA is digested by using the RNase-free DNase Set (Qiagen, Inc.). After the first (RW1) wash step, 10 μl of DNAse (27 Kunitz units) and 70 μl of RDD Buffer are mixed; the mixture is added directly onto the column and incubated at room temperature for 15 min. The Qiagen RNeasy isolation protocol is completed as directed, continuing from the second 350 μl (RW1) wash. The RNA is eluted from the column in 50 μl of RNAse-free water and stored frozen at $-20°$ until use.

Probe Preparation

Fluorescent cDNA Synthesis from Total RNA

Global gene expression profiling in prokaryotes lagged behind its counterpart in eukaryotes in part because of the absence of 3′-mRNA polyadenylated tails on which to prime reverse transcription of cDNA with oligo (dT). Furthermore, the currently available set of 4290 unique 3'-ORF-specific primers do not hybridize at equal efficiency and their complementary sequences show widely different degradation rates. Although priming cDNA synthesis with random hexamers suffers from the drawback that much of the label is incorporated into more abundant rRNA and tRNA species, others have shown that priming cDNA synthesis with random hexamers rather than 3'-ORF-specific primers is required for accurate measurement of gene expression levels in bacteria.[3] Total RNA (6 μg) is first annealed to 12 μg of random hexamers (Gibco-BRL) in 22 μl for 10 min at 70°, and then for an additional 10 min at room temperature. The RNA is then labeled with 0.05 mM, Cy3-dCTP or Cy5-dCTP (Amersham Pharmacia Biotech) in a 40 μl reaction containing 0.4 units of Superscript II-Reverse Transcriptase (Gibco-BRL), 50 mM Tris-HCl (pH 8.3), 75 mM KCl, 3 mM MgCl$_2$, 0.1 mM each of dATP, dGTP, dTTP, 0.05 mM of dCTP, and 0.01 M dithiothreitol (DTT) at 42° for 1 hr. All manipulations are conducted henceforth in amber microcentrifuge tubes under dim light because of the photosensitivity of Cy3 and Cy5 dyes. cDNA synthesis is terminated by heating at 95° for 5 min. Residual RNA is hydrolyzed at 37° for 10 min following the addition of 2 μl of 5 M NaOH. The reaction is then neutralized with 3 μl of 5 M HCl and 5 μl of 1 M Tris-HCl (pH 7.2). Labeled cDNA is purified to remove unincorporated nucleotides using a PCR Purification Kit (Qiagen, Inc.) according to manufacturer's instructions with the exception that the bound cDNA is washed three times with PE buffer before elution with 60 μl EB buffer.

Synthesis of Fluorescent Genomic DNA

Genomic DNA is isolated from *E. coli* MG1655 by a standard procedure[12] and approximately 1.5 μg is digested in a 50 μl reaction with the restriction endonuclease *Dpn*II (New England Biolabs, Beverly, MA) for 2 hr at 37°. The resultant DNA fragments are purified using a PCR Purification Kit (Qiagen, Inc.) and mixed with 6 μg of random hexamers (Operon Technologies, Inc., Alameda, CA) in a total volume of 24 μl. The DNA is denatured by heating at 94° for 10 min prior to annealing on ice for 10 min. Fluorescent labeling of genomic DNA (0.75 μg) is accomplished in a 60 μl reaction containing 50 mM Tris-HCl (pH 7.2), 10 mM MgSO$_4$, 0.1 mM dithiothreitol, 5 μM each of dATP, dGTP, dTTP, 2.5 μM dCTP, 50 μM of either Cy3-dCTP or Cy5-dCTP (Amersham Pharmacia Biotech), and 15 units of the Klenow fragment of DNA polymerase I (Promega) for 2.5 hr at room temperature. The labeled DNA probe is subsequently purified twice using a PCR Purification Kit (Qiagen, Inc.) before drying in a speed vacuum.

[12] T. K. Van Dyk and R. A. Rossen, *in* "Methods in Molecular Biology: Bioluminescence Methods and Protocols," Vol. 102 (R. A. LaRossa, ed.), p. 85. Humana Press, Inc., Totowa, NJ, 1998.

Hybridization and Washing Conditions

One of the key advantages to the use of cDNA-based technology is that the large DNA fragments on the microarray allow for high-stringency hybridization. Arrayed slides are placed in 2-propanol for 10 min, then in boiling ultrapure water for 5 min, and are dried under a stream of nitrogen gas prior to prehybridization. The slides are prehybridized in 3.5× SSC (Gibco-BRL), 0.2% sodium dodecyl sulfate (SDS) (Gibco-BRL), and 1% bovine serum albumin (Sigma) for 20 min at 60° to block nonspecific binding of the probe. The slides are then rinsed five times in distilled, deionized water at room temperature and twice in 2-propanol before drying under a nitrogen stream. Control and experimental cDNA probes equally labeled with 30–100 pmol of Cy dye at an efficiency of about 0.4 Cy dye/dCTP incorporated, eqivalent to 0.1 Cy dye/nucleotide of cDNA, are mixed, dried down, and resuspended in 30 μl of hybridization solution containing 5× SSC, 0.1% SDS, 50% formamide (Sigma), and 200 μg/ml denatured salmon sperm DNA (Sigma). The probes are denatured for 5 min at 95° and briefly spun in a microcentrifuge. As illustrated in Fig. 1, the mixed probes are added to the array by capillary action under a coverslip (Corning) and incubated overnight at 37° in a sealed hybridization chamber containing a small reservoir of water to maintain humidity. After hybridization, the coverslips are removed by gentle shaking in 2× SSC, 0.1% SDS buffer at 42° and then the slides are washed once in 2× SSC, 0.1% SDS buffer for 5 min at 42°, once in 0.1× SSC buffer for 5 min at room temperature, and three times at room temperature in 0.1× SSC buffer for 2 min. Once the coverslip has been removed, the remaining washes are performed in opaque containers (e.g., 50-ml conical tubes wrapped in aluminum foil) to prevent bleaching of light-sensitive Cy dyes. The slides are then dried under a stream of nitrogen gas and transported in opaque slide boxes until scanned.

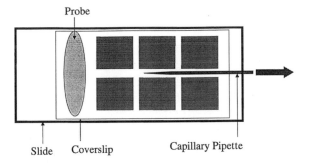

FIG. 1. Method for applying probe to DNA microarray. A slide is placed on Kimwipe paper. Equal amounts of Cy3- and Cy5-labeled hybridization probes are mixed and applied near the top of the slide. A tapered glass capillary pipette is placed at the bottom of the slide. A coverslip is first applied at the top of the slide just above the liquid drop containing the probe. Then the capillary pipette (VWR, West Chester, PA) is slowly removed from under the coverslip to disperse the probe by capillary action.

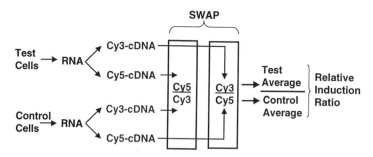

FIG. 2. Experimental strategy. The Cy dyes used to label the control and experimental probes are "swapped" on duplicate slides and the four normalized control signals (from duplicate spots of each PCR product on the two slides) and the four normalized experimental signals are separately averaged and used to determine the relative induction ratio for each gene.

Correcting for Cy Dye Bias

The Cy dye used to label the control and experimental cDNA probes can skew the results.[6] This dye bias can be observed on log–log scatter plots of the normalized signal intensities (see below) from control and experimental samples. Dye bias is visualized as a displacement in the distribution of the data scatter away from the diagonal toward either the abscissa (x axis) or ordinate (y axis), especially at low signal intensities. To minimize this effect, each dye is used to separately label the control and experimental probes. The probes are "swapped" on duplicate slides and the four normalized control signals (from duplicate spots on each of the two slides) and the four normalized experimental signals are separately averaged and then used to determine the relative induction ratio for each gene, as illustrated in Fig. 2. Averaging the "swapped" normalized signals from each transcript minimized skewing and decreased data scatter.[6] We routinely run replicate "swaps" on the same RNA sample to evaluate data quality.

Data Acquisition

Hybridization is quantified with a confocal laser scanner (Molecular Dynamics) that differentiates the Cy3 and Cy5 fluorescent signals and produces separate TIF images. Cy5 signals are measured following excitation at 633 nm with a 675B emission filter before Cy3 signals are measured following excitation at 532 nm with a 575B emission filter. The pixel density for each spot is measured with Array Vision 4.0 software (Imaging Research, Inc., Ontario, Canada). The local mean background of adjoining, nonspotted regions of the slide (N_j) is subtracted from the fluorescence intensity (I_j) associated with each spotted gene (j). These readings ($R_j = I_j - N_j$) are then normalized to the median of all background-subtracted signals on the slide, to permit dye-to-dye and slide-to-slide comparisons. Each gene is present in duplicate on each slide and the average of at least eight data points (duplicate Cy3 signals and duplicate Cy5 signals on each of two "swapped" slides) is used to determine the relative induction ratio for each gene. These data are exported to Excel (Microsoft) spreadsheets for further data manipulation and analyses.

Data Analysis

Even a simple experiment using replicate "swaps" generates an enormous amount of data. Manual compilation of this data is tedious, time consuming, and prone to error. We have developed a Gene Array Data Analysis program that acquires the normalized signals in Excel spreadsheets, averages normalized control and experimental signals from replicate "swaps," calculates the relative induction ratio for each gene, runs statistical analysis of the data, sorts the data by relative induction ratio, and identifies induced or repressed genes according to user-defined confidence limits. A scatter plot, which compares the normalized transcript levels for each gene under control and experimental conditions, is then generated to visualize the global expression changes that have occurred within the cell. As illustrated in Fig. 3 with GeneSpring software (Silicon Genetics, Redwood City, CA), most genes typically cluster along a diagonal line which passes through the origin of the scatter plot and represents a relative induction ratio of 1.0, meaning that their expression is unaffected by the conditions tested. We considered genes with relative induction ratios greater or less than the 99.5% confidence limits based on the standard deviation of all ratios to be significantly induced or repressed, respectively. These data points are clearly observed outside of the clustered region on the scatter plot.

The signals from a single microarray are subject to some variability. Potential sources of variation may include slide chemistry, spotting consistency, cell sampling, RNA isolation, cDNA labeling efficiency, hybridization efficiency, dye-specific bleaching, image processing, and the discrepancy between the fluorescence from different Cy dyes and the mRNA level. More reliable gene expression data can be obtained by pooling the data from multiple replicates, resulting in fewer false positives and negatives.[13] Another advantage is that the use of multiple replicates reduces the data scatter, particularly at low signal intensities.

The noise present in high-complexity DNA microarray hybridization also requires rigorous statistical analysis to determine the quality of the results. The significance of the differential gene expression measurements cannot be assessed simply by the magnitude of the relative induction ratio derived from control and experimental conditions.[3] For example, T-test p values for genes with small relative expression ratios can be much lower than for genes with larger expression ratios.

[13] M.-L. T. Lee, F. C. Kuo, G. A. Whitmore, and J. Sklar, *Proc. Natl. Acad. Sci. U.S.A.* **97,** 9834 (2000).

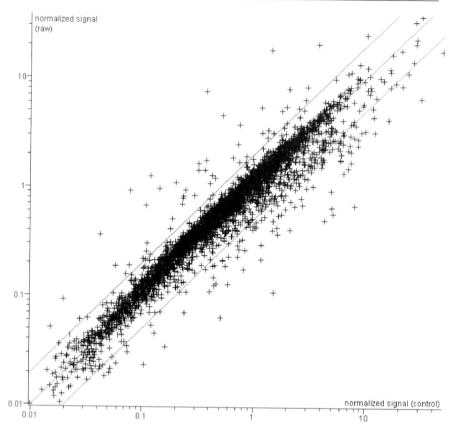

FIG. 3. Typical scatter plot of DNA microarray data. Note that the use of multiple replicates results in narrow data scatter even at very low signal intensities. The central line represents an induction ratio of 1 while the flanking lines indicate the 99.5% confidence limits.

Thus, it is inaccurate to classify genes as differentially expressed solely on the basis on an arbitrarily chosen relative induction ratio, i.e., 2-fold. A linear analysis of variance (ANOVA) for a statistically designed experiment using the micro-array protocols described here has shown the variance between replicate "swaps" on the same RNA sample, including all steps from cDNA labeling through data acquisition, to be statistically insignificant (results not shown). However, RNA isolation was found to be the single largest source of variation, contributing up to 80% of the total variance between replicate RNA preparations from the same cell sample. Others using macroarrays have observed that this variance accounted for an approximately 2.5-fold greater false-positive rate than that expected by chance alone, and could be minimized by averaging the data across RNA preparations.[3] We recommend averaging the relative induction ratios from independent RNA samples. The ANOVA may also be used to confirm that the data used to classify genes as induced or repressed is reliable and reproducible at greater than the 99.5% confidence interval.

Another way to analyze gene expression data is with pattern-recognition hierarchical cluster analyses that use standard statistical algorithms to identify genes with similar expression patterns, regardless of the level of their expression.[14,15] Using this approach, genes can be clustered into discrete sets that appear to reflect related functions or common regulatory networks. These analyses are especially valuable when comparing gene expression profiles from different experiments. Such methodology has been used to increase the number of known *E. coli* stimulons by categorizing ORFs with no previously assigned function into regulatory and motif classes.[14]

RNA Abundance

To convert normalized equivalent readings (ER) into measures of transcript abundance (AB), a further correction based on the hybridization signal arising from an equimolar concentration of all transcripts is needed. The surrogate for this correction factor is the fluorescence intensities generated from hybridization with fluorescent copies of genomic DNA.[6] The fluorescence intensities from hybridization with RNA-derived probes are corrected using fluorescent intensities arising from genomic DNA-derived probes. The abundance of each transcript is determined by dividing the normalized equivalent reading of a genomic DNA-derived sample into the normalized equivalent reading from the RNA-derived sample ($AB = \text{norm } ER_{\text{transcripts}}/\text{norm } ER_{\text{genome}}$).

Validation of DNA Microarray Results

The array methods described here provide a global view of gene expression in response to genetic or environmental changes. Other methods, including real-time RT-PCR, primer extension, Northern hybridization, and transcriptional gene fusions, can be used to measure transcriptional changes. Studies have been performed that demonstrate agreement between microarray and primer extension results for the *oxyR* regulon,[16] between microarray and RT-PCR results for identification of the *sdiA* regulon,[9] and between microarray and *lux* fusion results for identification of the mitomycin C stimulon as well as the pleiotropy of a *rpoC* mutation.[17] These alternative methods also provide a means to identify potential false positives from microarray results.

[14] M. B. Eisen, P. T. Spellman, P. O. Brown, and D. Botstein, *Proc. Natl. Acad. Sci. U.S.A.* **95,** 14863 (1998).

[15] S. Tavaoie, J. D. Hughes, M. J. Campbell, R. J. Cho, and G. M. Church, *Nat. Genet.* **22,** 281 (1999).

[16] M. Zheng, X. Wang, L. J. Templeton, D. R. Smulski, R. A. LaRossa, and G. Storz, *J. Bacteriol.* **183,** 4562 (2001).

[17] T. K. VanDyk, Y. Wei, M. K. Hanafey, M. Dolan, M. G. Reeve, J. A. Rafalski, L. B. Rothman-Denes, and R. A. LaRossa, *Proc. Natl. Acad. Sci. U.S.A.* **98,** 2555 (2001).

Conclusion

The "genomics revolution" has opened a new dimension of microbial physiology. With 36 microbial genomes sequenced and more than 100 others in progress, a focus of research is shifting away from the individual gene and reactions in a metabolic pathway to the complex interactions among metabolic pathways and the regulatory networks that determine the behavior of the entire cell. Yet, translating this enormous wealth of sequences into meaningful functional information represents a great challenge facing the biological sciences today. High-density gene arrays are now providing important insights into gene function and microbial physiology and are fast becoming an essential tool to link information from genomics, bioinformatics, proteomics, metabolite analyses, and flux analyses into a cohesive "snapshot" of cellular metabolism.

[14] *Neisseria* Microarrays

By Colin R. Tinsley, Agnès Perrin, Elise Borezée,
and Xavier Nassif

Introduction

The sequencing of microbial genomes opens up new avenues to the understanding of bacterial pathogenesis. The genome sequence provides the organization of thousands of genes, both previously confirmed and putative. However, in addition to its unprecedented size and complexity, this wealth of new data poses many more questions than it answers. For the two sequenced strains of *Neisseria meningitidis,* Z2491[1] and MC58,[2] 30 and 25% of the gene products are assigned no known function. Because a weak homology with a protein of demonstrated function from another bacterium does not necessarily mean that the protein will perform the same role in *N. meningitidis,* a considerable proportion of the genome

[1] J. Parkhill, M. Achtman, K. D. James, S. D. Bentley, C. Churcher, S. R. Klee, G. Morelli, D. Basham, D. Brown, T. Chillingworth, R. M. Davies, P. Davis, K. Devlin, T. Feltwell, N. Hamlin, S. Holroyd, K. Jagels, S. Leather, S. Moule, K. Mungall, M. A. Quail, M. A. Rajandream, K. M. Rutherford, M. Simmonds, J. Skelton, S. Whitehead, B. G. Spratt, and B. G. Barrell, *Nature* **404,** 502 (2000).
[2] H. Tettelin, N. J. Saunders, J. Heidelberg, A. C. Jeffries, K. E. Nelson, J. A. Eisen, K. A. Ketchum, D. W. Hood, J. F. Peden, R. J. Dodson, W. C Nelson, M. L. Gwinn, R. DeBoy, J. D. Peterson, E. K. Hickey, D. H. Haft, S. L. Salzberg, O. White, R. D. Fleischmann, B. A. Dougherty, T. Mason, A. Ciecko, D. S. Parksey, E. Blair, H. Cittone, E. B. Clark, M. D. Cotton, T. R. Utterback, H. Khouri, H. Qin, J. Vamathevan, J. Gill, V. Scarlato, V. Masignani, M. Pizza, G. Grandi, L. Sun, H. O. Smith, C. M. Fraser, E. R. Moxon, R. Rappuoli, and J. C. Venter, *Science* **287,** 1809 (2000).

remains to be deciphered. Furthermore, the interactions of these gene products in global regulatory networks, the relays underlying the bacterial response to changes in its environment during the disease process, are for the most part very poorly understood.

The results of genome sequencing have underscored the large number of genes possessed even by "simple" organisms. There are too many open reading frames and potential regulatory interactions for comprehensive analysis by traditional molecular biological methods. Hence new analytical tools are needed to allow the scientist to take advantage of the new data. DNA array technology has grown in response to this need, and also depends on the sequence data itself. The concept was developed earlier, however. For example, in 1993, Chuang and co-workers[3] probed an array of some 400 overlapping lambda clones covering the *Escherichia coli* chromosome, bound to nylon membranes, with ^{32}P-labeled cDNA to reveal changes in expression under different growth conditions. Since then, genome sequence data have permitted the development of arrays of thousands of PCR products corresponding to the entire chromosomal complement of genes.

The underlying principle (and indeed the experimental method) is essentially that of Southern blotting,[4] with the advantage that an ordered array of DNA fragments attached to a solid support, each one corresponding to defined positions on the chromosome allows the direct relation of a signal to its corresponding gene. In addition it is possible to analyze simultaneously the relative concentrations of each component in a complex mixture of nucleic acids (e.g., chromosomal DNA, or total RNA representing the transcriptional response of the organism—the transcriptome). It is this parallel analysis of the entire genetic complement of an organism which is the strength of the technique, particularly in the case of RNA expression analysis. DNA arrays provide information on the simultaneous changes in transcription levels of each of the genes of an organism. In 1997 DeRisi and co-workers[5] demonstrated coordinate changes in gene expression that accompany the shift from fermentation (anaerobic glucose utilization) to respiration (aerobic ethanol utilization) in yeast. Although functional molecules are usually proteins rather than RNA, which may be translated at different rates into proteins of which half lives may vary depending on the state of the cell, limited studies in *Escherichia coli*[6] and in *Saccharomyces cerevisiae*[7] have shown a reasonably good correlation between mRNA and protein levels.

---

[3] S. E. Chuang, D. L. Daniels, and F. R. Blattner, *J. Bacteriol.* **175,** 2026 (1993).

[4] E. M. Southern, *J. Mol. Biol.* **98,** 503 (1975).

[5] J. L. DeRisi, V. R. Iyer, and P. O. Brown, *Science* **278,** 680 (1997).

[6] S. M. Arfin, A. D. Long, E. T. Ito, L. Tolleri, M. M. Riehle, E. S. Paegle, and G. W. Hatfield, *J. Biol. Chem.* **275,** 29672 (2000).

[7] B. Futcher, G. I. Latter, P. Monardo, C. S. McLaughlin, and J. I. Garrels, *Mol. Cell. Biol.* **19,** 7357 (1999).

Another use of DNA arrays, which lends itself to analysis of pathogenicity in *Neisseria,* is genome comparison. Here the arrays are used to test the hybridization of the chromosomal DNA of a test strain with each gene of a reference strain. The presence or absence of each gene in the test strain is read directly from the array results. In the case of an organism whose genome has been entirely sequenced, this method is a simpler and more comprehensive approach than subtractive hybridization methods[8,9] for the identification of sequence differences between closely related organisms. By testing a panel of strains or species, associations may be revealed between the presence or absence of genes or chromosomal regions and particular phenotypes (e.g., the ability of *Neisseria meningitidis* to cause meningitis in contradistinction to other commensal species). Hence new insights into pathogenicity may be revealed.

Microarray Formats

The DNA array consists of an ordered series of "spots" of DNA corresponding to defined positions on the chromosome, bound to a solid support. This support may be a nylon membrane of dimensions of the order of 10 cm or treated glass microscope slides. The choice of support will determine the techniques used for genomic or transcriptome comparison.

In general DNA arrays on nylon membranes are hybridized with ^{33}P-labeled nucleic acids. The hybridization signal is read using a phosphorimager which produces a computer-readable image of the spot intensity (Fig. 1) and the intensity of each spot is then determined by computer analysis of the image. The presence or absence of a gene in a test strain or the change in mRNA expression in response to a given stimulus is measured by comparing the intensity of the same spot on membranes hybridized with nucleic acid prepared from the test and control bacteria.

For DNA arrays on glass slides[10,11] the two nucleic acid preparations to be compared are labeled with different fluorophores (usually Cy3 and Cy5), mixed in same amounts and reacted with the arrays. Laser confocal microscopy is used to determine the relative intensity of hybridization of the two samples with each spot, and hence the relative abundance of the corresponding nucleic acid species in the samples. DNA arrays printed on glass slides do save some time in that they are directly read without prior exposure to the phosphorimager screen, and the miniaturization facilitates parallel investigations of multiple conditions. However,

---

[8] N. Lisitsyn, N. Lisitsyn, and M. Wigler, *Science* **259,** 946 (1993).

[9] C. R. Tinsley and X. Nassif, *Proc. Natl. Acad. Sci. U.S.A.* **93,** 11109 (1996).

[10] M. Schena, D. Shalon, R. W. Davis, and P. O. Brown, *Science* **270,** 467 (1995).

[11] M. A. Behr, M. A. Wilson, W. P. Gill, H. Salamon, G. K. Schoolnik, S. Rane, and P. M. Small, *Science* **284,** 1520 (1999).

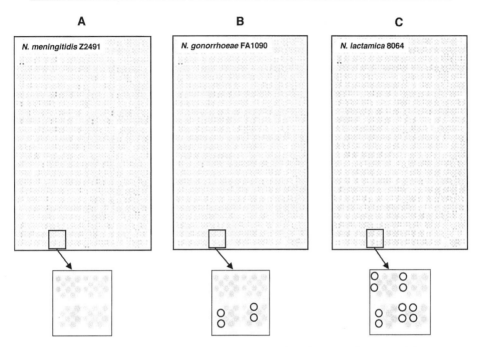

FIG. 1. DNA arrays on nylon membranes. Membranes (8 × 11 cm) are shown containing spots (5 × 5 cells gridded in a 384-well microtiter format) corresponding to each gene of *N. meningitidis* Z2491 in duplicate. Membranes were reacted respectively with DNA from the reference strain Z2491, from *N. gonorrhoeae,* and from *N. lactamica.* A selected region of each hybridization is shown magnified, with nonreacting amplicons circled. Hence genes absent from both *N. gonorrhoeae* and *N. lactamica* are potentially meningococcus-specific, whereas those absent from *N. lactamica* but present in *N. gonorrhoeae* are pathogen-associated.

nylon membranes are more sensitive when the quantity of RNA is limited,[12] as is the case in the study of the transcriptional response of *Neisseria* in biological experiments in which large batch cultures are not feasible. Otherwise the quality of the results seems to be similar. In a comparison of nylon membrane- versus glass slide-based arrays, Richmond and co-workers[13] reported that the glass slides gave more reproducible results and recommended that membranes be hybridized sequentially with the samples to be compared to minimize intermembrane differences. On the other hand, Arfin and co-workers[6] found that the variations between membranes had much less effect on reproducibility than did fluctuations in RNA

[12] F. Bertucci, K. Bernard, B. Loriod, Y. C. Chang, S. Granjeaud, D. Birnbaum, C. Nguyen, K. Peck, and B. R. Jordan, *Hum. Mol. Genet.* **8,** 1715 (1999).

[13] C. S. Richmond, J. D. Glasner, R. Mau, H. Jin, and F. R. Blattner, *Nucleic Acids Res.* **27,** 3821 (1999).

levels in different preparations from cells grown under the same conditions. The production and use of arrays bound to glass slides have been reviewed elsewhere (see, for example, Ref. 14) and this article will concentrate on the use of genomic DNA arrays on nylon membranes.

Preparation of Arrays

Preparation of Target DNA

Several strategies for the design of the target DNA "spots" of the array have been used. The use of PCR-amplified M13 clones generated during a sequencing project[15] bypasses the need for a full annotation of the genome. However, for sequenced and annotated genomes such as *Escherichia coli, Bacillus subtilis, Helicobacter pylori,* and *Neisseria meningitidis,* it is possible to design oligonucleotides to amplify the predicted genes. Another strategy is to use arrays of oligonucleotides bound to glass slides, 20 to 30 bases long, corresponding to portions of the open reading frames, as has been described for yeast,[16,17] *E. coli,*[18] and *Streptococcus pneumoniae.*[19] For nylon membrane arrays of *Neisseria meningitidis* we have used PCR-amplified DNA targets (amplicons) based on the raw sequence data and, after annotation of the genomes, based on the predicted genes. The former strategy has the advantage of circumventing possible errors in the annotation of the genomes, which would give misleading results. Furthermore, there is an advantage in providing target DNAs of a uniform size, in that the smaller ORFs (open reading frames) would give low signals which are difficult to interpret. However, there will be a loss of precision where two genes are present on the same amplicon.

In general we have used DNA arrays based on the unannotated sequence for genome comparison in the search for relatively large species and strain-specific regions, and ORF-based arrays for transcriptional analysis, taking advantage of the genome annotations to design oligonucleotide primers to amplify each predicted gene. Resources for publicly available pathogenic *Neisseria* genome sequence information are listed in Table I.

[14] M. B. Eisen and P. O. Brown, *Methods Enzymol.* **303,** 179 (1999).

[15] R. Rimini, B. Jansson, G. Feger, T. C. Roberts, M. de Francesco, A. Gozzi, F. Faggioni, E. Domenici, D. M. Wallace, N. Frandsen, and A. Polissi, *Mol. Microbiol.* **36,** 1279 (2000).

[16] D. J. Lockhart, H. Dong, M. C. Byrne, M. T. Follettie, M. V. Gallo, M. S. Chee, M. Mittmann, C. Wang, M. Kobayashi, H. Horton, and E. L. Brown., *Nat. Biotechnol.* **14,** 1675 (1996).

[17] L. Wodicka, H. Dong, M. Mittmann, M. H. Ho, and D. J. Lockhart, *Nat. Biotechnol.* **15,** 1359 (1997).

[18] D. W. Selinger, K. J. Cheung, R. Mei, E. M. Johansson, C. S. Richmond, F. R. Blattner, D. J. Lockhart, and G. M. Church, *Nat. Biotechnol.* **18,** 1262 (2000).

[19] R. Hakenbeck, N. Balmelle, B. Weber, C. Gardes, W. Keck, and A. de Saizieu, *Infect. Immun.* **69,** 2477 (2001).

TABLE I
SOURCES OF DATA FOR PATHOGENIC *Neisseria* GENOME SEQUENCING PROJECTS

Species	Strain	Complete	Annotated	Sequencing center	Internet access
N. gonorrhoeae	FA1090	+	+	University of Oklahoma	www.genome.ou.edu/gono.html
N. meningitidis	Z2491	+	+	The Sanger Centre, UK	www.sanger.ac.uk/Projects/N_meningitidis
N. meningitidis	MC58	+	+	The Institute for Genomic Research/Chiron Corporation	www.tigr.org
N. meningitidis	FAM18	+	−	The Sanger Centre, UK	www.sanger.ac.uk/Projects/N_meningitidis

The similarity of the 5′ ends of many genes is not generally a problem in terms of specificity of amplification. Hence oligonucleotide primers can be designed to amplify entire genes or fragments of larger ones. In order to produce roughly similar quantities of the amplified product corresponding to each gene, amplicons should not be much longer than 1.5 kb. Several programs are available for the design of primers.[20–22] The pathogenic *Neisseria* species do, however, pose particular problems in the choice of regions for amplification. The genomes contain a large number of multiply repeated elements,[1] ranging from the ubiquitous 10-base pair (bp) DNA uptake sequence[23] (GCCGTCTGAA; about 1 copy per kilobase) to the Correia sequences[24] (a family of elements, generally about 150 bp in length, terminating in 26 bp partial inverted repeats, and present about 200 times in the chromosome), several multicopy insertion sequences, and gene and chromosomal duplications.[1,2,25] There are gene families such as that of the surface-exposed Opa proteins, which consists of several (three in Z2491, four in MC58) variant genes having conserved ends between which are inserted "cassettes" of variable sequence.[26,27] The pilin genes are another family in which there is a single transcribed gene, *pilE,* and a number of silent, truncated genes, *pilS* (eight in the sequenced meningococci), which donate their sequence information to *pilE*

[20] www-genome.wi.mit.edu/cgi-bin/primer/primer3_www.cgi

[21] promoter.ics.uci.edu/Primers/ORF.htm

[22] G. Raddatz, M. Dehio, T. F. Meyer, and C. Dehio, *Bioinformatics* **17,** 98 (2001).

[23] S. D. Goodman and J. J. Scocca, *Proc. Natl. Acad. Sci. U.S.A.* **85,** 6982 (1988).

[24] F. F. Correia, S. Inouye, and M. Inouye, *J. Biol. Chem.* **263,** 12194 (1988).

[25] N. J. Saunders, A. C. Jeffries, J. F. Peden, D. W. Hood, H. Tettelin, R. Rappuoli, and E. R. Moxon, *Mol. Microbiol.* **37,** 207 (2000).

[26] A. Stern, M. Brown, P. Nickel, and T. F. Meyer, *Cell* **47,** 61 (1986).

[27] T. D. Connell, W. J. Black, T. H. Kawula, D. S. Barritt, J. A. Dempsey, K. Kverneland, Jr., A. Stephenson, B. S. Schepart, G. L. Murphy, and J. G. Cannon, *Mol. Microbiol.* **2,** 227 (1988).

in another relatively frequent genetic rearrangement.[28] Furthermore, some genes appear to be associated with cassettes of similar but not identical sequence downstream that could by recombination change the coding information in their 3' ends, e.g., *fhaB* and *mafB*. These repeated sequences may cause problems in the PCR reactions (one must avoid primers containing uptake or Correia sequences) and/or in interpretation of the hybridization results. If it is decided to include insertion sequences and gene families in the array, they should be placed separate from the main array to avoid their potentially high intensities interfering with the reading of adjacent spots. This also applies to the ribosomal RNA genes, which are present four times in *Neisseria,* as they indeed exist in multiple copies in most bacteria.

The amplicons are produced by PCR using each of the primer pairs and chromosomal DNA of the reference strain (see Table I) as template. The optimum quantity of template DNA to use per reaction, in order to give sufficient product and at the same time to avoid background reaction with residual chromosome bound to the membranes, should first be determined empirically. For the meningococcus, with a chromosome of 2.2 Mb, we have used 5 ng of DNA in each 50 μl. In each case, the quantity and quality of the PCR products should be controlled by agarose gel electrophoresis. There should be a good quantity (between 200 and 300 ng/μl) of a single band of the correct size. Failed PCR reactions should be repeated under different conditions (e.g., annealing temperatures, magnesium concentrations) or with redesigned primers. In general one can expect to achieve a success rate of about 97%. [13,29] Typical PCR conditions per 50 μl reaction are as follows: 10 mM Tris-HCl (pH 8.5 at 25°), 50 mM KCl, 1.8 mM MgCl$_2$. 0.1 μM both primers, 5 ng template chromosomal DNA, 50 μM each of the four deoxynucleotide triphosphates, 0.25 units thermostable polymerase. Amplification is carried out for 35 cycles of 95° for 1 min, 50° for 30 sec, 72° for 1 min, followed by one cycle of 72° for 7 min. Note that some enzymes may require a preincubation at 95° for activation.

Printing of Arrays

The PCR products are precipitated and redissolved to equalize their concentrations, or more simply spotted directly onto positively charged nylon membranes. This step is performed by a robot. A single nylon membrane (8 × 11 cm) is sufficient to contain PCR products corresponding to each of the about 2200 meningococcal genes in duplicate in addition to positive and negative controls. For loading, the membranes are prewetted in denaturing solution (0.5 M NaOH, 100 mM NaCl) and 100 nl of the raw PCR products is spotted robotically as five successive volumes of 20 nl. The procedure up to this point, and in particular the printing, may be carried out by any number of commercial enterprises offering these services,

[28] P. Hagblom, E. Segal, E. Billyard, and M. So, *Nature* **315,** 156 (1985).
[29] A. Perrin, D. Talibi, X. Nassif, and C. R. Tinsley, unpublished results.

or a robotic microarraying device could be purchased. After spotting, the arrays are neutralized by washing in water and then in phosphate-buffered saline (PBS) (the DNA may be cross-linked to the nylon by exposure to ultraviolet radiation, but this is not necessary with positively charged nylon membranes. The arrays are stored dry. The choice of control spots is determined by the investigator. Positive controls might be a series of dilutions of the total chromosome, or of the ribosomal RNA genes. Care should be exercised in the interpretation of these spots (see subsequent section on normalization), but a considered placement of positive control spots on the membrane will help in the alignment of the grid necessary for the computer software to identify the spots. Negative controls can be chosen from among antibiotic resistance markers, which might then serve as positive controls for deliberately mutated strains, or eukaryotic genes such as β-globin (see section on normalization).

A useful control for the quality of the array is to probe with radiolabeled DNA from the reference strain. In each case spot reactivities should be sufficiently greater than the background to allow changes in their intensities to be recorded with confidence.

Preparation and Labeling of Probes

For studies of the distribution of chromosomal sequences between strains, DNA is extracted and purified by standard method,[30,31] then labeled by random-primed incorporation of $[\alpha\text{-}^{33}P]dCTP$. (Phosphorus-33 is more expensive than phosphorus-32, but its lower energy emission results in more clearly defined spots on imaging.) The reactions (20 μl) contain 100 ng of chromosomal DNA in 7 μl of water, 2 μl of random hexamer primers (2 mg/ml), 3 μl of a mixture of dATP, dGTP, and dTTP (1 mM each), 5 μl of $[\alpha\text{-}^{33}P]dCTP$ (10 μCi/μl, 4 μM), and 2 μl of 10 × DNA polymerase (Klenow fragment) buffer (New England Biolabs). Heat at 100° for 2 min and then cool on ice. Add 1 μl (5 units) of Klenow enzyme (NEB) and incubate at 37° for 30 min.

For transcriptome studies, the quality of the RNA is of paramount importance. Bacterial RNA is notoriously subject to degradation after lysis of the cells. It is equally important to extract RNA representative of the growth conditions and minimally affected by the extraction procedure. In this respect the *Neisseria* species are relatively easy to lyse. We use a hot acid–phenol method in which the bacteria are lysed at 65° in the presence of RNase inhibitors. Bacteria grown in suspension under defined conditions are briefly centrifuged to pellet the cells, resuspended in

[30] J. Sambrook, E. F. Fritsch, and T. Maniatis, "Molecular Cloning: A Laboratory Manual." Cold Spring Harbor Laboratory Press, Cold Spring Harbor, NY, 1989.
[31] F. M. Ausubel, R. Brent, R. E. Kingston, D. D. Moore, J. G. Seidman, J. A. Smith, and K. Struhl (eds.), "Current Protocols in Molecular Biology," Vol. 1. Greene Publishing Associates and Wiley-Interscience, New York, 1989.

500 μl of medium, and added to tubes at 65° containing 700 μl of Trizol reagent (Life Technologies) and ~50 μl of acid-washed glass beads. This is vortexed vigorously and placed immediately on dry ice. Chloroform (200 μl) is added and the contents mixed by vortexing. The tubes are centrifuged at 15,000g for 15 min at 4°; the aqueous phase is removed and reextracted once before precipitating by addition of 1 volume of 2-propanol. The RNA pellet is washed with 75% (v/v) ethanol, dried, and redissolved in RNase-free TE buffer (10 mM Tris-HCl, 1 mM EDTA-Na, pH 8.0). Meningococci from 1 ml of culture having an OD$_{600}$ of 1 contain about 10 μg of RNA. We have found that 1 μg of RNA is sufficient for the reaction of at least two arrays. A further consideration is that the amount of RNA which can be obtained from typical biological experiments may be limiting, and protocols may have to be modified accordingly. Because of the relatively greater abundance of RNA compared to DNA, and the specificity of reverse transcriptase, it is generally not necessary to perform a DNase step before marking.

Radiolabeled cDNA is prepared essentially as for the chromosomal DNA above, but using higher concentrations of nucleotides in accordance with the higher K_m of reverse transcriptase. The RNA is heated to 70° for 5 min to destabilize secondary structure in the mRNA, before addition of the reverse transcriptase. The reactions (25 μl) contain up to 2 μg of RNA in 8 μl (less RNA may be used; see above), 2.5 μl of 10 × reverse transcriptase buffer, 2.5 μl of random hexamer primers (2 mg/ml), 2.5 μl of a mixture of dATP, dGTP, and dTTP at 3.3 mM each, 2.5 μl of 100 mM dithiothreitol, 5 μl of [α-^{33}P]dCTP (10 μCi/μl, 4 μM), and 2 μl (200 units) of reverse transcriptase (Superscript II reverse transcriptase kit, Life Technologies).

Alternatively, the priming of cDNA synthesis may be directed by oligonucleotides complementary to each ORF (e.g., those used for fabrication of the array). However, in agreement with others,[6] we have found that the results are inferior, probably because of large differences in the efficiency of labeling of different messages. However, this technique may be useful in cases where there is considerable contamination with eukaryotic RNA, e.g., human cell binding assays.

After labeling the chromosomal DNA or RNA, separate the labeled probe from the unincorporated radioactivity by gel filtration and perform a rough estimation of the efficiency of incorporation (by holding the tubes at a fixed distance from a Geiger counter). If two labeled DNAs to be compared are of very different specific activities, normalization of the membranes will be difficult.

Hybridization of Labeled Probe to Membranes

For reaction with the membrane, the probe is diluted in 400 μl of TE, denatured at 100° for 2 min, then diluted in hybridization buffer, and reacted as for standard Southern blot. For simplicity, we use the system of Church and Gilbert.[32]

---

[32] G. M. Church and W. Gilbert, *Proc. Natl. Acad Sci. U.S.A.* **81**, 1991 (1984).

For prehybridization and hybridization, the buffer is 0.5 M sodium phosphate, pH 7.2, containing 1 mM EDTA and 7% sodium dodecyl sulfate (SDS) to block nonspecific binding. Subsequent washing of the membranes is carried out at 65° in 40 mM sodium phosphate, pH 7.2, containing 1 mM EDTA, and 0.1% SDS. The length of time of the hybridization should be such that no signals become saturated; however, using the above quantities of reagent, the relative intensities of spots does not change between hybridizations for 8, 16, or 32 hr.[33] If visualization is to be achieved using labeled streptavidin to recognize biotinylated probe molecules, the membranes must be thoroughly washed in sodium phosphate/EDTA to remove the SDS before equilibration in the buffer used for enzymatic detection.

After reading, the membranes may be dehybridized by heating for 5 min at 100° in 0.5% SDS, and reused at least 5 times.

Reading of Array Results

The hybridized and washed arrays are wrapped in thin plastic film and exposed to a phosphorimager cassette for about 24 hr in order to record an image of the radioactive spots. (Care should be taken to avoid pockets of liquid. More importantly, the membranes must not to be allowed to dry out, since this will fix the label and prevent subsequent washing and reuse.) The image is read by the phosphorimager, and the associated software subsequently produces a computer-readable, pixel image. In order to define the positions of each spot, several interactive programs are available (Table II) (note also the nonradioactive technology used by Ang and co-workers). Grids are placed over the arrays corresponding to the repeat motif, in a preliminary attempt to locate each spot. (Careful positioning of the relatively intense positive control spots will help the computer algorithm to find the best grid.) The user can then manually modify the grid to better define the positions of each imperfectly localized spot. The reaction of each spot is measured as the average intensity of the corresponding pixels. The software will allow the user to substract an overall background noise, defined by an unused area of the membrane, or a local background, taken from around the spot in the case of uneven background over the membrane. Both methods have their drawbacks, and we generally do not subtract a background at this stage. Visual inspection of the image and comparison of duplicate spots will normally be sufficient to point out potential noise problems.

Interpretation of Hybridization Results

Normalization of Readings

The computer output will be the signal intensity of each spot, together with its grid reference and designation given by the user (e.g., ORF number, gene

[33] A. Perrin, S. Bonacorsi, E. Carbonnelle, D. Talibi, P. Dessen, X. Nassif, and C. R. Tinsley, submitted, 2002.

TABLE II
APPARATUS AND SOFTWARE FOR READING DNA ARRAYS

Apparatus and software	Supplier (location)	Nucleic acid labeling[a]	Refs.[b]
Image reading			
STORM Phosphorimager 820 (840 or 860)	Molecular Dynamics (Sunnyvale, CA)	33p	(1–5)
Biospace Imager	Biospace Measures (Paris, France)	33p	(6)
PowerLook 3000 Scanner	Umax (Hsinchu, Taiwan)	Biotin–dUTP, HRP–streptavidin	(7)
Fujix BAS2000 phosphorimager	Fuji (Japan)	33p	(8)
Interpretation			
DNA array vision	Research Imaging, Inc. (St. Catharines, Ontario, Canada)	33p	(9)
Imaging software, version 1.62	NIH (Bethesda, MD)	Biotin-dUTP, HRP–streptavidin	(7)
XDotsReader	COSE (Dugny, France)	33p	(3)
Visage HDG Analyzer	R. M. Lupton, Inc. (Jackson, MI)	33p	(8)
ImageQuant v.4, v.5	Molecular Dynamics	33p	(1,4,5)
HD grid analyzer	Genomic Solutions, Inc.	33p	(6)

[a] HRP, Horseradish peroxidase.

[b] Key to references: (1) C. S. Richmond, J. D. Glasner, R. Mau, H. Jin, and F. R. Blattner, *Nucleic Acids Res.* **27,** 3821 (1999); (2) R. Rimini, B. Jansson, G. Feger, T. C. Roberts, M. de Francesco, A. Gozzi, F. Faggioni, E. Domenici, D. M. Wallace, N. Frandsen, and A. Polissi, *Mol. Microbiol.* **36,** 1279 (2000); (3) A. Perrin, S. Bonacorsi, E. Carbonnelle, D. Talibi, P. Dessen, X. Nassif, and C. R. Tinsley, submitted (2002); (4) H. Tao, C. Bausch, C. Richmond, F. R. Blattner, and T. Conway, *J. Bacteriol.* **181,** 6425 (1999); (5) J. Walker and K. Rigley, *J. Immunol. Methods* **239,** 167 (2000); (6) F. Bertucci, K. Bernard, B. Loriod, Y. C. Chang, S. Granjeaud, D. Birnbaum, C. Nguyen, K. Peck, and B. R. Jordan, *Hum. Mol. Genet.* **8,** 1715 (1999); (7) S. Ang, C. Z. Lee, K. Peck, M. Sindici, U. Matrubutham, M. A. Gleeson, and J. T. Wang, *Infect. Immun.* **69,** 1679 (2001); (8) C. N. Arnold, J. McElhanon, A. Lee, R. Leonhart, and D. A. Siegele, *J. Bacteriol.* **183,** 2178 (2001); (9) S. M. Arfin, A. D. Long, E. T. Ito, L. Tolleri, M. M. Riehle, E. S. Paegle, and G. W. Hatfield, *J. Biol. Chem.* **275,** 29672 (2000).

name). Data analysis can then be carried out using any standard spreadsheet, e.g., Microsoft Excel. After averaging duplicate spots, it is necessary to normalize each reading with respect to the membranes because the efficiency of labeling and the kinetics of hybridization may not be constant between the two samples to be compared.

For chromosome comparisons, the genomic plasticity of the *Neisseria* species, which often leads to gene duplication, gives preference to normalization based on the totality of membrane reactivity, rather than attempting to define a gene present and constant in all species. Thus, for an amplicon on a second membrane (or resulting from a hybridization with a probe derived from the chromosome of a second strain) the normalized intensity:

$$I_{norm} = I \times m$$

where m, the normalization factor is calculated as:

$$m = \Sigma_{membrane1}(I) / \Sigma_{membrane2}(I)$$

However, the ratio of intensities for a gene absent in one of the strains evidently will not be representative of the global ratio of reactivities on the membrane, so the approximation is valid only if the chromosomal differences are small. If the genomic differences are larger, normalization should be performed avoiding those genes which are absent in the strain to be tested.

These amplicons can easily be identified and removed from the calculation. Figure 2 shows a comparison between the meningococcus Z2491, used for production of the membrane, and the gonococcus FA1090, whose sequence has also been determined. Figure 2A is a scatter plot of the intensity of reaction of each amplicon on membranes probed with meningococcal and with gonococcal DNA. For the majority of spots, the hybridization signal after reaction with the gonococcus is proportional to that after reaction with the meningococcus. However, a subset of spots (black, Fig. 2A) show low reactivity with the gonococcus independent of their reaction with meningococcal DNA. Comparing the ratio of reactivity of each amplicon with Z2491 and FA1090 to the computer-calculated degree of homology (Fig. 2B) shows that these do indeed correspond to sequences absent from the gonococcus. In order to remove these data from the calculation of the normalization factor, we first reject all those for which the value on hybridization with gonococcus FA1090 differs by more than 50% from that dictated by the least squares fit line of proportionality for the whole data group, then all those for which the value differs by more than 20% from that dictated by the line of proportionality for the data remaining after the first stage. This procedure effectively removes the group of amplicons corresponding to nonhomologous sequences. After normalization, the data are represented as a histogram (Fig. 2C), allowing the choice of a cutoff value by visual inspection—here 0.94. It is apparent that there are two classes of sequences: those common to the two species and those specific to the meningococcus. Comparing these results with a computer-assisted (TblastN; protein to translated DNA[34]) comparison of the genes of Z2491 and FA1090 (Fig. 3) shows that the membranes are capable of distinguishing those genes absent from the test strain.

[34] S. F. Altschul, T. L. Madden, A. A. Schaffer, J. Zhang, Z. Zhang, W. Miller, and D. J. Lipman, *Nucleic Acids Res.* **25,** 3389 (1997).

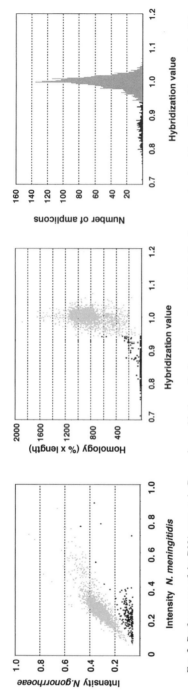

FIG. 2. Performance of the DNA arrays: 1. Comparison with theoretical gene homology. (A) Relation between the signal intensities (arbitrary units) of the amplicons from membranes hybridized with *N. gonorrhoeae* FA1090 and *N. meningitidis* Z2491. The signals are roughly proportional for the majority of spots. However, there is also a distinct group (black) which gives low values on reaction with ng DNA, independent of their reactivity with Z2491. (B) Low-reacting amplicons correspond to sequences absent in the gonococcus. The graph shows the relation between the calculated hybridization values (ratio of the logarithm of normalized intensity) and the degree of homology (% homology × length of the longest continuous region of homology identified by BlastN, with no gap of longer than 100 bases, i.e., which could react with a single labeled probe molecule, of about 300 bases) (J. Sambrook, E. F. Fritsch, and T. Maniatis, "Molecular Cloning: A Laboratory Manual." Cold Spring Harbor Laboratory Press, Cold Spring Harbor, NY, 1989). The amplicons fall generally into two groups: those giving a value of more than 0.94 and having more than about 200 bases in common with Z2491 in their longest region of homology and those giving a ratio of less than 0.94 and having fewer than 200 common bases. These latter are the same amplicons which are represented in black in (A). (C) Identification of strain-specific sequences: histogram classing the amplicons with respect to their hybridization values. The abscissa shows the ratio of the logarithms of the hybridizations since this value increases more regularly with the degree of homology that does the direct ratio of hybridization intensities, though this evidently does not affect the conclusions. As demonstrated in (B), the peak around unity corresponds to common sequences, and the second peak of lower-reacting amplicons to sequences absent in the gonococcal strain.

200

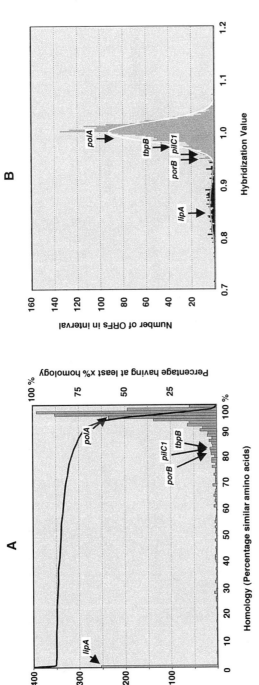

FIG. 3. Performance of the arrays: 2. Ability to predict the presence of functional homologs. (A) Homology of predicted proteins between *N. meningitidis* and *N. gonorrhoeae.* Gonococcal homologs to predicted ORFs from the genome sequence of *N. meningitidis* are identified by a TblastN search against the gonococcal genome. Generally ORFs either have high homology (e.g., PolA, a DNA polymerase), or do not exist (LipA is a capsular biosynthesis enzyme) in the gonococcus. The few that show intermediate homology include those encoding proteins known to be of common function but variable sequence between strains. PilC1 is an adhesin; PorB (the major porin) and TbpB (the iron scavenging, transferrin binding protein) are surface-exposed antigens. (B) Ability of the arrays to distinguish homologous genes from those absent in the test strain. The data are clearly divided into two groups, and in this case, since the majority of genes either are very homologous or show no homology, the distributions can be approximated as Gaussian curves, the cutoff of 0.94 being situated more than 2 standard deviations from both means. The sensitivity of a single comparison is applicable only to genes completely absent from the test strain, rather than to genes of intermediate homology. Reactivities are somewhat less well separated into two peaks in comparisons with the less closely related *N. lactamica.* In both cases reliability is augmented by performing comparisons with multiple strains.

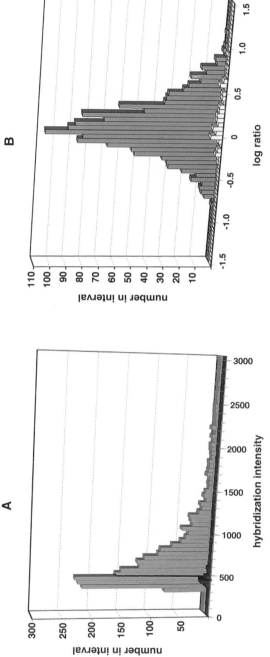

FIG. 4. Analysis of transcriptional levels. (A) Distribution of measured reactivities; comparison with selected control spots. RNA prepared from *N. meningitidis* Z2491 was used to probe the membranes, and the reactivities measured by the phosphorimager are represented as histograms for those amplicons corresponding to Z2491 ORFs (light gray) and those corresponding to ORFs having no homology with strain Z2491 (negative controls; dark gray). (B) Distribution of ratios of gene expression between bacteria grown in rich or in synthetic media (dark gray bars). Ratios were calculated for genes expressed in at least one of the conditions—i.e., giving signals greater than the mean of the negative controls. The set of ribosomal protein genes (light gray) is seen to be overexpressed in the rich medium where the bacteria are growing faster.

The choice of method for the normalization of DNA arrays used in the study of transcription poses somewhat different problems, although the aim is still to control for differences in labeling efficiency and hybridization kinetics. It is usual again to normalize with respect to the sum of the signals from the membrane, on the basis that differences in transcription will affect only a small number of genes. The use of rRNA genes is not recommended, because the proportion of rRNA to mRNA varies substantially depending on the growth phase of the bacteria.[35] Another possibility is to spike the labeling mixture with a small quantity (0.2 ng $\cong$ 0.2% of transcript abundance, assuming that 2 μg of total RNA contains 100 ng of mRNA) of mRNA corresponding to a gene with no homology in the *Neisseria* (e.g., rabbit β-globin—Life Technologies). A series of dilutions of the corresponding cDNA spotted onto the membranes may then serve as standards.

Subtraction of a background level of reactivity is necessary. In comparing patterns of transcription of the reference strain *N. meningitidis* Z2491 under different conditions, we have used as negative controls spots corresponding to those genes of *N. meningitidis* MC58 which are shown by a BlastN analysis to have no homology with the chromosome of Z2491. Hence there is no transcript corresponding to these amplicons. Figure 4A shows the data for reaction of an array with labeled RNA taken from *N. meningitidis* grown in liquid medium. The background spots (40 genes in duplicate) give an assumed Gaussian distribution of reactivities, and the cutoff may be taken as the mean of these negative control spots. Such an approach depends on the choice of appropriate negative control genes. For comparison of the transcriptome corresponding to two different growth conditions the levels of transcription are compared after normalization and subtraction of the background. In order to avoid spuriously high or negative ratios for genes which give low hybridizations in one of the two conditions the results are corrected after subtraction of the background so that the lowest value attributed to any spot is about 1/10 of the average value (50 for the experiment presented). The ratio of the values for each amplicon is calculated, and a gene is considered over- or underexpressed when the ratio is greater than 2. Figure 4B shows the distribution of genes expressed differentially in cells grown in synthetic or rich medium. As has been described for *Escherichia coli*,[36,37] genes involved in protein synthesis are overexpressed in rich medium, consistent with a faster rate of cell growth. Thus the differential expression of genes may be studied in response to various laboratory situations mimicking the *in vivo* disease conditions encountered by the bacteria, e.g., iron deprivation, contact with human cells, and growth in serum.

[35] H. Bremer and P. P. Denis, in "*Escherichia coli* and *Salmonella typhimurium:* Cellular and Molecular Biology" (F. C. Neidhardt *et al.,* eds.), pp. 1527. ASM Press, Washington, D.C., 1987.

[36] H. Tao, C. Bausch, C. Richmond, F. R. Blattner, and T. Conway, *J. Bacteriol.* **181,** 6425 (1999).

[37] Y. Wei, J. M. Lee, C. Richmond, F. R. Blattner, J. A. Rafalski, and R. A. LaRossa, *J. Bacteriol.* **183,** 545 (2001).

Interpretation of Array Data

DNA array data are inherently imprecise, each membrane giving only one or two values for each gene. Significance of the results will depend on the number of measurements, which can be accumulated by the following methods: (1) repeating the experiment several times with the same or similar strains; (2) studying groups of genes (e.g., those involved in energy production or iron acquisition), rather than single genes; or (3) in transcriptome studies, comparing the evolution of mRNA levels over several time points.

For genomic comparisons, the hybridization results used as reference (those of the strain Z2491 used for production of the array) are the average of four independent experiments. However, for comparison of the genetic complements of different strains, the strength of the evidence is based on comparison with a large number of different isolates and the association of the presence of a gene with a species or with a hypervirulent clone of meningococcus. The choice of strain is important for making comparisons. Until recently the number of meningococcal strains of known genetic derivation was limited, and this remains the case with *N. lactamica* and the more problematic *N. gonorrhoeae*. However, the multilocus enzyme electrophoresis (MLEE) system[38,39] and the multilocus sequence typing (MLST)[40,41] have permitted the elucidation of clonal relationships between meningococci (both disease and carrier isolates) on a global scale. Hence isolates can be chosen for study based on their clonal group and relation to disease-causing potential.

There are, however, some caveats concerning the interpretation of the results. The basis for comparative genomics is that potential determinants of the differential pathogenesis will be present, for example, in all virulent strains of *N. meningitidis* and absent from all *N. gonorrhoeae* and *N. lactamica*. Several such genes exist, and some, for example, the capsule biosynthesis gene cluster, have been shown to be important in the disease process. However, the situation may be considerably more complicated. First, two gene products may be interchangeable, so that the virulence phenotype would be dependent on the possession of one or the other, but not necessarily both. This situation will be difficult to recognize without knowledge of the functions of the genes. Second, the presence of a gene does not imply automatically the elaboration of the corresponding protein. Several genes expressed in *N. meningitidis* are nonfunctional pseudogenes due to mutation in

[38] R. K. Selander, D. A. Caugant, H. Ochman, J. M. Musser, M. N. Gilmour, and T. S. Whittam, *Appl. Environ. Microbiol.* **51**, 873 (1986).

[39] D. A. Caugant, *APMIS* **106**, 505 (1998).

[40] M. C. Maiden, J. A. Bygraves, E. Feil, G. Morelli, J. E. Russell, R. Urwin, Q. Zhang, J. Zhou, K. Zurth, D. A. Caugant, I. M. Feavers, M. Achtman, and B. G. Spratt, *Proc. Natl. Acad. Sci. U.S.A.* **95**, 3140 (1998).

[41] mlst.zoo.ox.ac.uk

N. gonorrhoeae (e.g., those for the porin *porA* and the γ-glutamyltranspeptidase *ggt*). Third, phenotypic differences between strains or species may result equally from differences in transcriptional regulation as from the presence or absence of a gene. In this case comparison of the transcriptomes may be more useful although the choice of growth conditions (e.g., iron limitation) will be important.

Analysis of the transcriptional response will deal with the genes induced in response to a growth condition or with the sequence of changes in gene expression over time. In the first case reliability of the results will depend on repetition of the experiment. (In addition, any response of interest should be exhibited by several different strains.) In the absence of multiple repetitions, an observed twofold or greater difference in expression is often taken as signifying that the gene is in reality over- or underexpressed, and intuitively one would have more faith in higher observed ratios. However, Arfin and co-workers[6] noted that there is in fact no relation between the observed fold increase and the probability that this represents a real difference in expression for an individual gene. Nevertheless, since each experiment provides thousands of individual data, it is possible without extensive repetition to make general statements concerning a predefined class of genes, for example, the tendency of the components of the translational apparatus (about 100 genes) to be overexpressed in faster-growing bacteria[36,37] (cf. Fig. 4A).

Another technique is to follow the changes in transcript level over time. The relation of the *Neisseria* with the host is a dynamic interaction, not only during initial colonization and disease, but also during asymptomatic carriage, in which the bacteria may interchange between an extracellular and intracellular lifestyle.[42] Furthermore, laboratory models of meningococcus–host cell adhesion demonstrate a well-defined sequence of interactions.[43] In such experiments the expression (relative to a reference condition) of each gene is followed at successive time points, and one of a number of clustering methods (see Refs. 44–47) is used to group together genes having similar patterns of change in expression over time. Analysis of the genes within clusters identified on this basis, their promoter sequences, and the expression of their protein products[44,48] has demonstrated that these clusters do indeed define genes which may be coregulated or are part of the same response network. Such parallel transcriptional analysis represents a new and powerful tool

[42] R. J. Sim, M. M. Harrison, E. R. Moxon, and C. M. Tang, *Lancet* **356**, 1653 (2000).
[43] C. Pujol, E. Eugene, M. Marceau, and X. Nassif, *Proc. Natl. Acad. Sci. U.S.A.* **96**, 4017 (1999).
[44] M. B. Eisen, P. T. Spellman, P. O. Brown, and D. Botstein, *Proc. Natl. Acad. Sci. U.S.A.* **95**, 14863 (1998).
[45] U. Alon, N. Barkai, D. A. Notterman, K. Gish, S. Ybarra, D. Mack, and A. J. Levine, *Proc. Natl. Acad. Sci. U.S.A.* **96**, 6745 (1999).
[46] S. Tavazoie, J. D. Hughes, M. J. Campbell, R. J. Cho, and G. M. Church, *Nat. Genet.* **22**, 281 (1999).
[47] M. P. Brown, W. N. Grundy, D. Lin, N. Cristianini, C. W. Sugnet, T. S. Furey, M. Ares, Jr., and D. Haussler, *Proc. Natl. Acad. Sci. U.S.A.* **97**, 262 (2000).
[48] M. T. Laub, H. H. McAdams, T. Feldblyum, C. M. Fraser, and L. Shapiro, *Science* **290**, 2144 (2000).

and will surely lead to great progress in understanding the complex interaction of the *Neisseria* with the human host.

Conclusion

DNA array technology is a new and increasingly useful tool which allows simultaneous measurements to be made on thousands of genes—the complete genome in the case of bacteria. In the case of the *Neisseria* species two types of utilization may be considered. The first, comparative genomics, aims to identify genes responsible for the disease-causing potentials of different species or subgroups by association of the possession of these genes with a certain phenotype. Such studies have been reported for *Helicobacter pylori*[49] and Streptococcus species.[19] The *Neisseria* are particularly well suited to this kind of analysis because their close interspecific relationship[50,51] contrasts with their dramatically differing pathogenic potential. *Neisseria meningitidis* is the agent of cerebrospinal meningitis; the gonococcus *N. gonorrhoeae* usually causes a localized inflammation of the urogenital tract, gonorrhea; whereas a number of closely related commensal species, for example, *N. lactamica,* are harmless inhabitants of the nasopharynx. Hence genes that are present only in virulent *N. meningitidis* may determine bloodstream survival and crossing of the blood–brain barrier, whereas genes present in the two pathogens, but absent from the commensal *N. lactamica,* will be involved in the common aspects of infection—adhesion to and invasion of epithelial cells at the site of initial colonization. Initial studies, undertaken before the availability of the genome sequence, using physical subtraction techniques, brought to light a number of potential virulence factors[9,52,53] and DNA arrays have expanded the number of species-specific genes identified and deserving of further practical investigation.[33] In the light of reports that pathogenesis may be the result of the absence rather than the presence of DNA sequences[54] and that commensal bacteria may promote host tolerance,[55] it is worth noting that it may be equally informative to examine the distribution of genes associated with commensalism. As genome sequencing becomes progressively easier, the sequence of a well-characterized

[49] N. Salama, K. Guillemin, T. K. McDaniel, G. Sherlock, L. Tompkins, and S. Falkow, *Proc. Natl. Acad. Sci. U.S.A.* **97,** 14668 (2000).

[50] D. T. Kingsbury, *J. Bacteriol.* **94,** 870 (1967).

[51] C. Hoke and N. A. Vedros, *Int. J. Syst. Bacteriol.* **32,** 57 (1982).

[52] A. Perrin, X. Nassif, and C. R. Tinsley, *Infect. Immun.* **67,** 6199 (1999).

[53] S. R. Klee, X. Nassif, B. Kusecek, P. Merker, J.-L. Berett, M. Achtman, and C. R. Tinsley, *Infect. Immun.* **68,** 2082 (2000).

[54] A. T. Maurelli, R. E. Fernandez, C. A. Bloch, C. K. Rode, and A. Fasano, *Proc. Natl. Acad. Sci. U.S.A.* **95,** 3943 (1998).

[55] A. S. Neish, A. T. Gewirtz, H. Zeng, A. N. Young, M. E. Hobert, V. Karmali, A. S. Rao, and J. L. Madara, *Science* **289,** 1560 (2000).

strain, for example, *Neisseria lactamica,* should allow the design of future arrays containing genes of both pathogenic and commensal strains.

The second use of DNA arrays is the comparison of RNA transcription profiles, and it is perhaps here that the technology will advance our understanding the most. As discussed above, the interaction of *Neisseria* with the human host is a dynamic process, particularly in the case of disease. For example, the meningococcus adheres to the nasopharyngeal cells, spreads to colonize the epithelium, traverses the cellular barrier, comes into contact with serum where it is able to survive and multiply, and finally breaches the blood–brain barrier by crossing the endothelium at the meninges. Laboratory models exist for each of these stages. The DNA arrays and associated analysis programs now available represent a powerful new approach to the understanding of meningococcal and gonococcal pathogenicity, and in this light we should not forget that the techniques may be equally well applied to the response of the host to the bacteria.[56,57] We now have the ability to extract a wealth of new and exciting information from the study of these models of bacteria–host interaction.

[56] M. Diehn and D. A. Relman, *Curr. Opin. Microbiol.* **4,** 95 (2001).

[57] D. B. Wells, P. J. Tighe, K. G. Wooldridge, K. Robinson, and D. A. Ala' Aldeen, *Infect. Immun.* **69,** 2718 (2001).

[15] Acquisition and Archiving of Information for Bacterial Proteomics: From Sample Preparation to Database

 By STUART J. CORDWELL

Introduction

The rapid expansion of proteomics as a tool for the study of complex biological systems has led to a concurrent improvement in technologies and methodologies for the separation and subsequent analysis of purified proteins. The standard proteome experiment involves cellular lysis and protein solubilization, a separation step [typically two-dimensional gel electrophoresis (2-DGE)], visual comparisons of these gels corresponding to altered biological states, and characterization of proteins of interest by mass spectrometry (MS). Without doubt standardization both within and across proteomics laboratories has been difficult to achieve because of the myriad combinations of solutions, suppliers, and techniques. This in turn has meant there is an irreproducibility factor associated with the technology. The inherent biological nature of protein rather than DNA samples is also a contributing factor, meaning that laboratories must first optimize a given protocol for any new biological system, and that this protocol may be different for other biologically derived samples. Since the beginning of the genome "gold rush," bacterial species have become particularly amenable to proteomic approaches.[1] Identification resources are common and genome expression is small enough to make 2-DGE a high-resolution approach. However, it is important to understand that several classes of proteins are currently incompatible with this process. They include basic ($pI > 10.0$), low abundance, and hydrophobic (Kyte–Doolittle > 0.20) proteins. These classes of proteins may best be studied utilizing complementary approaches including multidimensional liquid chromatography combined with mass spectrometry[2] and isotope-coded affinity tags.[3] The aim of this article is to provide a set of standardized methods, and some common variations, for proteomics experiments involving 2-DGE and mass spectrometry, from sample preparation to protein identification and interpretation of the final results in a microbial context.

Sample Preparation

The most critical phase in any proteomic project is the means by which protein mixtures are liberated from within whole cells and the solutions used to solubilize the majority of total protein species while reducing common contaminants that can inhibit the isoelectric separations. In microbial sample preparation, the important steps involve efficient cellular lysis by physical means including French press, tip-probe sonication, or disruption with zirconium or glass beads prior to solubilization with 2-DGE compatible sample buffers. Once cellular contents have been released the next most critical factor is the removal of salts, nucleic acids, and carbohydrates that may interfere with the separation process leading to poorly focused or streaky 2D gels. For many cell types this is as simple as centrifuging the final lysate to remove insoluble debris. However, in other cases additional steps must be taken to remove particular contaminants. For example, the outer wall and thick peptidoglycan layer of *Streptococcus mutans* can be removed with mutanolysin. Nucleic acids are removed by the addition of an excess of endonuclease which digests large nucleic acids into smaller fragments that pass rapidly through the SDS–PAGE second dimension gel. We have also found that antiprotease cocktails are unnecessary for protein mixtures solubilized in solutions containing Tris and urea. Where necessary, commercially available mixtures can be utilized; however, when used in excess these may result in additional protein spots visible on the gel

[1] S. J. Cordwell, A. S. Nouwens, and B. J. Walsh, *Proteomics* **1**, 461 (2001).
[2] M. P. Washburn, D. Wolters, and J. R. Yates III, *Nat. Biotechnol.* **19**, 242 (2001).
[3] S. P. Gygi, B. Rist, S. A. Gerber, F. Turecek, M. H. Gelb, and R. Aebersold, *Nat. Biotechnol.* **17**, 994 (1999).

image. The following methods are for solubilizing proteins from whole cells in a single step.

Method for Whole-Cell Protein Lysate

1. Pellet cells from culture.
2. Wash three times in low salt sample washing buffer or phosphate-buffered saline (PBS). Following the final wash, invert and tap the tube against a dry tissue to remove excess salt-containing buffer.
3. Freeze-dry the cells overnight.
4. Weigh 10 mg dry weight cells and add 1 ml of multiple surfactant solution [5 M urea, 2 M thiourea, 2 mM tributylphosphine (TBP), 2% CHAPS, 2% sulfobetaine 3–10, 40 mM Tris, 0.2% carrier ampholytes, and 0.002% bromphenol blue dye]. The TBP can be replaced by 100 mM dithiothreitol (DTT).
5. Vortex until cells are in solution (less than 1 min at maximum output).
6. Sonicate the cells using a tip-probe sonicator. This step is sample dependent and critical. Sonicate in replicate cycles of 30 sec with 1 min on ice between cycles. Longer bursts of sonication will lead to sample heating resulting in artifactual modifications on proteins visible as multiple spots on the final 2D gel image. Typically, for *Escherichia coli, Pseudomonas aeruginosa,* or *Caulobacter crescentus* (gram-negative bacteria) 90% lysis is achieved in two cycles; for gram-positives, acid-fast, or spore-forming bacteria multiple replicate rounds of sonication are required to achieve a similar degree of disruption.
7. Add 150 U of endonuclease (this is an excess which allows function in a highly denaturing solution for up to 20 min).
8. Centrifuge the lysate at 12,000g for 15 min at 15° to remove insoluble material.
9. Remove the supernatant to a fresh tube. The amount to be loaded depends on the sample type. However, the dynamic range is broad, i.e., for a 17 cm pH 4–7 IPG strip the equivalent of 250 μg to 2 mg starting dry weight can be loaded by passive rehydration (see below).

Sequential Extraction

Bacteriological factors play a significant role in determining the optimum conditions for physical disruption; for example, gram-negative cells are generally easier to disrupt than gram-positive cells. From 10 mg dry weight starting material, we have determined that 1 min of tip-probe sonication will result in approximately 90% protein recovery from *Escherichia coli, Pseudomonas aeruginosa,* and several other gram-negative species. However, because of a thick peptidoglycan layer in gram-positive bacteria, species such as *Staphylococcus aureus* and *Mycobacterium tuberculosis* may only provide 10% recovery following the same degree of disruption. The "perfect" disruption method is highly sample dependent and must be optimized prior to experimental analysis. These issues are critical when performing "sequential" solubilization methods[4,5] aimed at recovering sets of increasingly hydrophobic proteins across a series of 2D gels. If cells are not efficiently broken in the initial extract then the following extracts, rather than being "enriched" in hydrophobic proteins, will be repeat whole-cell lysates in alternative sample buffers, thus showing very little difference. We have determined that for several bacterial species, a three-step extraction is best performed with the following protocol, where extract 2 is replaced by extract 3 and the final extract is performed in SDS–PAGE sample buffer in conjunction with 1D PAGE (Fig. 1).

The following protocol describes the sequential extraction of a preparative amount of E. coli cells. However, we have demonstrated this procedure for a number of cells including C. crescentus, P. aeruginosa, and Helicobacter pylori. The sonication time must be altered according to the properties of a given sample.

Procedure

1. Weigh 10 mg of lyophilized cells and add 1 ml of 40 mM Tris (this volume can be increased if cells are too concentrated as the sample is collected via methanol precipitation). Vortex strongly and tip-probe sonicate for 4×30 sec, with 1 min on ice between each cycle.

2. Add 150 U of endonuclease and mix. Leave for 20 min.

3. Centrifuge at 12,000g for 15 min at 15°.

4. Collect the supernatant and keep the pellet.

5. Place the supernatant in a 50-ml centrifuge tube and add ice-cold methanol to a final volume of 40 ml. Place at −80° for a minimum of 2 hr to precipitate proteins.

6. Centrifuge at 12,000g for 30 min at 4°. Carefully remove and discard supernatant and to the pellet add 1 ml of standard sample solution (8 M urea, 4% CHAPS, 100 mM DTT, 40 mM Tris, and 0.2% carrier ampholytes). This is extract 1.

7. Wash the pellet from step 4 twice in 40 mM Tris to remove any contaminating Tris-soluble proteins and centrifuge at 12,000g for 15 min.

8. Add 1 ml of standard sample solution, vortex strongly, and tip-probe sonicate for 4×30 sec, as above.

9. Add 150 U of endonuclease, as above.

10. Centrifuge at 12,000g for 15 min at 15°.

11. Collect the supernatant and remove to a fresh 1.5-ml centrifuge tube. This is extract 2.

[4] M. P. Molloy, B. R. Herbert, B. J. Walsh, M. I. Tyler, M. Traini, J.-C. Sanchez, D. F. Hochstrasser, K. L. Williams, and A. A. Gooley, *Electrophoresis* **19**, 837 (1998).
[5] S. J. Cordwell, A. S. Nouwens, N. M. Verrills, D. J. Basseal, and B. J. Walsh, *Electrophoresis* **21**, 1094 (2000).

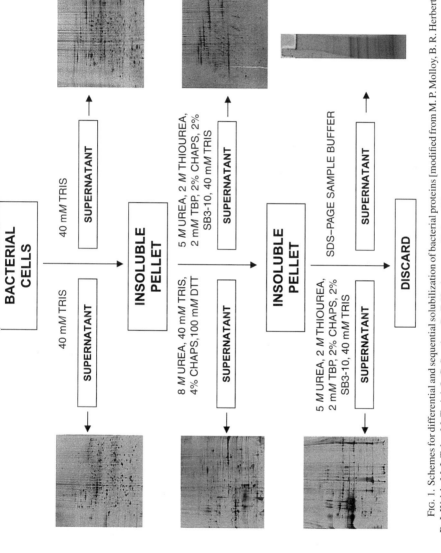

FIG. 1. Schemes for differential and sequential solubilization of bacterial proteins [modified from M. P. Molloy, B. R. Herbert, B. J. Walsh, M. I. Tyler, M. Traini, J.-C. Sanchez, D. F. Hochstrasser, K. L. Williams, and A. A. Gooley, *Electrophoresis* **19**, 837 (1998)]. Cells are subjected to rounds of physical disruption in each of the listed buffers followed by centrifugation of insoluble material.

12. Wash the pellet twice in standard sample solution to remove soluble proteins.

13. Add 500 μl of multiple surfactant solution to the pellet, vortex strongly, and sonicate for 4 × 30 sec.

14. Centrifuge at 12,000g for 15 min at 15°.

15. Collect supernatant and mark as extract 3.

Alternatively, perform step 6 with multiple surfactant solution and add SDS–PAGE sample buffer to the remaining insoluble pellet (step 13) prior to 1D SDS–PAGE.

Membrane Proteins

Bacterial membrane proteins are of critical importance in processes leading to pathogenicity. These include mediating host–pathogen interaction, efflux of antibiotics, import of nutrients, and evasion of the immune response. Proteome projects are therefore often centered on discovery of new vaccine targets or therapeutics by examining fractions of proteins enriched in membrane proteins. This method works very well for arraying proteins from gram-negative membranes.[6,7] For gram-positives, success is again dependent on efficient cellular lysis and removal of intact cells prior to precipitation of membranes in sodium carbonate.

Enrichment of Bacterial Membrane Proteins

This procedure has been modified from Ref. 6 and applied to the separation of membrane proteins from *P. aeruginosa*.[7] It should be noted that such protocols only provide an enrichment and can never yield 100% purity. In the study by Nouwens *et al.*,[7] all the identified proteins were predicted to contain at least one transmembrane spanning region (TMR). Furthermore, the large majority of the identified proteins were predicted to be hydrophilic with hydrophobic TMR. Results suggest that 2-DGE will not be amenable to the separation of highly hydrophobic proteins with currently available buffers.

Procedure

1. Suspend 40 mg dry weight cells in 5 ml of PBS containing 150 U of endonuclease. Sonicate in a tip-probe sonicator for 4 × 30 sec with 1 min on ice between each round of sonication.

2. Centrifuge the lysate at 6000g for 15 min at 15° to pellet unbroken cells. If the pellet is large remove the supernatant to a fresh tube and repeat steps 1 and 2 until there are few unbroken cells.

---

[6] M. P. Molloy, *Anal. Biochem.* **280**, 1 (2000).

[7] A. S. Nouwens, S. J. Cordwell, M. R. Larsen, M. P. Molloy, M. Gillings, M. D. P. Willcox, and B. J. Walsh, *Electrophoresis* **21**, 3797 (2000).

3. Pool the supernatants and add 5–6 volumes of ice-cold 0.1 M sodium carbonate. Stir the mixture for 1 hr at 4°.

4. Centrifuge at 100,000g for 1 hr at 4°.

5. Remove the supernatant and wash the membrane pellet twice in 40 mM Tris or PBS to remove any contaminating proteins or cells.

6. Add 1 ml of membrane-specific sample solution [5 M urea, 2 M thiourea, 40 mM Tris, 1% tetradecanoylamidopropyldimethylammoniopropane sulfonate (ASB-14), 2 mM TBP, and 0.2% carrier ampholytes] and vortex at maximum output.

7. Sonicate briefly for 30 sec.

8. Centrifuge at 20,000g for 12 min at 4° to pellet insoluble membrane material and remove the supernatant to a fresh 1.5-ml centrifuge tube.

Extracellular Proteins

Bacteria can also use secreted proteins such as proteases or lysins to induce pathogenic processes. Therefore, the examination of extracellular or culture supernatant (CSN) proteins is critical to understanding such events. The protocol given here is best utilized for organisms not grown in complex media containing serum or serum proteins as these will contaminate bacterial proteins and their abundance will override the bacteria-derived pattern.

Enrichment of Extracellular Proteins

1. Grow culture to required growth phase/optical density and centrifuge to pellet whole cells (10,000g for 15 min at 4°).

2. Filter the supernatant through a 0.2-μm filter (this should be performed with the culture supernatant on ice to reduce proteolytic activity).

3. Precipitate the proteins with 20% (w/v) trichloroacetic acid (TCA) on ice for 30 min with mixing.

4. Centrifuge at 17,000g for 30 min at 4°.

5. Wash the proteins three times in ice-cold methanol.

6. Add 1 ml of multiple surfactant solution to the final protein pellet.

Isoelectric Focusing

This article will only consider 2-DGE applications utilizing immobilized pH gradients (IPGs) in the first dimension. Commercially available dry strips come in several formats that should be considered prior to 2-DGE experiments. We find it most useful to use wide-range gradients (e.g., pH 3–10) for an initial screen and then combined sets of mid-range pH gradients (pH 3–6, 4–7, 5–8, 7–10) for high resolution separations. Narrow-range pH gradients (e.g., pH 4–5) can be used to enrich for low abundance proteins allowed by the increased loading capacity to separating area available.[5] IPG strips are uniformly dried when supplied and must be rehydrated in sample buffer either with or without the presence of sample. For many bacterial protein mixtures we have determined that "passive" rehydration is the optimum method for applying sample to the first dimension gel. In this case, lysates (diluted to 250 μg per gel for analytical loads, and up to 1–2 mg for preparative loads) are applied to the gel surface of the dry strip and allowed to reswell the strip for a minimum of 6 hr. Plastic strip trays are now commercially available; however, 2-ml disposable plastic pipettes are also useful. The alternative is to rehydrate the strip in an equal volume of sample buffer alone and add sample to a cup at either end of the IPG strip. For basic pH strips (pH 6–11, 9–12, 7–10 gradients) cup-loading at the anodic end appears to be essential to provide reproducible, high-quality patterns (Fig. 2). Isoelectric focusing (IEF) of proteins in IPG strips follows several different methods. Most involve the stepwise increase of voltages from an initial low voltage to remove current-inducing salts

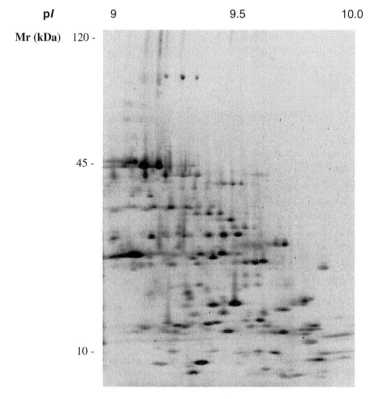

FIG. 2. SYPRO Ruby stained 2D gel of *H. pylori* proteins separated on a prototype pH9–10 immobilized pH gradient (Bio-Rad, Hercules, CA).

to high voltage for efficient and rapid protein separation. The total kVh used to focus proteins can be between 30 and 100 kVh depending on sample, choice of IPG strip, and the sample load. Some IEF systems provide an internal "ramp" on voltage to reach a preset total of kVh rather than via a stepwise procedure.

Equilibration

Post-IEF, the IPG strips must be equilibrated in SDS buffer prior to SDS–PAGE. SDS coats the proteins giving them the same net charge and allowing separation to occur in the second dimension on the basis of molecular mass. The equilibration procedure is mostly performed in two steps: first, a reduction and then an alkylation step using dithiothreitol (DTT) and iodoacetamide (IAA), respectively. We use the following one-step protocol (incubation period of 20 min), where DTT is replaced by TBP, and IAA is replaced by acrylamide monomer. The acrylamide monomer allows subsequent mass spectrometry searches to be performed with the cysteine-acrylamide option.

IPG Strip Equilibration

Immerse each IPG strip (gel-side exposed) in equilibration solution.

6 M Urea
2% SDS
1× Tris-HCl Gel Buffer (pH 8.8)
20% Glycerol
5 mM TBP
2.5% (v/v) Acrylamide solution

The strip is incubated in this solution, with shaking, for up to 20 min. The strip is then placed on top of the second dimension slab gel and a hot solution of 0.5% agarose in 1× Tris/glycine running buffer (192 mM glycine, 0.1% SDS, 24.8 mM Tris, pH 8.3) layered on top. A trace amount of dye (0.001% bromphenol blue) is added to this solution.

SDS–PAGE

Following equilibration, the IPG strip must be applied to the top of an SDS–PAGE slab gel.[8] The pore size of the slab depends on the application. For example, for high mass proteins pore sizes down as low as 4–7.5%T (total concentration of acrylamide monomer plus cross-linker) may be utilized. However, at this level

---

[8] B. J. Walsh and B. R. Herbert, *Methods Mol. Biol.* **112**, 245 (1999).

tensile strength is poor and gel handling may become difficult. To visualize the greatest number of proteins in the mass range 10–150 kDa, gradient gels of 8–18% can be used.

The following solutions can be used in a gradient pourer to cast six 8–18% slab gels for SDS–PAGE using a PROTEAN II Multi-Cell (Bio-Rad, Hercules, CA).

For 8% (200 ml):

> 40 ml 5× Tris/HCl Gel Buffer (1.875 M Tris, pH 8.8)
> 40 ml 40% Acrylamide solution [40% (w/v) acrylamide, 1% (w/v) piperazine diacrylamide (PDA)]
> 120 ml Ultrapure water

For 18% (200 ml):

> 40 ml 5× Tris/HCl Gel Buffer
> 90 ml 40% Acrylamide solution
> 70 ml 50% Glycerol

To both solutions add:

> 35 μl TEMED
> 360 μl 10% Ammonium persulfate (APS)

The solutions can be degassed prior to the addition of these reagents; however, this is not essential (degassed solutions will polymerize more rapidly). When the gels have been poured, water-saturated isobutanol should be layered on top of the gel solution in each glass plate assembly to ensure an even gel front. Allow at least 6 hr for complete polymerization.

Staining 2D Gels

Staining methods usually depend on the type of gel experiment performed. Traditionally, analytical protein loads were visualized by silver staining that modified proteins so that further characterization could not be performed, whereas preparative loaded gels were stained with Coomassie blue, compatible with further characterization but with only a fraction of the sensitivity of silver stains. With the dramatic improvement in sensitivity of mass spectrometry, the need for "analytical" and "preparative" gels has been removed. Furthermore, advances in fluorescent staining have reduced the need for labor intensive silver staining processes. Fluorescent dyes such as SYPRO Ruby (Molecular Probes, Eugene, OR)[9]

[9] W. F. Patton, *Electrophoresis* **21,** 1123 (2000).

are compatible with mass spectrometry, although silver stains not containing glu-taraldehyde can also be utilized.[10] We perform a "double staining" procedure, where gels are stained using fluorescent dyes and then scanned, followed by a colloidal Coomassie overlay that results in a visual image with the same sensitivity and resolution as the original fluorescent scan (Fig. 3).

Method for Staining 2D Gels

1. Place the gel in fixative containing 40% methanol, 10% acetic acid and leave with shaking for a minimum of 1 hr.

2. Remove the fixer and add SYPRO Ruby (Molecular Probes). Leave the gel immersed in stain overnight.

3. The appearance of protein spots can be monitored by periodic viewing on an "Orange/Blue" light box or UV scanner.

4. Remove the stain (stain can be reused several times with only a slight decrease in performance) and immerse the gel in destain [10% (v/v) methanol, 7% (v/v) acetic acid] for a minimum of 1 hr.

5. Scan the gel with a high-performance fluorescence scanning system.

6. Gels can be left for several days in destain.

Double Staining

1. Remove the destain and immerse the gel in colloidal Coomassie blue (CBB) G-250 (17% ammonium sulfate, 3% phosphoric acid, 0.1% Coomassie G-250, 34% methanol).

2. Leave the gel overnight with shaking at room temperature.

3. Remove the stain (CBB G-250 cannot be reused) and add destain (1% acetic acid) to enhance detection.

Image Analysis

Stained gel images from differing biological conditions must be visually compared to determine which proteins are up- or down-regulated as a result of biological stimuli, and therefore to determine which proteins need to be further characterized. Image analysis is generally performed using a commercial package [e.g., PD-Quest (Bio-Rad), Melanie (Swiss Institute for Bioinformatics, Geneva, Switzerland), Phoretix 2-D (NonLinear Dynamics, Newcastle-upon-Tyne, UK), and Z3 (Compugen, Tel Aviv, Israel)]. An excellent rule-of-thumb prior to performing image analysis is to visually inspect the images for obvious differences and then to make a decision on whether the gels are (a) reproducible (and thus worthy of "real" comparison; (b) of a sufficient quality (i.e., protein spots are well

[10] A. Shevchenko, M. Wilm, O. Vorm, and M. Mann, *Anal. Biochem.* **68,** 850 (1996).

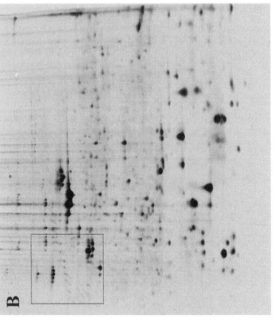

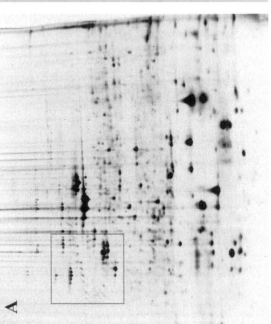

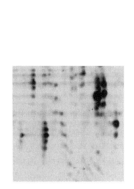

FIG. 3. (A) SYPRO Ruby stained 2D gel of *H. pylori* proteins separated on a pH 5–8 immobilized pH gradient and scanned using a Molecular Imager Fx (Bio-Rad); (B) the same gel following double staining with colloidal Coomassie blue G-250 and scanned using standard densitometry. Boxed areas are shown magnified below.

focused rather than streaks, etc.); and (c) contain differences. The answers to such questions allow the researchers to decide whether image analysis is warranted. When using image analysis care must be taken to ensure that differences are reproducible across gel sets and experiments. There is no set number of replicates that ensure the accuracy of the results; however, somewhere between 2 and 6 gel replicates is the accepted minimum. A second, replicate biological experiment and gel sets should also be performed where possible.

Spot Cutting

For high-throughput applications several commercial "spot cutters" are now available; however, for many laboratories that do not apply proteomics either on a day-to-day or at a genomic level, such instruments are outside the scale of the necessary requirements. Following "double" staining of 2D gels, spots are best excised using a sterile scalpel blade. Users should be aware of self-contamination of protein samples by human hair and skin. Although a background level of keratin in air-borne dust particles is always present, this can be minimized by performing gel cutting manipulations in a laminar flow hood, or if this is not available through the use of masks, gloves, and nonshedding laboratory coats. As mass spectrometry becomes more sensitive the need for negative pressure environments in which to perform manipulations will become a greater issue.

Sample Preparation for Mass Spectrometry

The second major phase in a discovery-based proteome project involves the identification and further characterization of gel-purified proteins via mass spectrometry. Typically, excised gel spots are treated with a proteolytic agent (e.g., trypsin) to produce peptide fragments that are used to generate peptide-mass mapping (PMM) data via matrix-assisted laser desorption/ionization (MALDI) mass spectrometry, or for electrospray ionization (ESI) MS applications including MS/MS sequencing and analysis of potential posttranslational modifications. The generation of peptides is most readily performed in-gel rather than following separation to an inert membrane. This is because latest generation mass spectrometers are more sensitive than staining techniques for 2D gels, and the blotting step generally results in a 15–50% loss of proteins as well as a complete loss of certain protein types. For MALDI-MS, the choice of matrix is generally based on the size of the generated peptides and for PMM applications we believe that α-cyanohydroxycinnamic acid is the most versatile. This article includes two alternative procedures for generating peptides from gel-purified proteins.

Sample Preparation for MALDI-PMM (I)

Multiple enzymatic digests are best performed in a 96- or 384-well plate.

1. Wash the CBB G-250 stained gel pieces in 50 mM NH$_4$HCO$_3$ (pH 7.8)/ 100% acetonitrile (60 : 40) for 1 hr. Repeat this procedure if CBB G-250 dye remains in the gel piece.
2. Dry gel pieces by vacuum centrifugation for approximately 20–30 min.
3. Add 15 μl trypsin (12 ng/μl of modified sequencing grade trypsin) in 50 mM NH$_4$HCO$_3$ solution to each gel piece.
4. Incubate at 4° for 1 hr, allowing the gel piece to rehydrate in trypsin solution without proteolytic activity.
5. Remove any remaining trypsin solution and add 15–20 μl of 50 mM NH$_4$HCO$_3$.
6. Incubate overnight at 37° with gentle shaking.

Sample Preparation for MALDI-PMM (II)

1. Wash the CBB G-250 stained gel pieces in 25 mM NH$_4$HCO$_3$, 50% acetonitrile at 37° with shaking for 10 min.
2. Remove excess solution and perform step 2 as above.
3. Add 8 μl of trypsin (15 ng/μl in 25 mM NH$_4$HCO$_3$) and incubate overnight at 37°.
4. Add a further 8μl of extraction solution (50% acetonitrile, 1% trifluoroacetic acid) and sonicate for 20 min in a water-bath sonicator.

For both method (I) and (II):

1. Place 1 μl of sample onto the MALDI target plate.
2. Add 1 μl of matrix (α-cyano-4-hydroxycinnamic acid, 10 mg/ml in 70% acetonitrile, 1% TFA) to the top of the sample droplet.
3. Allow the droplet to dry prior to MALDI-TOF MS.

Peptide Purification

The quality of mass spectrometry data depends greatly on the amount of common contaminants such as salts, the level of autolysis of the proteolytic enzyme, and background levels of interfering proteins such as keratins, as well as on the overall concentration of the original starting material. To acquire spectra with a high signal-to-noise ratio, chromatographic resins packed into low bed volume pipette tips can be used to bind and wash peptides free of noise-inducing salts. The bound peptides are eluted in a very low volume (<1μl) of matrix solution to concentrate the mixture onto the MALDI target (Fig. 4). Conversely, elution can

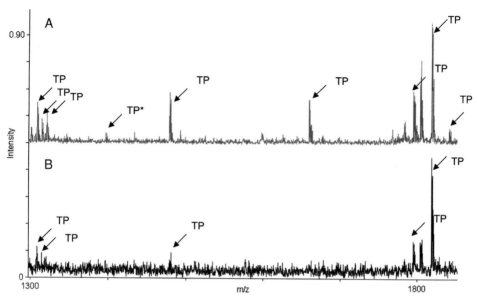

FIG. 4. MALDI-PMM mass spectrum of tryptic digest peptides derived from *Mycoplasma hyopneumoniae* Elongation Factor Tu. (A) Mass spectrum derived following concentration and desalting of 5 μl of peptide mix on a C_{18} Zip-Tip (Millipore, Bedford, MA). Nine peptides could be matched in the mass range 1300–1850 Da. (B) Mass spectrum derived from dried-droplet method using 1 μl of peptide mix. Five peptides could be matched in the mass range 1300–1850 Da. TP, tryptic peptide; TP*, tryptic peptide with one missed cleavage site.

be performed in low volumes of solvent and loaded directly into a capillary needle prior to ESI-MS. These manipulations can all be performed using prepackaged Zip-Tips (Millipore, Bedford, MA) loaded with either C_4 (for large peptides or whole proteins) or C_{18} (for lower mass peptides) chromatography resins. Other groups have shown that resins of many different properties loaded into thin diameter gel-loading tips can provide even greater sensitivity for eluted peptides prior to MALDI or ESI-MS.[11] Essentially, the technique described below is applicable to either type of tip.

Concentration and Desalting of Peptides using Zip-Tips or Micro-Columns

1. Activate the tip or column by washing through 20 μl of 100% acetonitrile.
2. Add 10 μl of 5% formic acid to the top of the column. Add an aliquot of the peptide mixture to be concentrated and desalted (for low abundance proteins this is typically the entire volume) to the column. Finally add another 10 μl of 5% formic acid on top of the sample.

[11] O. N. Jensen, M. R. Larsen, and P. Roepstorff, *Proteins* **Suppl. 2,** 74 (1998).

3. Using a modified syringe, apply air pressure to the top of the column, allowing the sample to pass slowly through the column.

4. Wash the bound peptides with 5% formic acid.

5. Add 0.7 μl of matrix solution to the top of the column and add finger pressure to spot 3–4 droplets of matrix/sample onto the MALDI target plate. The majority of the peptides will elute in the first and second droplets.

6. For electrospray (ESI) applications elute from the column in 1 μl of solvent (50% methanol, 1% formic acid) directly into a microcapillary compatible with the ESI-MS instrument.

Interpretation of MALDI-PMM Data

Acquisition of MALDI-MS PMM data results in a final spectrum taken from an average number of laser shots at the target plate. These single spectra are summed to achieve the final averaged spectrum. Masses are normally scanned over the range 800–3500 Da because lower mass signals are often due to matrix contaminants and poor database specificity of peptide products in this region. Higher masses are generally difficult to recover after cleanup as they bind tightly to the chromatography column and generally result in lower signals in the mass spectrometer. Modern mass spectrometers utilize sophisticated software to acquire data, calibrate, generate peak lists corresponding to peptide fragments, and search chosen databases for data output. Calibration is usually performed via internal fragments from trypsin autolysis peaks at each end of the mass range (e.g., 842.5 Da and 2211.1 Da). Care should be taken to examine all peptide masses—common contaminating peaks include peptide methylation (+14 Da), methionine sulfoxide (+16 Da), sodium adducts (+22 Da), and cysteine alkylated with acrylamide from the 2D gel process (+72 Da). Known keratin and trypsin peaks should also be removed. However, when adding these parameters to the database search profile, specificity may also be reduced. After the peak list is generated and the database is interrogated, a list of candidate matches is generated, based on the number of matching peptide masses within the peak list, and the percentage sequence covered by those matching peptides. Here it is important to remember that proteins with different properties may return erroneous results. For example, if no molecular mass cutoff is used, high mass proteins with many matching peptides derived by sheer coincidence, but with poor overall sequence coverage, may be found higher on a list of potential matches. A fraction of proteins may also contain very few arginine or lysine residues compatible with tryptic digestion. Although this is a low percentage (probably less than 5% for any given microbial genome), proteins that cannot be identified following PMM analysis should be subjected to ESI-MS/MS. Database output may also be misleading when cleavage products are identified on 2D gels (Fig. 5). For proteins appearing as "trains" of spots on 2D gels, "comparative" PMM can be performed. Mass spectra derived from the protein spots of interest are visually compared to determine which peptides are conserved between isoforms of the same protein, and which are unique. Phosphopeptides may ionize poorly using reflectron-MALDI scans; thus potentially posttranslationally altered peptide fragments can often be detected as a peak which disappears in the modified versus nonmodified spectrum. Scanning in linear-MALDI mode may aid in resolving modified peptides using MS. The presence of phosphopeptides

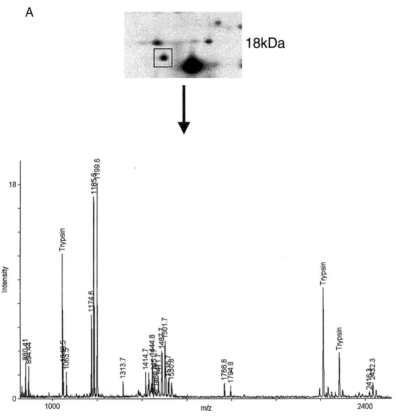

1	MRRSFLKTIG	LGVIALFLGL	LNPLSAASYP	PIKNTKVGLA	LSSHPLASEI	GQKVLEEGGN	AIDAAVAIGF	ALAVVHPAAG
81	NIGGGGFAVI	HLANGENVAL	DFREKAPLKA	TKNMFLDKQG	NVVPKLSEDG	YLAAGVPGTV	AGMEAMLKKY	GTKKLSQLID
161	PAIKLAENGY	AISQRQAETL	KEARERFLKY	SSSKKYFFKK	GHLDYQEGDL	FVQKDLAKTL	NQIKTLGAKG	FYQGQVAELI
241	EKDMKKNGGI	ITKEDLASYN	VKURKPVVGS	YRGYKIISMS	PPSSGGTHLI	QILNVMENAD	LSALGYGASK	NIHIAAEAMR
321	QAYADRSVYM	GDADFVSVPV	DKLINKAYAK	KIFDTIQPDT	VTPSSQIKPG	MGQLHEGSNT	THYSVADRUG	NAVSVTYTIN
401	ASYGSAASID	GAGFLLNNEM	DDFSIKPGNP	NLYGLVGDA	NAIEANKRPL SSMSPTIVLK NNKVFLVVGS PGGSPIITTV
481	LQVISNVIDY	NMNISEAVSA	PRFHHQULPD ELRIEKFGMP ADVKDNLTKM GYQIVTKPVM GDVNAIQVLP KTKGSVFYGS
561	TDPPKEF

FIG. 5. (A) Identification via MALDI-PMM of an 18 kDa C-terminal fragment of γ-glutamyltrans-peptidase (Ggp) from *H. pylori*. (B) Identification via MALDI-PMM of a 50 kDa N-terminal fragment of the same protein. Highlighted sequence shows the peptides covered in the PMM scan. Ggp is processed posttranslationally into a large and small subunit [C. Chevalier, J. M. Thiberge, R. L. Ferrero, and A. Labigne, *Mol. Microbiol.* **31**, 1359 (1999)].

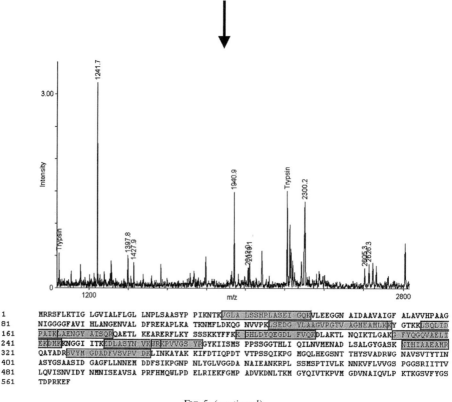

FIG. 5. (*continued*)

can be confirmed via on-target dephosphorylation using low volumes of alkaline phosphatase applied to the matrix dried droplet.[12]

Interpretation of ESI-MS/MS

Electrospray mass spectrometry (ESI-MS) has been used as an alternative to traditional Edman sequencing for obtaining sequence information from peptides. This technique is referred to as tandem mass spectrometry (MS/MS). The peptide

[12] M. R. Larsen, G. L. Sørensen, S. J. Fey, P. M. Larsen, and P. Roepstorff, *Proteomics* **1**, 223 (2001).

mixture is desalted and then applied to the mass spectrometer through a narrow-bore capillary. Traditionally ESI-MS is performed using a triple quadrupole analyzer, in which the initial quadrupoles are used to focus the ions, and the last to scan a fixed mass area. In MS mode the final quadrupole is used to scan the m/z (mass:charge) area for total peptides in the solution. In MS/MS mode, ions with a characteristic m/z can be selected in the first quadrupole. These ions are then collided with an inert gas (typically argon) in the second quadrupole, and the fragment ions thus produced are recorded in the third quadrupole, resulting in the generation of a fragment ion spectrum. Peptides preferentially fragment at amide bonds; thus the fragment ion spectrum contains a series of specific ion signals that differ in mass by one amino acid residue. Series of complementary N- and C-terminal sequence ion signals (b and y ions, respectively) can be observed, often resulting in complete sequence information of the isolated peptide (Fig. 6). Doubly and triply charged ions are preferentially selected for MS/MS since they fragment more readily than singly charged species. Highly unstable bonds in peptides, such as the bond between a phosphate group and serine or threonine amino acids, tend to break very easily in tandem MS/MS, resulting in additional ion signals corresponding to the loss of this modification from each of the peptides generated in the ladder sequence.

Data Storage

For high-throughput facilities performing 100 or more 2D gels and 1000–5000 PMM per week, an integrated LIMS (Laboratory Information Management System) approach is necessary. Proteomics-specific systems are beginning to reach the market, most notably the WorksBase software from Bio-Rad. Such systems allow chemical, IPG strip, and precast gel batch numbers to be cataloged, as well as processing information from the biological perspective including gel images, image analysis, cut spot annotation, and identification information from mass spectrometry. For example, quality control on isoelectric focusing gel experiments can be followed remotely, and therefore any problems arising during the course of these experiments can be immediately identified. For batch processing, high numbers of gels can be bar-coded and cataloged following image analysis, spots given unique identifiers, and the information from the mass spectrometer made readily searchable through an external index. Where lower throughput is necessary, 2D gels should be numbered sequentially for ease of identification (e.g., by using a sequential number system placed manually at the bottom of gel plates prior to pouring the second dimension gel). Spots cut from these gels should also be numbered sequentially or with the x,y well number corresponding to the position of that spot within a standard 96- or 384-well plate. Plates should therefore also have a unique identifier. Following imaging, 2D gels can be stored in airtight plastic bags with 1–5 ml of solution containing 1% or less sodium azide

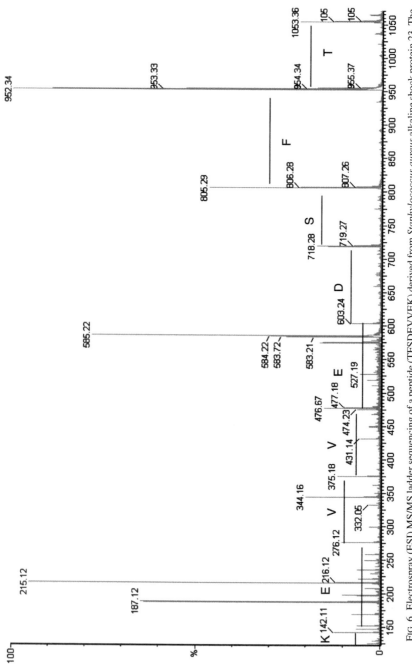

FIG. 6. Electrospray (ESI) MS/MS ladder sequencing of a peptide (TFSDEVVEK) derived from *Staphylococcus aureus* alkaline shock protein 23. The *y* ion series is shown.

(this prevents contamination and thus makes the gel amenable to further analysis several months, or even years, later). The majority of mass spectrometry manufacturers supply high-level software for searching, viewing, and interpreting both PMM and MS/MS data in a highly automated fashion. Such data can be viewed in a 96-well format (or other commercial format of choice) and can therefore be linked to the original plate and gel annotation system.

Biological Interpretation

The final step in the proteomics process involves the biochemical understanding of the results achieved within both a genomic and biological thought process, thus leading to sensible further experimentation. The researcher should ask: why was this protein affected by the experimental stimulus? Is this a secondary effect? How can the results be confirmed? Proteomics utilizing the separation and resolution of 2-DGE and the sensitivity of mass spectrometry is a useful tool for detecting proteins of interest in conjunction with thoughtful experimental design and for purifying those proteins for microcharacterization. New methodologies, technologies, and reagents should aid in standardizing this new field, thus leading to improvements in reproducibility across laboratories. Finally, improvements in cellular fractionation and enrichment, coupled with 2-DGE and mass spectrometry, will enable a greater percentage of expressed protein gene products to be identified and related to a cellular location. This should lead to the characterization of bacterial pathogenic determinants, virulence factors, and vaccine candidates, in turn providing a significant return to the world health community.

Acknowledgments

The author thanks colleagues at the Australian Proteome Analysis Facility, particularly Derek Van Dyk, Amanda Nouwens, Cassandra Vockler, Martin Larsen, and Brad Walsh. I also thank my collaborators: Christine Jacobs (Yale University) and Lucy Shapiro (Stanford University), Steven Djordjevic (Elizabeth Macarthur Agriculture Institute), Stuart Hazell (University of Southern Queensland), and Andrew Harris (University of New South Wales). All electrophoresis equipment and reagents were supplied by Bio-Rad Laboratories. Mass spectrometers and MS software were supplied by Micromass Ltd.

[16] Proteomic Analysis of pH-Dependent Stress Responses in *Escherichia coli* and *Helicobacter pylori* Using Two-Dimensional Gel Electrophoresis

By JOAN L. SLONCZEWSKI and CHRISTOPHER KIRKPATRICK

Introduction

As bacterial genomes are completed, we need to determine which of the genes predicted by annotation actually produce RNA and protein products, and under what regulation. Traditionally, such questions have been approached one gene at a time, by techniques such as Northern analysis, reporter gene fusion, and immunological detection. A different strategy is that of global and array-based methods which query a large number of genes at once.

The advantage of the global approaches, principally DNA microarray analysis of transcription of the genome[1] and two-dimensional gel electrophoresis (2D gels) of proteins in the "proteome," the total protein content expressed by a genome,[2,3] is that they reveal patterns of coordinate expression of numerous genes. Furthermore, genomic and proteomic methods may reveal particular genes and proteins whose expression might not have been tested under prevailing assumptions. Global approaches do miss a number of important genes, particularly those specifying RNA products and those whose changes in expression are small. Nonetheless, they have made important contributions to analysis of coordinate responses to environmental stresses such as heat shock and anaerobiosis, as well as virulence regulation.

The 2D gel analysis of proteins was developed by O'Farrell[2] and pursued extensively by Neidhardt, VanBogelen, and colleagues to explore stress response in *Escherichia coli*.[3] Advances in gel technology and protein identification, coupled with the availability of genomic sequence, have increased the attractiveness of 2D gels and made the technique accessible even to small laboratories. For a survey of methods, see Ref. 4. Here we summarize the principal options for 2D gel analysis and then outline a procedure which proves most fruitful in our laboratory. Current developments in our procedures are maintained on-line.[5]

[1] H. Tao, C. Bausch, C. Richmond, F. R. Blattner, and T. Conway, *J. Bacteriol.* **181,** 6425 (1999).

[2] P. H. O'Farrell, *J. Biol. Chem.* **250,** 4007 (1975).

[3] R. A. VanBogelen, K. Z. Abshire, A. Pertsemlidis, R. L. Clark, and F. C. Neidhardt, *in* "*Escherichia coli* and *Salmonella typhimurium*" (F. C. Neidhardt, R. I. Curtiss, C. C. Gross, J. L. Ingraham, and M. Riley, eds.), 2nd Ed., Chap. 115, p. 2067. ASM Press, Washington, D.C., 1996.

[4] M. R. Wilkins, K. L. Williams, R. D. Appel, and D. F. Hochstrasser (eds.), "Proteome Research: New Frontiers in Functional Genomics." Springer, Berlin, 1997.

[5] C. Kirkpatrick and J. L. Slonczewski, www2.kenyon.edu/depts/biology/slonc/labtools/2d_method. html, updated 08-08-02.

Options for 2D Gel Analysis

Experimental Design

To study protein profiles in response to a given environmental shift, there are two basic types of experimental design: kinetic shift and steady-state response. A kinetic experiment examines change in the rate of synthesis of given proteins following a sudden shift in an environmental factor such as temperature, oxygenation, or carbon source. The proteins must be pulse-labeled with a radiolabel, usually [^{35}S]methionine. The need for radiolabel limits the choice of culture conditions to defined media, either minimal medium or highly supplemented minimal medium minus methionine.[6] A steady-state experiment compares the total amount of various proteins synthesized under different long-term growth conditions.[7] Cells can be cultured under almost any conditions, even colonies scraped off plate agar.

The advantages of a kinetic experiment are that (a) radiolabel is highly sensitive, enabling detection of proteins at the lowest concentrations and (b) rapid changes in protein synthesis are detected. The advantages of a steady-state experiment are (a) the use of complex growth media, including methionine and other growth factors and (b) the ability to identify actual proteins observed, based on N-terminal sequence or spectroscopic analysis. In our experience, the steady-state approach is more likely to yield protein patterns specific to the conditions tested, rather than universal stress proteins which appear under almost any rapid change.

In testing environmental conditions, workers often neglect the role of pH. Both *Escherichia coli* and *Helicobacter pylori* possess the substantial metabolic capability to change the pH of their medium during growth. Growth on glucose is known to induce rapid acidification, but growth in peptide-rich media such as Luria broth causes alkalinization, up to pH 9 or higher.[8] A number of regulatory effects attributed to anaerobiosis or stationary phase have since been shown by our laboratory to be caused by shift in pH.[9] To control pH during growth to log phase or beyond, a buffer of appropriate dissociation constant (pK_a) must be included (Table I).

Gel System: Isoelectric Focusing and SDS–PAGE

A number of different gel systems are now available. For typical results of different approaches, Fig. 1 shows an example in which the first dimension (1-D) separation was performed through tube gel isoelectric focusing (IEF) of *E. coli* proteins and the second dimension (2-D) gel was stained with Coomassie blue.

[6] L. Lambert, K. Abshire, D. Blankenhorn, and J. L. Slonczewski, *J. Bacteriol.* **179,** 7595 (1997).

[7] D. Blankenhorn, J. Phillips, and J. L. Slonczewski, *J. Bacteriol.* **181,** 2209 (1999).

[8] J. L. Slonczewski, T. N. Gonzalez, F. M. Bartholomew, and N. J. Holt, *J. Bacteriol.* **169,** 3001 (1987).

[9] J. L. Slonczewski and J. W. Foster, *in* "*Escherichia coli* and *Salmonella typhimurium:* Cellular and Molecular Biology" (F. C. Neidhardt, R. I. Curtiss, C. C. Gross, J. L. Ingraham, and M. Riley, eds.), 2nd Ed., Chap. 96, p. 1539. ASM Press, Washington, D.C., 1996.

TABLE I
SULFONATE BUFFERS

Buffer	Chemical name	pKa[a]	Approximate pH range
HOMOPIPES	Homopiperazine-N,N'-bis-2(ethanesulfonic acid)	4.55	4.0–5.0
MES	2-(N-Morpholino)ethanesulfonic acid	5.96	5.5–6.5
PIPES	Piperazine-N,N'-bis(2-ethanesulfonic acid)	6.66	6.0–7.0
MOPS	3-(N-Morpholino)propanesulfonic acid	7.01	6.5–7.7
TES	N-[Tris(hydroxymethyl)methyl]-2-aminoethanesulfonic acid	7.16	6.8–8.2
TAPS	N-[Tris(hydroxymethyl)methyl]-3-aminopropanesulfonic acid	8.11	7.5–8.5
CAPSO	3-(Cyclohexylamino)-2-hydroxy-1-propanesulfonic acid	9.43	9.0–10.0
CAPS	3-(Cyclohexylamino)-1-propanesulfonic acid	10.08	9.5–10.5

[a] At 37°.

Figure 2[10] presents H. pylori proteins separated on an Amersham IPGphor strip nonlinear pH 3–10, silver stained, and Fig. 3 shows E. coli proteins separated on IPGphor pH 4–7, silver stained, composite images overlaid to reveal differences between two growth conditions.

The first-dimensional separation of proteins is performed in an isoelectric focusing gradient (IEF), which separates proteins based on their isoelectric point. The pH gradient is formed in a polyacrylamide gel containing multiply charged ampholines. Apparatus for automated pouring has greatly simplified the generation of IEF gels in narrow tubes (for example, the 1D system from Genomic Solutions). An even more convenient alternative is that of fixed IEF gels, in which the gradient of ampholines is polymerized into the gel as it is poured (Amersham IPG-phor; Bio-Rad, Hercules, CA). The lyophilized gel is provided on a strip which is rehydrated directly with the protein sample.

At present, the main advantage of tube gels is that they may enable detection of proteins of higher molecular weight (above 100 kDa). In our experience, however, detection of larger proteins is similar with both tube gel and fixed IEF systems. The fixed IEF systems have the advantages of generating more uniform gradients, easier application by novice users, and the availability of a wider range of pH gradients. Both tube gel and fixed IEF systems are limited in their ability to separate proteins of high hydrophobicity, such as integral membrane proteins.

The second-dimension gel system is sodium dodecyl sulfate–polyacrylamide (SDS–PAGE).[11] New forms of acrylamide such as Duracryl (Genomic Solutions)

[10] J. L. Slonczewski, D. J. McGee, J. Phillips, C. Kirkpatrick, and H. L. T. Mobley, *Helicobacter* **5,** 240 (2000).

[11] P. H. O'Farrell, *J. Biol. Chem.* **250,** 4007 (1975).

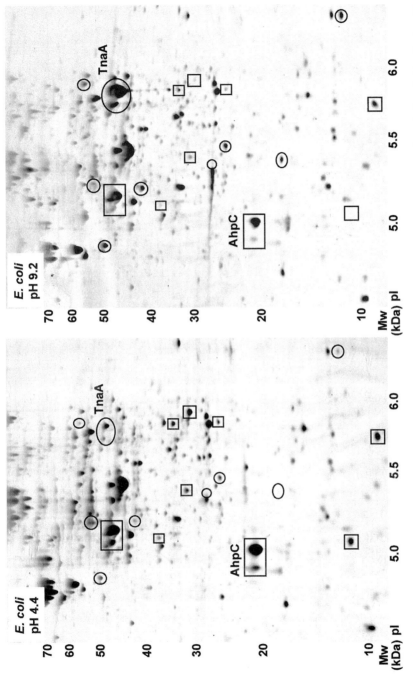

FIG. 1. *Escherichia coli* W3110 proteins following growth in acid or base. First dimension was run through a tube gel, ampholines 3–10, effective pH gradient of pH 4.5–6.5. The 2D slab gel was stained with Coomassie blue. Growth media contained LBK buffered with 100 m*M* HOMOPIPES, pH 4.4, or 100 m*M* AMPSO, pH 9.2, respectively. (○) Proteins were induced at high pH; (□) proteins were repressed. Horizontal axes represent approximate p*I* values of separated proteins; the vertical axes represent molecular masses in kDa. Modified from Ref. 7 with permission.

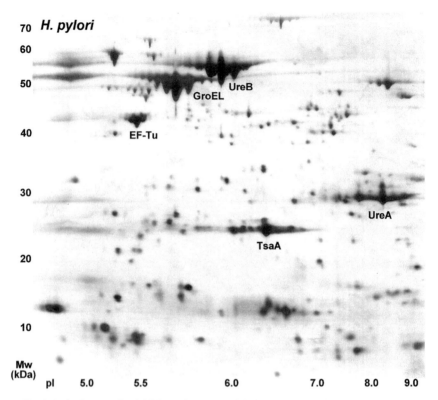

FIG. 2. *Helicobacter pylori* 26695 proteins expressed during growth on buffered *Brucella* agar with 12% O_2. The 1D gel was IPG strip, nonlinear pI range of pH 3–10. The 2D gel was silver stained. Horizontal axes represent approximate pI values of separated proteins; the vertical axes represent molecular masses in kDa.

enable production of gels with remarkable handling strength, over a wide range of acrylamide concentration to optimize the size range of proteins observed.

Identification and Quantification of Proteins

The protein stains most commonly used today for 2D gels are silver stain or Coomassie blue. Silver stain is more sensitive than Coomassie blue (in our experience it reveals about three times as many proteins), although its chemical response varies from protein to protein. It may actually not stain a few proteins detected by Coomassie blue. Both stains are highly quantitative with respect to varying amounts of a given protein.

If a single protein is of interest, its identification can be confirmed by its absence in a genetic deletion strain. This approach is not practical, however, for screens of large numbers of unknown protein spots. Gel position is also helpful, but rarely provided absolute identification, except for well-characterized spots. The

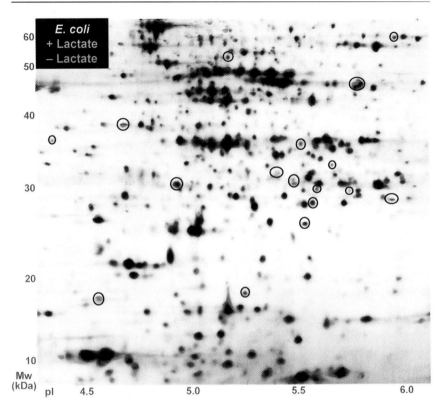

FIG. 3. *Escherichia coli* W3110 proteins expressed during growth with or without 50 m*M* L-lactate. The 1D gel was IPG strip, p*I* range of pH 4–7. The 2D gel was silver stained. Growth media contained LBK buffered with 50 m*M* MOPS and 50 m*M* TES at pH 6.7. Three gels of each condition are combined digitally to form a composite image. The two composite images are overlayed with color coding for spots that appear on one gel set but not the other. Pink indicates protein is induced in lactic acid; green indicates the protein is repressed. Horizontal axes represent approximate p*I* values of separated proteins; the vertical axes represent molecular masses in kDa.

technique of choice for protein identification today is MALDI-TOF spectroscopy, although some proteins, particularly those of molecular mass <15 kDa, still require N-terminal peptide sequence analysis.

A complication often observed is the appearance of a given protein in several different spots, most commonly in a "train" of spots over a range of p*I*. See, for example, TnaA in *E. coli* (Fig. 1) and UreA in *H. pylori* (Fig. 2[10]). The reasons for multiple spots remain unclear, although they probably arise from a combination of posttranslational modifications *in vivo* as well as chemical modifications that occur during sample preparation.

To compare different growth conditions, the size of a given protein spot can be compared between patterns of protein spots from gels performed on samples grown under different conditions. In the older literature, a protein was considered to be "induced" if it appeared in two out of three gels from independently grown cultures of an experimental condition, compared with gels from a control growth condition (see for example Ref. 6). More recently, computer image analysis has been used to subtract background from the protein pattern and normalize the spot pixel densities against the total protein density. This approach suffers from high errors associated with the background count, and from the skewing of total protein count by a few proteins whose high concentration overloads the relatively narrow range of image intensity available for current scanners; see for example the protein pattern of *H. pylori,* dominated by the greatly overloaded spots of UreB and GroEL (Fig. 2). Computer software such as Compugen Z-3 replaces background subtraction with a spot quantitation algorithm that is relatively insensitive to background and to overexposure, as well as a normalization algorithm that compares the histogram of all proteins visualized between two gel patterns.

Procedure for 2D Gel Experiment

Growth of Bacteria

Escherichia coli. E. coli W3110 is cultured in modified Luria broth with 100 mM KCl replacing NaCl (LBK), to avoid the toxicity of sodium ion in cultures grown at high pH. Buffers of appropriate pK_a are included (Table I) at 100 mM, pH adjusted with KOH. Sulfonate buffers are preferred to amine-based buffers such as Tris, whose deprotonated form can cross the cell membrane and elevate intracellular pH.[12]

Overnight cultures of *E. coli* grown at 37° are diluted 500-fold into 12 ml buffered medium per 125-ml baffled flask; for early log-phase growth, 1000-fold dilution is preferred. Cultures are aerated by rotating at 125–175 rpm (aerobic growth) or rotating slowly in closed tubes (anaerobic growth). Four different replicates are run, each from a separate overnight culture. The final OD_{600} is recorded, and each culture is resuspended at 4° in 1 ml of unbuffered growth medium or saline solution (to avoid buffer interference with the IEF gradient). The cells are spun again, and the pellet stored at −80°.

Helicobacter pylori. H. pylori 26695 is grown in buffered yeast tryptone medium (HPYT, Ref. 10) or on *Brucella* agar plates with oxygen regulation (T. Seyler and R. Maier, unpublished, 2000). Oxygen is maintained at 12% by incubation in a Forma Scientific incubator (CO_2, water-jacketed, with oxygen sensors nitrogen gas injection). The advantage of growth in liquid medium is that the cell density can be controlled. On the other hand, plate-grown cells appear healthier, show cleaner protein content, and allow for better control of oxygen concentration.

---

[12] D. R. Repaske and J. Adler, *J. Bacteriol.* **145,** 1196 (1981).

Protein Preparation

Bacterial protein preparation is based on the protocol of Genomic Solutions, developed by VanBogelen.[12a] All solutions are made in water purified to resistivity 18 MΩ-cm on a Milli-Q UFPlus Reagent grade water purification system (Millipore, Fisher). *Note:* Urea contained in sample buffers needs to be of the highest electrophoresis grade, stored under desiccation. If vertical streaks appear in the second-dimension gels, replace all urea solutions.

Buffers

Sample Buffer 1: 0.3% (w/v) Sodium dodecyl sulfate (SDS), 200 mM dithiothreitol (DTT), 28 mM Tris-HCl, 22 mM Tris base. Aliquot the mixture to 1 ml volumes and store at $-80°$.

Sample Buffer 2 : 24 mM of a 1.5 M Tris base stock, 476 mM of a 1.5 M Tris-HCl stock, 50 mM of a 1.0 M MgCl$_2$ stock, 1 mg/ml DNase I, 0.25 mg/ml RNase A. Mix the water and first three reagents and chill on ice before adding the DNase I and RNase A. Aliquot the mixture to 100 μl volumes and store at $-80°$.

Rehydration Solution: 8.0 M urea, 2.0% (w/v) CHAPS, trace amounts of bromphenol blue. Aliquot the mixture to 2.5 ml and store at $-20°$. Just prior to use, add 7 mg of DTT and 12.5 ml IPG buffer (same pH range of the IPG strip) to each 2.5 ml aliquot of rehydration stock solution. [*Note:* Sample Buffer 3 (Genomic Solutions) can also be used, although we find better results with Rehydration Solution.]

To prepare protein sample: Sample buffer 1 is added to each of the samples, then mixed by pipetting up and down until the pellet has completely dissolved. Add a volume (in μl) of sample buffer 1 equal to the product of 6.15 × the culture volume (ml) × the final OD$_{600}$ reading after growth of the culture. Mix by pipetting until the pellet has completely dissolved, then boil for 3.5 min in a locking-lid microfuge tube. Place on ice for 10 min, then add Sample Buffer 2 (1/10 × volume added of Sample Buffer 1), and vortex. Place on ice, then add Rehydration Solution (4 × volume Sample Buffer 1). Immediately mix by pipetting, then store at $-80°$. Note that when samples are taken for gel runs, the remainder can be returned to $-80°$ and stored for several weeks to run additional samples.

1D Gel Isoelectric Focusing

For the IPGphor Immobiline strip, there are a range of choices. We find that the range from pH 4 to pH 7 covers a majority of the proteins of interest in *E. coli*, and it is the best choice for beginning investigation. *H. pylori*, however, shows a greater

[12a] R. A. VanBogelen and F. C. Neidhardt, *Proc. Natl. Acad. Sci. U.S.A.* **87**, 5589 (1990).

proportion of alkaline proteins; we therefore use the nonlinear pH 3–10 range. The nonlinear gradient spreads out proteins in the range from pH 4 to pH 7 in which the majority of proteins still appear, but this also enables visualization of proteins up to p*I* 9 or 10 (Fig. 2). We use 18-cm strips; smaller or larger lengths are available.

Procedure. Strip holders are prepared by washing with detergent to remove residual protein and are rinsed thoroughly with 18 MΩ-cm water. Protein samples are set up according to the IPGphor procedure (Amersham Biosciences). For 10 protein samples, use two 2.5-ml aliquots of Rehydration Stock solution. To each aliquot add 7 mg of DTT and 12.5 μl IPG Buffer (same pH range as the IPG strip) per aliquot. Add 325 μl of rehydration solution to each of 10 microcentrifuge tubes. Of each protein sample, 25 μl (approximately 50 μg protein) is added to the rehydration solution, for a total of 350 μl. (*Note:* For Western blot and protein identification, use 150 μl of protein sample and 200 μl rehydration solution.)

Sample/rehydration solution is slowly applied to the center of each strip holder. The protective plastic cover is removed from the IPG strip and, with the gel side down, the anodic (pointed) end of the gel strip is lowered onto the anodic (pointed) end of the strip holder. Anodic end first, the gel is lowered into the solution, to coat the entire gel, then lifted, and lowered to ensure complete wetting. Make certain that both electrodes are covered with the gel. To each gel strip holder, 0.75–1.0 ml of PlusOne IPG DryStrip cover fluid (Amersham Pharmacia Biotech) is applied, coating the entire IPG strip. Strip holder covers are then placed on each strip holder, with the protrusions facing the gel, to provide complete contact between the gel and electrode. The gels are loaded and run on the IPGphor isoelectric focusing system, and set for 14 hr rehydration, 1 hr at 500 V, 1 hr at 2000 V, and 5 hr at 8000 V.

2D SDS–PAGE Gel Electrophoresis

Buffers and Reagents

1.5 *M* Tris Blend: 190.8 g TRIZMA preset crystals (tris[hydroxymethyl] aminomethane and Tris hydrochloride) pH 8.8 (Sigma, St. Louis, MO), in 1 liter. Store at 4°.

SDS equilibration buffer: 50 m*M* Tris Blend, 6 *M* urea, 30.0% (v/v) glycerol, 2.0% (w/v) SDS, trace amounts of bromphenol blue. Aliquot to 25.0 ml and store at −20°. Just prior to use add 0.25 g DTT.

10× Tris/Gly/SDS Running Buffer: 10 g SDS, 30.28 g Tris base, 144.13 g glycine, and 18 MΩ-cm water to 1 liter.

Slab Solution (11.5% acrylamide): 465 ml Duracryl (Genomic Solutions), 300 ml 1.5 *M* Tris blend, 436 ml 18 MΩ-cm Water, 12.3 ml 10% SDS, 0.618 ml TEMED, 3.04 ml of fresh 10% ammonium persulfate.

Top Chamber Running Buffer: 2 liter 18 MΩ-cm Water plus 250 ml 10× Tris/Gly/SDS running buffer.

Bottom Chamber Running Buffer: 10 liter 18 MΩ-cm Water plus 1.1 liter 10× Tris/Gly/SDS running buffer.

Procedure. Gel plates are assembled and poured as indicated (Genomic Solutions). To obtain a gel approximately 22 cm square, glass plates (28 cm × 23 cm) are cleaned and assembled, separated by spacer sheets. The beveled edge side of each of two plates is cleaned with 70% (v/v) ethanol, then assembled with spacers and glue stick. The 2D casting chamber is then covered and leveled.

The slab solution is poured into the filling chamber. When most of the bubbles have risen, the clamp is opened and the plates are filled to approximately 0.5 cm from the top. 0.65 ml 18 MΩ-cm water is applied evenly with the pipette tip pointed at a slant directly into the gel, starting from the middle outward; then another 0.65 ml is applied out toward the other side. This needs to be done rapidly for all 12 gels, but not so fast that the buffer projects too far into the gel. After polymerization (30 min) the gels are overlaid with 0.5× Tris/Gly/SDS running buffer. Gels can be left at room temperature overnight. The coolers for the gel running chambers are set to 14° and allowed to equilibrate.

After polymerization, the slab gels are removed, and excess gel is rinsed off with warm water. The gels are fitted into latex gaskets. When the gel sandwiches are loaded, the bottom chamber running buffer is drained just above the electrodes and the top running buffer is added to just above the top of the plates. Each gel is equilibrated for 10 min. Two aliquots of equilibration buffer are thawed, and 0.25 g DTT is added to each. Ten ml equilibration buffer is pipetted into five extrusion trays. Two 1D gels are placed into each tray and moved periodically to completely wet each gel in the buffer. After 10 min, tweezers are used to place the 1D gel into the beveled edges of the 2D sandwiches. With the plastic back of the gel toward the back of the chamber, the gel is firmly positioned into place with a gel installation tool, avoiding bubbles. The remaining top running buffer is carefully poured into the corner of the top chamber. The 2D gels are run at 500 V for 4–4.5 hr, or until the dye has reached roughly 0.5 cm to the bottom of the gel.

After the blue dye has reached within 1 cm of the bottom edge, the slab gel sandwich is lifted from the running chamber, and the top plate is removed with a metal spatula. Each gel is notched from 1 to 5 according to its position in the 2D chamber.

To silver stain on the next day, 1 liter of fixative (recipe under Silver Stain) is poured into each of two 7.8-liter storage containers, followed by the 2D gels. Each container is labeled 1 or 2 to indicate what chamber the gels are from, then rotated gently for 30 min and placed in a cold room overnight.

To clean the glass plates: Place overnight in a tank of Micro detergent solution, in racks made out of test-tube holders to keep the glass plates apart. The next day, rinse plates for 2 hr in a bucket with a line from the sink directing water from the bottom upward. Place in test-tube racks to dry.

Silver Staining of 2D Gels

The following procedure is modified from Amersham Biosciences (Piscataway, NJ) Silver Staining Kit, optimized for 22-cm slab gels and subsequent MALDI-TOF analysis.

Reagents (to Stain 10 Gels)

Fixative: 800 ml 95% (v/v) Ethanol, 200 ml glacial acetic acid, and 1000 ml 18 MΩ-cm water to a 2-liter flask.

Sensitizing Solution: 600 ml 95% Ethanol, 80 ml 5%(w/v) sodium thiosulfate, 136 g sodium acetate, and 18 MΩ-cm water to 2 liter.

Silver Stain: 0.75 g Silver nitrate to 300 ml 18 MΩ-cm water. Repeat this for as many gels to be stained.

Developing Solution: Add 7.5 g sodium carbonate to 300 ml. Repeat this for as many gels to be stained. Add 120 μl 37% (v/v) formaldehyde just prior to use.

Stop Solution: 29.2 g Disodium ethylenediaminetetraacetic acid (EDTA) to 2 liter.

Preserve Solution: 450 ml 95% Ethanol, 69 ml 87% (v/v) glycerol, and 981 ml 18 MΩ-cm water.

Procedure. The gels are carefully transferred from the fixative into two separate storage containers, each with half of the Sensitizing Solution, then rotated for 30 min. The gels are washed four times by transferring them into new containers, filled with enough 18 MΩ-cm water to cover the gels, and rotated. The first three washes are done for 5 min and the last one for 10 min. This step is important in removing excess Sensitizing Solution from the gel, which can add background.

Transfer each gel of a set of five to a developing tray containing silver stain, then rotate five trays gently for 20 min. (Handling of individual gels is best accomplished by folding a gel over into quarters, transfering with a glove, then immediately unfolding and opening out the entire gel in the new solution.) Transfer each gel into a wash container for two 2-min washes, then 1-min, followed by a quick rinse to remove excess silver stain, then transfer to trays containing Developing Solution. Rotate trays until each gel has developed all protein spots but avoided darkening of background. Transfer each developed gel to a container containing half the Stop Solution. Shake gently for 10 min.

Wash three times for 5 min, then transfer into two containers with half of the Preserve Solution each, and rotate gently for 1 hr. Each gel is placed into a 1 gallon plastic bag labeled with the strain information, protein preparation date, and the run date. Gels are scanned against a white background, taking care to roll out bubbles using a test tube.

Silver-stained gels may be stored for months at 4–10°. To identify protein, a gel spot can be cut out with a clean razor or spatula, then placed in a microfuge tube with 2 μl water and shipped to a MALDI-TOF laboratory. *Note:* During the entire process of gel and spot handling, gloves must be changed frequently to avoid keratin contamination of protein samples.

Coomassie Blue Staining of 2D Gels

Although less sensitive than silver stain, a few proteins show up only with Coomassie blue stain.

Reagents

   100% TCA (Trichloroacetic acid): Add 210 ml 18 MΩ-cm water to a 500 g bottle of TCA.
   2× Stock: For 4 liter add 80.0 ml 100% TCA, 80.0 ml 0.5% Coomassie blue, 600 ml glacial acetic acid, and 18 MΩ-cm water to 4 L.

Procedure. Place up to five gels in a container of 500 ml 2× stock and 500 ml 95% ethanol. Rotating gently overnight. Pour off solution and add mixture of 500 ml 2× stock and 500 ml 18 MΩ-cm water. Rotate gently for 15 min. Pour off and add mixture of 500 ml 2× stock and 500 ml 7.5% acetic acid. Rotate gently for 15 min. Gels are bagged, labeled, and scanned as for the silver staining procedure (above).

Western Blotting

Most proteins now are identified by MALDI-TOF of silver-stained gels, but occasionally proteins, especially of low molecular weight, cannot be identified by this method; in this case, N-terminal sequence identification may be performed. The gels must be blotted immediately onto a nylon membrane, which is then stained with Coomassie blue. The procedure below is based on the Type II Investigator Graphite Electroblotter (Genomic Solutions).

Buffers and Reagents

   Anode Buffer 1: 36.3 g Tris base, 100 ml methanol, and 18 MΩ-cm water to 1 liter. Store at room temperature for up to 1 week.
   Anode Buffer 2: 12.1 g Tris base, 100 ml methanol, and 18 MΩ-cm water to 1 liter. Store at room temperature for up to 1 week.
   Cathode Buffer: 3.03 g Tris base, 3 g 6-amino-*n*-caproic acid, 200 ml methanol, and 18 MΩ-cm water to 1 liter. Store at room temperature for up to 1 week.
   Coomassie Stain: 500 ml methanol, 100 ml acetic acid, 1 g Coomassie Brilliant Blue R-250, and 18 MΩ-cm water to 1 liter.
   Destain: 500 ml methanol, 100 ml acetic acid, and 18 MΩ-cm water to 1 liter.

Procedure. For each gel, 5 sheets of 3-mm chromatography paper and 1 sheet of Millipore (Bedford, MA) Immobilon-P 0.45 μm transfer membrane are cut into squares 9.25 inches by 9.25 inches. Soak gel in Cathode Buffer for 5 min, with gentle rotating. Two sheets of chromatography paper are saturated in Anode Buffer 1 and placed on the anode. One sheet of chromatography paper is saturated in Anode Buffer 2 and placed on top of the stack. One sheet of Immobilon-P membrane is soaked in 100% methanol, then saturated in Anode Buffer 2, and placed on top. The gel equilibrated in Cathode Buffer is placed on top, then covered with a dialysis membrane sheet soaked in Cathode Buffer. Two sheets of chromatography paper are wet in Cathode Buffer and placed on top of the stack. Bubbles are removed by rolling a glass pipette over the stack.

Two gels can be run simultaneously, the two stacks separated by a piece of dialysis membrane soaked in Cathode Buffer. The cathode of the electroblotter is placed on top of the stack and connected to the power supply. Run power for 30 to 60 min, depending on whether to optimize recovery of smaller or larger proteins. (Proteins below 20 kDa tend to run through the stack after 60 min, whereas larger proteins may not fully transfer after only 30 min.)

Rinse filters in 100% methanol and rotate in Coomassie blue stain for 30 min. Transfer to Destain, rotating until spots appear. Dried protein blots can be stored up to a month without significant degradation of protein. Excise each protein spot and wash 3 times in 10% methanol, for 2 min each, to remove residual glycine from the 2D gel running buffer. Samples are sent to an N-terminal sequence facility (we use the Molecular Structure Facility at the University of California at Davis). As few as seven amino acid residues of sequence may be needed to generate a match with the bacterial genome through SwissProt or GenBank.

Quantitative Analysis Using Compugen Z3

Compugen Z3 (version 2.0) enables quantification of protein spots based on pixel density, with a sophisticated algorithm that detects spot differences more reliably than methods based on background subtraction. Furthermore, pairwise comparisons between protein spots on multiple replicate gels are computed automatically, enabling a level of statistical analysis not possible previously. We use Z3 to perform pairwise comparisons of gel images from two different sets of three independent replicate cultures.[13,14]

The gel images are loaded into Z3 to be analyzed and converted into layered views containing the image of a comparative gel overlaying the image of a reference gel. We use gels from three independent cultures at each of two growth conditions (Fig. 3). It is best to select gel images that exhibit comparable amounts of overall protein and intensity of stain, although Z3 compensates for modest differences. Relative spot densities are computed by comparing the nonsaturated pixels of the spot on each of two gels and fitting by linear regression. Normalization of overall protein content is performed automatically by comparison of the overall histograms of spot ratios (the "differential expression histograms") between two gels.

[13] C. Kirkpatrick, L. M. Maurer, N. E. Oyelakin, Y. Yontcheva, R. Maurer, and J. L. Slonczewski, *J. Bacteriol.* **183,** 6466 (2001).
[14] L. M. Stancik, D. M. Stancik, B. Schmidt, D. M. Barnhart, Y. N. Yoncheva, and J. L. Slonczewski, *J. Bacteriol.* **184,** 4246 (2002).

TABLE II

Differential Expression Ratios for Pairwise Comparisons of Protein Spot Densities from Reference versus Comparative Gel Images[a]

Protein ID No.	1A–1B	1A–1C	1B–1C	2A–2B	2A–2C	2B–2C	1A–2A	1A–2B	1A–2C	1B–2A	1B–2B	1A–2C	1C–2A	1C–2B	1C–2C	Geometric mean	LDE ±SD
1001							Um+	Um+	Um+	Um+	Um+	Um+	Um+	Um+	Um+	Um+	Um+
1002							4.56	3.60	2.60	5.13	2.77	6.99	2.91	5.02	3.21	3.88	0.59 ± 0.1
1003	0.56						0.70	0.31	0.22	0.31	0.32	0.24	0.44	0.38	0.28	0.34	−0.47 ± 0.
1004	0.39						0.58	0.63	0.52	1.02	1.66	1.50	1.01	.90	0.51	0.85	−0.07 ± 0.
1005	Um+		2.61			0.60	2.73	Um+	3.59	1.01	0.84	Um+	.72	1.12	0.76	—	—

[a] Reference gel images (1A, 1B, 1C) and comparative gel images (2A, 2B, 2C).

Um+ indicates spots on Gel 2 lacking a matching spot on Gel 1.

DE is the differential expression ratio of the comparative to the reference spot density.

LDE is the average of the $\log_{10}$ of all differential expression ratios.

Table II shows typical differential expression ratios (DE values) for proteins from pairwise comparisons of gel images from two different sets of three independent replicate cultures. The reference gel image is designated gel 1 and the comparative gel image is designated gel 2. For each protein, there are three Gel 1–Gel 1 ratios; three Gel 2–Gel 2 ratios; and nine Gel 2–Gel 1 ratios. We consider a protein to exhibit significant differential expression if its DE ratios are greater than 1.5 (50% induced) or less than 0.67 (30% repressed) for at least 7 of 9 Gel 2–Gel 1 layered views (Table II). Dashed cells indicate control gel comparisons showing insignificant difference. Protein 1001 is "unmatched," appearing only in the comparative growth condition. Protein 1002 is induced nearly fourfold; protein 1003 is repressed about threefold. The last two rows in Table II show typical DE values of proteins showing no consistent pattern of induction or repression.

Because expression values represent ratios between conditions, we perform logarithmic conversion and represent their distribution as the mean log ratio of the DE values (LDE). Proteins induced in the comparative gel show a positive LDE; proteins repressed show a negative value of LDE. The LDE values can be plotted as a function of the growth conditions.[14]

An alternative approach to analysis and presentation is to generate a composite gel image for each experimental condition, then to overlay the two composite images (Fig. 3). Proteins induced in the comparative composite gel appear pink, whereas proteins repressed appear green. We use the composite gel comparison to present an overall global picture of the spot differences.[14]

Acknowledgment

This work was supported by Grant MCB 9982437 from the National Science Foundation.

[17] Mycobacterial Proteomes

By HANS-JOACHIM MOLLENKOPF,*[†] JENS MATTOW,[†] ULRICH E. SCHAIBLE, LEANDER GRODE, STEFAN H. E. KAUFMANN, and PETER R. JUNGBLUT

Introduction

The term proteome is used to describe the protein complement of an organism or biological compartment at a certain time under certain conditions.[1] The protein

* Corresponding author.
† Equal contribution to the manuscript.

[1] V. C. Wasinger, S. J. Cordwell, A. Cerpa-Poljak, J. X. Yan, A. A. Gooley, M. R. Wilkins, M. W. Duncan, R. Harris, K. L. Williams, and I. Humphery-Smith, *Electrophoresis* **16,** 1090 (1995).

complement reflects the genetic information and in addition biological and environmental influences, which results in diversification by co- and posttranslational protein modifications of the primary translation products.[2] Moreover, one open reading frame (ORF) can give rise to more than one protein species as a result of differential pre-mRNA splicing. Global changes in composition or abundance of proteins can be studied by proteome analysis. The classical proteome approach comprises protein separation by two-dimensional gel electrophoresis (2-DE) and protein identification and characterization by mass spectrometry (MS) and/or N-terminal sequencing by Edman degradation. Electrophoretic protein separation may be replaced by liquid chromatography (LC), which requires digestion of the protein mixture into peptide fragments before separation.[3] After LC the peptides are identified by data-dependent electrospray ionization mass spectrometry (ESI-MS). The major advantages of proteome analysis include the identification of co- and posttranslational modifications as well as the study of global changes in protein composition between different biological or environmental situations, e.g., virulent versus attenuated pathogen strains or in health versus disease by comparative proteome analysis. The high-resolution protein separation methods, together with the development of rapid and highly sensitive MS methods for protein identification and characterization and the rapidly growing DNA and protein databases, have paved the way for high-throughput proteome analysis.

Identification of gel-separated proteins by matrix-assisted laser desorption/ionization MS (MALDI-MS) and ESI-MS has been reviewed.[4] MALDI-MS peptide mass fingerprinting (PMF)[5-8] has been shown to be a method for sensitive high-throughput identification of gel-separated proteins. For PMF analysis, the protein spot of interest is excised from a preparative 2-DE gel and digested, usually by enzymatic proteolysis with trypsin or by chemical cleavage. The resulting peptides are desalted, concentrated, and subsequently analyzed by MALDI-MS (or ESI-MS) to determine their masses. The experimentally measured peptide masses are compared with peptide mass maps created by *in silico* cleavage of the proteins in a protein database. In general only a few accurately measured peptide masses (with a mass error < 30 ppm) are sufficient to identify proteins with high confidence.[9,10] Although MALDI-MS PMF is an effective tool for high-throughput protein

[2] P. R. Jungblut, B. Thiede, U. Zimny-Arndt, E.-C. Müller, C. Scheler, B. Wittmann-Liebold, and A. Otto, *Electrophoresis* **17,** 839 (1996).

[3] A. J. Link, J. Eng, D. M. Schieltz, E. Carmack, G. J. Mize, D. R. Morris, B. M. Garvik, and J. R. Yates, *Nat. Biotechnol.* **17,** 676 (1999).

[4] S. P. Gygi and R. Aebersold, *Curr. Opin. Chem. Biol.* **4,** 489 (2000).

[5] W. J. Henzel, T. M. Billeci, J. T. Stults, S. C. Wong, C. Grimley, and C. Watanabe, *Proc. Natl. Acad. Sci. U.S.A.* **90,** 5011 (1993).

[6] P. James, M. Quadroni, E. Carafoli, and G. Gonnet, *Biochem. Biophys. Res. Commun.* **195,** 58 (1993).

[7] M. Mann, P. Hojrup, and P. Roepstorff, *Biol. Mass Spectrom.* **22,** 338 (1993).

[8] D. J. C. Pappin, P. Hojrup, and A. J. Bleasby, *Curr. Biol.* **3,** 327 (1993).

[9] H. W. Lahm and H. Langen, *Electrophoresis* **21,** 2105 (2000).

[10] K. Gevaert and J. Vandekerckhove, *Electrophoresis* **21,** 1145 (2000).

identification, it has its shortcomings. It is often insufficient to reliably identify low molecular mass proteins and multiple proteins in single spots, especially at very low protein levels. Furthermore, the identification of unpredicted proteins which are not included in protein databases requires additional protein sequence information because databases containing nucleotide sequences cannot be searched with PMF information alone.[9,10] In these cases MALDI-MS PMF should be complemented by other protein analytical techniques that generate sequence information, e.g., postsource decay (PSD) MALDI-MS[11] or nanoelectrospray ionization tandem mass spectrometry (nESI-MS/MS),[12] for unequivocal protein identification.

Background

Pulmonary tuberculosis due to infection with the intracellular bacterial pathogen *Mycobacterium tuberculosis* is the major cause of morbidity and mortality by a single bacterial pathogen, resulting in 1.7–2.2 million reported deaths annually worldwide and an estimated infection rate of one-third of the world population (World Health Organization, 2000). Increasing numbers of multidrug-resistant *M. tuberculosis* strains and coinfection with HIV largely account for the recent resurgence of the tuberculosis epidemics. The increasing prevalence of tuberculosis makes this disease a major focus for vaccine and drug research and for the development of novel diagnostic tools.[13] This research is no longer restricted to single gene driven approaches but also includes systematic research strategies such as proteome, transcriptome, and genome analysis.

A vaccine against tuberculosis was created as early as 1927 by Calmette and Guérin. This so-called Bacillus Calmette–Guérin (BCG) vaccine is an attenuated strain of *M. bovis* which is still in use today. BCG has been administered to more than 3 billion people worldwide, being the most widely used live vaccine. In children BCG vaccination can prevent miliary and meningeal forms of tuberculosis, but protection against pulmonary tuberculosis in adults is highly variable and generally incomplete.[14] To date the reason for the attenuation of BCG has not been analyzed in detail. Hence comparative proteome analysis represents a major method for the detection and identification of differences between virulent and attenuated mycobacteria not predicted by genome research.

The complete genomic DNA sequence of *M. tuberculosis* strain H37Rv was published in 1998 by Cole *et al.*[15] and the DNA sequences of the clinical isolate *M. tuberculosis* CDC1551 and *M. leprae* strain *TN* have also been completely sequenced (www.tigr.org). Several other mycobacterial species are also currently

---

[11] B. Spengler, D. Kirsch, R. Kaufmann, and E. Jaeger, *Rapid Commun. Mass Spectrom.* **6**, 105 (1992).
[12] M. Mann and M. Wilm, *Anal. Chem.* **66**, 4390 (1994).
[13] S. H. E. Kaufmann, *Nat. Med.* **6**, 955 (2000).
[14] P. E. Fine, *Lancet* **346**, 1339 (1995).
[15] S. T. Cole *et al.*, *Nature* **393**, 537 (1998).

being sequenced (*M. bovis, M. bovis* BCG, *M. avium, M. paratuberculosis*) or their sequencing has been initiated (*M. microti, M. ulcerans*).[16]

Analysis of Mycobacterial Proteomes

We have used 2-DE for protein separation in combination with MS for the identification and characterization of gel-separated proteins in order to systematically analyze the proteomes of different virulent (*M. tuberculosis* H37Rv and Erdman) and attenuated (*M. bovis* BCG Chicago and Copenhagen) strains. Emphasis is on the identification of *M. tuberculosis* specific proteins (missing in BCG preparations), as we consider these proteins to represent putative virulence factors and potential antigens for diagnosis of and vaccination against tuberculosis. As a basis for further investigations we have established a mycobacterial proteome database,[17] which is available via Internet (www.mpiib-berlin.mpg.de/2D-PAGE). The different 2-DE patterns from the mycobacterial strains mentioned above and from different biological compartments of the bacteria—whole cell preparations (CP) and culture supernatants (CSN)—have been introduced into this database.[18] The separation of mycobacterial CP by 2-DE resulted in silver-stained protein patterns comprising approximately 1800 distinct protein spots (Fig. 1). Silver-stained 2-DE patterns of mycobacterial CSN comprise approximately 800 protein spots (Fig. 2). To date, we have analyzed over 1300 protein species from mycobacterial 2-DE patterns by MS with an identification efficiency >90%. In order to minimize experimental error and to ensure reproducibility of our proteome data we rely on triplicate analysis of independent sample preparations. CP preparations of all mycobacterial strains examined show a high density of spots in the acidic range, whereas in the basic range, spot density is clearly reduced. This observation is in agreement with the theoretical distribution of the predicted proteins of *M. tuberculosis* H37Rv.[19] The genomes of members of the *M. tuberculosis* complex, including the four mycobacterial strains examined to date, are highly conserved.[20] Consistent with this, our studies confirmed that the vast majority of protein spots have counterparts with identical electrophoretic mobility in all mycobacterial strains investigated. However, we also detected clear differences between different mycobacterial strains including variations in spot intensity, in the presence or absence

[16] A. S. Pym and R. Brosch, *Genome Res.* **10,** 1837 (2000).

[17] H. J. Mollenkopf, P. R. Jungblut, B. Raupach, J. Mattow, S. Lamer, U. Zimny-Arndt, U. E. Schaible, and S. H. E. Kaufmann, *Electrophoresis* **20,** 2172 (1999).

[18] P. R. Jungblut, U. E. Schaible, H. J. Mollenkopf, U. Zimny-Arndt, B. Raupach, J. Mattow, P. Halada, S. Lamer, K. Hagens, and S. H. E. Kaufmann, *Mol. Microbiol.* **33,** 1103 (1999).

[19] B. L. Urquhart, S. J. Cordwell, and I. Humphery-Smith, *Biochem. Biophys. Res. Commun.* **253,** 70 (1998).

[20] S. Sreevatsan, X. Pan, K. E. Stockbauer, N. D. Connell, B. N. Kreiswirth, T. S. Whittam, and J. M. Musser, *Proc. Natl. Acad. Sci. U.S.A.* **94,** 9869 (1997).

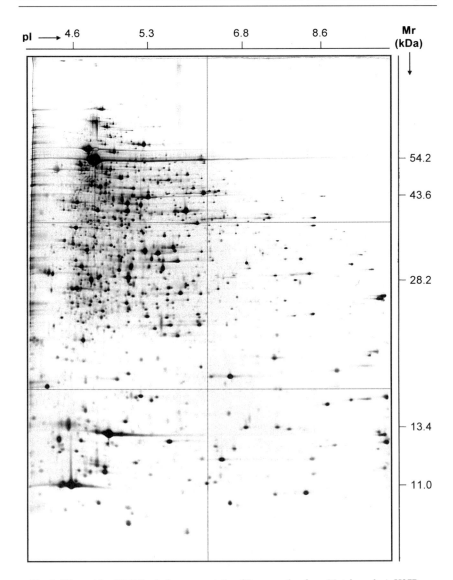

FIG. 1. Silver-stained 2-DE gel of a representative CP preparation from *M. tuberculosis* H37Rv.

of particular spots or in the electrophoretic mobility of protein species.[18,21] Comparing the proteomes of *M. tuberculosis* H37Rv and *M. bovis* BCG Chicago we initially detected as many as 31 variant protein species.[18] In a more recent study we systematically compared 2-DE patterns of whole cell preparations of virulent

[21] J. Mattow, P. R. Jungblut, E.-C. Müller, and S. H. E. Kaufmann, *Proteomics* **1,** 494 (2001).

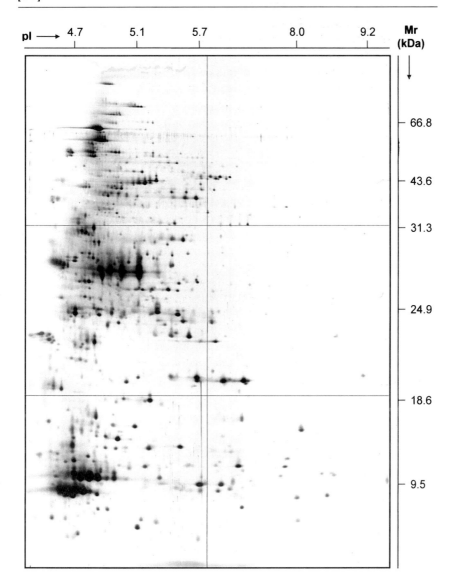

FIG. 2. Silver-stained 2-DE gel of a representative CSN preparation *M. tuberculosis* H37Rv.

(*M. tuberculosis* H37Rv, Erdman) and attenuated (*M. bovis* BCG strains Chicago, Copenhagen) mycobacterial strains.[21] This study resulted in the identification of 40 protein species that were unique for the attenuated *M. bovis* BCG strains and 56 protein species only observed for the virulent *M. tuberculosis* strains. Forty-four of the observed *M. tuberculosis*-specific protein species have been analyzed by MS and 32 of them have been identified.

In another investigation we focused on the identification and characterization of 190 acidic, low molecular mass cellular protein species (with a pI in the range of pH 4 to 6 and a M_r ranging from 6000 to 15,000) of *M. tuberculosis* H37Rv by MALDI- and ESI-mass spectrometry. These represent about 1/10 of all cellular protein species of *M. tuberculosis* H37Rv.[22] This investigation not only led to the identification of proteins with assigned putative functions, but also facilitated the identification of numerous proteins that had only been predicted at the DNA level ("conserved hypothetical proteins"; "unknown proteins"). We also identified six proteins that had not been predicted by the *M. tuberculosis* H37Rv genome project.[15] For five of these proteins identical homologs have been described for the *M. tuberculosis* clinical isolate CDC1551.[23] These findings clearly illustrate that proteomics plays a vital role in complementing genomic investigations.

Betts *et al.*[24] compared the protein composition of *M. tuberculosis* H37Rv with that of *M. tuberculosis* clinical isolate CDC 1551 and detected within 1750 protein spots only 17 differences confirming the high similarity between members of the *M. tuberculosis* complex at the proteome level.[20] Monahan *et al.*[25] found that six *M. tuberculosis* proteins were present with higher abundance inside macrophages as compared to growth in culture medium. All of the mycobacterial proteins identified in 2-DE patterns up to the year 2000 were presented by Rosenkrands *et al.*[26] Hendrickson *et al.*[27] reported on identification of an *M. tuberculosis* protein by MS as a serological marker of tuberculosis. The importance of genomics and proteomics of *M. tuberculosis* in the postgenomic age has been reviewed by Domenech *et al.*[28]

Materials

Mycobacterial Strains

M. tuberculosis H37Rv is isolated from human lung and was purchased from the American Type Culture Collection (ATCC; Manassas, VA). *M. tuberculosis* Erdman is a gift from William Jacobs (New York, NY). The vaccine strain *M. bovis* BCG Copenhagen (Danish 1331) is purchased from Statens Serum Institute (SSI;

[22] J. Mattow, P. R. Jungblut, E.-C. Müller, and S. H. E. Kaufmann, *Proteomics* **1**, 494 (2001).

[23] P. E. Jungblut, E.-C. Müller, J. Mattow, and S. H. E. Kaufmann, *Infect. Immun.* **69**, 5905 (2001).

[24] J. C. Betts, P. Dodson, S. Quan, A. P. Lewis, P. J. Thomas, K. Duncan, and R. A. McAdam, *Microbiology* **146**, 3205 (2000).

[25] I. Monahan, J. Betts, D. Banerjee, and P. Butcher, *Microbiology* **147**, 459 (2001).

[26] I. Rosenkrands, A. King, K. Weldingh, M. Moniatte, E. Moertz, and P. Andersen, *Electrophoresis* **21**, 3740 (2000).

[27] R. C. Hendrickson, J. F. Douglass, L. D. Reynolds, P. D. McNeill, D. Carter, S. G. Reed, and R. L. Houghton, *J. Clin. Microbiol.* **38**, 2354 (2000).

[28] P. Domenech, C. E. Barry III, and S. T. Cole, *Curr. Opin. Microbiol.* **4**, 28 (2001).

Copenhagen, Denmark) and *M. bovis* BCG Chicago (Tice 1) is purchased from ATCC.

Mycobacterial Culture Media and Growth Conditions

Bacterial cultures are grown in Middlebrook 7H9 broth (Difco, Detroit, MI) supplemented with Middlebrook albumin–dextrose–catalase (ADC) enrichment (Difco) or in Sauton minimal medium. Sauton minimal medium is prepared according to the following formulation: 0.4% L-asparagine, 0.2% citric acid, 0.05% $MgSO_4 \cdot 7H_2O$, 0.05% K_2HPO_4, 0.005% ferric ammonium citrate, 0.05% $ZnSO_4 \cdot 7H_2O$ is dissolved by the addition of 6% (v/v) glycerol in warm H_2O. The pH is adjusted by NaOH either before autoclaving to pH 6.8 or before sterile filtration to pH 7.0. The sterile medium is aliquoted into Erlenmeyer flasks and can be stored at 4° for approximately 6 weeks. Mimicking of intracellular environment can be achieved by the alteration of culture conditions in Sauton medium, e.g., by adjustment of the medium pH, variation of the iron concentration, or the addition of H_2O_2 or NO donors such as *S*-nitrosoglutathione (GNSO), nitroso-*N*-acetylpenicillamine (SNAP), SIN-1, NOC-7, or nitroprusside to the culture.[29,30]

Mycobacterial seed lot stocks are cultured after mouse passage on Middlebrook 7H11 agar supplemented with oleic acid–albumin–dextrose–catalase (OADC) enrichment 1339 (Difco) and subsequently in Middlebrook 7H9 medium containing ADC enrichment. Stocks are washed twice with phosphate buffered saline (PBS) without Ca^{2+}, and maintained in 10% glycerol at $-70°$ until use.

Methods

Preparation of Cellular Proteins (CP)

1. Inoculate 100 ml Middlebrook 7H9 culture medium with a total of 2×10^8 mycobacteria from seed lot.

2. Culture mycobacteria for 6–8 days at 37° with agitation to a cell density of $1–2 \times 10^8$ ml^{-1} or an optical density of 600 nm (OD_{600}) between 0.8 and 1.

3. Pellet mycobacterial cells at 4,000g for 15 min at 4°.

4. Wash pellet twice with 100 ml cold PBS without Ca^{2+}.

5. Resuspend pellet in 1 ml PBS and transfer into a microcentrifuge screw cap tube with lid gasket.

6. Centrifuge at 10,000g for 15 min at 4° to pellet bacteria.

7. To avoid proteolytic degradation 1 μl of an equal mixture of the proteinase inhibitors TLCK (stock 100 mg/ml), pepstatin A (stock 50 mg/ml), leupeptin (stock

[29] T. R. Garbe, N. S. Hibler, and V. Deretic, *Mol. Med.* **2**, 134 (1996).
[30] D. K. Wong, B. Y. Lee, M. A. Horwitz, and B. W. Gibson, *Infect. Immun.* **67**, 327 (1999).

100 mg/ml), and E64 (stock 25 mg/ml) diluted in dimethyl sulfoxide (DMSO) is added to the cell pellet.

8. Lyse cells by sonication using a Sonifier 250 and a sonification cup (Branson, Cincinnati, OH) with 100% output and 50% interval for 10 min.

9. Continue with sample preparation for 2-DE.

Sample Preparation for 2-DE

1. Gradually add urea to a final concentration of 9 M to the sonicate (108 mg urea per 100 μl sonicate).

2. Add dithiothreitol (DTT, stock concentration 1.4 M) to a final concentration of 70 mM, ampholytes (Servalytes 2–4, stock concentration 40%), and Triton X-100 to a final concentration of 2% each to completely denature and reduce the sample (10 μl 1.4 M DTT, 10 μl 40% ampholyte stock, and 4 μl 100% Triton X-100 per 100 μl sonicate).

3. Keep sample for 30 min at room temperature and stir occasionally.

4. Centrifuge at 10,000g for 15 min at 16°.

5. Centrifuge again at 100,000g for 30 min at 16° to completely clear sample using a tabletop ultracentrifuge (Beckman, Optimax TLX, Palo Alto, CA).

Preparation of Culture Supernatant (CSN)

1. Inoculate 100 ml Middlebrook 7H9 preculture with a total of 2×10^8 mycobacteria.

2. Culture mycobacteria for 12 days at 37° with agitation.

3. Pellet mycobacterial cells at 4000g for 15 min at 4°.

4. Wash pellet three times with cold PBS without Ca^{2+}.

5. Culture washed mycobacteria in 1 liter Sauton medium for 21 days at 37° with continuous slow shaking.

6. Harvest culture supernatant (CSN) by filtering twice through 0.2 μm Millipore (Bedford, MA) filters (5 cm diameter).

7. Add sodium deoxycholate to a final concentration of 0.015% at room temperature with shaking.

8. Incubate 10 min at room temperature.

9. Add trichloroacetic acid (TCA) to a final concentration of 10%, shaking the sample continuously.

10. Incubate for 60 min at 4°.

11. Centrifuge at least 4000g for 15 min at 4° to precipitate CSN proteins and decant supernatant.

12. Wash precipitate twice with 100 ml cold acetone.

13. Resuspend pellet in 500 μl H_2O and add 10 μl 1 M Tris pH 9.

14. Transfer material into a microcentrifuge screw cap tube with lid gasket and determine the wet weight of the sample.

15. Sonicate precipitate using a Sonifier 250 and a sonication cup with 100% output and 50% interval for 10 min.

16. Continue with sample preparation for 2-DE.

Preparation of Membrane Fraction (MF) and Somatic Fraction (SF)

1. Inoculate 100 ml Middlebrook 7H9 preculture with a total of 2×10^8 mycobacteria.

2. Culture mycobacteria for 12 days at 37° with agitation.

3. Pellet mycobacterial cells from 100 ml culture at 4000g for 15 min at 4°.

4. Wash pellet 3× with cold PBS without Ca^{2+}.

5. Culture washed mycobacteria in 1 liter Sauton medium for 21 days at 37° with continuous slow shaking.

6. Pellet mycobacterial cells from 1 liter culture at 4000g for 15 min at 4°.

7. Wash pellet 2× with cold PBS without Ca^{2+}.

8. Resuspend the pellet in 1 ml PBS and transfer into a tube filled with glass beads (FastRNA Blue Kit; BIO101, Quantum Appligene, Heidelberg, Germany).

9. Lyse the bacteria with FASTprep FP120 apparatus (BIO101, Savant Instruments, Holbrook, New York) using the following setting: speed 6.5 for 45 sec.

10. Centrifuge the lysate at 10,000g for 30 min at 4°.

11. The supernatant represents the somatic protein fraction (SF) which in our hands gave similar 2DE pattern as CP preparations.

12. Wash pellet 3× with cold PBS without Ca^{2+}.

13. Resuspend pellet in 1 ml H_2O.

14. Add to a final concentration 70 mM dithiothreitol (DTT, stock concentration 1.4 M), 2% ampholytes (Servalytes 2–4, stock concentration 40%), and 2% Triton X-100 to completely denature and reduce sample.

15. Isolate the membrane proteins with the FASTprep FP120 apparatus using the following setting: speed 6.5 for 45 sec.

16. Centrifuge the lysate at 10,000g for 30 min at 16°.

17. Take supernatant and centrifuge at 100,000g for 1 hr at 16° to completely clear the membrane function.

18. Gradually add urea to a final concentration of 9 M.

Preparation of Mycobacteria of Phagosomal Origin

A. Isolation and Infection of Murine Bone Marrow-Derived Macrophages (BMMO)

1. Isolate and differentiate murine BMMO as described previously.[31]

2. Plate and cultivate 10^9 BMMO in T75 tissue culture flasks ($\sim 10^7$/flask).

[31] S. Sturgill-Koszycki, P. L. Haddix, and D. G. Russell, *Electrophoresis* **18,** 2558 (1997).

3. Infect BMMO at a multiplicity of infection (MOI) of 10 : 1 with *M. tuberculosis* or *M. bovis* BCG.

4. Incubate for 2 hr at 37° with 7% (v/v) CO_2.

5. Wash infected BMMO with PBS at 37°.

6. Incubate for 2–6 days at 37° with 7% (v/v) CO_2.

7. Wash infected BMMO with PBS at 37°.

B. Preparation of Mycobacteria of Phagosomal Origin[31–33]

8. Scrape infected BMMO with lysis buffer (20 mM HEPES, 8.55% sucrose, 1 μl protease inhibitor mix as described in preparation of CP step 7, pH 6.5).

9. Adjust to 10^8 cells ml^{-1} with lysis buffer.

10. Lyse cells by passing 15 times through a 28-gauge syringe. Monitor cell lysis microscopically during the procedure. Note that for *M. tuberculosis,* cell lysis must be performed under a laminar air flow. We use 50-ml Falcon tubes with a hole in the screw cap to fit a 20-ml syringe. The gap between syringe and lid is sealed with Parafilm.

11. Stop lysis at a ratio of 9 : 1 between nuclei and intact cells, as controlled by microscopy. Slides with *M. tuberculosis* lysates should be kept in a Parafilm-sealed petri dish when taken out of the hood for microscopy.

12. Centrifuge cell lysate 3 times by postnuclear spins at 120g for 8 min at 4°; collect and pool supernatants.

13. Wash pellet with lysis buffer; add washes to the supernatant.

14. Layer 3 ml supernatants on top of a 12%/50% (2 ml each) sucrose step gradient.

15. Centrifuge for 45 min at 450g at 4°.

16. Recover 1.5 ml at the 12%/50% interphase using a syringe.

17. Control purity of phagosomes by phase-contrast microscopy.

18. Dilute phagosomes in 200 mM HEPES containing 0.005% noxidet P-40 (NP-40).

19. Recover mycobacteria by centrifugation for 15 min at 4,000g at 4°.

20. Resuspend in 20 mM HEPES containing 0.05% NP-40.

21. Transfer to a microcentrifuge screw cab tube with lid gasket.

22. Wash bacterial cell pellet twice with 0.6 M KCl containing 0.05% NP-40.

23. Wash bacterial cell pellet twice with 0.05% NP-40.

24. Lyse bacterial cells by sonication using a Sonifier 250 and a sonication cup with 100% output and 50% interval for 10 min.

25. Continue with sample preparation for 2-DE.

[32] U. E. Schaible, S. Sturgill-Koszycki, P. H. Schlesinger, and D. G. Russell, *J. Immunol.* **160,** 1290 (1998).

[33] G. Dietrich, U. E. Schaible, K. D. Diehl, H. J. Mollenkopf, S. Wiek, J. Hess, K. Hagens, S. H. E. Kaufmann, and B. Knapp, *FEMS Microbiol. Lett.* **186,** 177 (2000).

An alternative protocol for the preparation of mycobacteria of phagosomal origin has been described.[25]

Two-Dimensional Electrophoresis (2-DE)

The high resolution 2-DE technique that we use for mycobacterial proteome analysis combines nonequilibrium pH gradient electrophoresis (NEPHGE) carrier ampholyte isoelectric focusing (IEF) with sodium dodecyl sulfate–polyacrylamide gel electrophoresis (SDS–PAGE). For estabishing the technology it is strongly recommended to start with exactly the same protocol as worked out in detail by Klose and Kobalz.[34] As the description of the entire procedure would fill more than 15 pages, here we only describe the main conditions of our system. The size of the 2-DE gels is 23 × 30 cm. IEF is performed in rod gels containing 9 M urea, 3.5% acrylamide, 0.3% piperazine diacrylamide and a total of 4% carrier ampholytes pH 2–11. Protein samples are applied at the anodic side of the IEF gels and focused under NEPHGE conditions (8870 Vh). For analytical and preparative analysis 0.75 and 1.5 mm thick IEF gels are used respectively. For analytical investigations 60–100 μg protein is loaded; for preparative experiments we use up to 900 μg protein. SDS–PAGE is performed in gels containing 15% acrylamide using the IEF gels as stacking gels. For analytical and preparative analysis proteins are visualized by either silver-staining or Coomassie Brilliant Blue (CBB) G250-staining, respectively.

Silver Staining

Silver staining is performed as described by Jungblut and Seifert.[35] In order to obtain optimal results all solutions should be prepared immediately before use. Unless otherwise stated all steps are performed at room temperature with shaking in a volume of 1 liter.

1. Incubate gel for 2 hr in 50% ethanol, 10% acetic acid.
2. Store gel overnight at 4° in this solution.
3. Incubate gel for 2 hr in 30% ethanol, 0.5% glutardialdehyde, 4.1% sodium acetate, 0.2% sodium thiosulfate.
4. Wash gel twice for 20 min in 4 liter H_2O.
5. Incubate gel for 30 min in 0.1% silver nitrate, 0.01% formaldehyde.
6. Wash gel for 15 sec in H_2O.
7. Incubate gel for 1 min in 2.5% sodium carbonate.
8. For protein visualization incubate gel for 4–6 min in 2.5% sodium carbonate, 0.01% formaldehyde.

[34] J. Klose and U. Kobalz, *Electrophoresis* **16**, 1034 (1995).
[35] P. R. Jungblut and R. Seifert, *J. Biochem. Biophys. Methods* **21**, 47 (1990).

9. To stop the developing process incubate gel for 20 min in 0.05 M disodium ethylenediamine tetraacetate dihydrate (EDTA), 0.02% thimerosal.

10. Dry the stained gel.

Coomassie Brilliant Blue (CBB) G-250 Staining

The CBB G-250 staining is performed as described by Doherty et al.[36] All steps are performed at room temperature under shaking.

1. Incubate gel overnight in 1 liter 50% methanol, 2% o-phosphoric acid.

2. Wash gel 3× for 30 min in 2 liter H_2O.

3. Incubate gel for 1 hr in 1 liter 34% methanol, 2% o-phosphoric acid, 17% ammonium sulfate.

4. After 1 hr of incubation add 0.66 g CBB G-250 (final concentration: 0.066%) and incubate the gel for another 5 days. Do not dissolve the powdery CBB G-250 before adding it.

5. Wash gel for 1 min in 1 liter 25% methanol.

6. Wash gel for 1 min in 1 liter H_2O.

7. Store the stained gel in a plastic bag at 4°.

Sample Preparation and MALDI-MS

The procedure described here is a modification of the methods described earlier.[37,38]

1. Excise spot of interest from preparative CBB-stained 2-DE gel.

2. Incubate the gel material for 20 min at 30° in a microcentrifuge tube in 500 μl 100 mM Tris pH 8.5 in 50% acetonitrile (ACN) with shaking.

3. Discard the supernatant.

4. For equilibration add 500 μl 50 mM ammonium bicarbonate buffer pH 7.8 in 5% ACN and incubate for 30 min at 30° with shaking.

5. Discard the supernatant and dry the gel material using a Speed Vac concentrator for approximately 30–45 min. The resulting gel material should not be totally dry.

6. Dissolve 20 μg lyophilized trypsin (Promeg, Madison, WI) according to the manufacturers instructions in 100 μl 50 mM resuspension buffer (50 mM acetic acid).

[36] N. S. Doherty, B. H. Littman, K Reilly, A. C. Swindell, J. M. Buss, and N. L. Anderson, *Electrophoresis* **19**, 355 (1998).

[37] A. Otto, B. Thiede, E.-C. Müller, C. Scheler, B. Wittmann-Liebold, and P. Jungblut, *Electrophoresis* **17**, 1643 (1996).

[38] S. Lamer and P. R. Jungblut, *J. Chromatogr. B. Biomed. Sci. Appl.* **752**, 311 (2001).

7. Add 1 μl of the resulting trypsin solution (containing 0.2 μg trypsin) and 50 μl digestion buffer (50 mM ammonium bicarbonate buffer pH 7.8 in 5% ACN) to the dried gel material. Make sure that the gel material is covered by liquid. The trypsin solution and the digestion buffer should be mixed before use.

8. Incubate the gel material overnight at 37° with shaking.

9. Centrifuge at 4000g for 1 min.

10. Add 50 μl 0.1% trifluoroacetic acid (TFA).

11. The resulting peptide mixture is subsequently desalted and concentrated using a ZipTip$_{C18}$ pipette tip (Millipore, Eschborn, Germany) according to the manufacturer's instructions. In the final step of the procedure elute peptides from the ZipTip$_{C18}$ pipette tip with 5 μl 50% ACN.

12. Mix 2 μl of the resulting peptide solution with 2 μl of an aqueous saturated matrix solution (2% α-cyano-4-hydroxycinnamic acid in 30% ACN, 0.3% TFA).

13. Apply 2 μl of the resulting solution to the MALDI-MS sample template and let it dry.

14. Perform MALDI-MS measurements.

We perform MALDI-MS measurements in the reflectron mode of a time-of-flight MALDI mass spectrometer *Voyager Elite* (Perseptive, Framingham, MA) with delayed extraction using the following parameters: 20 kV accelerating voltage, 70% grid voltage, 0.050% guide wire voltage, 100 ns delay, and a low mass gate of 500, we summarize 256 laser shots to obtain mass spectra.

We obtain mass accuracy in the range of 30 ppm by internal calibration using synthetic peptides with known molecular mass as internal markers or by internally calibrating the spectra by use of matrix specific peaks or peptide mass peaks that occur due to trypsin autolysis.

Peptide Mass Fingerprint (PMF)

As described in the introduction, PMF protein identification is based on comparing peptide masses derived from a protein digest with peptide mass maps created by *in silico* cleavage of proteins in a protein database. There are several search programs, e.g., MS-FIT (prospector.ucsf.edu/ucsfhtml3.4/msfit.htm), PeptIdent (www.expasy.ch/tools/peptident.html), Mascot (www.matrixscience.com/cgi/index.pl?page = ../home/html), and peptide Search (www.mann.embl-heidelberg.de/peptidesearchpage.html) and protein databases, e.g., the protein database from the National Centre for Biotechnology Information (NCBI; www.ncbi.nlm.nih.gov:80/entrez/query.fcgi?db = Protein) and the protein database of the Swiss Institute of Bioinformatics (SwissProt; www.expasy.ch/sprot) that can be used in order to perform the database searches. We perform protein identification by searches in the NCBI and SwissProt protein databases. Searches are performed using the program MS-FIT, reducing the proteins of the databases to mycobacterial proteins and a M_r range estimated from 2-DE $\pm 20\%$, allowing a mass tolerance of 0.1 Da. If no proteins match, the M_r window is extended. Partial enzymatic cleavages leaving two cleavage sites, acetylation of N termini of proteins, pyroglutamate formation at N-terminal glutamine of peptides, oxidation of methionine, and modification of cysteine by acrylamide should be considered in the database searches. We regard a protein as identified if the matched peptides cover at least 30% of the complete protein sequence. Assignments with a sequence coverage (SC) < 30% are only accepted if at least the three most intense peaks of the mass spectrum match with a database sequence.

Conclusions

The proteome reflects the functional status of an organism or a cell in response to environmental or extracellular stimuli, and hence serves as a valuable complement to the genome. The techniques described here form the basis for standardized proteome investigations, and together with the standard 2-DE patterns in the 2-DE database (www.mpiib-berlin.mpg.de/2D-PAGE) they represent a prerequisite for future comparative analyses. Complementary proteome technologies and information on the interactions between proteins and other molecules will provide a better understanding of the cell at the molecular level.

Acknowledgments

The authors thank Ursula Zimny-Arndt and Stephanie Lamer for excellent technical assistance, Dr. Helen Collins for critical reading of the manuscript, and Lucia Lom-Terborg for secretarial help. Stefan H. E. Kaufmann acknowledges financial support by DFG, BMBF, WHO, EC, Chiron Behring, and Fonds Chemie.

[18] Proteomic Analysis of Response to Acid in *Listeria monocytogenes*

By LUU PHAN-THANH

Introduction

A proteomic approach is adopted to study the response to acid in *Listeria monocytogenes*. This response is materialized by up-regulation of the synthesis of a number of cellular proteins. These are separated by two-dimensional electrophoresis and identified by mass spectrometric analysis. This article describes the methodological steps and detailed techniques aiming at identifying and characterizing these acid-induced proteins in order to gain insights into the mechanism by which *L. monocytogenes* responds to acid.

In spite of ever-improving hygienic and sanitary measures taken in agroalimentary sectors, the food-borne pathogen *Listeria monocytogenes* still raises concerns for public health in many industrialized countries. Much research effort has been done in the past decade to delineate the genes involved in the virulence and pathogenesis of this gram-positive bacterium. Its capabilities to survive in hostile conditions are, however, not much studied, at least to the depth of regulatory mechanisms. Among the harsh conditions of the environment unfavorable for *Listeria* growth, acidity is perhaps the most encountered in both natural habitats and infected host systems (acid rain, industrial and agricultural wastes, fermented food and feed products, gastric secretions, phagocytosomal vacuoles of the host cells, etc.). When facing such adverse conditions, *Listeria*, like other organisms, evolves mechanisms that permit it to resist the adversity and eventually to adapt for survival. The phenomenon of acid tolerance in other bacteria has been studied for many years, mostly in gram-negative bacteria, especially in *Escherichia coli*[1-3] and Salmonella typhimurium,[4,5] much less in gram-positive bacteria,[6,7] and only recently in *Listeria*.[8,9]

The lowest pH that *L. monocytogenes* can resist depends on the strain and the kind of acid present.[9] Adaptation to an intermediary nonlethal acidic pH increases the bacterial resistance to a subsequent lethal acid stress. This capability of acid resistance is also growth phase-dependent. In the stationary phase *L. monocytogenes* exhibits a certain natural acid tolerance. At the same external pH, weak organic acids exert a more deleterious effect on *Listeria* than strong inorganic acids do. In an acidic medium the metabolism of *L. monocytogenes* decreases: many physiological processes slow down and fewer proteins are synthesized. At the same time, however, *L. monocytogenes* cells increase the synthesis of a number of indispensable proteins, which help them to resist the acidity. To gain insights into the mechanisms of response to acid stress, it is necessary to identify these acid-induced proteins and to determine the role they may play in the process. In spite of the difficulty with reproducibility, two-dimensional electrophoresis (2DE) is the method of choice for protein separation because it is the

[1] D. Poynter, S. J. Hicks, and R. J. Rowbury, *Lett. Appl. Microbiol.* **3**, 117 (1986).
[2] K. P. Stim-Herndon, T. M. Flores, and G. N. Bennet, *Microbiology* **142**, 1311 (1996).
[3] E. W. Hickey and I. N. Hirshfield, *Environ. Microbiol.* **56**, 1038 (1990).
[4] S. Bearson, B. Bearson, and J. W. Foster, *FEMS Microbiol. Lett.* **147**, 173 (1997).
[5] J. Lin, I. S. Lee, J. Frey, J. S. Slonczewski, and J. W. Foster, *J. Bacteriol.* **177**, 4097 (1995).
[6] B. Poolman, A. J. Driessen, and W. N. Konings, *Microbiol. Rev.* **51**, 498 (1987).
[7] A. Hartke, S. Bouché, J. C. Giard, A. Benachour, P. Boutibonnes, and Y. Auffray, *Curr. Microbiol.* **33**, 194 (1996).
[8] J. D. Mark, P. J. Coote, and C. P. O'Byrne, *Microbiology* **142**, 2975 (1996).
[9] L. Phan-Thanh and A. Montagne, *J. Gen. Appl. Environ. Microbiol.* **44**, 183 (1997).

only method actually capable of resolving in one experiment a complex mixture of proteins such as cellular proteins. The differential analysis of cellular proteins in acidic and neutral conditions is rendered possible by radioactive labeling of proteins and the use of specialized computer software. With the advent of mass spectrometric techniques such as matrix-assisted laser desoption/ionization time of fly (MALDI-TOF) and electrospray tandem mass spectrometry (ESI MS/MS) the protein identification can be done much more rapidly than with using classical methods and with the amounts of protein obtained from spots on a 2DE gel. This article describes such a proteomic approach with methodological steps and technical details.

Bacterial Culture

Media

Brain heart infusion broth (Difco, Detroit, MI) is used for preculture. This is a rich medium excellent for growth of *Listeria,* but it is of an undefined composition and protein-rich, and hence not suited for radioactive labeling. A synthetic medium of chemically defined composition has been designed[10] for *Listeria* both for culture and metabolic labeling. Its composition is given in Table I.

Growth Conditions

The pathogenic *L. monocytogenes* EGD strain from stock agar is first inoculated and grown at 37° with agitation in 10 ml of brain heart infusion broth for several generations (7–8 hr), allowing the bacteria to recover from previous stress and damage. An inoculum from this preculture is then diluted into 100 ml of synthetic medium so that there are about 10^5 cells/ml. The culture proceeds at 37° with agitation to mid-exponential phase (optical density measured at 620 nm is about 0.2). The cells are harvested by centrifugation (3000g for 10 min) and submitted to acid treatment.

Acid Stress and Adaptation

The harvested bacteria are acid-stressed by transferring into a fresh synthetic medium that has been previously adjusted to the desired pH with HCl (use 1 N and 0.1 N HCl). Acid stress consists of treatment with a lethal pH (e.g., pH 3.5). Acid adaptation is carried out by treatment with an intermediary nonlethal acidic pH (e.g., pH 5.0–5.5) for a relatively long period (2–5 hr). The conditions of acid stress (duration of stress imposition and pH value) should be chosen so that they generate a strong enough effect without killing a high proportion of the cells at the end of

---

[10] L. Phan-Thanh and T. Gormon, *Int. J. Food Microbiol.* **35,** 91 (1997).

TABLE I
COMPOSITION OF CHEMICALLY DEFINED MEDIUM FOR
Listeria GROWTH

Component	Concentration (per liter)
KH_2PO_4	6.56 g
$Na_2HPO_4 \cdot 7H_2O$	30.96 g
$MgSO_4 \cdot 7H_2O$	0.41 g
Ferric citrate	88 mg
Glucose	10 g
L-Glutamine	0.6 g
L-Leucine	0.1 g
DL-Isoleucine	0.1 g
DL-Valine	0.1 g
DL-Methionine	0.1 g
L-Arginine hydrochloride	0.1 g
L-Cysteine hydrochloride	0.1 g
L-Histidine hydrochloride	0.1 g
L-Tryptophan	0.1 g
L-Phenylalanine	0.1 g
Adenine	2.5 mg
Biotin	0.5 mg
Riboflavin	5 mg
Thiamin hydrochloride	1 mg
Pyridoxal hydrochloride	1 mg
p-Aminobenzoic acid	1 mg
Calcium pantothenate	1 mg
Nicotinamide	1 mg
Thioctic acid (α-lipoic acid)	5 μg

the treatment. These conditions depend on the bacterial strain, its physiological state, and the nature of the acid used. For example, with *Listeria* EGD strain grown to mid-exponential phase, about 70% of the cells survive after 30 min of stress at pH 3.5 (in the presence of HCl).

Radioactive Labeling and Protein Extraction

The metabolic labeling of bacterial proteins is performed by incorporation of $[S^{35}]$methionine and $[S^{35}]$cysteine in the labeling medium. About 4×10^9 cells harvested at mid-exponential phase are put in 4 ml of labeling medium that has been previously adjusted to the desired pH and prewarmed to 37°. The labeling medium is derived from the chemically defined growth medium in which methionine and cysteine are omitted. The labeling is carried out by adding 400 μCi of EXPRE^{35}S^{35}S labeling mixture (NEN-Dupont, Boston, MA) to the bacteria suspension, allowing the labeling to proceed at 37° for 30 min, and stopping by chilling on ice. For acid stress (pH 3.5), the labeling is started 15 min after the pH shift, the time thought to be necessary for the external pH to exert an action on the synthesis machinery inside the cell. For acid adaptation, the labeling is started 2 hr after bacterial adaptation to pH 5.5. The labeling in control condition (neutral pH) lasts 5 min, the metabolism at neutral pH being much more active than that at acidic pH values. Following labeling, the bacteria are washed three times with buffered physiological saline, centrifuged at 3000g for 10 min, and placed in a 3-ml microtube with 260 μl of sonication buffer, which is composed of 140 mg of Tris base, 200 μl of Biolyte pH 3–10 (Bio-Rad, Hercules, CA) or IPG-buffer pH 3–10 (Amersham Pharmacia Biotech, Uppsala, Sweden), and 1 tablet of protease inhibitor cocktail (MiniComplete, EDTA-free, Boehringer, Mannheim, Germany), and dissolved in 25 ml of ultrapure water obtained with a Milli-Q system (Millipore, Bedford, MA). The bacterial suspension is then sonicated with a Branson sonifier (Danbury, CT) 4 times for 1 min on ice using a microtip at power level 3. Nucleic acids in the mixture are digested for 15 min at room temperature with 5 μl of a DNase and RNase solution that is prepared by dissolving 1 mg of DNase and 5 mg of RNase (Sigma, St. Louis, MO) in 0.5 ml of 20 mM Tris buffer containing 0.5 M Mg^{2+}. Solid urea (250 mg) and thiourea (70 mg) are added and agitated to complete dissolution before adding 36 μl of 50% (w/v) 3-[(3-cholamidopropyl)dimethylammonio]-1-propane sulfonate (CHAPS), 30 μl of 20% (w/v) Nonidet P-40 (NP-40), and 12 μl of 200 mM tributylphosphine stock solution in 2-propanol. The mixture is left at room temperature for 30 min with intermittent vortexing. The soluble protein is finally separated from insoluble material by centrifugation at 20,000g for 30 min and stored at $-70°$ in aliquots until use. An aliquot of 50 μl is saved for determination of protein concentration. Two 5-μl aliquots of each sample are used to measure the incorporation of radioactivity. In a microtube containing 60 μl of cold 12% trichloroacetic acid (TCA), add 5 μl of protein extract and 50 μl of 1 mg/ml bovine serum albumin (BSA) as carrier, mix, and leave the mixture on ice for 15 min, centrifuge at 20,000g for 15 min, and wash the precipitate with 2 × 100 μl of cold 10% TCA, measure the radioactivity after resuspending the precipitate in 1 ml of scintillation liquid.

Protein Determination of Samples Destined for Two-Dimensional Electrophoresis

The concentration of protein in bacterial extracts destined for 2DE cannot be directly determined by the current methods of protein determination (Lowry, Biuret, Bradford, or bicinchoninic acid) due to interference by combinatory effects of urea, detergents, carrier ampholytes, and reducing compounds present in sample-solubilizing buffer. A method modified from that of Bradford[11] allows direct

---

[11] L. S. Ramagli and L. V. Rodriguez, *Electrophoresis* **6**, 559 (1985).

estimation of protein in 2DE samples over a range of 0.5 to 50 μg. A simpler method with Advanced Protein Assay Reagent from Cytoskeleton (Denver, CO) gives a linear standard curve in the range of concentrations from 0.5 to 4 mg/ml. The concentrations of the protein extracts prepared from *Listeria* cells for 2DE purposes are usually in that range. All dilutions of protein standard solutions are made in solubilizing buffer, which is also used for the blank assay.

Solubilizing Buffer

Mix together 2080 μl of sonication buffer (see above), 2000 mg of urea, 560 mg of thiourea, 290 μl of 50% CHAPS, 2400 μl of 20% NP-40, and 100 μl of 200 mM tributylphosphine, and adjust to 4.8 ml with Milli-Q water.

Assay

In 5-ml tubes dilute 10 μl of the solutions to be quantified (samples, standard solutions, and blank) to 1000 μl with Milli-Q water. Add 200 μl of concentrated ADV01 reagent. Mix and measure the absorbance at 590 nm in 1-ml cuvettes. The slope of the standard curve depends on the standard protein used. Ovalbumin as standard gives rather overestimated results for *Listeria* protein extracts, whereas ovine serum albumin has the tendency to give an underestimation.

Two-Dimensional Electrophoresis

Two-dimensional electrophoresis is very sensitive to the physicochemical parameters of the microenvironment in which the proteins find themselves. The quality and reproducibility of the gels are crucial for the detection and differential analysis of stress proteins. To ensure a high degree of reproducibility, care should be taken to keep at minimum the variability of these physicochemical parameters between samples and during different steps of the process. In spite of all precautions, however, variations inherent in the technique are often unavoidable. Therefore, a minimum of five gels are routinely run for each of the samples and the entire experiment should be repeated two times. The medium, the physiological state of bacteria, and the timing of the labeling as well as the conditions under which the proteins are extracted and separated by 2DE all affect the bacterial 2D protein patterns.

First Dimension

The proteins are separated in the first dimension according to their charge on Immobiline-based gels, using 18-cm IPG DryStrips of pH 4–7 and pH 6–11 (Amersham Pharmacia Biotech) that covers almost the whole pH range of *Listeria* proteins (17-cm IPG ReadyStrips from Bio-Rad can also be used with equivalent results).

The first dimension is performed at $20°$ in a Multiphor II apparatus (Amersham Pharmacia Biotech) as described in the manufacturer's instructions.

Protein samples are incorporated in the rehydration step. The rehydration buffer is composed of $6 M$ urea, $2 M$ thiourea, 2% (w/v) CHAPS, $2 mM$ tributylphosphine, and 2% (v/v) IPG buffer pH 4–7 for DryStrips of pH 4–7 (or 0.5% IPG buffer pH 6–11 for DryStrips of pH 6–11). To reduce the electroendoosmotic effect with the DryStrips of pH 6–11 the rehydration buffer should contain an additional 10% 2-propanol. The sample is mixed with rehydration buffer to a total volume of $450 \mu l$ for each DryStrip (this volume should be totally absorbed by the IPG strips). For analytical gels, the protein sample should be about $50 \mu g$ (or 4×10^6 cpm) and the same amounts of protein (or cpm) should be used for all the samples to be analyzed.

Comment

1. Incorporation of samples in the rehydration step is more advantageous than sample loading with minicups, since it permits loading large sample volumes and avoids the problem of protein precipitation at the place on the gels under the loading cups.

2. Noncharged tributylphosphine is used as a reducing agent instead of dithio-threitol or dithioerythritol, since these compounds produce charged molecules that will migrate to the anode during the first dimension separation, causing possible precipitation of proteins due to disulfide reformation.

The electric current applied should not exceed $70 \mu A$ per IPG strip. The voltage is set at low values (100–500 V) for the first 5 hr before being gradually increased to the maximum value, so that a total product of 160 kV.hr is obtained for the pH 4–7 IPG strips and 200 kV.hr for the pH 6–11 IPG strips. At the end of the first dimension separation, immediately store the IPG strips in sealed plastic bags at $-80°$ until the second dimension separation.

Second Dimension

The second dimension separation is performed in a Hoefer DALT vertical unit (Amersham Pharmacia Biotech) with 10 gels run simultaneously using Lämmli SDS electrophoresis buffers. Gels used are 12.5% polyacrylamide and are cast without SDS in the gel formula to prevent eventual formation of minipockets of unpolymerized polyacrylamide. Before transferring onto the gel plates, the IPG strips are first equilibrated for 15 min in equilibration buffer (put each strip in a 5-ml blunt-end plastic pipette filled with 3 ml of buffer, seal the pipettes with flexible Parafilm, and agitate on a rocking tray). The equilibration buffer is composed of 14 g urea, 4 g thiourea, 1 g SDS, 15 g glycerol, and some grains of bromphenol blue, dissolved in 50 ml of $50 mM$ Tris buffer pH 8.8; just before use, add tributylphosphine ($20 \mu l$ of $200 mM$ TBP stock solution per ml of equilibration buffer). This equilibration step is followed by a 15 min acetylation step, in which TBP in the equilibration buffer is replaced by iodoacetamide (2 mg per ml of buffer). The IPG strips are fixed to the gel plates with a hot solution of 0.6% pure agarose in electrophoresis running buffer. The second dimension separation is performed overnight at constant amperage (26 mA/gel) until the front line reaches 1 cm from the end of the gels.

Determination of Molecular Mass and Isoelectric Point

Although a protein cannot be identified solely on the basis of its molecular mass and isoelectric point (pI), these parameters are useful in excluding many proteins in databases during identification procedures.

To determine molecular mass and pI, protein standards of known mass and pI are coelectrophoresed with bacterial proteins. These gels are stained with Coomassie blue or silver nitrate before being submitted to autoradiography. The molecular mass and pI standards are from Pharmacia and Bio-Rad; carbamylated protein standards are especially useful for pI determination. Use radioactive ink for marking the spots for protein standards on the dried gels (dilute 5 μl of EXPRE^{35}S^{35}S in 2 ml of India ink). A standard curve of the logarithm of molecular mass as a linear function of migration distances in the second dimension is established using a function in the software package of 2D gel analysis system. For pI estimation, a pI grid can be created within the gel image analysis system from the theoretical pI values of known protein spots.

Autoradiography

Following 2D electrophoresis, the radioactive gels are fixed in 50% (v/v) ethanol containing 2% (v/v) phosphoric acid (85% concentrated) for at least 2 hr, washed in 20% ethanol 2X for 30 min, vacuum-dried, and exposed to Biomax Kodak film for 14 days. Autoradiography is preferred to fluororadiography, the latter method yielding enhanced intensity but diminished resolution.

Computer-Aided Analysis of 2-DE Gels

Software developed for 2D gel image analysis is now available commercially: for example, Melanie and PDQuest from Bio-Rad and ImageMaster Elite from Amersham Pharmacia Biotech. They all provide the basic analysis of 2D gels. We use Kepler, a sophisticated software from Large Scale Biology (Rockville, MD). The autoradiograms, CB- or silver-stained gels are scanned with a Kodak Eikonik 1412 CCD camera. Gel comparisons are made against a master image that is a synthetic image containing all protein spots of all of the gels to be analyzed. Only consistent protein spots are retained for analysis. A protein spot is

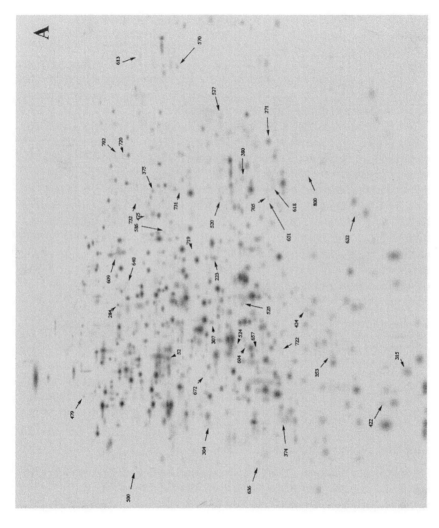

FIG. 1. (continued)

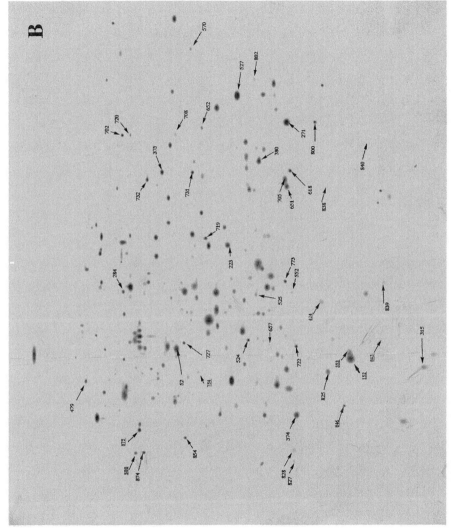

FIG. 1. *(continued)*

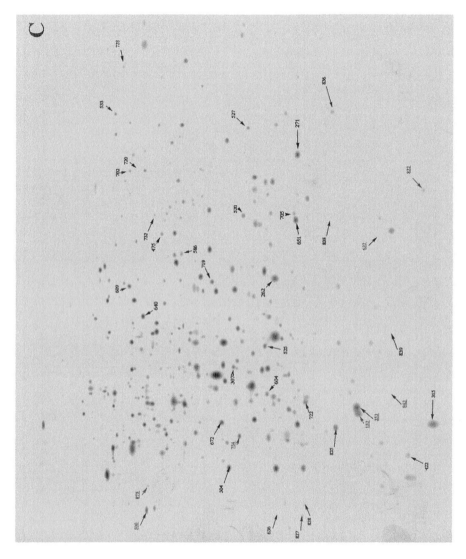

FIG. 1. (continued)

considered consistent when it is present at least in $n-1$ gels out of n gels for a given sample. Although the same amounts of radioactivity are loaded, corrective calculations of spot intensity (integrated intensity) are made to compensate for eventual discrepancies in protein sample loading or in the duration of film exposure. The intensity ratio of a defined spot in the gels from two experiments indicates the change in expression level of the corresponding protein under the two conditions. This ratio is called induction ratio when the stress condition is compared to the control condition.

Figure 1 shows the proteins induced by acid stress at pH 5.5 and pH 3.5. If a twofold up-regulation is chosen as minimal threshold for consideration, 39 and 49 proteins are found to be induced by pH 3.5 and pH 5.5, respectively (Table II). These proteins can be grouped into two categories. The first category (majority) consists of constitutive proteins, the expression of which is up-regulated by the acidic conditions. The second category (minority) consists of novel proteins, which are the proteins absent at neutral pH and uniquely synthesized at acidic pH values. Twenty-three proteins are found to be induced both at pH 5.5 and pH 3.5. These results suggest that when *Listeria* cells encounter an acidic environment, they respond by up-regulating the synthesis of a number of proteins, which actually participate in the process of resistance to acid. When the bacterial cells face a more severe acidity, they need to synthesize additional proteins for survival.

Micropreparative 2D Electrophoresis

For micropreparative purpose, samples of nonradioactive bacterial proteins are loaded on the gel in amounts of 200 μg for silver staining and 600–800 μg for Coomassie blue staining. The focusing time is longer than in the case of analytical gels: 270 kVh for IGP strips of pH 4–7 and 325 kVh for IPG strips of pH 6–11. The second dimension gel plates are 1.5 mm thick.

Comment. Samples up to 2 mg may be loaded on the gel, but only for preparing low abundance proteins. Sample amounts that are too large give rise to aggregation of protein spots and worsened resolution.

Colloidal Coomassie Blue Staining

Following the second dimension separation, the gels are fixed in 50% (v/v) ethanol containing 2% (v/v) phosphoric acid for at least 2 hr (conveniently overnight). Then the gels are washed 3 times for 20 min with distilled water,

---

FIG. 1. 2DE autoradiograms of cellular proteins from *L. monocytogenes* synthesized at different pH values: (A) pH 7.2; (B) pH 3.5; and (C) pH 5.5. First dimension: pH 4–7, acidic end at the left. Second dimension, SDS–PAGE with 12% T, low MW at the bottom. The acid-induced proteins are indicated by arrows with identification numbers (their molecular mass and p*I* are given in Table II). Reprinted with permission from L. Phan-Thanh and F. Mahouin, *Electrophoresis* **20**, 2214 (1999).

TABLE II

PROTEINS INDUCED IN *L. monocytogenes* BY pH 3.5 (ACID STRESS) AND pH 5.5 (ACID ADAPTATION)

Identification number	Induction ratio[a]	Molecular mass (kDa)	p*I*	Induced at pH
52	2.6	43.8	5.76	3.5
152	3.2	18.2	5.70	3.5 and 5.5
223	3.4	37.1	6.28	3.5
262	4.1	30.2	6.31	5.5
271	2.7 and 2.2	28.2	6.79	3.5 and 5.5
284	2.4	53.1	6.12	3.5
304	4.8	38.0	5.40	5.5
307	2	37.8	6.03	5.5
315	2.5 and 8.5	12.6	5.65	3.5 and 5.5
353	4.4 and 8.3	18.2	5.47	3.5 and 5.5
374	6.6	25.0	5.40	3.5
375	3.9	48.5	6.54	3.5
380	3.5	31.0	6.60	3.5
422	3.8	14.1	5.46	5.5
424	3.6	22.4	6.08	3.5
475	2.8	50.0	6.44	5.5
479	6.5	75.8	5.49	3.5
500	13.3 and 15.4	54.9	4.95	3.5 and 5.5
520	3.1	37.0	6.50	5.5
524	2.8	33.1	5.88	3.5
525	2.9 and 6.3	31.6	6.11	3.5 and 5.5
527	14.6 and 7.7	36.3	6.84	3.5 and 5.5
552	2.6	27.8	6.17	3.5
570	3.1	35.7	6.94	3.5
586	4.4	47.9	6.37	5.5
604	2.8 and 5.6	30.2	5.90	3.5 and 5.5
609	11	56.0	6.28	5.5
618	3.3	26.9	6.58	3.5
632	2.4	16.8	6.43	5.5
636	4.8	29.5	4.96	5.5
640	12.8	50.2	6.22	5.5
651	8.2 and 9.3	26.1	6.48	3.5 and 5.5
652	3.9	42.2	6.77	3.5
657	3.0	29.8	5.84	3.5
672	19.6	39.6	5.60	5.5
695	2.7	37.1	6.15	3.5
702	3.8 and 8.6	50.3	6.65	3.5 and 5.5
705	9.1 and 3.6	23.5	6.26	3.5 and 5.5
708	6.0	47.9	6.76	3.5
719	6.6 and 11	41.7	6.29	3.5 and 5.5
720	7.5 and 8.5	50.9	6.76	3.5 and 5.5
722	4.8 and 11.4	27.0	5.85	3.5 and 5.5
727	6.0	47.1	5.84	3.5

TABLE II (*continued*)

Identification number	Induction ratio[a]	Molecular mass (kDa)	p*I*	Induced at pH
728	3.2	54.6	6.65	5.5
731	4.4 and 5.4	45.7	6.55	3.5 and 5.5
732	7.0 and 6.7	48.9	6.54	3.5 and 5.5
754	22.8 and 9.5	42.1	5.59	3.5 and 5.5
773	5.7	27.4	6.21	3.5
783	2.7	35.5	5.30	3.5
800	7.5	23.7	6.81	3.5
802	8.7 and 3.8	32.0	6.88	3.5 and 5.5
822	Novel	13.7	6.72	5.5
825	Novel	22.1	5.60	3.5 and 5.5
827	Novel	25.3	4.95	3.5 and 5.5
828	Novel	25.3	5.10	3.5 and 5.5
836	Novel	22.3	6.87	5.5
838	Novel	22.6	6.53	3.5 and 5.5
839	4.5	16.0	6.20	3.5
840	Novel	17.8	6.75	3.5
843	Novel	15.9	6.29	3.5 and 5.5
846	Novel	19.7	5.44	3.5
854	Novel	44.7	5.15	3.5
874	5.2	56.9	4.98	3.5
875	Novel	51.2	4.86	3.5 and 5.5

[a] The induction ratio given are means of the values from several gels. When two ratios are given, they correspond to induction at pH 3.5 and 5.5, respectively. Reprinted with permission from L. Phan-Thanh and F. Mahouin, *Electrophoresis* **20**, 2214 (1999).

incubated for 20 min in 15% (w/v) ammonium sulfate containing 2% phosphoric acid, and finally stained for 24–48 hr in colloidal Coomassie blue which is prepared by mixing 1 volume of ethanol with 4 volumes of a CB G-250 stock suspension. To prepare CB stock solution, add with vigorous agitation 1 g of Coomassie Brilliant Blue G-250 to 1 liter of 15% ammonium sulfate containing 2% phosphoric acid. After staining, wash the gels briefly in 20% ethanol to remove any stain precipitates which adhere to the gel surfaces.

Silver Staining

The silver staining techniques which use covalent fixing agents such as glutaraldehyde are not suited for the subsequent analysis of proteins by mass spectrometry. Mild fixatives should be used instead.

Second dimension gels should contain sodium thiosulfate incorporated in gel solution before casting (1.25 g per liter) to avoid posterior impregnation and washings. Following second dimension electrophoresis, the gels are fixed overnight in 50% (v/v) ethanol containing 2% (v/v) phosphoric acid and washed 3 times for

30 min with distilled water before being impregnated for 1 hr in a solution of silver nitrate (2 g/liter) containing formaldehyde (250 μl/liter). Color development ensues by agitating the gels in a solution of K_2CO_3 (40 g/liter), $Na_2S_2O_3$ (10 mg/liter), and formaldehyde (250 μl/liter). The development solution must be changed when it turns cloudy. When the small spots become visible, stop the development reaction by transferring the gels to a solution of 1% (v/v) acetic acid. After 1 hr the gels are washed several times with water and stored in 20% ethanol containing 2% (w/v) glycerol at 4°.

In-Gel Digestion of Proteins for Mass Spectrometric Analysis

Care must be taken during gel handling to avoid contamination by proteases and other proteins, especially skin keratin (wear washed powder-free gloves; clean staining trays, etc., with detergent).

Most stress proteins are hardly visible or not visible on CB- or silver-stained gels. Therefore a number of specific identifiable spots seen both on CB or silver-stained gels and radiolabeled gels must be chosen to serve as landmarks for locating the stress proteins. One or two micropreparative 2D gels normally yield sufficient protein for identification by mass spectrometry. Modern mass spectrometers are capable of giving strong signals with 50–100 fmol of protein.

The steps in the in-gel digestion procedure are as follows. The spots chosen are excised from the micropreparative gel, transferred into siliconized microtubes, and cut into small pieces with a scalpel (about 1 mm × 1 mm; do not crush the gel). A protein-free gel piece used as a negative control is also treated in the same manner throughout the process.

Destaining

If the gel has been CB-stained, the stain is removed by washing three times with a solution of 25 mM ammonium bicarbonate pH 8.0 and 50% acetonitrile (of sufficient volume to immerse the gel pieces). If the gel has been silver-stained, the stain is removed with a destaining solution freshly prepared by mixing 1 volume of 30 mM potassium ferricyanide and 1 volume of 100 mM sodium thiosulfate. The destaining is followed by several washes with 25 mM ammonium bicarbonate and a final wash with of 25 mM ammonium bicarbonate/50% acetonitrile (the gel turns opaque white). Use a centrifuge to pull down droplets after each agitation and use gel-loading pipette tips to remove the supernatant after each wash (to avoid sucking away tiny pieces of gel). Dry the gel pieces in a Speed Vac for 30 min.

Reduction–Alkylation

The gel pieces are left to swell in 10 mM dithiothreitol/25 mM ammonium bicarbonate before being incubated at 56° for 45 min. After the supernatant is discarded, an equivalent volume of acetonitrile is added, which is then rapidly replaced with a volume of 55 mM iodoacetamide/25 mM ammonium bicarbonate. Then incubate the gel pieces and supernatant in the dark for 30 min at room temperature. After discarding the supernatant, the gel pieces are washed with 25 mM ammonium bicarbonate, dehydrated with 100% acetonitrile, and dried in the Speed Vac.

Digestion

Rehydrate the dried gel pieces for 45 min at 4° (on ice) in 25 mM ammonium bicarbonate buffer containing 5 mM CaCl$_2$ and 12.5 ng/μl trypsin (sequencing grade, Boehringer, Mannheim, Germany). Add only enough solution so that it is totally absorbed by the gel (too much free enzyme in solution can result in greater enzyme autolysis). The digestion continues at 37° overnight (if necessary, add enough 25 mM ammonium bicarbonate buffer to immerse the gel pieces).

Peptide Recovery

Extract the supernatant (if the gels are dried, adding some water may facilitate this). Reserve 1 μl for direct MALDI analysis. Perform two or three extractions with a suitable volume of 50% acetonitrile/5% trifluoroacetic acid (about twice the volume necessary to immerse the gel pieces). The supernatants are pooled in siliconized microtubes, concentrated in a Speed Vac, and brought back up to 25 μl with 50% acetonitrile/5% TFA (a high concentration of TFA serves to minimize adsorptive sample loss). The amount of volatile salts can be decreased by adding water and reducing the volume using a Speed Vac. Repeat this step as many times as needed. Although the MALDI-TOF technique is relatively tolerant to salts, high amounts of salts may interfere with the efficiency of the matrix, and hence with the ionization of peptide fragments. Store the recovered peptides at below −20° until needed.

Protein Identification by Analysis of Peptide Mass Fingerprints Using Matrix-Assisted Absorption/Desorption Ionization–Time of Flight Mass Spectrometry (MALDI-TOF)

Peptide mass data are acquired using a Voyager Elite mass spectrometer from Applied Biosystems (Framingham, MA) equipped with a delayed extraction device and a time-of-flight analyzer.

A 2-μl aliquot of each tryptic digest is desalted using a C$_{18}$ ZipTip (Millipore) and diluted directly onto the MALDI sample target plate with 2 μl of 2,5-dihydroxybenzoic acid as matrix. Mass spectra are recorded in positive reflector mode. External calibration is performed using des-Arg[1]-bradykinin M$_y$ 904.46 and ACTH M$_y$ 2465.20. The list of peptide masses in a digest mixture as determined by MALDI is compared to theoretical lists produced *in silico* (by computer) based on databases through one of several programs available on the Internet. The most used program is MS-FIT from the University of California at

San Francisco (prospector.ucsf.edu/msfit.htm). Mass search programs can also be locally installed on the computer-operating MALDI-TOF machine and linked to the main protein databases such as those from NIH's National Center for Biology Information (www.ncbi.nlm.nih.gov), SwissProt (www.expasy.ch/sprot), and European Bioinformatics Institute (ftp://ftp.ebi.ac.uk/pub/databases/Peptide-Search). The search results for identification depend on such parameters as species, molecular weight and pI range, number of missed cleavages, mass tolerance, and possible amino acid modifications. The matching output is a list of proteins ranked by the number of peptides shared with the unknown protein (MOWSE score). The protein that gives rise to the largest number of "hits" is in principle the best candidate for correct identification. Peptide mass accuracy and number of matched peptides as well as the percent of coverage of the identified protein are the usual criteria used to evaluate the confidence of the identification. One should, however, bear in mind that peptide mass fingerprinting rarely yields all expected peptides from a protein, because large or hydrophobic peptides either are not extracted or are poorly extracted from the gel, others are not ionized efficiently in the mass spectrometer, and some sites may not be cut by the enzyme as expected. Nonspecific enzymatic cleavages are also commonplace. Furthermore, peptides from a protein that do not match those from the database may carry posttranslational or artifactual modifications or may result from truncations.[12] Because the genome sequence of *L. monocytogenes* is only partially characterized, cross-species identification is necessary in most cases. Wherever possible, preference is given to *Bacillus subtilis,* the gram-positive bacterium phylogenetically closest to *Listeria* among the microorganisms whose genome has been sequenced and better characterized. Given the evolutionary relatedness of protein sequences between species, the genome/proteome data from well studied organisms can be used for the identification and understanding of proteins from other species. Figure 2 reproduces a typical mass spectrum of the tryptic digest of an acid-induced protein (spot 702). The search results through MS-FIT program leads, at highest MOWSE score, to a *Listeria* protein of unknown function and highly similar to the glycine betaine ABC transporter (ATPase homolog). The proportion of matched masses is 9/31 and the protein coverage 30%. For reasons cited above, this identification needs to be confirmed by another protein attribute such as a protein sequence tag,[13] which can be obtained using a mass spectrometer equipped with a postsource decay device or a collision cell.

---

[12] A. A. Gooley and N. H. Packer, *in* "Proteome Research: New Frontiers in Functional Genomics" (M. R. Wilkins, K. L. Williams, R. D. Appel, and D. F. Hochstrasser, eds.), p. 65. Springer-Verlag, Berlin, 1997.

[13] M. R. Wilkins and A. A. Gooley, *in* "Proteome Research: New Frontiers in Functional Genomics" (M. R. Wilkins, K. L. Williams, R. D. Appel, and D. F. Hochstrasser, eds.), p. 35. Springer-Verlag, Berlin, 1997.

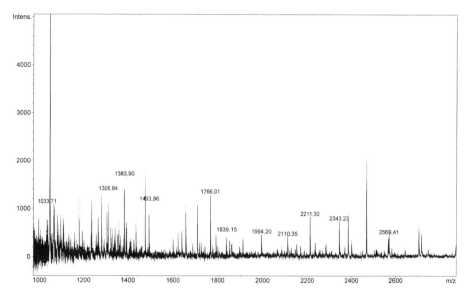

FIG. 2. MALDI mass spectrum of the tryptic digest of spot 702.

Comments. More traditional protein identification techniques such as immuno-blotting, Edman partial sequencing of the protein and of internal peptides, amino acid composition, or comigration analysis of known and unknown proteins can provide additional protein attributes which help to achieve high-confidence identification. These techniques, although effective, are time-consuming and labor-intensive and are not suited for a high-throughput proteomic approach.

Alternatively, one may increase the identification confidence by performing on a duplicate sample another peptide mass fingerprinting with a second proteolytic enzyme, such as chymotrypsin.

Sequence Tag Determination by Electrospray Ionization Tandem Mass Spectrometry (ESI MS-MS)

The samples from the spots not identified or ambiguously identified by MALDI are prepared for ESI MS-MS analysis using a Micromass Q-ToF hybrid tandem mass spectrometer equipped with a nanospray electrospray ionization device and a time-of-flight analyzer.

A 5-μl aliquot of each sample is desalted and concentrated on Poros R2 (Applied Biosystems) packed in the end of a gel loader tip and eluted into a nanovial in 2 μl of 1:1 methanol:0.2% aqueous formic acid (to have a protein concentration about 1 pmol/μl) before being introduced into the electrospray source on the tip of a microcapillary.

In the MS mode of operation, the infusion of the tryptic mixture into the Q-ToF allows mass measurements to be made for all of the individual peptides. This is followed by specific selection of individual peptides and their subsequent MS-MS fragmentation to generate sequence data. Figure 3 reproduces the typical mass spectrum in MS-MS mode of Q-ToF fragmentation of a sample (tryptic digest of spot 728). The peptide at m/z 771.34 has been selected for MS-MS fragmentation. The product ion scan from the doubly charged ions at m/z 771.34 shows, after deconvolution, both y''_n (C-terminal) and b_a (N-terminal) sequence ions (Fig. 4), from which the sequence DVVIVAGNVATAEGAR is determined using the integrated MassLynx software. Using this sequence tag as part of the query specification, the database search against *Listeria* genome results in a protein of 488 amino acid residues whose function is unknown and which has a homology with inosine-monophosphate dehydrogenase of *Bacilus subtilis*. The theoretical values of its molecular mass and isoelectric point (53 kDa and p*I* 6.5) are in accordance with those estimated experimentally (54.5 kDa and p*I* 6.6). The entire sequence of this protein is:

MWETKFAKEGLTFDDVLLVPAKSDVLPNDVDLSVEMAPSLKLNVPIWSA
GMDTITEAKMAIAARQGGIGVVHKNMSIEQQAEEIEKVKRSESGVIIDPF
YLTPDHQVFAAEHLMGKYRISGVPIVNNEKERKLVGILTNRDLRFISDYDT
VIKDVMTKENLVTAPVGTTLKQAEQILQKHRIEKLPLVDEAGILKGLITIK
DIEKVIEFPNSAKDKHGRLLAAAVGITNDTFVRVEKLIEAGVDAIVIDTAH
GHSAGVINKISEIRQTFK**DVVIVAGNVATAEGAR**ALFEVGVDIVKVGIGPG
SICTTRVVAGVGVPQITAIYDCATVAREFGKTIIADGGIKYSGDIVKALAAG
GNAVMLGSMLAGTDESPGETEIFQGRQFKTYRGMGSLAAMEHHGSKDR
YFQADAKKLVPEGIEGRVPYKGSVADIIFQLVGGIRSGMGYTGSPDLRHLR
EEAAFVRMTGAGLRESHPHDIQIKKEAPNYSIS

Conclusions

To identify and characterize the proteins induced by acidic conditions, the proteomic approach using two-dimensional electrophoresis combined with mass spectrometric analysis rationally offers the best solution: 2D electrophoresis allows analyzing differentially expressed proteins under different stress conditions and mass spectrometry allows high-throughput identification of great numbers of proteins that no other technique can afford to do. The major difficulty with this proteomic approach lies, however, in 2D electrophoresis. Despite its unparalleled resolving power, 2D electrophoresis still has problems with reproducibility and at least for the time being it is rather a matter of craftsmanship than a precisely controlled technology. As for mass spectrometry, it has rapidly become the method of choice, unchallenged, for protein identification. For organisms whose genome is sequenced and well characterized, peptide mass fingerprinting using MALDI mass spectrometry may provide a complete solution. However, in practice, because

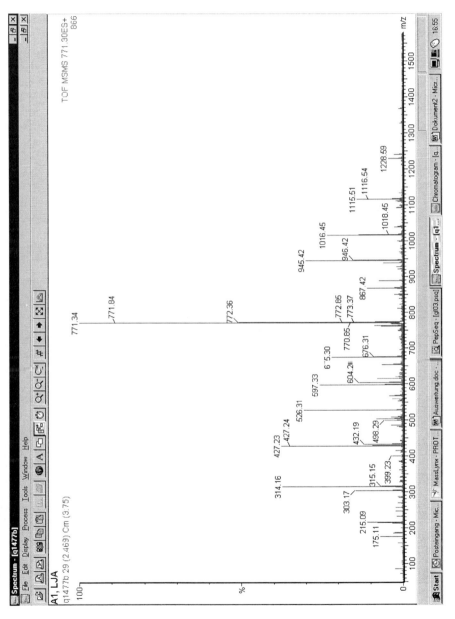

FIG. 3. ESI MS-MS mass spectrum of the tryptic digest of spot 728. The peptide at *m/z* 771.34 is selected for MS-MS fragmentation.

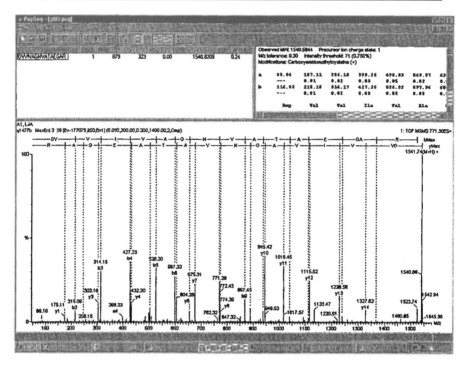

FIG. 4. MS-MS product ion spectrum (after deconvolution) of the doubly charged ions at *m/z* 771.34 from the tryptic digest of spot 728, from which the sequence is determined.

of poor protein digestion or too small an amount of extractable peptides from the 2D gel, proteins cannot always be identified unambiguously by MALDI MS. Electrospray ionization tandem mass spectrometry (ESI-MS-MS) offers a powerful complementary tool capable of providing sequence data from individual peptides generated by the enzymatic digestion of protein, thus increasing the confidence level of protein identification. The sequence tags not only serve to identify proteins, they can also be used for cloning the genes with which regulation experiments can be done to verify the effect of the stress and to confirm the role these genes play in the process of resistance to the stress.

[19] Proteome Analysis of *Chlamydia pneumoniae*

By BRIAN BERG VANDAHL, SVEND BIRKELUND, and GUNNA CHRISTIANSEN

Introduction

Today 32 bacterial genomes are available through NCBI.[1] Genome sequencing has clearly provided major new insights in the molecular biology of these bacteria, but genes are turned on and off and synthesized proteins may possibly be processed, modified, or even exported. The protein composition of a living bacterium cannot be predicted from the genome sequence alone. Direct investigation of the total content of protein species in a cell is the task of proteomics. Proteomics as approached today in most laboratories involves two steps: Separation of proteins according to their isoelectric point and molecular weight by two-dimensional polyacrylamide gel electrophoresis (2D PAGE) and subsequent identification of those proteins.[2] The method of choice for large-scale identification is mass spectrometry (MS) but all identification methods known from one-dimensional electrophoresis apply. Bacteria are especially well suited for proteome analysis, because of the availability of high quality genomes, the relatively small size of the genomes, and the low level of functional redundancy.

Chlamydia Pneumoniae

Chlamydia pneumoniae primarily infects epithelial cells of the respiratory tract causing pneumonia and bronchitis. A correlation between *C. pneumoniae* and atherosclerosis has been suggested. The determinants of pathogenesis are only poorly understood. All *Chlamydia* species are obligate intracellular gram-negative bacteria sharing a characteristic biphasic developmental cycle in which they alternate between two morphologically and metabolically distinct forms, the infectious elementary bodies (EB) and the replicative reticular bodies (RB). The EB are small rigid bodies about 300 nm in diameter which are characterized by having a high content of cross-linked cysteine-rich proteins in the outer membrane and a nucleus that is condensed by histone-like proteins. An infection begins with the attachment of EB to a host cell. On attachment the EB are internalized by endocytosis which is thought to be induced by the EB themselves. Following ingestion, the EB are transformed into RB, which have a permeable membrane and normal bacterial DNA. The RB divide by binary fission inside the phagosome, which is called the chlamydial inclusion. The chlamydiae stay inside the inclusion

---

[1] www.ncbi.nlm.nih.gov:80/PMGifs/Genomes/org.html

[2] M. J. Dunn and J. M. Corbett, *Methods Enzymol.* **271**, 177 (1996).

throughout their intracellular life. At about 48 hr postinfection (hpi), the RB begin differentiation back into EB and about 72 to 96 hpi a new generation of infectious EB is released on disruption of the host cell.[3]

Proteome Studies of Chlamydia

A few proteome studies of *C. trachomatis* have been published in recent years. Immunoreactive *C.trachomatis* proteins recognized by sera from 17 patients suffering from genital inflammatory disease have been identified by Sanchez-Campillo *et al.*[4] Immunological sensitization in chlamydial pathogenicity had been thought to be sustained by *C. trachomatis* GroEl, which is a homolog to human heat shock protein 60 (HSP60), but Sanchez-Campillo *et al.* identified other conserved bacterial proteins as antigens.[4] As part of a comparative study of *C. trachomatis* serovars, Shaw *et al.* identified a difference in the tryptophan synthase α-subunit which can explain difference in pathogenesis between serovars.[5] Other studies include identification of early proteins and iron-response proteins.[6,7] In 1996 a preliminary analysis of the total *C. trachomatis* proteome was published, but only very highly abundant proteins were identified.[8]

Considerations Regarding Proteomics on Chlamydia pneumoniae

The first *C. pneumoniae* genome was published in 1999 by Kalman *et al.*[9] and today three *C. pneumoniae* genomes are available. Clearly, much knowledge has been acquired about the life of *Chlamydia* by genomics, but we expect proteomics to prove especially valuable since the *Chlamydia* species are obligate intracellular and no transformation system exists which can be used in the investigation of gene expression. Furthermore, no information on the regulation of proteins during the developmental cycle can be obtained by genomics. Transcript analysis reveals which genes are transcribed into RNA. However, RNA is not necessarily translated into protein and information on posttranslational modification, processing, and transportation can only be achieved by looking at the protein level directly.

[3] J. W. Moulder, *Microbiol Rev.* **55,** 143 (1991).
[4] M. Sanchez-Campillo, L. Bini, M. Comanducci, R. Raggiaschi, B. Marzocchi, V. Pallini, and G. Ratti, *Electrophoresis* **20,** 2269 (1999).
[5] A. C. Shaw, G. Christiansen, P. Roepstorff, and S. Birkelund, *Microbes Infect.* **2,** 581 (2000).
[6] A. G. Lundemose, S. Birkelund, P. M. Larsen, S. J. Fey, and G. Christiansen, *Infect.Immun.* **58,** 2478 (1990).
[7] J. E. Raulston, *Infect. Immun.* **65,** 4539 (1997).
[8] L. Bini, M. Sanchez-Campillo, A. Santucci, B. Magi, B. Marzocchi, M. Comanducci, G. Christiansen, S. Birkelund, R. Cevenini, E. Vretou, G. Ratti, and V. Pallini, *Electrophoresis* **17,** 185 (1996).
[9] S. Kalman, W. Mitchell, R. Marathe, C. Lammel, J. Fan, R. W. Hyman, L. Olinger, J. Grimwood, R. W. Davis, and R. S. Stephens, *Nat. Genet.* **21,** 385 (1999).

Furthermore, the EB are described as being transcriptionally inactive. To elucidate which investigations of *C. pneumoniae* it would be reasonable to approach by proteomics, we found that a global analysis was required and thus initiated such a work.

Methods for Construction of Proteome Reference Map on *Chlamydia pneumoniae*

Construction of a proteome reference map of EB proteins is described in Vandahl *et al.*[10] In future studies proteins can be identified by comparison to this reference map which has been made available on the Internet (www.gram.au.dk). Proteins are radiolabeled to facilitate selective visualization of bacterial proteins by inhibition of host cell expression during the labeling period. In this way, the expression pattern in specific periods of time during the developmental cycle can be investigated without the need for purification of bacteria, which is a major advantage as purification of the fragile RB is cumbersome compared to purification of EB. To ensure that radioactivity is incorporated into proteins synthesized at different times during infection, a number of labeling periods are introduced throughout the developmental cycle before purification of EB for the reference map.

Cultivation and Radioactive Labeling of Chlamydia Proteins

HEp-2 cells (ATCC, Manassus, VA) are cultivated in RPMI 1640 (Gibco-BRL, Gaithersburg, MD) with 10% heat inactivated fetal calf serum and 10 μg/ml gentamicin in a 5% CO_2 and 85% humidity atmosphere at 37°. Cells detected free of *Mycoplasma* are cultivated in 3.5 cm in diameter (i.d.) wells. Three ml of medium containing 100,000 cells per ml is added to each well the day before infection. Semiconfluent monolayers of cells are infected with one inclusion forming unit of *C. pneumoniae* (VR1310, ATTC) per cell by 30 min of centrifugation at 34° at 1000g. The medium for the centrifugation step is 50% phosphate buffered saline (PBS) and 50% infection medium (RPMI1640 with 10% fetal calf serum, 2 μg/ml gentamicin, 1 μg/ml cycloheximide, and 0.5% glucose). After centrifugation the medium is changed to infection medium and the temperature is lowered to 34°. The efficacy of the infection is checked by indirect immunofluorescence microscopy of infected cells cultivated on glass coverslips using antibodies specific for *C. pneumoniae*. Labeling is performed by changing the medium to methionine/cysteine free RPMI 1640 containing 40 μg/ml cycloheximide and 100 μCi/ml [^{35}S]methionine/cysteine Promix (Amersham, England). Cells are harvested in lysis buffer immediately after labeling periods of 2 hr to visualize the proteins expressed during the labeling period. The amount of incorporated radioactivity is

---

[10] B. Vandahl, S. Birkelund, H. Demol, B. Hoorelbeke, G. Christiansen, J. Vandekerckhove, and K. Gevaert, *Electrophoresis* **22**, 1204 (2001).

determined by TCA precipitation of 10 μl sample followed by scintillation counting of the precipitate. Using the method described above to label *C. pneumoniae* infected cells at 48–50 hpi will yield a total of approximately 10^7 cpm from a 3.5 cm i.d. well.

Preparation of Elementary Bodies

Cultures used for purification of EB have the medium changed back into normal medium after the 2 hr labeling period and until purification of EB. Different wells are labeled at different points in time before infected cells are scraped off in cold PBS at 72 hpi and purification of EB is performed in principle according to Schachter and Wyrick.[11] The labeled samples are supplemented with unlabeled samples since purification requires a certain amount of material. For a standard purification we use 600 square cm of infected cell culture which yields about 1 mg of protein from purified EB. Samples in PBS are pooled and sonicated briefly to rupture host cells and cell debris is removed by slow speed centrifugation ($350g$) for 10 min. The purificentation of EB is obtained by two steps of ultracentrifugation through discontinuous gradients made from Visipaque (Nycomed, Norway) and HEPES. For the lowest gradient in the first centrifugation step we use a high density layer ($\delta = 1.20$), which cannot be penetrated by the bacteria since collection at the bottom of the centrifuge tube can result in clotting of bacteria and contaminants. Above the Visipaque/HEPES layer of $\delta = 1.20$ we use a layer of 50% sucrose and on top a Visipaque/HEPES layer of $\delta = 1.12$. The bacteria are collected from the top of the bottom layer after 1 hr of centrifugation at $10^5 g$ at $5°$ using a Beckman SW-28 rotor. The collected bacteria are sonicated briefly and then treated with RNase and DNase before centrifugation through Visipaque/HEPES gradients of $\delta = 1.13, \delta = 1.15$, and $\delta = 1.16$ for 2 hr at $10^5 g$ at $5°$. The EB are collected from top of the $\delta = 1.15$ layer and further purified by mixing with HEPES and 30 min of centrifugation at 28,000 rpm. The purity of the EB preparation is verified by electron microscopy.

Two-Dimensional Polyacrylamide Gel Electrophoresis

Sample Preparation

A crucial step in 2D PAGE is the sample preparation in which major problems have been concerned with limited entry into the gel of high molecular weight proteins, highly hydrophobic proteins, and basic proteins. The most widely used solubilization procedure is that based on O'Farrell[12] using a mixture of urea, detergent, reducing agent, and carrier ampholytes. However, progress has been made

[11] J. Schachter and P. B. Wyrick, *Methods Enzymol.* **236,** 377 (1994).
[12] P. H. O'Farrell, *J. Biol. Chem.* **250,** 4007 (1975).

by the introduction of new solubilizing agents. Thiourea has been shown to supplement urea in a way that facilitates improved solubilization of hydrophobic proteins, sulfobetains have proved superior to 3-[(3-chloramidopropyl)dimethylammonium]-1-propane sulfonate (CHAPS) as detergent in certain cases, and tributylphosphine (TBP) is a stronger alternative to dithioerythritol (DTE) for reduction of disulfide bridges; see Gorg *et al.*[13] for an excellent description of 2D PAGE methods. In our proteomic investigations of *C. pneumoniae* we have focused on the best possible solubilization of outer membrane proteins and we currently use a lysis buffer based on Harder *et al.*[14] containing 7 M urea, 2 M thiourea, 4% (w/v) CHAPS, 40 mM Tris base, 65 mM DTE, and 2% (v/v) Pharmalyte 3-10. To further improve the recovery of outer membrane proteins, samples are presolubilized in a minimal amount of 1% sodium dodecyl sulfate (SDS), 50 mM Tris–HCl pH 7.0, sonicated and boiled for 5 min. By subsequent addition of lysis buffer the SDS is displaced. In our hands there is no marked improvement by using TBP instead of DTE on *Chlamydia* samples. It is important that the lysis buffer is prepared fresh before use or alternatively stored in aliquots at $-70°$ and only thawed once. Cell lysates in lysis buffer must be kept below $37°$ to avoid artifactual carbamylation of proteins.

Electrophoresis

Precast polyacrylamide gel strips with an immobilized pH gradient (IPG) are used for separation of proteins according to their pI in the first dimension. A major advantage of IPGs compared to the use of carrier ampholytes is the high reproducibility which makes interlaboratory comparison possible.[15] Furthermore, the osmotic drift against the cathode is limited in an IPG which means that a good resolution can be obtained also for basic proteins, and the focusing time can be prolonged so that the amount of sample can be increased. We apply samples to the IPG by rehydration. Rehydration under application of a low voltage (10–50 V), which has been made possible by introduction of the IPGphor (Amersham Pharmacia, Piscataway, NJ), has markedly improved the recovery of especially high molecular weight proteins.[16] High protein load and long focusing time can cause precipitation of salts and migration of water at the ends of the IPG. We have experienced good results by placing lightly hydrated paper strips between the IPG and the electrode. The osmotic drift is reduced and accumulating ions are absorbed by the paper. The electrode pads are placed immediately after rehydration of the IPGs and changed when the voltage reaches 1000 V.

---

[13] A. Gorg, C. Obermaier, G. Boguth, A. Harder, B. Scheibe, R. Wildgruber, and W. Weiss, *Electrophoresis* **21**, 1037 (2000).

[14] A. Harder, R. Wildgruber, A. Nawrocki, S. J. Fey, P. M. Larsen, and A. Gorg, *Electrophoresis* **20**, 826 (1999).

[15] J. M. Corbett, M. J. Dunn, A. Posch, and A. Gorg, *Electrophoresis* **15**, 1205 (1994).

[16] X. Zuo and D. W. Speicher, *Electrophoresis* **21**, 3035 (2000).

A total of 750 μg *Chlamydia* protein is loaded on each strip for identification purposes and the amount of labeled sample is 3,000,000 cpm as determined by scintillation counting. This amount of *Chlamydia* protein, comprising maximally 1000 different proteins, focuses well using the following procedure on the IPGphor: 350 μl of the lysis solution containing 750 μg protein is used to rehydrate a nonlinear pH 3–10 strip (Amersham Pharmacia) at 21°. Rehydration is carried out for 14 hr at 30 V. Focusing is obtained by application of 250 V for 1 hr, 500 V for 2 hr, 1000 V for 1 hr, 3000 V for 1 hr, and 8000 V for 12 hr. The current is limited to 50 mA per strip. Following first dimension, strips are equilibrated in 6 *M* urea, 2% SDS, 30% glycerol (87%), and 2% DTE and subsequently in a solution where iodoacetamide replaces DTE. Equilibration is performed in a reswelling tray (Amersham Pharmacia) using 2.5 ml of each solution per strip, which in our hands are the sufficient but required amounts to obtain reduction and alkylation of cysteines. Second dimension is carried out on polyacrylamide gels which are cast at least 4 hr before use. We use 9–16% linear gradient gels; no stacking gel is required.

Visualization

To visualize proteins, all conventional staining procedures apply but detection of radiolabeled proteins has certain advantages. Radiolabeling is sensitive and does not interfere with subsequent identification methods, and furthermore, pulse/chase studies can be performed. Preparative gels for MS are washed for 30 min in double distilled H$_2$O before they are vacuum dried onto chromatographic paper. Comparative gels are fixed in a solution containing 10% acetic acid and 25% 2-propanol for 30 min and treated with Amplify (Amersham) for another 30 min before drying. The protein load is limited to 200,000–300,000 cpm as measured by scintillation counting when Amplify is used. Using this load, gels are exposed to Kodak Biomax-MR X-ray films for 7 days at −70°. Before exposure, the gels are dried by heating in vacuum. To ensure that the gels do not crackle, the temperature is increased to 75° within 1 hr and kept at 75° for 5 hr. Autoradiographs and dried gels are aligned using radioactive ink markers, and spots for MS analysis are excised by cutting through a transparent copy of the autoradiograph. To avoid keratin contamination during excision this step is performed in a laminar flow bench and only disposable equipment is used for handling the samples.

Identification

Mass spectrometry (MS) has become the most widespread identification tool in proteome projects since it allows for a high throughput. By MS the mass of molecules converted to gas phase ions can be measured with an accuracy as good as 50 ppm[17] which corresponds to less than 0.1 Da for small peptides.

---

[17] O. N. Jensen, A. V. Podtelejnikov, and M. Mann, *Anal. Chem.* **69,** 4741 (1997).

The masses of a collection of peptides resulting from enzymatic digestion of a protein, the so-called peptide mass fingerprint (PMF), can be used to search a protein database which is theoretically digested with the same enzyme, or sequence information (peptide sequence tags, PSTs) can be obtained by fragmentation of peptides.[18,19]

Gel Resolution

For identification purposes we use narrow pH gradients such as pH 4–7 and pH 6–11 to minimize the number of overlapping spots and thereby achieve the best possible purity of samples. The number of gels can be reduced in comparative studies by using one nonlinear 3–10 IPG. About 650 protein spots can be resolved using these IPGs for separation of *C. pneumoniae* samples. In nonlinear pH 3–10 gels of *C. pneumoniae* EB, the spot of highest density accounted for 45,000 ppm and the spot of lowest density unambiguously identified by use of the MelanieII software accounted for 10 ppm. This gives a dynamic range of 4500-fold. The protein identified from the spot of lowest intensity accounted for 38 ppm when corrected for methionine content which compared to a corrected value of 62,000 ppm for the highest abundant protein gives a dynamic "identification" range of 1500-fold. Even if it is estimated that no protein is lost during the procedures, the protein amount in the lowest abundant spot was less than 500 fmol. To date a total of 275 proteins have been identified representing 170 genes from literally all protein categories. An image from labeling 55–57 hpi is shown in Fig. 1.

Applications of Bacterial Proteomics

Besides mapping the total protein content in a developmental stage such as the *Chlamydia* EB, proteome analysis can be used in a variety of investigations. We have concentrated on time-dependent expression as exemplified in Vandahl *et al.*,[20] but the effect of all conditions which can be controlled in the laboratory such as temperature changes, heat shock, acidic shock, partial pressure of oxygen, amino acid depletion, or the effect of different drugs or cytokines can be investigated. Identification of immunoreactive proteins is facilitated by reacting 2D immunoblots with patient sera; the total content of the outer membrane complex can be determined or secreted proteins can be identified in the supernatant of bacteria growing in suspension. The remainder of this article will concern a method to identify proteins secreted from obligate intracellular bacteria.

---

[18] M. Mann, P. Hojrup, and P. Roepstorff, *Biol. Mass. Spectrom.* **22,** 338 (1993).

[19] K. Gevaert and J. Vandekerckhove, *Electrophoresis* **21,** 1145 (2000).

[20] B. Vandahl, K. Gevaert, H. Demol, B. Hoorelbeke, A. Holm, J. Vandekerckhove, G. Christiansen, and S. Birkelund, *Electrophoresis* **22,** 1697 (2001).

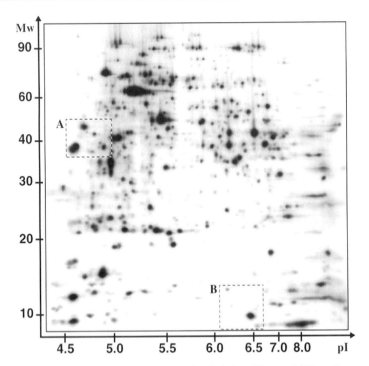

Fig. 1. Autoradiograph of 2D PAGE protein profile of whole cell lysate of *Chlamydia pneumoniae* infected cells radiolabeled 55–57 hr postinfection. The boxed areas (A) and (B) are enlarged in Fig. 3A and 3C, respectively.

Secreted Pathogenicity Factors

The chlamydiae are protected from the host cell by the inclusion membrane throughout the intracellular development. By separating itself from the host cell the bacterium escapes the degradation and presentation of bacterial peptides as T-cell antigens at the surface of the infected cell. Secreted effector proteins, however, will be present in the cytoplasm of the host cell, and hence these proteins will be important pathogenicity factors and vaccine candidates. No *Chlamydia* proteins have been identified in the cytoplasm of the host cell, but key members of a type III secretion system have been identified in the genome sequence and in our proteome analysis and there is evidence that the apparatus is indeed functional.[21] The type III secretion apparatus spans the periplasmic space of gram-negative bacteria and involves a protrusion from the outer membrane which is able to penetrate cell membranes. In the case of *Chlamydia* this system may be operational

[21] K. A. Fields and T. Hackstadt, *Mol. Microbiol.* **38**, 1048 (2000).

both prior to infection by delivering bacterial proteins across the cell membrane of a future host cell and during infection by transporting proteins across the inclusion membrane. *Chlamydia* synthesize proteins which are inserted into the inclusion membrane, but the specific function of these proteins is not known. Evidence has been provided that several inclusion membrane proteins can be secreted through the type III secretion system.[22] Type III secreted effector proteins contain no known sequence motif or structural motif by which these can be predicted. Furthermore, proteins secreted from different organisms may carry out a variety of different functions, meaning that proteins secreted from *Chlamydia* will not necessarily be homologous to proteins secreted from other organisms. The structural components of the type III secretion apparatus are located in several clusters in the *Chlamydia* genome meaning that genes encoding proteins secreted through the type III system cannot be easily identified based on their genomic localization alone. Because of the fragility of RB it would be difficult to prepare host cell cytoplasm without contamination from intracellular *Chlamydia* proteins, which, immediately, would be required for the isolation of secreted proteins.

Proteome-Based Method for Identification of Candidate Secreted Proteins

The proteome reference map has been based on purified EB meaning that the protein profile contains only bacterial intracellular proteins. In the protein profile from whole cell lysate of infected cells both bacterial intracellular proteins and all proteins present in the host cell will be present. However, only bacterial proteins will be visualized by autoradiography if host cell protein synthesis is inhibited during labeling. The additional protein spots visualized in gels from whole cell lysates of infected cells compared to gels from EB must therefore contain secreted proteins, RB-specific proteins, or proteins which have been processed or modified in a way that has changed their position in the period of time from labeling to the purification of EB. RB-specific proteins can be discarded as secretion candidates by examination of protein profiles of purified RB. When searching for secreted proteins by protein profile comparison it is important that identification of proteins from the gels of whole cell lysates can be achieved despite the eukaryotic background, as secreted proteins of course cannot be enriched by purification of the bacteria. An example of identification of a protein which is absent from EB has been given in Vandahl *et al.*[20] The bacterial protein spot was excised from a gel loaded with whole cell lysate and as expected several human proteins were present in the same spot. After HPLC fragmentation of a tryptic digest of the protein it was possible to carry out postsource decay (PSD) analysis on single peptides which resulted in secure identification of a *C. pneumoniae* protein. Further analysis revealed that this particular protein was absent from EB because it was processed

---

[22] A. Subtil, C. Parsot, and A. Dautry-Varsat, *Mol. Microbiol.* **39,** 792 (2001).

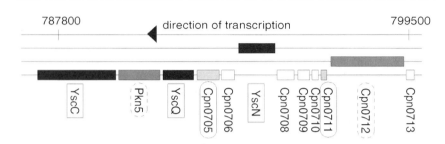

FIG. 2. Region of the *Chlamydia pneumoniae* genome containing three type III secretion apparatus proteins (boxed), two earlier suggested candidate secreted proteins (encircled, dashed line), and two new secretion candidates (encircled, solid line). The genomic segment is redrawn from the Web site of the Chlamydia Genome Project (*Chlamydia*-www.berkeley.edu:4231/).

within 2 hr after synthesis. An alternative to PSD as identification method could have been MS/MS.

Preliminary Results

At least 90 protein spots which are clearly present in gels from whole cell lysate of infected cells labeled 55–57 hpi are absent from or significantly reduced in gels from purified EB labeled at different times throughout the developmental cycle including 55–57 hpi. The reference map from purified EB includes the identification of 15 of these proteins. Ten proteins are likely to be RB specific proteins which will be confirmed upon purification of RB. Five proteins are hypothetical proteins including two proteins (Cpn0705 and Cpn0711) which are encoded by genes located close to the type III component genes *yscC, yscQ,* and *yscN* and the genes for the suggested effector proteins Pkn5[23] and Cpn0712[24] (Fig. 2).[9] It can be expected that genes located in proximity to type III clusters will include secreted effector proteins. The proteins Cpn0705 and Cpn0711 were found in very limited amounts in EB, whereas major spots were found at the position of Cpn0705 and Cpn0711 when comparing to the gels loaded with whole cell lysates labeled from 55–57 hpi (Fig. 3). These findings can be interpreted as the proteins Cpn0705 and Cpn0711 have been secreted, but they may also be RB specific or present in EB in a processed form. If these proteins are secreted they do not necessarily carry out a function in the host cell cytoplasm but could be inserted into the inclusion membrane as a part of the type III apparatus. Based on their localization in the genome we suggest that they are either type III secreted effector proteins or members of

[23] A. Subtil, A. Blocker, and A. Dautry-Varsat, *Microbes Infect.* **2,** 367 (2000).
[24] T. L. Yahr, A. J. Vallis, M. K. Hancock, J. T. Barbieri, and D. W. Frank, *Proc. Natl. Acad. Sci. U.S.A.* **95,** 13899 (1998).

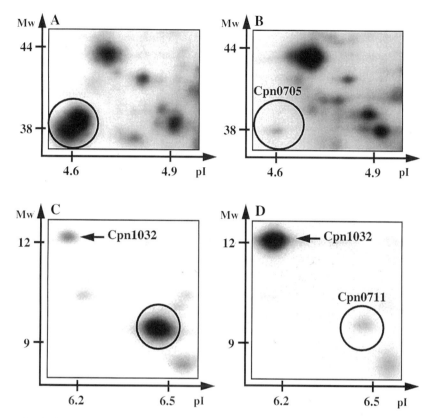

FIG. 3. Comparison of 2D PAGE protein profiles of whole cell lysates of infected cells labeled from 55–57 hpi and harvested 57 hpi (A and C) to purified EB labeled at 6–8, 12–14, 24–26, 36–38, 42–44, 48–50, 54–56, 60–62 hpi and harvested 72 hpi (B and D). The encircled spots in (B) and (D) were identified from EB gels. The corresponding areas in the whole cell lysate gel are encircled in (A) and (C). Note that two spots are detected within the circle in (A). The pI and molecular weight values are determined by comparison to the EB reference map (available at www.gram.au.dk).

the type III secretion apparatus. If they are part of the secretion apparatus they must be located in the inclusion membrane since they are absent from EB. In both cases these proteins are candidate pathogenicity factors. The subcellular location of the proteins will be further analyzed by immunofluorescence microscopy with antibodies raised against these. Two protein spots are present in the whole cell lysate gel (Fig. 3A) at the position of Cpn0705, but only one spot can be detected in the purified EB gel (Fig. 3B). The protein causing the additional spot in the whole cell lysate gel is one of the 75 candidate secreted proteins which await identification.

The described proteome comparison is a novel method to identify candidate secreted proteins for obligate intracellular bacteria. Identification and further investigation of the proteins differing between whole cell lysates and purified bacteria will reveal important *Chlamydia* pathogenicity factors and vaccine candidates and contribute to the understanding of the intracellular life of *Chlamydia*.

Acknowledgments

We are greatly indebted to Prof. Joel Vandekerckhove, Dr. Kris Gevaert, and co-workers at Ghent University, Belgium, for performing mass spectrometry identification.

[20] Enrichment and Proteomic Analysis of Low-Abundance Bacterial Proteins

By MICHAEL FOUNTOULAKIS and BÉLA TAKÁCS

Not all proteins of an organism are expressed at the same level at a certain time point. Whereas some species, such as heat shock proteins or elongation factors, are present in as many thousands of copies per cell, others, such as transcription factors, exert their functions at very low concentrations, sometimes just a few copies per cell. In the small sample volume (about 10–500 μl), which is usually used for proteomic analysis, the majority of the expressed proteins are not present in sufficient quantities to be visualized and identified. Therefore, proteins present in low copy numbers cannot be readily detected during the analysis of total proteins. The study of the low copy number gene products may be more interesting in investigating biological events than their high abundance counterparts.

Proteomics is a technology-based science which studies in a high throughput mode changes in the levels, the modifications, and the interactions of proteins that result from specific diseases or from exposure to external factors, such as toxic agents. Proteomics usually employs two steps: (*i*) the separation of a protein mixture by biochemical methods, usually by two-dimensional (2D) electrophoresis and (*ii*) the identification of the separated proteins by various analytical techniques, mainly by mass spectrometry.[1] The major advantage of 2D electrophoresis is that it enables the simultaneous separation and visualization of thousands of unknown protein forms. No other method can do that at the present time. There are several other proteomics-related technologies, such as protein arrays, that are presently in the developmental stage.

[1] M. Fountoulakis, "Encyclopedia of Separation Science, II/Electrophoresis," p. 1356. Academic Press, London, 2000.

Two-dimensional electrophoresis comprises two steps (dimensions): (i) separation of the proteins on the basis of differences in their net charge, called isoelectric focusing (IEF), which is usually performed on immobilized pH gradient (IPG) strips, and (ii) separation of the focused proteins on the basis of differences in their molecular masses, which is performed in sodium dodecyl sulfate (SDS)–polyacrylamide gels. Two-dimensional electrophoresis has certain limitations: (i) only the major components of a protein mixture can be visualized (this is also valid for the other separation approaches) and (ii) the detection of the low and high molecular mass proteins, as well as the basic and hydrophobic proteins, is inefficient.

Samples representing the total protein mixture have usually been analyzed and only the abundant, hydrophilic components have been visualized. These proteins could be solubilized with reagents compatible with isoelectric focusing (IEF), for example, urea and CHAPS. Such an analysis provides a limited image of the proteome, which is insufficient for the detection of the majority of proteins. In a 2D gel, where about 1 mg of total bacterial proteins have been resolved, 1000–3000 protein spots can be detected, using Coomassie blue staining. The spots represent the products of only 200–300 different genes. Other gene products, not visualized, are most likely expressed at levels too low for detection or they cannot be identified because of limitations of the current technology: they are too small, too large, basic, or hydrophobic. Here we will discuss protein enrichment approaches prior to the analysis and we will limit the proteomic analysis to the use of 2D electrophoresis.

Protein Enrichment Prior to 2D Electrophoresis

Analysis is usually successful if the protein is present in a sufficient amount so that the corresponding spot is visible in a polyacrylamide gel after staining with Coomassie blue. In general, for the detection of a protein, three prerequisites must be fulfilled: (1) The protein should be available in a sufficient amount in the protein mixture prior to 2D electrophoresis (otherwise enrichment is necessary, for example, by chromatography). (2) The protein should be brought into solution with IEF-compatible reagents and kept in solution during the first dimensional separation (for proteins difficult to solubilize strong solubilizing agents should be employed, such as SDS). (3) The protein should belong to that category of proteins separable by 2D electrophoresis, i.e., it should have average pI values usually between pH 4 and 10 and a molecular mass between 10 and 120 kDa and it should not be strongly hydrophobic (otherwise variations of 2D electrophoresis or an alternative separation procedure should be used).

For the enrichment of low-abundance bacterial proteins from crude extracts, biochemical protein enrichment methods are usually employed. These methods separate the original protein mixture into simpler fractions, each containing a lesser

TABLE I
ENRICHMENT AND ANALYSIS OF LOW-ABUNDANCE BACTERIAL PROTEINS

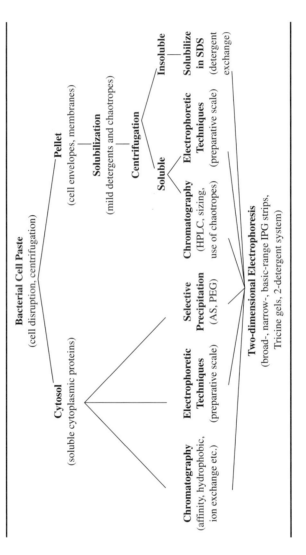

number of total proteins in comparison with the starting material. This increases the likelihood of detecting low-abundance proteins. Two approaches are usually employed: (1) separation of the mixture into cytosolic and membrane protein fractions and (2) protein enrichment from larger volumes, after the separation of the cytosolic and membrane fractions, by fractionation of the protein mixture, using selective fractionation, chromatographic steps, or electrophoretic procedures.

Table I shows the general scheme followed for the enrichment of low-copy-number bacterial proteins. Here we describe the enrichment of low-abundance proteins of the bacteria *Haemophilus influenzae* and *Escherichia coli. Haemophilus influenzae,* a gram-negative bacterium and a major causative agent of respiratory tract infections, represents a didactic model of the application of proteomics. Because the entire genome of the microorganism has been sequenced,[2] its proteome includes a relatively small number of possible gene products (approximately 1740), and it is easily cultivated in defined media. Similar procedures can be applied for other bacteria, such as *Escherichia coli.* The genome of this microorganism has also been completely sequenced and its genome includes approximately 4300 genes.[3]

Preparation of Cytosolic and Membrane Protein Fractions

The separation of the protein mixture into cytosolic and membrane fractions prior to the 2-D electrophoretic analysis is usually the first step to increase the chance of detecting low-copy-number proteins of either fraction. Cell envelope and membrane proteins usually represent approximately 15% of the total cellular proteins,[4] so that the probability of detecting a low-abundance membrane protein increases by about 10-fold if prefractionation is employed. In eukaryotic systems, the separation of additional organelles, such as mitochondria, is strongly recommended. Figure 1 shows the effect of separating the protein mixture into cytosolic and membrane fractions on the enrichment of selected membrane proteins. Analysis of the cytosolic fraction alone would not have resulted in the detection of three membrane proteins that are represented by weak spots (there is practically never

[2] R. D. Fleischmann, M. D. Adams, O. White, R. A. Clayton, E. F. Kirkness, A. R. Kerlavage, C. J. Bult, J. F. Tomb, B. A. Dougherty, J. M. Merrick, K. Kenney, G. Sutton, W. FitzHugh, C. Fields, J. D. Gocayne, J. Scott, R. Shirley, L.-I. Liu, A. Glodek, J. M. Kelley, J. F. Weidman, C. A. Phillips, T. Spriggs, E. Hedblom, M. D. Cotton, T. R. Utterback, M. C. Hanna, D. T. Nguyen, D. M. Saudek, R. C. Brandon, L. D. Fine, J. L. Fritchman, J. L. Fuhrmann, N. S. M. Geoghagen, C. L. Gnehm, L. A. McDonald, K. V. Small, C. M. Fraser, H. O. Smith, and J. C. Venter, *Science* **269,** 496 (1995).
[3] F. R. Blattner, G. Plunkett III, C. A. Bloch, N. T. Perna, V. Burland, M. Riley, J. Collado-Vides, J. D. Glasner, C. K. Rode, G. F. Mayhew, J. Gregor, N. W. Davis, H. A. Kirkpatrick, M. A. Goeden, D. J. Rose, B. Mau, and Y. Shao, *Science* **277,** 1453 (1997).
[4] H. R. Kaback, *Methods Enzymol.* **22,** 99 (1971).

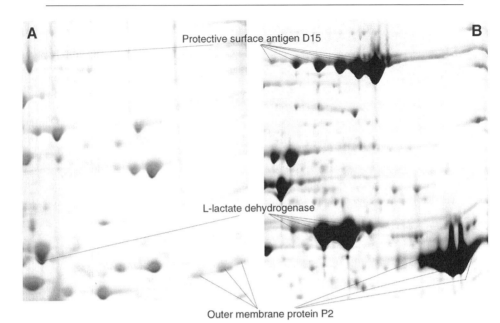

FIG. 1. Partial 2D gel images demonstrating the effect of protein sample fractionation prior to the 2D electrophoretic analysis on the visualization of spots representing membrane proteins. (A) Cytosolic fraction; (B) membrane fraction. The proteins of *H. influenzae* were separated as stated in the text. The cytosolic and membrane proteins were analyzed by 2D electrophoresis. The gels are stained with Coomassie blue.

a perfect separation of a protein mixture into cytosolic and membrane fractions). These same proteins are represented by very strong spots in the membrane fraction.

Procedure. *H. influenzae* (strain Rd KW20), is grown in a 10 liter fermentor. Typically about 70 g (wet weight) of cell paste is obtained from a 10 liter culture.[5] Cell pellets are kept frozen at −70° until use. For cell fractionation, 20 g (wet weight) of cell paste is suspended by gentle stirring in 25 ml of lysis buffer: 20 mM HEPES–OH, pH 8.0, containing 10% (v/v) glycerol, 150 mM NaCl, and 1 mM MgSO$_4$. Benzonase (Merck, Darmstadt, Germany) is added to 250 units/ml to hydrolyze DNA and RNA. The protease inhibitors, Trasylol (Bayer, Leverkusen, Germany) and ε-aminocaproic acid (Serva, Heidelberg, Germany), are added to 100 units/ml and 5 mM, respectively. Cells are broken in a precooled French pressure cell (SLM Instruments, Urbana, IL) at 20,000 lb/in^2. EDTA·Na$_2$ is added to 5 mM final concentration and the lysed cell suspension is centrifuged at 3000g for 20 min at 4° to sediment intact cells and debris. The supernatant is centrifuged at

---

[5] H. Langen, B. Takács, S. Evers, P. Berndt, H.-W. Lahm, B. Wipf, C. Gray, and M. Fountoulakis, *Electrophoresis* **21**, 411 (2000).

150,000g for 90 min at 4° to sediment cell membranes. The supernatant from this centrifugation is dialyzed two times against 1 liter of 5 mM HEPES–OH, pH 8.0, containing 10% glycerol and 1 mM EDTA·Na$_2$, at 4° for 18 hr to reduce the ionic strength. The dialyzate is filtered through a 0.45-μm pore-size membrane and used for analysis of the soluble cytoplasmic proteins or for further fractionation by chromatography. About 1.3 g of total cytosolic proteins are obtained from 20 g of cell paste.

The membrane pellet is resuspended in about 80 ml of 5 mM HEPES–OH, pH 8.0, containing 1 M LiCl, 10% (v/v) glycerol and 1 mM EDTA·Na$_2$, and pelleted at 30,000g for 60 min. The pellet is resuspended in 30 ml of a hypertonic solution: 5 mM HEPES–OH, pH 8.0, containing 20% (w/v) sucrose, 10% (v/v) glycerol, and 1 mM EDTA·Na$_2$. A rapid 10-fold dilution with this buffer without sucrose is performed to remove soluble components trapped in vesicular structures by osmotic shock. After centrifugation as above, the pellet is either used for 2D analysis of the cell membrane proteins or kept frozen at −70° until use or further fractionated into cell wall and cytoplasmic membrane components.[6] About 175 mg of membrane proteins are usually obtained from 20 g of H. influenzae cells.

For the 2D electrophoretic analysis, the membrane proteins need to be solubilized with the use of detergents or chaotropes. It is useful to analyze the preparation first by 1D electrophoresis in SDS–polyacrylamide gels. For SDS–PAGE, the membranes are usually solubilized in a buffer containing 0.1–2% SDS and a reducing agent. For the 2D electrophoretic analysis, the membranes should be dissolved in an IEF compatible solution, such as 8 M urea and 4% CHAPS or 7 M urea, 2 M thiourea and 4% CHAPS.[7] Nonionic or zwitterionic detergents and nonionic chaotropes are compatible with IEF, but they are not usually very efficient in dissolving certain difficult to solubilize proteins. The strong detergents, SDS and lithium dodecyl sulfate (LDS), are useful in the solubilization of most membrane proteins; however, they are ionic and need to be exchanged against zwitterionic detergents, such as CHAPS, before IEF.[8] Removal of SDS is difficult and most likely the free SDS and not the protein-bound detergent is removed.

Enrichment of Low-Abundance Proteins from Large Volumes

Selective Fractionation

The proteins in a mixture can be fractionated on the basis of their density by centrifugation or solubility, using ammonium sulfate, polyethylene glycol, or organic solvents. These classical procedures have been described in detail.[9] Precipitation

[6] B. Takács and J. Rosenbusch, J. Biol. Chem. 250, 2339 (1975).
[7] M. Chevallet, V. Santoni, A. Poinas, D. Rouquié, A. Fuchs, S. Kiefer, M. Rossignol, M. Lunardi, J. Garin, and T. Rabilloud, Electrophoresis 19, 1901 (1998).
[8] M. Fountoulakis and B. Takács, Electrophoresis 22, 1593 (2001).
[9] A. A. Green and W. L. Hughes, Methods Enzymol. 1, 67 (1955).

agents need to be removed prior to IEF by dialysis, ultrafiltration, or similar procedures.

Chromatographic Methods

Various chromatographic methods can be used to separate complex protein mixtures into simpler fractions on the basis of different binding principles and every approach adds a unique resolving power. The proteins are usually separated on the basis of affinity, using antibody, heparin, metal chelate, triazine dye, etc., chromatography; charge, employing ion-exchange chromatography, or chromato-focusing; hydrophobicity, and size. The choice of the chromatographic method best suited to fulfill the experimental requirements is essential for the success of the experiment. Sequential chromatographic steps can be even more powerful.

Here we describe four chromatographic steps used to enrich low-abundance cytosolic bacterial proteins from larger volumes: affinity chromatography on heparin,[10–12] ion-exchange chromatography on Polybuffer Exchanger (chromato-focusing),[13] hydrophobic interaction chromatography on TSK-phenyl column,[14] and hydroxyapatite chromatography.[15]

Heparin Chromatography. Heparin, a highly sulfated glycosaminoglycan, has an affinity for a broad range of nucleic acid-binding proteins, mainly growth factors, protein synthesis factors, and coagulation factors. The anionic sulfate groups of heparin also function as a high capacity cation exchanger. This step is especially effective if the main interest is the detection of nucleic acid binding proteins.

Procedure. Four ml of the cytosolic protein fraction (see above), in 5 mM HEPES–OH, pH 8.0, 10% (v/v) glycerol, and 1 mM EDTA $\cdot$ Na$_2$, containing approximately 70 mg of protein, is applied onto an 80-ml heparin–Actigel (Sterogene Bioseparations, Inc., Carlsbad, CA) column equilibrated in 20 mM HEPES–OH, pH 7.8, containing 10% glycerol and 1 mM EDTA. The column is washed with the same buffer until OD$_{280}$ reaches a background level. Unbound material is saved separately. Bound proteins are eluted with 10 column volumes of a linear gradient of 0–2 M NaCl in the same buffer. Ten-ml fractions are collected. Peak fractions are pooled according to the elution profile and concentrated by centrifugation at 2000g in Ultrafree-15 centrifugal filter devices, equipped with Biomax 10 K NMWL membranes (Millipore Corp., Bedford, MA), prior to the 1D and 2D electrophoretic analysis. The protein recovery in the salt-eluted fractions is approximately 40% of the input.

[10] M. Fountoulakis, H. Langen, S. Evers, C. Gray, and B. Takács, *Electrophoresis* **18**, 1193 (1997).

[11] M. Fountoulakis, B. Takács, and H. Langen, *Electrophoresis* **19**, 761 (1997).

[12] M. Fountoulakis and B. Takács, *Prot. Expr. Purif.* **14**, 113 (1998).

[13] M. Fountoulakis, H. Langen, C. Gray, and B. Takács, *J. Chromatogr. A* **806**, 279 (1998).

[14] M. Fountoulakis, M.-F. Takács, and B. Takács, *J. Chromatogr. A* **833**, 157 (1999).

[15] M. Fountoulakis, M.-F. Takács, P. Bernd, H. Langen, and B. Takács, *Electrophoresis* **20**, 2181 (1999).

Results. Figure 2 shows the 2D electrophoretic analysis of the starting material and the protein composition of eluted peak fractions from the heparin–Actigel column. Heparin chromatography is relatively easy to perform. In our hands, approximately 150 different proteins were identified in the pools collected from the column. About 50 of these are low-abundance proteins, which had not been found in the 2D gel of the starting material. Approximately 40% of the heparin-bound proteins are nucleic acid binding proteins, including ribosomal proteins and tRNA synthetases. Basic proteins, in particular, ribosomal proteins, can be identified using this step.[10,11]

Chromatofocusing. Chromatofocusing is an ion-exchange chromatographic step. The proteins are bound to the gel matrix, Polybuffer Exchanger, and are eluted with a specific buffer, Polybuffer, in the order of their decreasing isoelectric points. Proper choice of the pH of the equilibration and elution buffers and of the dimensions of the column can result in an efficient protein concentration and high resolution.[16–18] This step is usually chosen because the ion exchanger has a high protein binding capacity and can discriminate and enrich proteins with minor differences in their isoelectric points (pI values).

Procedure. About 200 mg of cytosolic *H. influenzae* proteins are dialyzed against 25 mM Tris–acetate, pH 8.3 buffer to remove salts. The dialyzate is filtered through a 0.22-μm pore-size membrane and applied onto a 187 ml (1.6 × 93 cm) Polybuffer Exchanger 94 (Amersham Biosciences, Uppsala, Sweden) column, equilibrated in 25 mM Tris–acetate, pH 8.3, at 1 ml/min. The column is washed with 200 ml of the same buffer and the proteins are eluted with Polybuffer (Amersham Biosciences), pH 5.0 (800 ml). The elution buffer is prepared by diluting a mixture of Polybuffer 96 and Polybuffer 74, 3 : 10 (v/v) 10-fold with H$_2$O and adjusting the pH to 5.0 with acetic acid. The column is further eluted with a linear gradient of increasing salt concentration from 0 to 2 M NaCl in 25 mM Tris–acetate, pH 8.3. Fractions of 8 ml are collected and pooled according to the elution profile. Each pool is concentrated to about 1 ml by centrifugation at 2000g in Millipore Ultrafree-15 devices with Biomax-10 membrane. Each pool is twice diluted 10-fold with 20 mM Tris-HCl, pH 8.0, and concentrated as above. The concentrates are analyzed by 1D and 2D electrophoresis.

Results. The chromatofocusing step, as described here, results in the detection of a low number of basic proteins (they probably precipitate on top of the column). Most fractions and in particular those eluted with salt contain mainly acidic proteins. No clear separation of the proteins based solely on their pI values is achieved. About 125 unique proteins were identified in the pools. Approximately 30 of them are low-abundance gene products not identified before. Mainly acidic proteins, in

[16] L. A. Æ. Sluyterman and O. Elgersma, *J. Chromatogr.* **150,** 17 (1978).

[17] L. Giri, *Methods Enzymol.* **182,** 380 (1990).

[18] M. Fountoulakis, E. J. Schlaeger, R. Gentz, J.-F. Juranville, M. Manneberg, L. Ozmen, and G. Garotta, *Eur. J. Biochem.* **198,** 441 (1991).

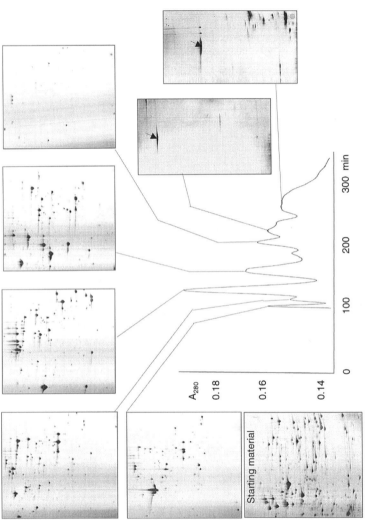

FIG. 2. Protein elution profile from the heparin–Actigel column. The cytosolic proteins of *H. influenzae* were fractionated by heparin chromatography. The proteins were eluted as described in the text and the absorbance was recorded at 280 nm. The proteins in the indicated peaks and the starting material applied onto the column were analyzed by 2D electrophoresis, on broad pH 3–10 IPG strips, except for the samples from the two last peaks which were analyzed on basic pH 6–11 IPG strips. The arrowheads indicate basic proteins enriched by heparin chromatography, which had not been detected in the starting material. The gels are stained with Coomassie blue.

the majority enzymes with a wide spectrum of catalytic activities, are enriched using this step.

Hydrophobic Interaction Chromatography. In hydrophobic interaction chromatography, proteins are separated on the basis of differences in their hydrophobicity. The proteins are adsorbed onto an uncharged matrix carrying hydrophobic groups, in the presence of salts. Elution is achieved by lowering the salt concentration.[19,20] Because of the various numbers of hydrophobic sites carried by proteins, hydrophobic interaction chromatography can efficiently fractionate complex protein mixtures.

Procedure. Five ml of cytosolic proteins of *H. influenzae,* containing about 90 mg of total protein, are dialyzed against 50 m*M* sodium phosphate, pH 7.0, containing 1 *M* ammonium sulfate and 1 *M* glycine. The dialyzate is centrifuged at 30,000*g* for 15 min at 23°. The soluble protein fraction is filtered through a 0.22-μm pore-size membrane (Millex-GV, Millipore) and applied onto a 28-ml (16 × 140 mm) TSK-phenyl 5-PW (TosoHaas, Montgomeryville, PA) column,[21,22] equilibrated in 50 m*M* sodium phosphate buffer, pH 7.0, containing 1 *M* ammonium sulfate and 1 *M* glycine. The column is washed with three column volumes of loading buffer and bound proteins are eluted with 20 column volumes of the same buffer without ammonium sulfate at 1 ml/min. Fractions of 5 ml are collected and pooled according to the elution profile. The pools are each concentrated to about 1 ml by centrifugation at 2000*g* in Millipore Ultrafree-15 devices with Biomax-10 membrane. The concentrates are analyzed by one- and two-dimensional gel electrophoresis.

Results. About 200 different proteins bound to the TSK-phenyl column were identified, about 30 of which are low-abundance proteins. This step mainly enriches proteins with catalytical activities as well as hypothetical proteins having a wide spectrum of p*I* values, including basic proteins.

Hydroxyapatite Chromatography. Hydroxyapatite chromatography for protein isolation has been introduced by Tiselius *et al.*[23] and has been systematically studied by Bernardi[24] and Gorbunoff.[25] The hydroxyapatite matrix carries positively charged (calcium) and negatively charged (phosphate) sites. Proteins bind either by nonspecific, electrostatic interactions between their positive charges and the general negative charge on the hydroxyapatite column, when the column is equilibrated with phosphate buffer, or by complexation of the carboxyl groups with

[19] S. Påhlman, J. Rosengren, and S. Hjertén, *J. Chromatogr. A* **131,** 99 (1974).
[20] N. C. Robinson, D. Wiginton, and L. Talbert, *Biochemistry* **23,** 6121 (1984).
[21] Y. Kato, T. Kitamura, and T. Hashimoto, *J. Chromatogr. A* **333,** 202 (1985).
[22] Z. El Rassi, A. L. Lee, and C. Horvath, *in* "Separation Process in Biotechnology" (J. A. Asenjo, ed.), p. 447. Marcel Dekker, New York, 1985.
[23] A. Tiselius, S. Hjerten, and O. Levin, *Arch. Biochem. Biophys.* **65,** 132 (1956).
[24] G. Bernardi, *Methods Enzymol.* **22,** 325 (1971).
[25] M. J. Gorbunoff, *Methods Enzymol.* **182,** 329 (1990).

TABLE II
PROTEIN ELUTION SCHEME AND BUFFERS USED FOR
CHROMATOGRAPHY ON HYDROXYAPATITE

Buffer	Composition	Eluent volume (column volume)
A	0.5 mM Sodium phosphate, pH 6.8, 5% glycerol	Until OD$_{280}$ reaches background level
B	50 mM MgCl$_2$, 5% glycerol	5
C	1.5 M MgCl$_2$, 5% glycerol	5
D	1.5 M NaCl, 5% glycerol	5
E	2.5 M NaCl, 5% glycerol	3
A	0.5 mM Sodium phosphate, pH 6.8, 5% glycerol	Until OD$_{280}$ reaches background level
F	500 mM Sodium phosphate, pH 6.8, 5% glycerol	16

the calcium sites on the column, as they are repelled electrostatically from the negative charge of the matrix.[25] The hydroxyapatite chromatographic step represents a standard protein purification procedure and the column binds a wide range of proteins by mechanisms different than the other separation techniques.

Procedure. The *Escherichia coli* cytosol fraction is prepared as described above for *H. influenzae.* The cytosolic fraction is dialyzed twice against 1 liter of buffer A (Table II) for 18 h at 4° in a Spectra/Por 3 dialysis membrane (Spectrum, Houston, TX) with a molecular weight cutoff of 3500. The dialyzate is filtered through a 0.22 μm pore-size membrane (Millex-GV, Millipore) and applied onto a 100 ml Macro-Prep Ceramic Hydroxyapatite (Type I, 20 μm) column (Bio-Rad, Hercules, CA) equilibrated in buffer A.

The steps in the chromatography on hydroxyapatite and the elution buffers used are given in Table II. The column is washed with buffer A at 1 ml/min until OD$_{280}$ reaches a background level. Ten-ml fractions are collected. Elution is continued with 5 column volumes of buffer B at 0.75 ml/min, then with 5 column volumes of buffer C, followed by 5 column volumes of buffer D. The column is then washed with buffer E and reequilibrated with buffer A. Acidic proteins are eluted with 16 column volumes of buffer F. The fractions are analyzed by SDS–PAGE and are pooled according to the analysis. The pools are concentrated to about 0.5 ml by ultrafiltration at 2000g at 23°. During concentration, the various buffers are exchanged against 5 mM HEPES–NaOH, pH 8.0, for the 2D electrophoretic analysis.

Results. Because of the complexity of the protein–hydroxyapatite interactions and in order to achieve a high resolution, various steps are carried out for the development of the column. Approximately 300 different proteins were identified in the pools collected from the column and about 100 of these are low-abundance

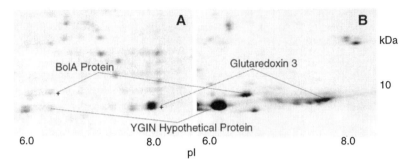

FIG. 3. Partial 2D gel images showing the enrichment of low molecular mass, low-abundance proteins by hydroxyapatite chromatography. Soluble proteins from *E. coli* were chromatographed on a hydroxyapatite column as described in the text. (A) Starting material; (B) proteins eluted with sodium phosphate, pH 6.8 (buffer A, Table II). The spots, representing three low molecular mass proteins enriched with this step, are shown.

proteins. Basic proteins, as well as many low molecular mass proteins, the proteolytic products of many proteins, and their full-length counterparts, are enriched by this method. Figure 3 presents examples of enrichment of three low molecular mass, low-abundance *E. coli* proteins using chromatography on hydroxyapatite.

Comments

1. The approaches described are usually performed in low-pressure columns. They can be scaled up for the fractionation of relatively large amounts and volumes of protein mixtures. Because in most cases, the goal of a chromatographic approach is the enrichment of a certain class of proteins (for example, nucleic acid-binding proteins) and not the isolation of a particular protein, it is not essential to obtain an optimal peak separation. Often the same protein is distributed in several fractions.

2. High-performance liquid chromatography (HPLC), for example, reversed-phase, is usually used with lower protein amounts. As in most other chromatographic steps, the elution solvents need to be exchanged against IEF-compatible solutions, i.e., salts or organic solvents have to be removed.

3. In protein comparison studies, the reproducibility of the chromatographic steps is essential. Reproducibility becomes more difficult with increasing protein amounts. On the other hand, a certain quantity of protein should be present in the fractions collected for the subsequent protein identification. Standard protocols and the use of automation can improve the reproducibility. The HPLC steps usually have good reproducibility. Because fewer chromatographic steps can be performed, the statistical analysis is usually based on fewer samples.

4. Size-exclusion chromatography of a crude extract does not usually result in an efficient protein fractionation, as the mixture is not resolved into distinct

peaks. Size-exclusion chromatography, however, can be performed in the presence of high concentrations of chaotropes, such as 8 M urea, to yield an efficient fractionation.

5. A significant overlap was observed between the proteins enriched by heparin chromatography, chromatofocusing, and hydrophobic interaction chromatography. An estimated 40% of the proteins enriched by hydrophobic interaction chromatography were also enriched by heparin chromatography and about the same percentage was also enriched by chromatofocusing. About 20% of the proteins enriched by hydrophobic interaction chromatography had also been enriched by heparin chromatography as well as chromatofocusing.[14] After having applied two or more fractionation methods, the use of additional chromatographic separations may not result in the detection of significant numbers of proteins not detected or enriched by the previous methods. This points to the limits of the chromatographic enriching techniques and suggests that a large percentage of the proteins not detected are expressed at very low levels or that enrichment or detection should be tried by other technologies.

Protein Enrichment by Electrophoretic Procedures

The electrophoretic methods comprise the separation of protein mixtures by (1) preparative polyacrylamide gel electrophoresis on the basis of protein size usually in the presence of ionic detergents or by (2) preparative isoelectrofocusing on the basis of protein charge in the presence of ampholines.

For preparative gel electrophoresis, we used the PrepCell system (Bio-Rad), following the instructions of the supplier. We used this system for the separation of eukaryotic proteins but similar conditions can be applied for bacterial proteins as well. The cylindrical separation gel is about 6 cm long and has an acrylamide concentration of 11%. The stacking gel is 2.5 cm long and has an acrylamide concentration of 4%. Ten to 50 mg of total proteins in about 4 ml of 50 mM Tris-HCl, pH 6.8, containing 25% glycerol and 1% LDS, is applied onto the stacking gel. The electrophoresis buffer is 0.198 M glycine and 25 mM Tris and contains 0.1% LDS. The electrophoresis is performed at 250 V and the proteins eluted from the gel are collected in 0.198 M glycine and 25 mM Tris, containing 0.1% CHAPS at 30 ml/h. Ten-ml fractions are collected and concentrated to about 0.2 ml by ultrafiltration. Excess salt and LDS are reduced by diluting the sample with a low ionic strength buffer without LDS. LDS is used because it can be removed from the proteins more easily in comparison with SDS, thereby interfering less with the 2D electrophoresis.[8] Preparative gel electrophoresis can often result in the enrichment of certain low-abundance and low molecular mass proteins, which cannot be achieved by other approaches. It has the disadvantage of a relatively low total protein recovery. We estimated that approximately 25% of the input is typically recovered.

Preparative IEF separates mixtures into less complex fractions according to the pI values of the components and facilitates the subsequent 2D electrophoretic analysis. We used the Rotofor system (Bio-Rad) according to the instructions of the supplier. This method may be compromised by serious precipitation problems especially for membrane proteins.

Protein Analysis by 2D Electrophoresis

Following the enrichment by any of the methods described above or similar approaches, the visualization of a protein is usually performed by 2D electrophoresis. Here we describe the general principles for performing 2D electrophoretic analysis. Technical details concerning buffers and other reagents can be found in Walsh and Herbert[26] or in Westermeier.[27]

Procedure. The 2D gel electrophoresis is performed essentially as reported earlier.[28] Protein samples of 0.1–1.0 mg in 20 mM Tris, 8 M urea, 4% CHAPS 10 mM 1,4-dithioerythritol, and a mixture of protease inhibitors [1 mM phenylmethylsulfonyl fluoride (PMSF) and 1 μg/ml of each pepstatin A, chymostatin, leupeptin, and antipain) are applied on immobilized pH gradient (IPG) strips using sample cups at both the basic and acidic ends of the strips. Focusing is initiated at 200 V. The voltage is gradually increased to 5000 V over 24 hr and kept at 5000 V for a further 24 hr. The second-dimensional separation is usually performed on 9–16% linear gradient Tris–glycine polyacrylamide gels or on other types of gels if necessary.[28,29] After protein fixation with 40% (v/v) methanol containing 5% (v/v) phosphoric acid for 12 hr, the gels are stained with colloidal Coomassie blue (Novex, San Diego, CA) for 48 hr. The molecular mass is determined by running standard protein markers shown at the right hand side of selected gels. The size markers (Gibco, Basel, Switzerland) cover the range of 10–200 kDa. The pI values are estimated according to information provided by the supplier of the IPG strips and from the average theoretical values of selected proteins. The gels are destained with H$_2$O and scanned in an Agfa DUOSCAN scanner. The images are processed using the Photoshop (Adobe) software. They are stored as tiff and jpeg formats.

Use of narrow pH range (pH 4–7 or pH 6–11) or very narrow pH range (for example, pH 5–6) IPG strips and application of a relatively large amount of protein (about 2 mg) may in certain cases lead to the detection of new spots. These spots could have overlapped with larger spots focusing at approximately the same

[26] B. J. Walsh and B. Herbert, rbams3115/Pages/2DPAGE/ABRFNews_2dpage.html (1998).
[27] R. Westermeier, *in* "Electrophoresis in Practice" (R. Westermeier, ed.), p. 215. VCH Verlagsgesellschaft, Weinheim, 1993.
[28] H. Langen, D. Röder, J.-F. Juranville, and M. Fountoulakis, *Electrophoresis* **18**, 2085 (1997).
[29] M. Fountoulakis, J.-F. Juranville, D. Röder, S. Evers, P. Berndt, and H. Langen, *Electrophoresis* **19**, 1819 (1998).

TABLE III
COMPOSITION OF SEPARATING AND STACKING PARTS OF TRICINE GELS

Stock solution	Separating gel (10.4% T, 6% C, 6.2M urea)	Stacking gel (5.4% T, 3% C)
49.5% T, 6% C acrylamide solution[a]	9.0 ml	—
48.5% T, 3% C acrylamide solution[b]	—	0.9 ml
Gel buffer[c]	7.5 ml	1.5 ml
Urea[d]	16 g	—
1 M Sodium thiosulfate	200 μl	40 μl
TEMED	25 μl	4.7 μl
10% APS	125 μl	23 μl
H$_2$O to a final volume of[e]	43 ml	8 ml

[a] The acrylamide stock solution contains 46.5 g of acrylamide and 3 g of bisacrylamide per 100 ml of solution. T indicates the total percentage acrylamide concentration (acrylamide and bisacrylamide) and C indicates the percentage concentration of the bisacrylamide in relation to the acrylamide. Both acrylamide stock solutions were kept at 4°.

[b] The acrylamide stock solution contains 47 g of acrylamide and 1.5 g of bisacrylamide per 100 ml of solution.

[c] The gel buffer consists of 3 M Tris-HCl, pH 8.45, containing 0.3% SDS.

[d] This gel type contains 6.2 M urea in the separating gel. The other gel type contains 3.1 M urea (8 g urea in a final volume of 43 ml).

[e] Volumes sufficient to prepare one gel of 180 × 180 × 15 mm casting dimensions.

position or are representative of low-abundance proteins and become visible due to the increase in the focusing area, the application of a large amount of protein, and the exclusion of high-abundance proteins.

Tricine Polyacrylamide Gels. For the detection of low molecular mass proteins, the use of Tricine gels is often more efficient than the Tris–glycine gels. The first dimensional separation is performed as described above. We prepare the Tricine gels according to a modified method of Schägger and von Jagow.[30] The gels consist of a lower, separating gel of 10.4% acrylamide and a 20-mm-high stacking gel of 5.4% acrylamide concentration. The separating gel contains urea at two different concentrations: either 3.1 or 6.2 *M*. The stacking gel does not include urea. The gels have the composition indicated in Table III. The solutions are degassed before gel preparation. The anode buffer is Tris–glycine SDS running buffer (25 m*M* Tris-base, 192 m*M* glycine, 0.1% SDS, final pH 8.3) (Novex, San Diego, CA) and the cathode buffer is Tricine SDS running buffer (100 m*M* Tris-base, 100 m*M* Tricine, 0.1% SDS, final pH 8.3) (Novex). Electrophoresis is performed in the PROTEAN II system (Bio-Rad) at 40 mA per gel for approximately 4 hr. Proteins are fixed with a solution of 40% (v/v) methanol containing 5% (v/v) phosphoric acid

[30] H. Schägger and G. von Jagow, *Anal. Biochem.* **166,** 368 (1987).

for 16 hr. The gels are stained either with colloidal Coomassie blue (Novex) for 48 hr or with silver. After removal of the stacking gel part, the Tricine gels have smaller dimensions (approximately $160 \times 160 \times 15$ mm) than the 9–16% gradient Tris–glycine gels (approximately $205 \times 180 \times 15$ mm).

Note: In order to reduce carbamylation of free amino groups on proteins, urea and thiourea are deionized by passing them over a mixed bed ion-exchange resin. The 2D electrophoresis buffer should be prepared fresh with deionized chaotropes and if not used the same day, be kept frozen at $-20°$ after preparation and thawed only once.

After separation and visualization in 2D gels, protein identification is usually performed by mass spectrometry. The most used and most sensitive approach is matrix-assisted laser desorption ionization mass spectrometry (MALDI-MS). Protein identification is based on the principle of peptide mass fingerprinting. Tryptic digestion is performed in a low-salt buffer.[31] Low molecular mass proteins usually do not give rise to a sufficient number of peptides to be identified by MALDI-MS. These proteins can be analyzed by electrospray techniques.

Comments. Often it is not sufficient to enrich a protein for a successful detection. Protein detection and identification have certain limitations.

1. Various chromatographic and other methods used result in the enrichment of both low- and high-abundance proteins. This is to be expected because the high- and low-abundance proteins bind to the column matrix based on the same principles. Certain high-abundance protein may represent up to 50% of the protein content in a particular fraction, thereby suppressing the low-abundance proteins and hindering their detection in 2D gels. The high-abundance proteins, for example, heart shock proteins, elongation factors, or certain housekeeping enzymes in bacteria, can be removed prior to 2D electrophoresis by affinity chromatography using specific antibodies.

2. The detection in 2D gels and identification of low-abundance proteins, already successfully enriched from dilute solutions, might be technologically limited because of their charge and size. Figure 4 shows approximately 300 *E. coli* proteins enriched by hydroxyapatite chromatography sorted according to their molecular mass and p*I* values. The majority of the proteins have average molecular mass and p*I* values. Only a few proteins were found with molecular masses below 10 kDa or above 70 kDa. No protein was detected with a p*I* below 4 and relatively few were detected with a p*I* value higher than 8. In such cases, alternative separation techniques may need to be employed, for example, Tricine gels for the detection of the low molecular proteins.[29,30] Figure 5B shows the successful detection of several low molecular mass proteins from *E. coli* on Tricine gels. The detection is not satisfactory on conventional Tris–glycine gels (Fig. 5A).

[31] M. Fountoulakis and H. Langen, *Anal. Biochem.* **250**, 153 (1997).

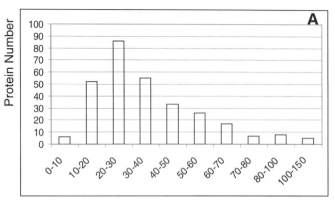

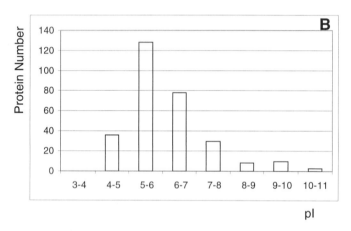

FIG. 4. *E. coli* proteins, bound to the hyrdoxyapatite column, sorted according to their theoretical molecular mass (A) and p*I* (B) values. Very few proteins with a molecular mass below 10 or higher than 70 kDa and p*I* values below 4 or higher than 8 were detected.

3. Another reason for unsuccesful detection of a protein might be related to the hydrophobicity and low solubility of certain proteins. A protein can only be visualized and analyzed if it can be brought into solution and kept in solution during the whole enrichment process and 2D electrophoretic analysis. In general, the solubility problem arises at two time points during 2D electrophoresis: (i) during the initial extraction step to solubilize membrane proteins (the cytosolic proteins are already in solution) using agents that are compatible with isoelectric focusing, such as urea and CHAPS, and (ii) during the performance of the first dimensional separation, when hydrophobic proteins could precipitate at their application positions. Poorly soluble proteins can be brought into solution with the use of strong

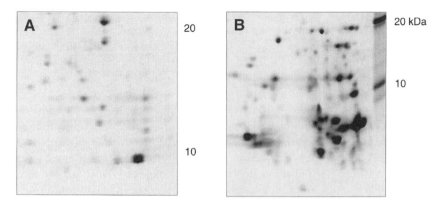

FIG. 5. Partial 2D gel images showing the detection of low-molecular mass cytosolic proteins of *E. coli* separated on a conventional 9–16% gradient polyacrylamide Tris–glycine gel (A) and on a Tricine gel (B). The gels are stained with Coomassie blue. A larger number of small proteins are detected in the latter case.

detergents and chaotropes, such as SDS. Figure 6 presents an example of visualization of membrane proteins of *H. influenzae* after solubilization with the strong detergent SDS. These proteins could not be solubilized with the IEF-compatible detergent CHAPS even in the presence of 7 *M* urea (Fig. 6A). The urea/CHAPS insoluble pellet was dissolved in 0.1% SDS, resulting in the successful visualization of three outer membrane proteins (Fig. 6B). Hydrophobic proteins with multiple transmembrane regions do not usually enter the immobilized pH gradient strips during the first dimensional separation and thus they cannot be detected. It seems that no single solubilizing agent is sufficient for the solubilization, resolution, and visualization of all the proteins of a proteome.

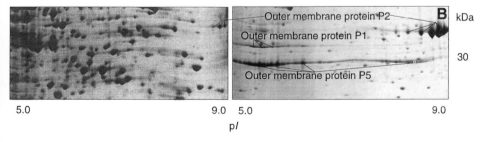

FIG. 6. Partial 2D gel images showing the total proteins of *H. influenzae* solubilized with 7 *M* urea and 2% CHAPS (A) and the proteins of the urea/CHAPS insoluble pellet solubilized with 1% SDS (B). The outer membrane proteins P1, P2, and P5 that could not be solubilized with the relatively mild agents urea and CHAPS were dissolved in SDS and visualized in the 2D gel (B).

4. Membrane proteins can also be separated in a discontinuous, two-detergent system.[5,32] In this system, the first dimensional separation is performed in 7.5% polyacrylamide gels in the presence of 250 mM benzyldimethyl-n-hexadecyl-ammonium chloride (16-BAC) and the second dimensional separation in 9–16% linear gradient gels in the presence of 0.1% SDS.

Conclusions. The proteomic analysis of bacterial proteins detects mainly the high-abundance, hydrophilic components of a mixture. For the detection of the low-abundance proteins, a separation of the complex mixture into fractions with fewer components is necessary. This can be achieved using biochemical approaches, such as centrifugation and chromatography.[33] A combination of separation techniques might be more efficient. Even if a protein has been sufficiently enriched, it is possible that it cannot be detected because of limitations in the separation and visualization methods, mainly in 2D electrophoresis. Therefore, in conjunction with the enrichment approaches, variations in 2D electrophoresis need to be evaluated, i.e., use of Tricine gels for the low molecular mass proteins, use of heavily loaded narrow pH range IPG strips, use of the two-detergent system, and use of different detergents and chaotropes for efficient protein solubilization.

In spite of the current limitations of a proteomic study, this is the only method for high-throughput qualitative and quantitative analysis of a protein mixture. Moreover, with the sequencing of genomes of about 30 microorganisms to date, many new proteins with no known functions have been deduced. These proteins are referred to as hypothetical proteins or predicted coding regions and some of them might be of biological interest. Many of these proteins are produced in low copy numbers and no information is available in the literature for their isolation. In our hands, a large number of hypothetical or unknown proteins have been identified following enrichment with the chromatographic steps described here. The analysis of the proteomes of higher organisms is anticipated to be more complex in comparison with the bacterial proteomes and will probably require a combination of many analytical approaches, comprising multiple enrichment steps and efficient separation and detection methods.

Acknowledgment

We thank Dr. S. Evers for critical reading of the manuscript.

[32] J. Hartinger, K. Stenius, D. Högemann, and R. Jahn, *Anal. Biochem.* **240,** 126 (1996).
[33] B. Takács, in "Encyclopedia of Analytical Chemistry" (R. A. Meyers, ed.), p. 5955. John Wiley and Sons, New York, 2001.

[21] Immunoproteome of *Helicobacter pylori*

By PETER R. JUNGBLUT and DIRK BUMANN

Introduction

The detection of antigens on blots from sodium dodecyl sulfate gels by reaction with antibodies revolutionized the visualization of antigens.[1] A secondary antibody directed against the first antibody class is conjugated with an enzyme, which catalyzes a staining reaction. This procedure, termed Western blotting or more precisely immunoblotting, is used in many life sciences laboratories and has led to the development of diagnostic kits for several pathogenic microorganisms. Because of the complexity of bacterial and parasite proteomes, one-dimensional separation is not sufficient to resolve all components of the organism. Within a band of a 1D gel many proteins may be present and an unknown number of these proteins are immunoreactive. Additionally, the exact position is difficult to standardize, leading to many uncertainties in the literature, as exemplified in the description of the p100 protein of *Borrelia burgdorferi*, which was characterized by molecular masses between 90 and 100 kDa and has a real M_r of 87,000, calculated from the sequence.[2] An increase in the resolution power of the separation system by two-dimensional electrophoresis (2-DE) offers a solution to most of these problems. With large gels it is possible to resolve up to 10,000 spots.[3] This improves the procedure to the level of a real proteome analysis. The set of antigens that reacts with sera of infected patients can be defined as the immunoproteome of a microorganism.[4] Comprehensive immunoproteome signatures can only be obtained from the analysis of large numbers of sera. An immunoproteome cannot be completed because of the unique immune reaction of each patient, but also because of differences between the proteomes of various isolates of a single species of microorganism. Complications because of cross-reactive antibodies that were induced by other microorganisms may be overcome by a large series of samples and controls using sera from patients without previous detectable contact to the investigated microorganism.

Here we describe how 2-DE techniques can be combined with various immunoblotting procedures and how such data sets from infected patients can be evaluated in a semiquantitative manner as a basis for the development of sensitive and specific diagnostic tests.

[1] H. Towbin, T. Staehelin, and J. Gordon, *Proc. Natl. Acad. Sci. U.S.A.* **76,** 4350 (1979).

[2] P. R. Jungblut, G. Grabher, and G. Stoffler, *Electrophoresis* **20,** 3611 (1999).

[3] J. Klose and U. Kobalz, *Electrophoresis* **16,** 1034 (1995).

[4] P. R. Jungblut, D. Bumann, G. Haas, U. Zimny-Arndt, P. Holland, S. Lamer, F. Siejak, A. Aebischer, and T. F. Meyer, *Mol. Microbiol.* **36,** 710 (2000).

Methods

Preparation of Helicobacter pylori Cell Proteins

All tested *H. pylori* isolates grow well in a microaerobic atmosphere (5% O_2, 15% CO_2, 85% N_2, v/v/v) on serum–agar plates[5] or in liquid cultures in bovine brain–heart infusion supplemented with 10% fetal calf serum. Division times in liquid cultures are in the range of 4 hr during exponential growth.

Harvested bacteria are washed twice in ice-cold phosphate-buffered saline (PBS) containing proteinase inhibitors [1 mM phenylmethylsulfonyl fluoride (PMSF), 0.1 μM pepstatin, 2.1 μM leupeptin, and 2.9 mM benzamidine] to minimize proteolytic degradation. The weight of the resulting bacterial pellet is determined. After resuspension in the same amount of distilled water, the bacteria are lysed by the addition of urea (final concentration 9 M), dithiothreitol (final concentration 70 mM), CHAPS (final concentration 1% w/v), and Servalyte pH 2–4 (Serva, Heidelberg, Germany) (final concentration 2% w/v).[4] This preparation method typically yields a protein concentration of about 15 mg ml^{-1}.

Two-Dimensional Electrophoresis (2-DE)

Carrier ampholyte isoelectric focusing combined with sodium dodecyl sulfate polyacrylamide gel electrophoresis as described by Klose and Kobalz[3] is the basis for the immunoblotting procedure presented here. 2-DE parameters that influence subsequent blotting yields include gel thickness, the ionic strength in the gel, and the amount of protein that is applied. The following blotting procedure has been optimized for gels that have a thickness of 1 to 1.5 mm and a SDS–PAGE with 15% acrylamide gels as a second dimension[6] with a sufficiently high ionic strength to promote the binding of eluted proteins to the hydrophobic blotting membrane. Between 10 and 1000 μg protein can be separated on large gels (23 cm × 30 cm).

Blotting

Originally, protein blotting was performed within a buffer tank ("wet blot").[1] For larger gels and less buffer consumption, a semidry blotting technique was introduced.[7] Today, polyvinylidene difluoride (PVDF) membranes (Immobilon, Millipore, Bedford, MA) or nitrocellulose membranes are used to immobilize proteins for immunostaining. The efficiency of protein elution from the gel and the binding of the eluted proteins to the blotting membrane determine the overall blotting yield.[8] Both factors depend on protein characteristics (size, hydrophobicity), gel parameters [sodium dodecyl sulfate (SDS) concentration, ionic strength],

[5] S. Odenbreit, B. Wieland, and R. Haas, *J. Bacteriol.* **178,** 6960 (1996).
[6] U. K. Laemmli, *Nature* **227,** 680 (1970).
[7] J. Kyhse-Andersen, *J. Biochem. Biophys. Methods* **10,** 203 (1984).
[8] P. Jungblut, C. Eckerskorn, F. Lottspeich, and J. Klose, *Electrophoresis* **11,** 581 (1990).

blotting buffer composition (methanol concentration, pH, ionic strength), the electric field, and the blotting time. In addition, elution is also influenced by the acrylamide composition of the gel, and protein binding is also influenced by the hydrophobicity of the blotting membrane. Because protein characteristics influence blotting yields, a single condition that allows an optimal transfer of all proteins that can be separated on 2-DE gels does not exist. The following conditions represent a compromise optimal for proteins within a size range of 25 to 60 kDa.[8] Possible modifications for larger or smaller proteins are also described.

Materials. Prepare in advance.

Semidry Transfer Cell: Trans-Blot SD (Bio-Rad, Munich, Pt/Ti electrodes); Amersham Pharmacia (Piscataway, NJ) (Hoefer Semiphor); Schleicher and Schuell (Keane, NH), Carboglass.

Gel-blotting paper (Schleicher and Schuell) PVDF membrane: Immobilon, Millipore, Bio-Rad, Gels: 1–1.5 mm thick 2-DE gels, Methanol, pro analysis.

Blotting buffer: 50 m*M* borate/20% methanol, pH 9.0: 3.09 g boric acid (Merck)+750 ml deionized water+200 ml methanol+18.6 ml 1 *N* NaOH. Fill with deionized water to 1000 ml

Deionized water, scissors, gloves, glass stick or pipette, dishes

Procedure. (Start 30 min before the end of the second dimension of 2-DE.)

Preparation of PVDF membrane: Dip membrane in methanol for 30 sec, shake in blotting buffer 3 × 5 min.

Preparation of blotting chamber: Wash the chamber with deionized water, graphite electrodes should stay damp until the beginning of the blotting process.

Five minutes before the end of the second dimension, three layers of filter paper are rolled individually on the anode (filter papers should extend the 2-DE gel by about 3 mm on each side). Air bubbles are expelled by rolling a glass pipette across the filter paper with firm pressure. The blotting membrane is carefully placed on top of the filter paper stack without trapping air bubbles (do not roll the membrane to avoid damage). The membrane must not dry until the gel is placed on it, but excess liquid between gel and membrane should also be avoided.

After the end of the second dimension, the 2-DE gel is rolled onto the blotting membrane. Three layers of filter paper are rolled individually onto the sandwich and each time the glass pipette is rolled across to remove air bubbles. Excess water is removed with a Kimwipe.

The blotting chamber is closed. For chambers with graphite electrodes, weights of about 1 kg are placed on the chamber to ensure firm contact between the membrane and the gel. Blotting is carried out at 1 mA/ cm^2 for 3 hr with an upper voltage limit set to 100 V. During blotting of a 55 cm^2 gel, the voltage typically increases from 6 V to 20 V.

Modifications for Large and Small Proteins. Proteins with high molecular mass (>50 kDa) tend to poorly elute from the gel. This problem can be overcome by decreasing the concentration of methanol to 10% and by using 0.5% SDS in the cathode buffer. Increasing the blotting time is only recommended in combination with SDS in the cathode buffer.

Proteins with low molecular mass (<20 kDa) may not efficiently adsorb to the membrane during their fast transfer. Hence, it is necessary to decrease the electric current to 0.5 mA and to promote binding to the membrane by increasing ionic strength with 100 mM borate and eventually raising the methanol concentration to 30%. The blotting time can also be shortened.

Immunostaining

Alternative A: Alkaline phosphatase/Naphthol and Fast Red

Solutions and buffers:

Tris/NaCl wash buffer: 100 mM Tris-HCl pH 7.5, 135 mM NaCl

PBS, pH 7.4: 20 ml 1 M KH$_2$PO$_4$ (6.81 g/50 ml distilled water), 80 ml 1 M K$_2$HPO$_4$ (17.41 g/100 ml distilled water), and 45 g NaCl are brought to 5 liter with deionized water.

PBS/Tween: 1 liter PBS, pH 7.4, and 3 g Tween 20

PBS/Tween/MP3%: 100 ml PBS/Tween and 3 g skim milk

PBS/Tween/MP1%: 100 ml PBS/Tween and 1 g skim milk

Substrate buffer: 50 ml of 0.8 M Tris–base (9.69 g Tris–base in 100 ml deionized water), 53.6 ml of 0.4 N HCl, and 4.1 ml 0.048 M MgCl$_2$ (1.016 g MgCl$_2$·6H$_2$O in 50 ml deionized water) are brought to 100 ml with deionized water (final pH of 8.0).

Development solution: Mix 1 : 1 a 0.4 mg naphthol/ml substrate buffer solution and a 6.0 mg Fast Red (Sigma, Munich)/ml substrate buffer solution. A mixture of 60 ml is sufficient for 4 blot membranes 7 cm × 8 cm. This mixture has to be freshly prepared directly before use in a light-protected beaker.

Procedure. Staining is performed at room temperature with continuous slight shaking. For each membrane a separate dish should be used.

(a) After blotting, wash the membrane 3 times in Tris/NaCl wash buffer for 10 min.

(b) Block for at least 1 hr (or overnight at 4°) in PBS/Tween/MP3%.

(c) Wash three times for 10 min in PBS/Tween.

(d) Incubate the blocked membranes with the first antibody or a serum (diluted in PBS/Tween/MP1%) for 1 hr or overnight at 4°. At least 0.2 ml antibody solution/cm^2 membrane area should be used. The dilution of the antibody or the serum can be optimized using 1D immunoblots.

(e) Wash three times for 10 min in PBS/Tween.

(f) Incubate with 0.2 ml/cm^2 secondary antibody–alkaline phosphatase conjugate (antibody directed against the antibody class and organism of the primary antibody) at room temperature for 1 hr. The antibody is diluted in PBS/Tween/MP1% and the dilution is optimized on 1D blots.

(g) Wash three times for 10 min in PBS/Tween followed by washing in tap water until no foaming occurs.

(h) Preincubate for 10 min in substrate buffer.

(i) The freshly prepared development solution (at least 0.2 ml/cm^2 membrane area) is added to the membranes. The staining is visually controlled and stopped by exchanging the development solution with deionized water. The membranes are dried on filter paper.

Alternative B: Alkaline Phosphatase/NBT and BCIP

Solutions and buffers

TBST (Tris-buffered saline/Tween 20): 20 mM Tris-HCl, pH 7.3; 154 mM NaCl, 0.05% Tween

Blocking solution: TBST + 5% bovine serum albumin (BSA) or 5% skim milk

AP-Buffer (alkaline phosphatase buffer): 100 mM Tris-HCl, pH 9.5; 100 mM NaCl; 5 mM MgCl$_2$

Staining solution: Add per 10 ml AP buffer 33 μl nitroblue tetrazolium (NBT) (Aldrich, Milwaukie, WI) and 66 μl 5-bromo-4-chloro-3-indolyl phosphate (BCIP) (Sigma, St. Louis, MO). This mixture has to be freshly prepared directly before staining.

Procedure. Staining is performed at room temperature with continuous slight shaking. A separate dish should be used for each membrane.

(a) Wash 10 min in TBST.

(b) Block 30 min in 0.15 ml/cm^2 blocking solution.

(c) Add primary antibody in optimized dilution in blocking solution. Incubate for 2 hr.

(d) Wash 5 times for 10 min in 0.5 ml TBST/cm^2 of membrane.

(e) Add 0.15 ml/cm^2 secondary antibody–alkaline phosphatase conjugate at an optimized dilution in blocking solution and incubate for 2 hr.

(f) Wash 5 times for 10 min in 0.5 ml TBST/cm^2 of membrane.

(g) Stain in freshly prepared staining solution using visual control as described above.

(h) Exchange the staining solution with deionized water and dry the membranes.

Alternative C: Peroxidase/Chemoluminescence

Materials. Western blot chemoluminescence reagent for ECL (enhanced chemiluminescence) immunstaining (NEN, Boston, MA), (alternatively, SuperSignal West Pico, Pierce, Bonn, Germany) can be used according to the manufacturer's instructions, Kodak (Rochester, NY) BioMax MR1 film, autoradiography film cassette, staining tray.

Procedure. Staining is performed at room temperature with continuous slight shaking. For each membrane a separate dish should be used.

(a) Dip PVDF membrane in methanol to moisten the surface.

(b) Block the surface with 0.3 ml/cm^2 5% skim milk in PBST buffer (2.03 g sodium dihydrogen phosphate hydrate, 14.4 g disodium hydrogen phosphate dihydrate, 85 g NaCl, 5 ml Tween 20 to 1 liter, dilute directly before use 1 : 10 with deionized water). Incubate for 1 hr at room temperature or overnight at 4°.

(c) Wash 3 times for 5 min each in 0.5 ml PBST/cm^2 of membrane.

(d) Incubate with 0.2 ml/cm^2 primary antibody (antibody, serum or plasma) diluted in incubation solution (1% BSA in PBST buffer) for 1 hr at room temperature.

(e) Wash 4 times for 15 min each in 0.5 ml PBST/cm^2 of membrane.

(f) Incubate with 0.2 ml/cm^2 diluted secondary antibody–horseradish peroxidase conjugate (antibody directed against the antibody class and organism of the primary antibody) for 1 hr at room temperature.

(g) Wash 4 times for 15 min each in 0.5 ml PBST/cm^2 of membrane.

(h) Mix equal volumes of enhanced luminol reagent and oxidizing reagent to obtain 0.125 ml/cm^2 membrane, add this solution to the wet membrane, and incubate for 1 min, remove solution, and blot with filter paper.

(i) Wrap the blot in a transparent foil avoiding any folds or air bubbles, fix the blot within a film cassette, place a film on the wrapped blot, expose for 15 sec to 15 min, and develop the film (Kodak BioMax MR1).

Stripping of Blotting Membrane

Sometimes it is preferable to use a single blot to test several sera. For this purpose, the antibodies are removed after staining by a procedure named "stripping" and a new serum may be applied to the same blot.[9] Usually this process can only be repeated 4 or 5 times.

---

[9] S. H. Kaufmann, C. M. Ewing, and J. H. Shaper, *Anal. Biochem.* **161**, 89 (1987).

Stripping of Nitrocellulose

Buffers

Stripping buffer: 62 mM Tris-HCl, pH 6.8, 2% SDS, 100 mM 2-mercaptoethanol (0.25 ml per cm^2 membrane area is required).
Wash buffer: PBS/Tween 0.05%

Procedure. Stripping is performed with continuous slight shaking in a hood.

(a) Wash membrane in wash buffer.
(b) Heat stripping buffer to 50° and incubate membrane for 30 min at 50°.
(c) Wash membrane 3 to 5 times for 10 min in wash buffer.

After blocking, the membrane can be immunostained with another antibody.

Stripping of PVDF

Buffers

Stripping buffer: 62.5 mM Tris-HCl, pH 6.7, 2% SDS, 100 mM 2-mercaptoethanol (0.25 ml per cm^2 membrane area are required).
Wash buffer: TBS/Tween: 20 mM Tris-HCl, pH 7.4, 0.5 M NaCl, 0.05% Tween 20

Procedure. Stripping is performed with slight shaking in a hood.

(a) Wash membrane in wash buffer.
(b) Heat stripping buffer to 50° and incubate membrane for 30 min at 50°.
(c) Wash 3 times for 10 min each in wash buffer.

After blocking, the membrane can be immunostained with another antibody.

Staining of Blotted Membrane with Coomassie Brilliant Blue R-250

Proteins may be detected by Coomassie Brilliant Blue (CBB) R-250 directly after blotting, or after immunostaining.[10] Staining with CBB R-250 after immunostaining helps to assign recognized antigens to spots, e.g., for subsequent identification by mass spectrometry. This protein counterstaining is possible despite the blocking of available membrane surface areas with unrelated proteins (such as those present in skim milk). The reversibly bound proteins of the blocking solution are removed from the surface of the membrane during the multiple washes of the immunostaining procedure, whereas the blotted proteins remain trapped inside the membrane.

[10] E. Zeindl-Eberhart, P. R. Jungblut, and H. M. Rabes, *Electrophoresis* **18,** 799 (1997).

Solutions

Staining solution: 0.1% Serva Blue R-250/40% methanol/10% acetic acid:
1 g Serva Blue R-250 + 400 ml methanol + 100 ml acetic acid p.a. (Merck)
+ 500 ml deionized water
Destaining solution: PVDF: 40% methanol/10% acetic acid: 400 ml methanol
+ 100 ml acetic acid p.a. (Merck) + 500 ml deionized water

Procedure:

Membrane is carefully shaken in staining solution for 5 min.
Membrane is carefully shaken in destaining solution three times for 5 min.
Hang the membrane up or place it on several layers of Kimwipes for drying.

Evaluation

Proteins that are recognized by test sera cannot be directly identified from the immunoblot. A parallel gel of the same sample that is stained with, e.g., Coomassie Brilliant Blue G-250 (CBB) may serve as the starting material for identification by peptide mass fingerprinting with MALDI mass spectrometry. In this case, the assignment of the recognized spot on the immunoblot to the corresponding spot on the CBB-stained gel is a critical point. Counterstaining of the immunoblot with Coomassie Brilliant Blue R-250 as described above greatly improves the reliability of such an assignment. Another possibility is to produce replica blots that can be directly superimposed by blotting in two directions.[11] In general, the accurate assignment of recognized antigens to gel spots is easier for larger gels with high resolution.

Strategies for Further Diagnostic Applications

Antigen detection aims primarily at development of diagnostic tools to detect an infection with a certain pathogen. A basis for this is the knowledge of the genetic variability of the pathogen and the geographic origin of the different strains or species causing disease manifestations. A first step is to elucidate antigens, which are specifically and consistently detected. Because this is labor intensive, an initial focus on one typical strain may be necessary. In addition to the diversity of the pathogen, host antibody responses are usually also highly diverse. A reduction in the number of required experiments can be obtained by the pooling of different sera. However, starting with pooled sera may result in misinterpretations because a single serum containing exceptionally abundant antibodies to certain antigens can dominate the result, while less abundant but consistent antibody specificities might

---

[11] K. E. Johannsson, *J. Biochem. Biophys. Methods* **13**, 197 (1986).

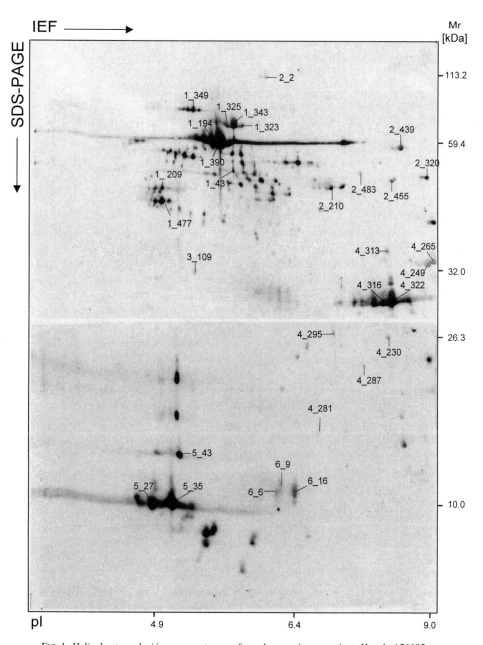

FIG. 1. *Helicobacter pylori* immunoproteome of an adenocarcinoma patient. *H. pylori* 26695 proteins were separated by a large gel 2-DE system (23 × 30 cm) and blotted by semidry blotting onto PVDF membranes. Antigens were visualized by the ECL system as described in the section on methods. Spots marked with numbers are already identified. The most prominent ones are 1_390, 60 kDa chaperonin GroeL (HP0010); 1_343, urease β subunit (HP0072); 1_477, EFTU (HP1205), 4_322 urease α subunit (HP0073), 5_35, 50S ribosomal protein L7/L12 (HP1199). The other identities may be found within the 2D-PAGE database (www.mpiib-berli.mpg.de/2D-PAGE).

escape detection. For more representative results that mimic the final diagnostic test conditions, different sera should be individually tested. From the collective results, the frequency and the semiquantitative strength of recognition can be calculated. Antigens that are frequently recognized by sera from infected individuals but rarely cross-reactive with noninfected control sera can be overexpressed and dotted onto test strips for simple serological assays. A minimum threshold number of recognized antigens can be defined to represent a positive result.

Helicobacter pylori Immunoproteome

The investigation of *H. pylori* immunoproteome revealed a high diversity of antigens. Within one large protein gel[4] (Fig. 1) 230 antigens were detected in the serum of a carcinoma patient. A systematic study using sera from 42 patients with active *H. pylori* infection, unrelated disorders, and gastric cancer resulted in a total of 310 antigenic protein species.[12] In this investigation small gels (7 cm × 8.5 cm) were used and proteins of *H. pylori* 26695 strain were screened for antigens. Of the 32 proteins most frequently recognized by *H. pylori* positive sera, 23 were confirmed from other studies and nine were newly identified. Three antigens of these were specifically recognized by *H. pylori* positive sera: the predicted coding region HP0231, serine protease HtrA (HP1019), and Cag3 (HP0522). Other antigens are recognized uniquely by sera from gastritis and ulcer patients, which may characterize them as candidate indicators for these disease manifestations. Proteomic data are stored in the 2D-PAGE database (www.mpiib-berli.mpg.de/2D-PAGE). Here, large 2-DE patterns of *H. pylori* 26695 and J99 are stored, containing about 1800 protein spots per strain, from which about 250 are identified for each strain. With the order "*Show all antigens*" the 32 most frequently detected antigens[12] are marked on the protein pattern and names of the proteins appear when the cursor is placed on the spots. By clicking on the spots, characteristics of the protein such as M_r, and pI, identification method, and accession numbers are available and links to sequence databases directly lead to protein and gene sequences.

---

[12] G. Haas, G. Karaali, K. Ebermayer, W. G. Metzger, S. Lamer, U. Zimny-Arndt, S. Diescher, U. B. Goebel, K. Vogt, A. B. Roznowski, B. J. Wiedenmann, T. F. Meyer, T. Aebischer, and P. R. Jungblut, *Proteomics* **2**, 313 (2002).

Section IV

Bacterial Perturbations of Eukaryotic Cell Cycle and Apoptosis

[22] Helicobacter pylori and Apoptosis

By EMILIA MIA SORDILLO and STEVEN F. MOSS

Helicobacter pylori is a gram-negative spiral-shaped bacterium that colonizes the stomach in at least half the world's population.[1] In most individuals, the presence of H. pylori does not appear to be associated with symptoms, despite evidence of an inflammatory reaction. However, chronic infection by H. pylori is clearly detrimental in some cases. H. pylori infection is the major cause of peptic ulcer disease and is associated with development of both gastric adenocarcinoma and gastric B-cell lymphoma of the mucosa-associated lymphoid tissue (MALT lymphoma). The mechanisms by which H. pylori infection causes disease are not completely understood. One potentially important pathogenic mechanism is the effect of H. pylori on cell turnover. Both proliferation and apoptosis of gastric epithelial cells are increased in association with H. pylori colonization,[2] and the balance between these may be critical in the development of both ulcer disease and gastric adenocarcinoma. The induction of epithelial apoptosis by H. pylori has been the subject of several recent investigations, but there is very little information thus far regarding the effect of H. pylori on the apoptotic response of inflammatory cells or lymphocytes. Although most studies have focused on H. pylori, the experimental approaches used to study H. pylori and apoptosis can also be applied to investigation of the interactions of related Helicobacter species and nonhuman tissue, or indeed, those between various other organisms and their target epithelia.

In general, apoptosis associated with H. pylori has been studied by one of two approaches: either by staining human and animal gastric tissue sections, or by coculturing H. pylori with gastric cells in vitro. The advantage of a coculture system is that the target cell and bacterium may be readily manipulated, thus enabling dissection of the pathways triggered by H. pylori. Furthermore, apoptosis can be reproducibly quantified. However, the coculture system is limited in some respects; for example, the relative contributions of bacterial factors versus the inflammatory reaction to a pathogenic outcome cannot be assessed.

In Vivo Methods for Human/Animal Studies

TUNEL: Terminal Deoxyuridine Nick End Labeling

H. pylori typically elicits an intense mixed inflammatory reaction in the gastric mucosa, including within the gastric epithelium. Because the appearance of an

---

[1] J. Parsonnet, *Infect. Dis. Clin. North Am.* **12,** 185 (1998).
[2] H. Shirin and S. F. Moss, *Gut* **43,** 592 (1998).

apoptotic gastric epithelial cell in hematoxylin and eosin (H&E)-stained sections may be difficult, if not impossible, to differentiate from the morphology of a neutrophil or a lymphocyte, apoptotic epithelial cells cannot be detected and counted in routinely stained sections. Thus, most studies of gastric apoptosis in tissue sections of human biopsy material or from animal models of *H. pylori* infection have depended on identification of fragmented DNA by the TUNEL method as a surrogate marker of apoptosis.[3–6] This *in situ* histochemical method was originally described by Gavrieli *et al.*[7] and is based on the ability of the enzyme terminal deoxynucleotidyltransferase to add labeled nucleotides to the many "sticky" 3′ DNA ends exposed in cells undergoing DNA fragmentation during apoptosis. A number of kits based on the TUNEL method are now commercially available. We adapted the originally described method specifically for human endoscopic biopsies in order to maximize the signal/noise ratio for good quality images for black and white photography and quantification.[8] In every batch of slides, it is important to use a control tissue to ensure that the method produces reproducible results. We use rat small intestine as a control and expect several epithelial cells at the tip of the villus to be TUNEL-positive, with rare positive cells at the base of the crypt (Fig. 1).

Materials. All materials are from Sigma (St. Louis, MO) unless stated otherwise.

> Proteinase K (make and freeze stocks at 1 mg/ml in deionized distilled water)
>
> TdT enzyme (Roche Applied Science, Indianapolis, IN) 250 U/ml
>
> TdT buffer [make as 5× concentrated: 30 mM Tris-HCl (pH 7.2), 140 mM sodium cacodylate, 1.25 mg/ml bovine serum albumin (BSA)]
>
> $CoCl_2$ 16.7 mM
>
> Termination Buffer (300 mM NaCl, 30 mM sodium citrate, pH 7.2)
>
> Digoxigenin-11–dUTP (Roche)
>
> 0.1 M Acetate buffer (0.2 M acetic acid plus 0.2 M sodium acetate in 1 : 20 ratio, then dilute with an equal volume deionized distilled water and adjust to pH 6)
>
> Nickel–diaminobenzene: Make fresh just before needed [mix 25 ml 5% nickel ammonium sulfate in 0.2 M acetate buffer, pH 6, with 24.3 ml water and

[3] S. F. Moss, J. Calam, B. Agarwal, S. Wang, and P. R. Holt, *Gut* **38**, 498 (1996).

[4] E. E. Mannick, L. E. Bravo, G. Zarama, J. L. Realpe, X. J. Zhang, B. Ruiz, E. T. Fontham, R. Mera, M. J. Miller, and P. Correa, *Cancer Res.* **56**, 3238 (1996).

[5] T. C. Wang, J. R. Goldenring, C. Dangler, S. Ito, A. Mueller, W. K. Jeon, T. J. Koh, and J. G. Fox, *Gastroenterology* **114**, 675 (1998).

[6] R. M. Peek, Jr., H. P. Wirth, S. F. Moss, M. Yang, A. M. Abdalla, K. T. Tham, T. Zhang, L. H. Tang, I. M. Modlin, and M. J. Blaser, *Gastroenterology* **118**, 48 (2000).

[7] Y. Gavrieli, Y. Sherman, and S. A. Ben-Sasson, *J. Cell Biol.* **119**, 493 (1992).

[8] S. F. Moss, L. Attia, J. V. Scholes, J. R. Walters, and P. R. Holt, *Gut* **39**, 811 (1996).

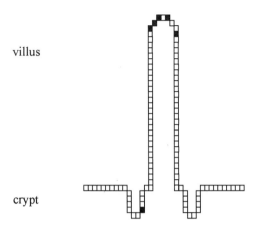

villus

crypt

FIG. 1. Expected distribution of TUNEL-positive epithelial cells (shown in black) in rat small intestine.

0.7 ml diaminobenzene (50mg/ml in water). Then add 100 mg D-glucose, 20 mg ammonium chloride, and 12.5 mg crude glucose oxidase; mix again]. Methyl green: [0.5% (w/v) in 0.1 M sodium acetate pH 4.4 (adjust pH with acetic acid), then filter (Whatman (Clifton, NJ) filter #3].

Note: Diaminobenzene (*p*-phenylenediamine) is a carcinogen and should be handled wearing protective gloves and under a fume hood with adequate ventilation. A respirator should be worn if there will be prolonged exposure. Additional information regarding safety measures for handling this and other hazardous chemicals is available (www.cdc.gov/niosh/ipcs and www.cdc.gov/od/ohs).

Methods

1. Rapidly fix tissue of interest (e.g., endoscopic biopsy) by placing fresh tissue in neutral buffered 10% (v/v) formalin within 30 sec and leave for 24–72 hr depending on tissue size (4–12 hr for biopsies).

2. Block tissue in paraffin; check for acceptable tissue orientation by staining a section with H&E.

3. Cut well-oriented sections at 4 μm onto Superfrost Plus microscope slides (Fisher Scientific, Pittsburgh, PA).

Perform all subsequent steps at 24° except where stated.

4. Deparaffinize sections by sequential soaking in xylene (5 min × 2), 100% ethanol (5 min × 2), 70% ethanol (5 min × 2), and deionized distilled water (5 min × 2).

5. Treat with proteinase K ($20\mu g/ml$) for 15 min.
6. Wash in deionized distilled water (5 min × 2).
7. Quench endogenous peroxidase activity with 2% H_2O_2 (v/v) for 5 min.
8. Wash in deionized distilled water (5 min × 2).

Steps 9–11 are performed by dropping reagents over entire tissue section on a flat slide.

9. Equilibrate in 1× TdT buffer for 10 min.
10. Blot off 1× TdT buffer and then add reaction mixture (50 μl per slide) comprising (for each slide): 10 μl 5× Tdt, 3μl 16.7 mM $CoCl_2$, 1 μl TdT enzyme (25 U), and 2 μl 10 nM digoxigenin-11–dUTP (Roche) and 34 μl deionized distilled water.
11. Incubate at 37° for 90 min in a humidified atmosphere.
12. Immerse in termination buffer 15 min.
13. Wash in deionized distilled water (5 min × 2).
14. Immerse in 2% (v/v) BSA in deionized distilled water.
15. Wash in deionized distilled water (5 min × 2).
16. Wash in phosphate-buffered saline (PBS) for 5 min.
17. Add anti-digoxigenin–POD Fab fragments (Roche), 1 : 300 in PBS for 30 min.
18. Wash in deionized distilled water (5 min × 2).
19. Rinse in 0.1 M acetate buffer, pH 6.
20. Immerse in freshly prepared nickel-diaminobenzene for 20 min.
21. Wash in deionized distilled water (5 min × 2).
22. Counterstain with methyl green for 2 min.
23. Wash in deionized distilled water (5 min × 2).
24. Rehydrate through 70% ethanol (1 min), 95% ethanol (1 min × 2), 100% ethanol (1 min × 2), xylene (1 min × 4).
25. Mount and coverslip slides.

Quantification Considerations

1. Because of the complex branching structure of gastric glands, they are rarely perfectly oriented. Examining nonadjacent sections separated by 50–100 μm within the same block will increase the number of suitable glands. We recommend counting at least 10 well-oriented glands per site (gastric antrum or corpus) to allow for reasonable counting reproducibility.[9] Also, at least two separate biopsies for each site per patient should be used, to avoid sampling errors from patchy *H. pylori* colonization.

---

[9] S. F. Moss, E. M. Sordillo, A. M. Abdalla, V. Makarov, Z. Hanzely, G. I. Perez-Perez, J. M. Blaser, and P. R. Holt, *Cancer Res.* **61,** 1406 (2001).

2. Choosing the best area to count is ideally done in collaboration with a histopathologist experienced in the interpretation of gastric pathology. Sections should be carefully chosen to avoid atrophy or intestinal metaplasia. This will avoid errors due to "comparison of apples with oranges," since, for example, apoptosis may be markedly reduced in intestinal metaplasia.[10] Areas of necrosis should also be avoided, since TUNEL does not differentiate between apoptosis and necrosis.

3. As for all histochemical interpretation, setting the threshold for positivity is subjective. This can be addressed by recounting 20% of cases at random to ensure acceptable reproducibility.

4. There is no standardized way to express results. Many investigators report the apoptotic index (the percentage of cells that are TUNEL positive) or the number of TUNEL positive cells per gland unit. We have found that the number of cells per gland and the gland density (number per linear mm) is not altered by *H. pylori* in human biopsies, so that either way of expressing the data would be satisfactory. Note that this may not be true for some animal models in which gland hyperplasia may occur with *H. pylori* infection.

5. Staining of serial sections for markers of proliferation (e.g., Ki67 or PCNA antigen in humans, PCNA or bromodeoxyuridine in animals) allows calculation of the proliferation: apoptosis ratio,[6,9] and comparison with H&E allows correlation with histological scores of inflammation.[11]

Other Methods

Many other methods have been tried to measure the effect of *H. pylori* on gastric epithelial cell apoptosis in biopsy material. These include the study of single epithelial cells isolated from gastric biopsies following treatment with EDTA and mechanical agitation,[12] the COMET assay, measurement of caspase activity, identification of single-stranded DNA as a marker of apoptosis, and measurement of the expression of apoptosis-associated proteins (such as p53 and the Bcl-2 family members) by histochemical methods. However, currently none of these methods has been extensively used or validated in gastric tissue.

In Vitro Methods

The interaction between *H. pylori* and gastric cells can also be investigated in coculture. A primary advantage of the coculture system is that single characteristics of the bacterial cells or their targets can be manipulated, and the interaction of the modified cells compared to that of the wild-type cells. A number of *H. pylori*

[10] I. A. Scotiniotis, T. Rokkas, E. E. Furth, B. Rigas, and S. J. Shiff, *Int. J. Cancer* **85,** 192 (2000).
[11] M. F. Dixon, R. M. Genta, J. H. Yardley, and P. Correa, *Am. J. Surg. Pathol.* **20,** 1161 (1996).
[12] X. Fan, H. Gunasena, Z. Cheng, R. Espejo, S. E. Crowe, P. B. Ernst, and V. E. Reyes, *J. Immunol.* **165,** 1918 (2000).

with mutations in one or more virulence genes have been constructed by insertional inactivation,[13–18] transposon shuttle mutagenesis,[19] and random insertion mutagenesis.[20]

However, there are concomitant limitations of the coculture system. In general, only one target cell type is represented, for example, gastric epithelial cells, in contrast to the many cell types that may be present *in situ*. Additionally, coculture experiments are generally performed with exponentially growing cells or cells that have been synchronized. Expression of cell cycle-dependent proteins and surface receptors will not be the same as that by confluent cells and may differ from expression in cells *in situ*. Similarly, adherence of bacteria to the target cell requires expression of surface molecules on each that may be differentially expressed during the cell cycle, or variably expressed in different media. Last, observations in coculture could potentially reflect metabolic consequences of coculture such as competition for nutrients or generation of lactic acid. Thus, the results of coculture experiments should be validated by comparison with *in vivo* findings in patient specimens or animal models whenever possible.

Choice of Epithelial Target Cell

The ideal target cell for study in a model *in vitro* system would be a nontransformed human gastric epithelial cell line, but none are currently available. Primary human gastric epithelial cells have been used,[21] but in limited fashion, since the cell yield from endoscopic biopsies is small, and the extensive manipulation required may itself trigger apoptosis. Thus, gastric cell lines have been used as target cells in most investigations. Because these cell lines are derived from gastric cancers, they are an imperfect model for the response of normal cells, although a limited comparison has suggested that some lines behave similarly to primary cultures.[21] We have used the AGS human gastric cancer cell line in most of our studies because it is readily available, is easily maintained, and is amenable to transient transfection. Furthermore, AGS cells share many properties with normal gastric epithelial cells. AGS cells have a wild-type p53

[13] R. L. Ferrero, V. Cussac, P. Courcoux, and A. Labigne, *J. Bacteriol.* **174,** 4212 (1992).
[14] T. L. Cover, M. K. R. Tummuru, P. Cao, S. A. Thompson, and M. J. Blaser, *J. Biol. Chem* **269,** 10566 (1994).
[15] M. K. Tummuru, T. L. Cover, and M. J. Blaser, *Infect. Immun.* **62,** 2609 (1994).
[16] M. Tsuda, M. Karita, K. Okita, and T. Nakazawa, *Infect. Immun.* **62,** 3586 (1994).
[17] S. A. Sharma, M. K. R. Tummuru, G. G. Miller, and M. J. Blaser, *Infect. Immun.* **63,** 1681 (1995).
[18] M. K. Tummuru, S. A. Sharma, and M. J. Blaser, *Mol. Microbiol.* **18,** 867 (1995).
[19] R. Haas, T. F. Meyer, and J. VanPutten, *Mol. Microbiol.* **8,** 753 (1993).
[20] J. J. E. Bijlsma, C. M. J. E. Vandenbroucke-Grauls, S. H. Phadnis, and J. G. Kusters, *Infect. Immun.* **67,** 2433 (1999).
[21] B. W. Wagner, J. Westermann, R. P. Logan, C. T. Bock, C. Trautwein, J. S. Bleck, and M. P. Manns, *Gastroenterology* **113,** 1836 (1997).

gene,[22] express MHC II,[12] and form monolayers to which *H. pylori* adhere readily. The MKN45 cell line also has wild-type p53, in contrast to the MKN28 and Kato III cell lines.

Cell Culture Conditions

We maintain the AGS human gastric epithelial cell line (American Type Culture Collection, Manassus, VA, CRL-1739) at 37° in 95% air/5% CO_2 (v/v) in 180-ml tissue culture flasks. The growth medium for these cells is Ham's F12 medium (available from several vendors) plus 10% (v/v) fetal bovine serum (FBS). Cells are fed every 2–3 days and split 1 : 4 when subconfluent. Refeeding after serum starvation will induce apoptosis of these cells.[23] AGS cells should be passed at least three times from frozen stocks before they are used in coculture.

General advice regarding cell culture techniques is available from the ATCC Web site (www.atcc.org). No antimicrobials should be added to the growth medium, because even agents that do not kill *H. pylori* may affect its growth. Scrupulous attention to sterile technique is required. It is helpful to prepare media and other components in small volumes, and to limit reuse of refrigerated components such as serum or trypsin to periods of a few days to 1 week at most. It may also be helpful to aliquot FBS after heat inactivation at 56° for 20 min into convenient volumes (usually 10 to 20 ml) in sterile Falcon snap-cap (BD Biosciences, Franklin Lakes, NJ) culture or conical centrifuge tubes, and freeze at −20° until use. Whenever possible, serum from the same lot should be used for replicate experiments. Prepared sterile 0.25% (w/v) trypsin–0.05% (w/v) EDTA solution in phosphate-buffered saline can also be defrosted and aliquoted in convenient volumes (e.g, 2 to 4 ml volumes in sterile snap-cap culture tubes) and frozen at −20° until use.

Note that water baths are a potential source of contamination in the cell culture laboratory. If a water bath is used to defrost or heat-inactivate FBS, the container should be wiped with ethanol and dried thoroughly before opening. The bath water should never be allowed to come in contact with the cap.

Bacterial Factors in Coculture Experiments

Biosafety. Biosafety level 2 practices are recommended for handling *H. pylori*. These practices are outlined in detail in guidelines published by the U.S. Department of Health and Human Services[24] and are available on the World Wide Web

[22] G. Chen, E. M. Sordillo, W. G. Ramey, J. Reidy, P. R. Holt, S. Krajewski, J. C. Reed, M. J. Blaser, and S. F. Moss, *Biochem. Biophys. Res. Commun.* **239,** 626 (1997).

[23] H. Shirin, E. M. Sordillo, S. H. Oh, H. Yamamoto, T. Delohery, I. B. Weinstein, and S. F. Moss, *Cancer Res.* **59,** 2277 (1999).

[24] U. S. Department of Health and Human Services, Public Health Service. Biosafety in Microbiological and Medical Laboratories. HHS Publication No. (CDC) 99-xxxx, 4th Ed., April 1999.

(www.cdc.gov). Because the major mode of transmission for this organism is thought to be fecal–oral or by direct contact with the gastric mucosa, there should be strict compliance with recommendations regarding handwashing and restriction on eating and drinking in the laboratory. Although aerosols are not thought to play a role in transmission of *H. pylori,* it is prudent to handle liquid cultures or bacterial suspensions in a biological safety cabinet.

Growth Characteristics and Expression of Virulence Factors. Certain bacterial characteristics or factors may not be expressed during growth on all media,[25] or expression may vary during exponential as compared to stationary phase growth. For example, metabolism of tryptophan by *Escherichia coli* is induced by growth on media containing blood, but not by growth on MacConkey agar (the indole reaction). Similarly, it is well recognized that glycosylation is influenced by the composition of the growth medium. It is interesting to speculate that growth on solid versus broth media might influence expression of characteristics that are induced by cell contact.

Gross changes in *H. pylori* are apparent as a morphologic change from curved rods during exponential growth to coccoid forms during the stationary and death phases.

This morphologic change is accompanied by changes in ultrastructure and in protein and antigen expression[26–28] and in their ability to induce changes in epithelial cell response such as interleukin-8 secretion.[29] In older cultures, corresponding to stationary and death phase growth, autolysis of some bacterial cells results in the release of potential virulence factors such as urease, which may adhere to the remaining viable organisms in the colony.[30]

The optimal conditions for growth and viability of *H. pylori* vary from strain to strain, and for each strain on various media. For example, we have found that on trypticase soy agar with 5% sheep blood incubated at 37° in 92% air/8% CO_2, the ATCC 49503 (60190, 88-23wt) strain approaches stationary phase in 48–72 hr, and its cagA-isogenic mutant in 72 hr. The ATCC 43504 strain requires about 96 hr to reach stationary phase under these conditions. By contrast, many fresh clinical strains require 5 or more days to mature on initial isolation, but the time required may shorten as these strains become laboratory-adapted after several passages. It is important to remember that clinical isolates should only be considered "fresh"

[25] C. Lindholm, J. Osek, and A. M. Svennerholm, *Infect. Immun.* **65,** 5376 (1997).

[26] L. Cellini, N. Allocati, E. DiCampli, and B. Dainelli, *Microbiol. Immunol.* **38,** 25 (1994).

[27] M. Benaissa, P. Babin, N. Quellard, L. Pezennec, Y. Cenatiempo, and J. L. Fauchere, *Infect. Immun.* **64,** 2331 (1996).

[28] J. G. Kuster, M. M. Gerrits, J. A. G. Van Strijp, and C. M. J. E. Vandenbroucke-Grauls, *Infect. Immun.* **65,** 3672 (1997).

[29] S. P. Cole, D. Cirillo, M. F. Kagnoff, D. G. Guiney, and L. Eckman, *Infect. Immun.* **65,** 843 (1997).

[30] B. E. Dunn and S. H. Phadnis, *Yale J. Biol. Med.* **71,** 63 (1998).

during the first few passages after isolation. To minimize variability, the growth medium and duration of culture for a given strain should be standardized from experiment to experiment in a replicate set.

Only a small inoculum is required for passage and maintenance of *E. coli* and many commonly studied, rapidly growing bacteria. In contrast, successful passage of *H. pylori* requires an adequate inoculum to grow well. In general, a loop rather than a needle should be used to take up several colonies that are streaked thickly. A requirement for large inocula also has been described in broth cultures, in which cultures with lower initial inocula of *H. pylori* have a longer "lag time" before initiation of exponential growth.[31]

Assessment of Inoculum Size. The bacterial dose is usually expressed as the ratio of bacterial cells to the target mammalian cell, and 10-fold increases in inoculum are compared (e.g., 1 : 1, 10 : 1, 100 : 1). The term colony-forming units (cfu) is used to describe the number of viable bacteria in the inoculum. The actual cfu per ml are determined by serial dilution (10-fold or 100-fold) and quantitative culture of the inoculum, but the results of quantitative cultures are not available immediately.

A rapid estimate of the inoculum dose can be made by comparison of the optical density of the inoculum suspension to a reference suspension, such as the McFarland barium sulfate turbidity standards (available from BD Biosciences, Franklin Lakes, NJ), either visually or by measurement of absorbance or percent transmittance, or by measurement with a nephelometer or photometer. This method is well established for preparation of bacterial suspensions used for antimicrobial susceptibility testing in a clinical setting.[32] McFarland estimates are most reliable at concentrations $\leq 10^9$ CFU/ml. The absorbance is usually measured at about 600 nm for bacteria suspended in broth and at 400–450 nm for bacteria in saline or phosphate-buffered saline.

A calibration curve can be constructed by plotting absorbance of bacterial suspensions made to match serial McFarland standards, against the cfu/ml determined by quantitative subculture. The calibration curve is specific to a given spectrometer and given diluent. Because variations in bacterial size and shape also affect the reproducibility of these quantitative estimates, the calibration curve is also specific to both the *H. pylori* strain and the age of the subculture. It is important to remember that both viable and dead bacteria contribute to the absorbance of the suspension. Similarly, light scattering and thus absorbance will differ for suspensions killed by formalin or heat treatment. Thus, when comparing the effect of killed and live bacteria, the absorbance of the suspension should be determined before heat or formalin treatment.

[31] X Jiang and M. P. Doyle, *J. Clin. Microbiol.* **38**, 1984 (2000).

[32] D. M. Carlberg, *in* "Antibiotics in Laboratory Medicine" (V. Lorian, ed.), Chap. 3. Williams & Wilkins, Baltimore, 1986.

Quantitative Culture

1. Lable (A through D) 4 sterile tubes that can be tightly capped (use a screw cap if available).

2. Pipette 5 ml sterile saline or PBS into each tube.

3. Use a pipettor and sterile tip to add 50 μl of the original suspension to tube to make a 10^{-2} dilution (dilution A).

4. Tighten the cap and invert the tube six times to mix thoroughly.

5. Add 50 μl of dilution A to tube B, tighten the cap, and invert 6 times to mix. This will be a 10^{-4} dilution of the original suspension (dilution B).

6. Continue the process for dilutions C and D.

For each dilution A through D:

7. Inoculate 100 μl of each dilution to a 5% sheep blood agar plate. Test each dilution in duplicate to increase the accuracy of the count.

8. Use a 10-μl loop or a glass rod to spread the inoculum evenly over the plate. Turn the plate 90 degrees and spread the inoculum again to ensure an even distribution.

9. Incubate at 37° in 92% air/8% CO_2.

10. Countable colonies will generally be visible after 2–4 days. Count colonies on plates that have between 10 and 200 colonies, and multiply by the dilution factor to determine the number of colony-forming units that were in the original suspension.

For example:

	Estimated cfu/ml	Expected colonies per 100 ul plated
Original (3.0 McFarland)	6 to 9×10^8	
A	6 to 9×10^6	TNTC[a] (6 to 9×10^5)
B	6 to 9×10^4	TNTC (6 to 9×10^3)
C	6 to 9×10^2	60 to 90
D	6 to 9	0 to 1

[a] Too numerous to count.

Preparing Bacterial Inoculum

Spontaneous mutations occur in all bacterial cultures, although the rate varies among strains and among individual genes and is influenced by physiologic conditions and the presence of stressors.[33] Because spontaneous mutations will

[33] J. L. Martinez and F. Bacquero, *Antimicrob. Agents Chemother.* **44**, 1771 (2000).

accumulate over time, it is very important to limit the number of times a bacterial strain is passed before it is used in an experiment. Conversely, bacterial strains, particularly after the stress of freezing, may require several passages on an enriched medium in order to enhance expression of various traits including virulence characteristics. For these reasons, we perform experiments with strains passed at least two and no more than eight times from frozen stocks.

Maintaining Stock Cultures

We maintain all bacterial stock cultures in either defibrinated sheep blood or in freezing medium consisting of brain–heart infusion broth supplemented with 5% fetal bovine serum and 20% glycerol. Storage media based on *Brucella* broth[34,35] and skim milk[35] have also been used successfully. Only pure cultures with >90% viability should be used for preparation of stocks. Stock cultures of *H. pylori* may be frozen indefinitely, but usually do not tolerate thawing and refreezing.

In Sheep Blood

1. Defibrinated sheep blood can be obtained in 50 to 100 ml volumes (PML, Tualatin, OR). On receipt, culture 1 to 2 ml to ensure the sheep blood was not contaminated during collection. If contaminated, discard.

2. Aliquot 1-ml volumes of the sheep blood in 1.8- or 2-ml sterile screw cap polypropylene freezing tubes (NUNC, Corning, or similar), date, and freeze at $-70°$ to $-90°$ until use.

3. Use a 1- or 10-μl loop to harvest several loopfuls of growth, and suspend evenly in the sheep blood.

4. Label the tube with the organism name and strain number, passage number and date, and freeze at $-70°$ to $-90°$.

In Freezing Medium. Prepare freezing medium by adding sterile glycerol to a final volume of 20% to brain–heart infusion (BHI) broth supplemented with 5% FBS.

Small volumes of freezing medium may be prepared as follows:

1. Add 0.5 ml heat-inactivated FBS to a sterile 15-ml centrifuge tube.

2. Add 7.5 ml of BHI broth to the tube.

3. Add 2 ml sterile glycerol and mix thoroughly by pipetting up and down several times.

The glycerol is very viscous, and will stick to the inside of the pipette.

[34] S. J. Liaw, L. J. Teng, S. W. Ho, and K. T. Luh, *J. Microbiol. Immunol. Infect.* **31,** 261 (1998).

[35] S. W. Han, R. Flamm, C. Y. Hachem, H. Y. Kim, J. E. Clarridge, D. G. Evan, J. Beyer, J. Drnec, and D. Y. Graham, *Eur. J. Clin. Microbiol. Infect. Dis.* **14,** 349 (1995).

The exact concentration of fetal bovine serum is not critical for this purpose. Thus, if BHI/5% FBS medium has been prepared for bacterial culture, the 20% glycerol may be added to the prepared medium. Unused freezing medium may be kept refrigerated for several weeks before use.

4. Pipette 2 or 3 ml of freezing medium into a sterile culture tube.

5. Use a sterile Dacron-tipped applicator (S/P Brand, Allegiance Healthcare Corp., McGaw Park, IL), a loop, or a sterile wooden applicator stick to harvest growth from a sheep blood agar plate. Suspend the bacteria in the freezing medium by rubbing the applicator swab, stick, or loop against the side of the tube to get a milky solution (approximately the density of a 2.0 or 3.0 McFarland).

6. Aliquot 1-ml volumes of the suspension into 1.8- or 2-ml sterile screw cap freezing tubes (NUNC or Corning). Sterile glass or plastic donut beads may be added.

7. Label the tubes with the organism name and strain number, passage number and date, and freeze at $-70°$ to $-90°$.

Reviving Frozen Stock Cultures

Subcultures. We subculture *Helicobacter* strains from frozen stocks to trypticase soy agar containing 5% sheep blood (TSAII, BBL, Becton Dickinson, Cockeysville, MD) and incubate at $37°$ in 5 to 10% CO_2 for a minimum of two and a maximum of eight passages. It may be easiest at first to subculture the entire tube of the stock culture. However, subculturing this volume is not usually necessary. If glass or plastic beads have been used, a single bead can be removed with a small loop and rolled across the blood plate, then dropped in BHI broth/5% FBS. Alternatively a small amount can be scraped from the frozen stock, or a sterile pipette can be used to obtain a core, which can be inoculated to the blood agar plate.

Isogenic mutant strains that were generated by insertional recombination of antibiotic resistance genes should be grown in the presence of the antimicrobial agent until one or two passages prior to the coculture experiments.

Preparation of Inocula for Coculture with Eukaryotic Cells

1. Calculate the required bacterial inoculum to achieve the desired proportion of viable bacteria or cfu per target cell.

2. Prepare bacterial suspensions from mature colonies of subcultures approaching stationary phase. Suspensions should be made to match the optical density of a known McFarland standard, and diluted in medium to the desired inoculum concentration. For example, we have found that bacterial suspensions from 48–72 hr subcultures of *H. pylori* ATCC 49503 made to match the absorbance of a 1.0 McFarland standard at 450 nm have averaged 2.3×10^8 cfu/ml when confirmed by quantitative cultures.

3. Use a sterile Dacron-tipped applicator to gently pick up bacterial colonies from the agar plate. Be careful to pick up the bacterial growth only, and not agar.

4. Rub the applicator swab against the side of the tube to make a suspension. If clumps are noted, these should be allowed to settle and the suspended organisms transferred to a new tube.

Assessment of Bacterial Virulence Factors

Viable vs Killed Organisms. Some bacterial virulence factors, such as toxins or LPS, may not require the organisms to be viable in order to produce disease. However, the method chosen to kill the bacterial cells may have effects other than the direct reduction of viability. For example, formaldehyde "fixes" cells by cross-linking proteins, and thus affects their availability for binding.

Isogenic Mutants. A number of isogenic mutants have been developed that allow the assessment of the role of specific gene products. For example, we have found that the wild-type 88-23 strain and its isogenic *cagA*⁻ mutant produce a similar apoptotic response and effect on the cell cycle in AGS cells.[22,23] In some laboratories, coculture experiments suggest the apoptotic response to *vacA*⁻ strains is reduced in comparison to wild-type strains.[36] Whenever possible, these results should be confirmed in an analogous mutant from a different wild-type strain or in clinical isolates.

Requirement for Cell Contact vs Soluble Factors. The requirement for cell contact has become an increasingly complex issue with the recognition that the type IV secretion system can transfer *H. pylori* products such as the *cag A* protein to the target cell.[37–39] It is not known if this type IV system is required for transfer of an apoptosis-inducing factor, or if this system acts to concentrate the transfer of apoptosis inducing factors, thus affecting the dose–response rate.

Our experiments using polycarbonate Transwell membranes (Corning Life Sciences, Acton, MA) suggest that direct bacterial target cell contact enhances the induction of apoptosis.[22]

Transwells used in coculture experiments should have a pore size (0.1 to 0.4 μm) that is adequate to prevent the movement of bacteria through the membrane. Cocultures using Transwells are also important to address any issues regarding competition for nutrients or other metabolic effects, since movement of these molecules would not be affected.

[36] R. M. Peek, Jr., M. J. Blaser, D. J. Mays, M. H. Forsyth, T. L. Cover, S. Y. Song, U. Krishnan, and J. A. Pietenpol, *Cancer Res.* **59**, 6124 (1999).

[37] E. D. Segal, J. Cha, J. Lo, S. Falkow, and L. S. Tompkins, *Proc. Natl. Acad. Sci. U.S.A.* **96**, 14559 (1999).

[38] S. Odenbreit, J. Puls, B. Sedlmaier, E. Gerland, W. Fische, and R. Haas, *Science* **287**, 1497 (2000).

[39] S. Censini, M. Stein, and A. Covacci, *Curr. Opin. Microbiol.* **4**, 41 (2001).

Analysis of Apoptosis in Coculture Experiments

A number of methods that can be used to evaluate apoptosis in cell culture systems have been described and have been reviewed in a volume of *Methods in Enzymology*.[40] Many of the reagents are available commercially as components, or even as complete kits, and are suitable for the analysis of bacterial–gastric cell coculture experiments. Various methods that have been used successfully to study apoptosis induced by *H. pylori* are as noted below.

Nuclear Dyes. Nucleic acid binding dyes such as propidium iodide, acridine orange, and fluorescent Hoechst dyes can be used to identify nuclei of cells exhibiting apoptotic morphology. This is a relatively easy assay method to approximate the percentage of cells undergoing apoptosis. However, the percentage of apoptotic cells may be underestimated if the floating cell population is not captured. Furthermore, the manual counting method is operator (counter) dependent and thus subjective.

Flow Cytometric Methods. Flow cytometric analysis allows precise and complete sampling of many thousands of cells and it is extremely useful for the quantification of apoptosis. As for all flow cytometric techniques, fixation and careful attention to filtering cells into single cell suspensions are critical, and familiarity with the analytical software is essential. It is important to collect cells that are floating in the cell culture (representing already apoptotic cells), as well as adherent cells (consisting of live cells and those undergoing the early stages of apoptosis prior to detachment from the substrate). A number of staining protocols have been described, including staining with fluorescein-conjugated Annexin V and propidium iodide, and several kits are available from commercial sources. We have found that dual staining with propidium iodide and Annexin V allows a reasonable separation of early apoptotic cells (which stain only with Annexin V) from later apoptotic cells (which exhibit dual staining due to some membrane leakage of propidium iodine) and necrotic cells (which stain only with propidium iodide).[22] A subjective element can be introduced into this analysis when the windows are set to sort these different fractions. Standard cell cycle analysis on propidium iodide stained cells can be performed once necrosis has been ruled out (because it is difficult to separate a necrotic from an apoptotic sub-G_1 fraction). Because cells in G_2 and M have twice the DNA content of cells in G_0/G_1, the G to M peak should be twice the G_0/G_1 channel position in a linear display and the sub-G_0/G_1 will correspond to the apoptotic population. Protocols and kits have also been developed for double staining with propidium iodide and a second marker such as TUNEL or bromodeoxyuridine, to assess proliferation and apoptosis simultaneously.

Electron Microscopy. This is a qualitative method only. Typical electron microscopic features of apoptosis such as vacuolation, membrane blebbing, and nuclear

---

[40] J. C. Reed, ed., "Apoptosis," *Methods Enzymol.* **322** (2000).

condensation and fragmentation, can be demonstrated in AGS cells after coculture with *H. pylori*.[21,22]

DNA Fragmentation Enzyme-Linked Immunosorbent Assays (ELISAs). These assays use DNA fragmentation as a surrogate marker for apoptosis. Kits are available from several commercial sources. The commercially available assays differ mainly in their specificity for the antigens exposed during the DNA fragmentation that typically accompanies apoptosis (such as novel histone epitopes, other DNA-associated proteins, or 3'sticky ends). These assays are conveniently designed for use in 96-well plates, and several groups have used them to measure *H. pylori*-induced apoptosis.[12,21,22,36] Approximately 10^4 cells are seeded per well in at least triplicate samples because there is some inevitable variation from well to well.

DNA Laddering. The fragmentation of DNA typical of apoptosis can also be detected by electrophoresis followed by ethidium bromide staining. However, this method is quantitative only, and some cell lines undergo DNA laddering more easily than others. Therefore it is of limited use in studies of *H. pylori*-induced apoptosis.

Metabolic Activity Assays. Tetrazolium salts such as MTT [3-(4,5-dimethylthi-azol-2-yl)-2,5-diphenyltetrazolium bromide] can be reduced to a colored product by metabolically active cells. These assays are commercially available in a 96-well plate format. Changes in the proportion of reduced to oxidized MTT have been used to monitor mitochondrial function. However, this method is subject to multiple limitations, since the level of reduced MTT can be altered by direct chemical interactions or pH, and by nonmitochondrial cytosolic dehydrogenase activity. Additionally, in coculture, the proportion of reduced MTT may reflect overall metabolic activity, not just that of the eukaryotic target cell. Furthermore, this assay does not differentiate between decreased proliferation and increased apoptosis. Assays of mitochondrial function using substrates other than MTT are available but they too have the inherent problems described above. Assays that utilize changes in growth curves and viability counts also present interpretative problems, as these parameters similarly are affected by changes in either proliferation or apoptosis.

Caspase Activity. The final common pathway of apoptosis is the activation of procaspase 3 to its active substrate, which subsequently degrades target proteins such as poly(ADP-ribose) polymerase and lamins. An assessment of caspase-3 activity is now widely used as a marker of apoptosis, but has been determined in only a few studies of *H. pylori*-induced apoptosis.[41,42] It is also possible to measure the activities of other caspases further upstream, including caspase 8 and caspase 9, by enzymatic and fluorimetric activity assays. Commercial kits supplied with specific inhibitors simplify these measurements. Qualitative data on

[41] A. Galmiche, J. Rassow, A. Doye, S. Cagnol, J. C. Chambard, S. Contamin, V. de Thillot, I. Just, V. Ricci, E. Solcia, E. Van Obberghent, and P. Boque, *EMBO J.* **19**, 6361 (2000).

[42] J. M. Kim, J. S. Kim, H. C. Jung, I. S. Song, and C. Y. Kim, *Scand. J. Gastroenterol.* **35**, 40 (2000).

specific caspase activation can also be obtained by changes in molecular mass after immunoblotting.

Expression of Apoptosis-Related Genes

It is critical that the cells used for these analyses be collected under the same conditions under which other parameters of apoptosis are assessed. In particular, cells should be in the logarithmic growth phase. We have measured the expression of members of the Bcl-2 family following standard protein extraction and immunoblotting.[22] Because some of the Bcl-2 family proteins may be membrane-associated the use of a strong detergent in the lysis buffer is recommended. However, there is so far no conclusive information on the expression of the Bcl-2 family or indeed other apoptosis-regulatory proteins, mitochondrial cytochrome *c* release, or the mRNA expression of apoptosis-related genes in *H. pylori*-induced apoptosis. Thus, whether any of these gene products will serve as suitable surrogate markers of apoptosis in coculture systems remains to be determined.

[23] Modulation of Apoptosis during Infection with *Chlamydia*

By JEAN-LUC PERFETTINI, MATHIEU GISSOT, PHILIPPE SOUQUE, and DAVID M. OJCIUS

Introduction

Cell death occurs through different mechanisms that can be distinguished morphologically and biochemically. The most common forms of cell death are necrosis and apoptosis.[1-3] Necrosis has also been referred to as accidental death and results from severe environmental disturbances. Apoptosis represents an active, orchestrated form of death that is initiated in response to physiological stimuli. The hallmarks of apoptosis are condensation of the nuclear chromatin and cytoplasm, activation of proteases (caspases) and endonucleases, loss of plasma membrane phosphatidylserine (PS) asymmetry, cleavage of the DNA into 200 base-pair oligonucleosomal fragments, and segmentation of the dying cell into membrane-bound apoptotic bodies. Intracellular microbes can also modulate apoptosis of the host cell, either inhibiting or promoting cell death, and it has been proposed that

---

[1] A. H. Wyllie, J. F. Kerr, and A. R. Currie, *Int. Rev. Cytol.* **68**, 251 (1980).
[2] A. H. Wyllie, *Brit. Med. Bull.* **53**, 451 (1997).
[3] P. Golstein, D. M. Ojcius, and J. D. Young, *Immunol. Rev.* **121**, 29 (1991).

METHODS IN ENZYMOLOGY, VOL. 358

Copyright 2002, Elsevier Science (USA).
All rights reserved.
0076-6879/02 $35.00

the persistence and pathogenesis of several pathogenic microbes may be related to their ability to dysregulate apoptosis.[4,5]

Caspases belong to a family of cysteine proteases containing at least 14 members.[6-8] In mammalian cells, caspases are present as inactive procaspases and are proteolytically processed into active caspases that cleave substrates after aspartate residues. Many of the substrates are proteins involved in maintaining cell structure or cell regulation. Specific caspase inhibitors, consisting of small tri- or tetrapeptides, have been developed that inhibit most types of apoptosis studied to date.[9]

Caspases can be activated through engagement of cell surface receptors such as Fas,[10] but they can also be activated following release of cytochrome c from mitochondria.[11] Dysfunction of mitochondria can lead to redistribution of cytochrome c from mitochondria to the cytosol, which can activate caspase-9 in the presence of dATP and the procaspase-processing factor, Apaf-1.[12] Caspase-9 in turn activates the effector caspase-3 that is responsible for many of the morphological and biochemical features of apoptosis.[6,8]

Mitochondria can be induced to release cytochrome c through a myriad of weak stress signals emanating from the interior of the cell, and they can also release cytochrome c following activation of caspases stimulated by ligation of surface receptors.[13] The integrity of the mitochondrial outer membrane and cytochrome c release are regulated by the Bcl-2 family of proteins, which consist of antiapoptotic factors such as Bcl-2 and Bcl-x_L, and proapoptotic proteins such as Bax and Bak.[14] Bcl-2 proteins prevent apoptosis by preventing the release of cytochrome c from mitochondria, whereas Bax stimulates release of cytochrome c, resulting in apoptosis.[15,16]

Nevertheless, caspase activation is not required for all types of cell death. In a growing number of cell death models, specific caspase inhibitors do not block

[4] J. C. Ameisen, J. Estaquier, and T. Idziorek, *Immunol. Rev.* **142,** 9 (1994).

[5] L.-Y. Gao and Y. Abu Kwaik, *Microbes Infect.* **2,** 1705 (2000).

[6] G. S. Salvesen and V. M. Dixit, *Cell* **91,** 443 (1997).

[7] D. W. Nicholson and N. A. Thornberry, *Trends Biochem. Sci.* **22,** 299 (1997).

[8] M. Los, S. Wesselborg, and K. Schulze-Osthoff, *Immunity* **10,** 629 (1999).

[9] M. Garcia-Calvo, E. P. Peterson, B. Leiting, R. Ruel, D. W. Nicholson, and N. A. Thornberry, *J. Biol. Chem.* **273,** 32608 (1998).

[10] S. Nagata, *Cell* **88,** 355 (1997).

[11] G. Kroemer, B. Dallaporta, and M. Resche-Rigon, *Annu. Rev. Physiol.* **60,** 619 (1998).

[12] P. Li, D. Nijhawan, I. Budihardjo, S. M. Srinivasula, M. Ahmad, E. S. Alnemri, and X. Wang, *Cell* **91,** 479 (1997).

[13] D. R. Green and J. C. Reed, *Science* **281,** 1309 (1998).

[14] J. M. Adams and S. Cory, *Science* **281,** 1322 (1998).

[15] S. Shimizu, M. Narita, and Y. Tsujimoto, *Nature* **399,** 483 (1999).

[16] T. Rossé, R. Olivier, L. Monney, M. Rager, S. Conus, I. Fellay, B. Jansen, and C. Borner, *Nature* **391,** 496 (1998).

apoptosis induced by proapoptotic stimuli,[17–19] and caspase activation is not sufficient for initiating apoptosis.[20] In addition, overexpression of Bax or Bak induces cell death without the involvement of caspases,[18,21] suggesting that mitochondrial dysregulation can trigger cell death in a caspase-independent manner, and that factors other than caspases can also mediate apoptosis.

Like mycobacteria and the herpes virus,[5,22] chlamydiae modify the apoptotic pathway in two opposing directions. *Chlamydia trachomatis* protects infected cells during early stages of the infection against apoptosis due to external stimuli, including staurosporine, etoposide, and anti-Fas antibody.[23] Apoptosis is apparently blocked through inhibition of cytochrome *c* release from mitochondria and subsequent caspase-3 activation.[23] In contrast, induction of host cell apoptosis has also been observed in macrophages and epithelial cells infected by *C. psittaci* during late stages of the infection, and the apoptosis requires intracellular bacterial replication.[24] *C. psittaci*-induced apoptosis may require secretion of a bacterial apoptotic factor, potentially via the type III secretion apparatus[25,26] and/or may be a stress response in the infected cell.[27] Consistent with the inactivation of caspases during infection,[23] apoptosis induced by *C. psittaci* does not involve the activity of caspase-3 or other known caspases.[24,28] Accordingly, this paper will describe methods used to measure apoptosis or inhibition of apoptosis during infection, focusing on techniques that reveal host cell morphological changes, caspase activation, mitochondrial membrane depolarization, cytochrome *c* release, and DNA fragmentation.

Chlamydia Strains Known to Modulate Host Cell Apoptosis

Epithelial cells or macrophages are cultured and infected with *Chlamydia* at multiplicities of infection ranging from 1 to 50.[23,24,29,30] *Chlamydia* strains known

[17] O. Deas, C. Dumont, M. MacFarlane, M. Rouleau, C. Hebib, F. Harper, F. Hirsch, B. Charpentier, G. M. Cohen, and A. Senik, *J. Immunol.* **161**, 3375 (1998).
[18] J. Xiang, D. T. Chao, and S. J. Korsmeyer, *Proc. Natl. Acad. Sci. U.S.A.* **93**, 14559 (1996).
[19] N. J. McCarthy, M. K. B. Whyte, C. S. Gilbert, and G. I. Evan, *J. Cell Biol.* **136**, 215 (1997).
[20] L. H. Boise and C. B. Thompson, *Proc. Natl. Acad. Sci. U.S.A.* **94**, 3759 (1997).
[21] J. G. Pastorino, S. T. Chen, M. Tafani, J. W. Snyder, and J. L. Farber, *J. Biol. Chem.* **273**, 7770 (1998).
[22] V. Galvan and B. Roizman, *Proc. Natl. Acad. Sci. U.S.A.* **95**, 3931 (1998).
[23] T. Fan, H. Lu, L. Shi, G. A. McCarthy, D. M. Nance, A. H. Greenberg, and G. Zhong, *J. Exp. Med.* **187**, 487 (1998).
[24] D. M. Ojcius, P. Souque, J. L. Perfettini, and A. Dautry-Varsat, *J. Immunol.* **161**, 4220 (1998).
[25] R.-C. Hsia, Y. Pannekoek, E. Ingerowski, and P. M. Bavoil, *Mol. Microbiol.* **5**, 351 (1997).
[26] K. A. Fields and T. Hackstadt, *Mol. Microbiol.* **38**, 1048 (2000).
[27] P. M. Bavoil, R.-C. Hsia, and D. M. Ojcius, *Microbiology* **146**, 2723 (2000).
[28] J. L. Perfettini, J. C. Reed, N. Israël, J. C. Martinou, A. Dautry-Varsat, and D. M. Ojcius, *Infect. Immun.* **70**, 55 (2002).
[29] J.-L. Perfettini, T. Darville, G. Gachelin, P. Souque, M. Huerre, A. Dautry-Varsat, and D. M. Ojcius, *Infect. Immun.* **68**, 2237 (2000).
[30] R. Coutinho-Silva, J. L. Perfettini, P. M. Persechini, A. Dautry-Varsat, and D. M. Ojcius, *Am. J. Physiol.* **280**, C81 (2001).

to inhibit or induce host cell apoptosis include *C. trachomatis* serovar L2, serovar C, serovar mouse pneumonitis (MoPn), lymphogranuloma venereum (LGV/L2), and *C. psittaci* strain guinea pig inclusion conjunctivitis (GPIC).[23,24,28-31] Infected cells are protected against apoptosis due to incubation with staurosporine, etoposide, tumor necrosis factor-α (TNFα), Fas antibody, and extracellular ATP.[23,30] Protection begins as early as 5 hr after infection.[23] Apoptosis due to the infection itself does not begin until later in the infection cycle, becoming manifest after approximately 20–40 hr of infection, depending on the *Chlamydia* strain.[24,29,31]

Evaluation of Nuclear Condensation and Segmentation by Cytofluorimetry

Condensation and segmentation of apoptotic nuclei can be measured conveniently and quantitatively by cytofluorimetry, using detergent-permeabilized cells stained with propidium iodide (PI).[32] This approach also allows one to identify infected cells simultaneously, by incubating the same samples with fluorescein isothiocyanate (FITC)-labeled anti-*Chlamydia* monoclonal antibody (MAb) (Argene, Varilhes, France) before adding the PI buffer.[24]

Buffers

Prepare 10× stock solutions of sodium tricitrate (0.1 g in 10 ml), Triton X-100 (100 μl in 10 ml), and 25 mg/ml PI. A "1× PI buffer" can then be prepared by adding:

> 1 ml Sodium tricitrate
> 1 ml Triton X-100
> 20 μl Propidium iodide
> 8 ml Distilled water

This solution may be stored in the dark at 4° for several weeks.

Procedure

Adherent cells are typically grown in 75-cm² tissue culture flasks (Costar) to 60 to 70% confluence, then incubated with chlamydiae in cell culture medium for various times at 37° in 5% (v/v) CO_2. Nonadherent cells may be incubated with chlamydiae in cell culture medium at 37° in 5% (v/v) CO_2. Cells (1–4 × 10⁶) are harvested, washed twice with cold phosphate-buffered saline (PBS), and resuspended in 0.5 to 1.0 ml "1× PI buffer" in 12 × 75-mm Falcon 2052 FACS tubes (Becton Dickinson, San Jose, CA). Before addition of "1× PI buffer," the cells may

[31] D. Gibellini, R. Panaya, and F. Rumpianesi, *Zentralblatt Bakteriol.* **288,** 35 (1998).
[32] R. S. Douglas, A. D. Tarshis, C. H. Pletcher, P. C. Nowell, and J. S. Moore, *J. Immunol. Methods* **188,** 219 (1995).

be fixed first with 4% (w/v) paraformaldehyde for 5 min, and then washed immediately with cold PBS. Data from 10,000 cells are typically collected in a FACScan flow cytometer with an argon laser tuned to 488 nm. As previously described,[33] the cytofluorimetry profile gives intense, well-defined fluorescent staining for viable cells at the G_1, S, or G_2 stages of the cell cycle, while the apoptotic cells appear in a much broader sub-G_0 region. Necrotic cells, cells at very advanced stages of apoptosis, or cell debris binds PI less efficiently and can be excluded from the analysis by placing an appropriate threshold level.

Comments

Both adherent cells and cells in the supernatant must be collected to measure apoptosis during infection due to the fact that apoptotic cells detach from growth substrate *in vitro*.[24] *In vivo,* apoptotic cells detach from neighboring cells and are subsequently removed by scavenger phagocytes.[1] To measure protection against apoptosis due to external ligands, adherent cells may be analyzed separately, or the extent of protection may be ascertained at early times of infection, before there is a high level of apoptosis due to the infection itself.[30]

Measurement of Nuclear Condensation by Fluorescence Microscopy

Prepare a stock solution of Hoechst 33258 (Sigma, St. Louis, MO) at 5 mg/ml in 20 mM HEPES, pH 7.4. Infected or uninfected cells grown on microscope coverslips are rinsed twice in PBS and fixed with 2.5% (w/v) paraformaldehyde, as previously described.[34] The cells are stained with 5 μg/ml Hoechst in PBS with 0.05% (w/v) saponin for 15 min at room temperature. After washing in PBS, stained samples are mounted on microscope slides as previously described[34] and observed with a fluorescence microscope. The Hoechst dye can be observed after excitation at 340–380 nm using a blue emission filter. Apoptotic cells have markedly condensed nuclei.

Alternatively, nuclei may be stained with 1 μg/ml 4′,6-diamidino-2-phenyl-indole (DAPI) (Molecular Probes, Eugene, OR) in PBS instead of Hoechst.[35]

Detection of DNA Fragmentation by Gel Electrophoresis

Infected or uninfected cells (2×10^6) are washed twice with PBS and centrifuged (270g for 5–10 min) at 4°. The cell pellet is lysed with 0.6% sodium dodecyl

[33] I. Nicoletti, G. Migliorati, M. C. Pagliacci, F. Grignani, and C. Riccardi, *J. Immunol. Methods* **139,** 271 (1991).

[34] D. M. Ojcius, F. Niedergang, A. Subtil, R. Hellio, and A. Dautry-Varsat, *Res. Immunol.* **147,** 175 (1996).

[35] T. Miller, L. A. Beausang, M. Meneghini, and G. Lidgard, *BioTechniques* **15,** 1042 (1993).

sulfate (SDS), 10 mM EDTA, 10 mM Tris, and 20 μg/ml RNase A, pH 7.5, for 1 hr at 37° in 2 ml. Two hundred microliters of 5 M NaCl is added to the lysis suspension, and the sample is incubated for 40 min on ice and then centrifuged for 30 min at 15,000g. The supernatant contains the DNA and is extracted with phenol–chloroform–isoamyl alcohol (25 : 24 : 1). The low molecular weight DNA is precipitated with ethanol. Samples containing approximately 3 μg of DNA per lane are separated by electrophoresis on a 1.8% agarose gel and revealed with ethidium bromide.

Detection of DNA Fragmentation by Fluorescence Microscopy

Samples for fluorescence microscopy are fixed with paraformaldehyde, incubated with antibodies, and mounted as described in detail elsewhere.[34] Apoptotic cell on coverslips are detected by enzymatic labeling of DNA strand breaks with the TUNEL technique[36] using the cell death detection kit from Roche Molecular Biochemicals (Indianapolis, IN). In these experiments, apoptotic nuclei appear in green (due to fluorescein-12-dUTP labeling of fragmented DNA). Infected cells can be identified simultaneously by revealing with unconjugated anti-*Chlamydia* MAb (Argene, Varilhes, France, 1 : 500 dilution) followed by incubation with Texas Red- or rhodamine-labeled anti-mouse immunoglobulin (Ig) polyclonal antibodies. Fluorescently labeled samples are examined by conventional fluorescence microscopy or confocal microscopy.

As most of the apoptotic cells are in suspension, cells from the supernatant may be incubated with the cell detection kit in Eppendorf tubes, and then centrifuged and mounted on coverslips. The adherent cells may be used by themselves to study protection against apoptosis.

Observation of Nuclear Condensation and Segmentation by Electron Microscopy

Both adherent cells and cells in suspension from samples infected for various times are fixed with 2.5% glutaraldehyde for at least 2 hr at room temperature or overnight at 4°. The fixed cells are centrifuged twice for 20 min at 20,000g in PBS, and the pellet is transferred in 2 ml of freshly prepared cacodylate buffer (5 mM CaCl$_2$, 5 mM MgCl$_2$, 0.2 M cacodylate, pH 7) into an Eppendorf tube. The tube is centrifuged and the pellet resuspended in 200 μl of 3% agar (Difco, Kansas City, MO) in cacodylate buffer at 45°. The warm agar is centrifuged 1 min at 20,000g and allowed to solidify at 4°. One ml of 1% OsO$_4$ in cacodylate buffer is added to the agar. After leaving for 1 hr at room temperature, the supernatant is removed and the pellet is incubated in a 1% uranyl acetate (Merck, Rahway, NJ) solution in cacodylate buffer for an additional hour at room temperature. The supernatant is

---

[36] Y. Gavrieli, Y. Sherman, and S. A. Ben-Sasson, *J. Cell Biol.* **119,** 493 (1992).

removed; the pellet is rinsed with water, and then dehydrated by rinsing with increasing concentrations (25, 50, 75, and 100%) of acetone. The preparations are then embedded in Epon. Ultrathin sections (50–100 μm) are prepared on a microtome and poststained with uranyl acetate and lead citrate for examination on an electron microscope.

Measurement of Phosphatidylserine (PS) Exposure on Surface of Dying Cells

In apoptotic cells, PS is translocated from the inner leaflet of the plasma membrane to the outer leaflet, exposing PS to the extracellular space. Annexin V is a phospholipid-binding protein that has a high affinity for PS and is therefore a convenient marker for identifying cells with surface-exposed PS. Fluorescein-conjugated Annexin V (Annexin V–FITC) labels apoptotic cells by binding to membrane PS, rendering apoptotic cells fluorescent.[37]

Annexin V–FITC: Staining reagent normally provided at a concentration of 30 μg/ml (R&D Systems, Minneapolis, MN).
Binding buffer: 10 mM HEPES/NaOH, pH 7.4, 140 mM NaCl, 2.5 mM CaCl$_2$. Store at 4°.

Procedure

Wash cells with PBS and resuspend in 1× binding buffer (1 × 10^6 cells/ml). Transfer 100 μl of the solution to a 5-ml culture tube. Add 10 μl of Annexin V–FITC. Incubate for 15 min at room temperature, being careful to protect the samples from light. Without washing the cells of excess reagents, add 400 μl of 1× binding buffer to each tube. Analyze by cytofluorimetry within 1 hr.

Comments

Annexin V–FITC may label both apoptotic and necrotic cells. The standard PI viability assay,[38] used to distinguish viable from nonviable cells, may be combined with Annexin V labeling to distinguish between apoptosis and necrosis. Viable cells with intact membranes exclude PI, while dead cells with damaged membranes are permeable to PI. Cells that stain with Annexin V–FITC and exclude PI are undergoing apoptosis. Cells that are labeled with both Annexin V–FITC and PI are undergoing necrosis or a late stage of apoptosis. Cells that do not stain with either Annexin V–FITC or PI are viable.

[37] I. Vermes, C. Haanen, H. Steffens-Nakken, and C. Reutelingsperger, *J. Immunol. Methods* **184,** 39 (1995).
[38] C. Matteucci, S. Grelli, E. De Smaele, C. Fontana, and A. Mastino, *Cytometry* **35,** 145 (1999).

Measurement of Caspase Activation and Caspase–Substrate Cleavage

Caspase Activation Observed by Cleavage of Chromogenic or Fluorogenic Substrates

Synthetic peptides with a reporter group attached on the C terminus of the peptide have been used to quantify caspase activity. Cleavage of the peptide substrate causes release of the fluorogenic or chromogenic probe, and the resulting increase in fluorescence can be measured in a spectrofluorimeter or spectrophotometer. Substrates that are specific for different caspases have been described.[39,40] Peptide substrates can be obtained from a number of commercial suppliers, including Alexis (San Diego, CA), Boehringer Mannheim (Indianapolis, IN), Calbiochem (San Diego, CA), and Enzyme Systems Products (Livermore, CA). The most common reporter groups are 7-amino-4-methylcoumarin (AMC) and 7-amino-4-trifluoromethylcoumarin (AFC). AMC can be detected with an excitation wavelength of 380 nm and emission at 460 nm, whereas AFC can be excited at 405 nm and its emission measured at 500 nm. AFC can also be detected colorimetrically because of an absorbance change at 380 nm. The most common colorimetric reporter is *p*-nitroanilide (pNA), which absorbs at 405–410 nm. In general, the fluorimetric assays are more sensitive than the colorimetric assays.

After various times of infection, whole cell extracts are obtained by centrifuging host cells and dissolving the pellet with lysis buffer containing 50 mM PIPES–NaOH (pH 7.5), 50 mM KCl, 5 mM EGTA, 2 mM MgCl$_2$, 1 mM dithiothreitol (DTT), 20 μM cytochalasin D, 1% CHAPS, 1 μg/ml leupeptin, 1 μg/ml pepstatin A, and 1 mM phenylmethylsulfonyl fluoride (PMSF) for 10 min at 4°. Samples are vortexed and sonicated in a cold water sonicator for 15 min. After centrifugation (13,000g at 4°), the pellet is discarded, the protein concentration measured by the Bradford assay, and supernatants stored at −20° until ready for use.

Lysates (25 μg of cytosolic protein) are diluted in 50 μl of buffer A [25 mM HEPES (pH 7.5), 5 mM MgCl$_2$, 5 mM EDTA, 1 mM EGTA, 1 mM PMSF, 1 mM DTT, 10 μg/ml pepstatin A, and 10 μg/ml leupeptin] to which are added 225 μl of buffer B (25 mM HEPES, pH 7.5, 0.1% CHAPS, 10 mM DTT, 1 mM PMSF) containing 100 μM fluorogenic substrate. Samples are incubated for 4 hr at 37°, and the reaction is terminated by addition of 1.225 ml of ice-cold buffer B. Cleavage of the substrate is quantified in a spectrofluorimeter.

[39] R. V. Talanian, C. Quinlan, S. Trautz, M. C. Hackett, J. A. Mankovich, D. Banach, T. Ghayur, K. D. Brady, and W. W. Wong, *J. Biol. Chem.* **272,** 9677 (1997).
[40] H. R. Stennicke and G. S. Salvesen, *J. Biol. Chem.* **272,** 25719 (1997).

Caspase Activation Observed by Western Blot Analysis

The status of caspase activation during *Chlamydia* infection can be detected by Western blot analysis,[23] because many commercially available antibodies against different caspases recognize both inactive, uncleaved caspases and active, cleaved caspases. For the downstream caspase, caspase-3, the inactive form has a molecular mass of 32 kDa, whereas the active form is smaller, at 17–20 kDa. Antibodies against the different caspases can be obtained from a growing number of suppliers, including Transduction Laboratories (Lexington, KY) and New England Biolabs (Boston, MA).

Infected or uninfected cells ($4–8 \times 10^6$) are lysed in 1 ml lysis buffer [20 mM HEPES, pH 7.5, 150 mM NaCl, 1 mM EGTA, and 1% Noxidet P-40 (NP-40)] with protease inhibitors (10 μg/ml aprotinine and 5 μg/ml pepstatin). Following incubation at 4° for 10 min, samples are centrifuged at 13,000 rpm in a microfuge for 20 min. Sixty μg of protein is subjected to SDS–PAGE at 60–70 V on a 15% gel. Proteins are transferred onto Hybond ECL (enhanced chemiluminescence) nitrocellulose membranes (Amersham Pharmacia Biotech, Piscataway, NJ) for 3 hr at 50–75 mA, and incubated with blocking solution containing 5% nonfat milk in PBST (1× PBS, 0.05% Tween 20) for 1 hr at room temperature. The membrane is then incubated with primary anti-caspase antibody in blocking solution at 4° overnight. To confirm equal loading, the membrane can be stained with GelCode Blue Stain Reagent (Pierce, Rockford, IL). The immunoreactive proteins are visualized using horseradish peroxidase-linked goat anti-mouse or anti-rabbit antibodies and enhanced chemiluminescence.

The same procedure can be followed to measure cleavage of the caspase-3 substrate, poly(ADP-ribose) polymerase (PARP), which is present as an intact 116 kDa protein or a cleaved 85 kDa form.[23]

Observation of Cytochrome *c* Release

In viable cells, the peripheral membrane protein cytochrome *c* is normally confined to the mitochondrial intermembrane space, where it participates in electron transport in the respiratory chain. In many cases of apoptosis, however, cytochrome *c* is released from the mitochondria to the cytosol, where it binds to Apaf-1, which can in turn bind to and activate procaspase-9.[12,41,42] Caspase-9 can then initiate a caspase cascade, activating downstream effector caspases such as caspase-3. Cytochrome *c* release can be measured by first separating cells into mitochondrial and cytosolic fractions and then revealing cytochrome *c* by Western blot, or directly by immunofluorescence (IF) microscopy, which gives a qualitative

[41] X. Liu, C. N. Kim, J. Yang, R. Jemmerson, and X. Wang, *Cell* **86,** 147 (1996).
[42] J. Yang, X. Liu, K. Bhalla, C. N. Kim, A. M. Ibrado, J. Cai, T. I. Peng, D. P. Jones, and X. Wang, *Science* **275,** 1129 (1997).

description of cytochrome c localization in an intact cell. The IF technique has the advantage that the distribution of cytochrome within a cell and the *Chlamydia* inclusion can be evaluated at the same time.

Western Blot Analysis of Subcellular Fractions

Cells are harvested in isotonic mitochondrial buffer (MB: 210 mM mannitol, 70 mM sucrose, 1 mM EDTA, 10 mM HEPES, pH 7.5) supplemented with protease inhibitors (100 μM PMSF, 10 μg/ml leupeptin, 2 μ/ml aprotinin), and broken by five passages through a 25G1 0.5 × 25 needle fitted on a 2-ml syringe. Samples are transferred to Eppendorf centrifuge tubes and centrifuged at 500g for 5 min at 4° to eliminate nuclei and unbroken cells. The supernatant is centrifuged at 10,000g for 30 min at 4° to obtain a pellet enriched in mitochondria. This supernatant is further centrifuged at 100,000g for 1 hr at 4° to give the final soluble fraction. The mitochondrial fraction is resuspended in MB supplemented with 1% Triton X-100. After the protein concentrations are measured by the Bradford assay, soluble and mitochondrial fractions (30 and 15 μg, respectively) are separated by SDS–PAGE (4–20%) and transferred to a nitrocellulose membrane. After blocking of nonspecific sites for 1 hr at room temperature with 5% nonfat milk in PBS supplemented with 0.2% Tween, the membrane is incubated overnight at 4° with purified anti-cytochrome c antibody (7H8.2C12, Pharmingen, San Jose, CA) diluted in PBS supplemented with 2.5% nonfat milk. To normalize for the amount of mitochondrial proteins loaded, the membrane can be stripped and reprobed with an antibody against cytochrome c oxidase subunit IV (COX-IV) or mitochondrial hsp70. The immunoreactive proteins are visualized using horseradish peroxidase-linked goat anti-mouse antibodies and enhanced chemiluminescence.

Detection of Cytochrome c by Fluorescence Microscopy

Cytochrome c release can also be observed by fluorescence microscopy, by incubating paraformaldehyde-fixed host cells on coverslips, prepared, incubated with antibodies, and mounted as described.[34] The intracellular distribution of cytochrome c can be revealed by incubating cells with anti-cytochrome c antibodies in the presence of 0.5% saponin. Anti-cytochrome c antibody 6H2.B4 (Pharmingen) works well on epithelial cells (e.g., HeLa cells) at a concentration of 10 μg/ml in PBS with 3% (w/v) BSA. Infected cells can be identified at the same time by revealing with unconjugated anti-*Chlamydia* antibodies followed by incubation with a secondary antibody conjugated to a different dye from that used for the secondary antibody against anti-cytochrome c. To identify condensed and fragmented nuclei of apoptotic cells, the cells can also be labeled with DNA-specific fluorochromes, such as Hoechst 33342, DAPI, or PI. These dyes can be added after cells have been incubated with the secondary antibody and washed to remove excess antibody. Fluorescently labeled samples are examined by conventional fluorescence microscopy or confocal microscopy.

Comments

The IF technique works best for adherent cells, which do not include most of the cells that may be dying through apoptosis. However, it is very useful for determining whether *Chlamydia* infection under different conditions renders host cells resistant to apoptosis dependent on cytochrome *c* release.

Measurement of Mitochondrial Membrane Depolarization

In many models of apoptosis, the transmembrane potential ($\Delta\Phi_m$) of the mitochondrial inner membrane decreases during early stages of apoptosis, preceding activation of caspases.[43] In intact cells, the change in $\Delta\Phi_m$ can be monitored using potentiometric fluorochromes that can be measured by cytofluorimetry or fluorescence microscopy. These lipophilic cations accumulate in mitochondria because of the inside-negative potential across the inner membrane. Commonly used fluorochromes include 3,3′-dihexyloxacarbocyanine iodide [$DiOC_6(3)$] (fluorescence in green)[44] and chloromethyl-X-rosamine (CMXros) (fluorescence in red),[43] which can be obtained from Molecular Probes (Eugene, OR).

Buffers

Prepare 40 μM $DiOC_6(3)$ or 1 mM CMXRos stock solutions in dimethyl sulfoxide (DMSO). Store in the dark at $-20°$. The mitochondrial uncoupler carbonyl cyanide *m*-chlorophenylhydrazone (CCCP), which disrupts $\Delta\Phi_m$, is used as a negative control. Store CCCP as a stock solution of 20 mM in ethanol.

Procedure

Collect infected or uninfected cells (5×10^6 cells in 0.5 ml PBS), and keep at $4°$ until ready for use. Stain by adding $DiOC_6(3)$ (final concentration of 20 nM) or CMXRos (final concentration of 100 nM), incubate at $37°$ for 15–20 min, and then place the cells again at $4°$. Without washing, measure the $\Delta\Phi_m$ by cytofluorimetry gating the side and forward scatter on normal-sized (viable) cells. $DiOC_6(3)$ can be monitored in FL1 and CMXRos in FL3. Use 100 μM CCCP as a negative control ($\Delta\Phi_m = 0$ mV).

Acknowledgments

Supported by funds from the Institut Pasteur, INSERM, and Université Paris 7. The authors are grateful to Dr. Philippe Kourilsky for encouragement and support.

[43] M. Castedo, T. Hirsch, S. A. Susin, N. Zanzami, P. Marchetti, A. Macho, and G. Kroemer, *J. Immunol.* **157**, 512 (1996).
[44] P. X. Petit, J. E. O'Connor, D. Grunwald, and S. C. Brown, *Eur. J. Biochem.* **220**, 389 (1990).

[24] Measurement of Pore Formation by Contact-Dependent Type III Protein Secretion Systems

By GLORIA I. VIBOUD and JAMES B. BLISKA

Introduction

Several gram-negative bacterial pathogens encode type III protein secretion systems that are activated on host cell contact.[1] These type III secretion systems (TTSSs) function to deliver protein toxins into or across membranes of host eukaryotic cells. The process by which TTSS toxins are delivered across host membranes is not well understood. Some TTSSs can introduce pores into plasma membranes of host eukaryotic cells or synthetic lipid bilayers. Pore formation by a TTSS is generally dependent on the same proteins that are essential for toxin translocation. A current model suggests that TTSS toxins move through a transmembrane channel formed by components of the TTSS, and that pore formation by a TTSS results from the opening of translocation channels to the extracellular environment.

In this article, we discuss methods that can be used to measure the pore-forming activity of TTSSs in host eukaryotic cells.

Hemolysis

Lysis of red blood cells (RBCs) is commonly used to assay pore formation by a variety of secreted toxins.[2] Interestingly, RBCs appear to be relatively permissive cell targets for TTSS-dependent pore formation. TTSSs capable of inducing RBC lysis *in vitro* include those encoded by *Shigella*,[3] *Yersinia*,[4] *Pseudomonas*,[5] and enteropathogenic *Escherichia coli* (EPEC).[6] Hemolysis assays are simple to perform and require little in the way of specialized equipment. The basic principle of the assay is as follows: equal volumes of bacteria and RBCs are mixed, gently centrifuged (see below), and incubated at 37° for varying lengths of time. The samples are then suspended and centrifuged to obtain a supernatant. Pore formation in the RBC membrane results in water influx, osmotic lysis, and release of hemoglobin. Hemoglobin released into the supernatant is detected spectrophotometrically at 540 nm.

[1] C. J. Hueck, *Microbiol. Mol. Biol. Rev.* **62,** 379 (1998).
[2] A. L. Lobo and R. A. Welch, *Methods Enzymol.* **235,** 667 (1994).
[3] P. J. Sansonetti, A. Ryter, P. Clerc, A. T. Maurelli, and J. Mounier, *Infect. Immun.* **51,** 461 (1986).
[4] S. Håkansson, K. Schesser, C. Persson, E. E. Galyov, R. Rosqvist, F. Homble, and H. Wolf-Watz, *EMBO J.* **15,** 5812 (1996).
[5] D. Dacheux, J. Goure, J. Chabert, Y. Usson, and I. Attree, *Mol. Microbiol.* **40,** 76 (2001).
[6] J. Warawa, B. B. Finlay, and B. Kenny, *Infect. Immun.* **67,** 5538 (1999).

Experiments performed with *Shigella* and *Pseudomonas* have examined the need for bacterium–RBC contact by omitting the centrifugation step. Hemolysis was found to be very inefficient in the absence of centrifugation.[3,5] Blocker *et al.* found that centrifugation facilitates hemolysis by forcing very close contact and extensive surface apposition of bacterium and RBC.[7] The suspension step could also play a role in hemolysis, but to our knowledge this has not been investigated. Contact between bacterium and RBC may be broken during suspension, leading to the exposure of translocation channels to the extracellular environment and the formation of pores.

A modification of the hemolysis assay in which RBCs are immobilized in cell culture dishes has been used to study pore formation by the TTSS of EPEC.[8] Bacteria are added to the wells and incubated at 37°. Supernatants are collected at different times after infection and analyzed for hemoglobin release. Interestingly, hemolysis of immobilized RBCs by EPEC can occur in the absence of a centrifugation or suspension step. Electron micrographs reveal that the EPEC TTSS produces a long filamentous organelle that mediates bacteria–RBC contact.[8] It is exciting to speculate that the tips of these organelles that contact the RBCs are connected to translocation channels, and that retraction or breakage of these conduits can result in pore formation.

Osmoprotection

Osmoprotection assays allow one to distinguish pore formation from other forms of membrane damage that can lead to osmotic lysis. The size of the pore may also be estimated by the use of osmoprotectants.[2] Osmoprotection assays have been used to investigate TTSS pore forming activity in RBCs and nucleated cells. When added to the incubation medium a high molecular weight sugar, dextran or glycol, can prevent water influx through pores and the resultant swelling. For osmoprotection to occur, the molecular diameter of the agent must be larger than the diameter of the pore, and the concentration of the agent must be high enough (usually 30 mM final concentration) to keep the external medium isotonic when compared to the cell interior. Compounds with a diameter smaller than the pore will equilibrate across the membrane and thus afford no protection from osmotic lysis. The size of the pore may be determined by testing a series of compounds of varying molecular diameter for osmoprotection.

Experiments of this type have been used to estimate the sizes of pores introduced into RBCs by the TTSSs of *Shigella*,[7] *Pseudomonas*,[5] and *Yersinia*.[4] In the case of *Yersinia*, 30 mM raffinose (diameter 1.2–1.4 nm) had no significant effect

[7] A. Blocker, P. Gounon, E. Larquet, K. Niebuhr, V. Cabiaux, C. Parsot, and P. Sansonetti, *J. Cell Biol.* **147**, 683 (1999).

[8] R. K. Shaw, S. Daniell, F. Ebel, G. Frankel, and S. Knutton, *Cell. Microbiol.* **3**, 213 (2001).

on hemolysis, while 30 mM dextran 4 (diameter 3–3.5 nm) offered significant protection.[4] Thus, the pore introduced into RBCs by the *Yersinia* TTSS was estimated to be 1.2–3.5 nm in diameter. A similar approach was used to estimate the *Pseudomonas* TTSS pore at 2.8–3.5 nm.[5] Pore size can be determined more accurately by measuring the incubation time required to achieve 50% hemolysis in the presence of different osmoprotectants. By this method the pore introduced into RBCs by the *Shigella* TTSS has been approximated at 2.5 nm.[7]

Pore Formation in Nucleated Cells

Although RBCs have proven to be extremely useful cell targets for analysis of pore-forming activity *in vitro,* they are unlikely to be relevant targets of TTSSs *in vivo.* Do TTSSs form pores in the types of nucleated cells that are encountered by bacterial pathogens *in vivo?* The answer is both yes and no. Wild-type strains of *Pseudomonas aeruginosa* have a pronounced pore forming activity on macrophages that is TTSS-dependent.[5] The TTSS of EPEC on the other hand does not appear to form pores in epithelial cells very efficiently if at all. How can this difference be explained? It may depend on the nature of the target cell and the role of the TTSS in pathogenesis. In the case of *Pseudomonas aeruginosa,* the pore-forming activity of the TTSS may be a defense against phagocytes. For EPEC, pore formation in epithelial cells may be an undesired side effect of the TTSS translocation process. Most TTSSs are probably designed to avoid pore formation during bacteria–host cell interaction. Only when the TTSS is perturbed in some manner does pore forming activity become evident. As an example, the plasmid-encoded TTSS in *Yersinia* displays pore-forming activity against macrophages or epithelial cell only when genes encoding certain secreted components of the system (Yops) have been inactivated by mutation.[4] Experiments indicate that specific Yops function within the host cell to prevent the formation of pores in the plasma membrane.[9]

Several techniques can be used to measure pore formation in the types of nucleated cells that are targets of TTSSs *in vivo.* These include very simple assays that measure release of cytoplasmic enzymes, to more complicated assays that measure uptake or release of fluorescent probes.

Release of LDH

Lactate dehydrogenase (LDH) is a ubiquitous and stable cytosolic enzyme that is released on cell lysis. The amount of LDH released into culture supernatants can be measured by a colorimetric reaction. In a coupled reaction LDH and a diaphorase convert a tetrazolium salt into a red formazan product. A convenient kit for performing LDH release assays can be purchased from Promega (Madison, WI)

---

[9] G. I. Viboud and J. B. Bliska, *EMBO J.* **20,** 5373 (2001).

(CytoTox 96 cytotoxicity assay). The amount of formazan product is measured spectrophotometrically by absorbance at 490 nm. A comparable but less convenient assay that utilizes radioactive detection measures release of ^{51}Cr.[10]

In a typical application, cultured cells are seeded into wells of a tissue culture plate the day prior to the experiment. Bacteria grown under appropriate conditions are washed and added to the wells to initiate infection. A brief centrifugation step may be used to bring the bacteria and host cells in contact. The length and multiplicity of infection should be varied to identify optimal conditions for pore formation. After incubation at 37° and 5% (v/v) CO_2, samples of culture medium from infected cells are collected. Wells containing uninfected cells are lysed by treatment with Triton X-100 or by a freeze/thaw cycle to obtain total LDH. Samples are centrifuged at 13,000 rpm for 10 min at 4°, and supernatants are then transferred to a 96-well plate. A culture medium background control can be added as a blank to correct for the absorbance contributed by the phenol red or the serum that may be present in the cell culture medium. Samples and controls are then incubated with the substrate for 30 min. After the reaction is stopped, the absorbance values are determined using a microplate reader. All values are normalized by the absorbance of the background control. Percent LDH release is calculated by dividing the amount of LDH released from infected cells by the amount of total LDH.

Because of its size (135 kDa), LDH is larger than the pores formed by TTSSs and can only be released by cell lysis. Osmoprotection experiments should be performed to confirm that TTSS-dependent release of LDH is in fact due to pore formation. If osmoprotection experiments are used, care should be taken to ensure that the compound used does not interfere with bacteria–host cell contact. In our own experiments, we found that bacteria–host cell contact was significantly reduced if dextran 6000 was present at 30 mM in culture wells before bacteria were added.[9] This problem could be bypassed by performing a brief centrifugation step and allowing bacteria–host cell contact to occur for 5 min before addition of the dextran 6000.

Release of BCECF

The formation of pores in host cells can also be measured by the release of small intracellular marker molecules. As an example, Neyt and Cornelis[11] measured release of a fluorescent probe from macrophages infected with *Yersinia*. For this assay, the host cells are loaded with BCECF AM [2',7'-bis(2-carboxyethyl)-5-(and-6-carboxyfluorescein), acetoxymethyl ester] (Molecular Probes, Eugene, OR) prior to infection. BCECF AM is a nonfluorescent compound that readily permeates cells. Esterases inside the cell convert BCECF AM to BCECF,

---

[10] J. D. Goguen, W. S. Walker, T. P. Hatch, and J. Yother, *Infect. Immun.* **51**, 788 (1986).

[11] C. Neyt and G. Cornelis, *Mol. Microbiol.* **33**, 971 (1999).

a membrane-impermeant fluorescent molecule of approximately 615 Da. Following infection, culture supernatants are collected and BCECF release is quantified by fluorescence measurements (excitation 490 nm and emission 520 nm). Supernatants from uninfected cells lysed with Triton X-100 (0.1%) are analyzed in parallel to determine total BCECF. As BCECF is small enough to pass through TTSS pores, release of this probe in the presence of an osmoprotectant is indicative of pore formation. The assay does require the use of a spectrofluorimeter, a type of equipment that may not be widely available.

Uptake of Fluorescent Molecules

The uptake of membrane-impermeable fluorescent dyes, such as propidium iodide, ethidium bromide (EtdBr), or Lucifer Yellow CH, by host cells can be used as an additional assay of pore formation. These experiments require the use of a fluorescence microscope. We have used a mix of two dyes that intercalate into nucleic acids, EtdBr and acridine orange (AO), to measure pore formation in host cells by *Yersinia*.[9] AO penetrates cells with intact membranes, whereas EtdBr selectively enters cells with perforated membranes. EtdBr is sufficiently small (394 Da) to pass through TTSS pores.

Cultured cells are prepared and infected as described above for the LDH release assay with the exception that seeding is done onto sterile glass coverslips. At the end of the infection, the coverslips are inverted onto a 5 μl drop of the dye mix (25 μg/ml ethidium bromide and 5 μg/ml acridine orange in PBS) placed on the surface of a glass slide. Dye uptake is practically instantaneous and therefore coverslips can be examined immediately by epifluorescence microscopy using a 40× or 60× objective. A filter set appropriate for detecting fluorescein is used to observe cells for AO staining. EtdBr staining is detected using a rhodamine filter set. Cells with intact membranes will appear green because of AO staining; those with damaged membranes will exhibit bright orange nuclear staining from the EtdBr. The percentage of cells with damaged membranes can thus be calculated by visual observation. Addition of an osmoprotectant can be used to confirm that uptake of EtdBr is due to pore formation. A major disadvantage of this method is that samples must be analyzed immediately after staining. The number of cells that take up EtdBr will increase over time after mounting of the coverslip. This staining technique is not compatible with fixation, as the fixation procedure will make the membranes permeable to the dyes.

Concluding Remarks

The methods and techniques described in this article can be used to gain basic information about the pore-forming activity of a TTSS. Assays that measure pore formation using semipurified or purified TTSS components and synthetic membranes have also been described[12–14] but have not been covered here because of space limitations. Many important issues remain to be addressed in this area. One of the most important issues is the source of the pores introduced by TTSSs. Are they the result of translocation channels that become exposed to the extracellular environment, or are they a result of a distinct process? The methods described here may also be applied to the discovery of pore-forming activities associated with other types of bacterial secretion systems. For example, Kirby *et al.* have obtained evidence for pore-forming activity in the type IV secretion system of *Legionella pneumophila.*[15]

[12] F. Tardy, F. Homble, C. Neyt, R. Wattiez, G. R. Cornelis, J. M. Ruysschaert, and M. Cabiaux, *EMBO J.* **18,** 6793 (1999).

[13] J. Lee, B. Klüsener, G. Tsiamis, C. Stevens, C. Neyt, A. P. Tampakaki, N. J. Panopoulos, J. Nöller, E. W. Weiler, G. R. Cornelis, J. W. Mansfield, and T. Nürnberger, *Proc. Natl. Acad. Sci. U.S.A.* **98,** 289 (2001).

[14] A. Holmström, J. Olsson, P. Cherepanov, E. Maier, R. Nordfelth, J. Pettersson, R. Benz, H. Wolf-Watz, and Å. Forsberg, *Mol. Microbiol.* **39,** 620 (2001).

[15] J. E. Kirby, J. P. Vogel, H. L. Andrews, and R. R. Isberg, *Mol. Microbiol.* **27,** 323 (1998).

[25] Interaction of Enteropathogenic *Escherichia coli* with Red Blood Cell Monolayers

By Stuart Knutton, Robert Shaw, and Gad Frankel

Introduction

Enteropathogenic *Escherichia coli* (EPEC), an important human infantile diarrheal pathogen, like numerous other animal and plant bacterial pathogens, employ a type III secretion system (TTSS) to inject virulence proteins directly into host cells.[1] The TTSS is a macromolecular protein complex which spans both bacterial membranes and has a short external needle structure through which proteins are secreted.[2] Pore-forming proteins secreted by the TTSS and inserted into the host cell membrane complete the secretion apparatus by producing a continuous pathway for protein translocation from the bacterial to the host cell cytosol.[1] Injection of proteins into host cells by *Shigella* and *Yersinia* species, has been correlated with their ability to cause a contact-dependent hemolysis of red blood cells (RBC), hemolysis in which the TTSS needle structure has to be brought into close contact

[1] C. J. Hueck, *Microbiol. Mol. Biol. Rev.* **62,** 379 (1998).

[2] A. Blocker, N. Jouihri, E. Larquet, P. Gounon, F. Ebel, C. Parsot, P. Sansonetti, and A. Allaoui, *Mol. Microbiol.* **39,** 652 (2001).

with the RBC membrane by centrifugation.[3,4] Contact-dependent hemolysis has also been reported for EPEC.[5] However, in addition to secreting pore-forming proteins, the EPEC TTSS also secrete EspA, a protein which is assembled into a filamentous structure that is believed to form a long extension to the TTSS needle.[6] Because it is these long EspA filaments that form the direct link between the bacterium and the host cell,[7] we predicted that close bacteria–host cell contact would not be required for EPEC to cause hemolysis and we confirmed this by demonstrating that EPEC cause total hemolysis of RBCs without the need for centrifugation.[8] RBCs are difficult to work with especially when quantitation is involved. So to overcome this problem and also to provide a system more suitable for microscopic examination of EPEC–RBC interaction, we developed an infection model based on the use of monolayers of RBCs attached to plastic or glass.[9]

Preparation of RBC Monolayers

RBCs have a negatively charged surface and so will bind to a positively charged substratum. RBCs will bind to tissue culture cell dishes but the attachment is not sufficiently strong to withstand the repeated washings required to perform bacterial infections. Consequently a stronger positively charged surface is required. This is achieved by coating the surface with poly(L-lysine). We found that small (~3 cm diameter) cell culture dishes are suitable for performing quantitative hemolysis assays; RBC monolayers formed on 10–13 mm diameter glass coverslips are suitable for microscopical studies.

Reagents

 Cell culture dishes (3 cm diameter)
 Washed glass coverslips placed in the wells of 24-well plates
 Phosphate-buffered saline (PBS)
 A 0.02% solution of poly(L-lysine) (70,000–150,000 molecular weight, Sigma, Poole, Dorset) in PBS
 Freshly drawn human blood
 HEPES-buffered Dulbecco's modified Eagle's medium (DMEM)(Sigma)

[3] P. Clerc, B. Baudry, and P. J. Sansonetti, *Ann. Institut Pasteur Microbiol.* **137A,** 267 (1986).
[4] S. Hakansson, K. Schesser, C. Persson, E. E. Galyov, R. Rosqvist, F. Homble, and H. Wolf-Watz, *EMBO J.* **15,** 5812 (1996).
[5] J. Warawa, B. B. Finlay, and B. Kenny, *Infect. Immun.* **67,** 5538 (1999).
[6] G. Frankel, A. D. Phillips, I. Rosenshine, G. Dougan, J. B. Kaper, and S. Knutton, *Mol. Microbiol.* **30,** 911 (1998).
[7] S. Knutton, I. Rosenshine, M. J. Pallen, I. Nisan, B. C. Neves, C. Bain, C. Wolff, G. Dougan, and G. Frankel, *EMBO J.* **17,** 2166 (1998).
[8] R. K. Shaw, S. Daniell, F. Ebel, G. Frankel, and S. Knutton, *Cell. Microbiol.* **3,** 213 (2001).
[9] T. Bachi, G. Eichenberger, and H. P. Hauri, *Virology* **85,** 518 (1978).

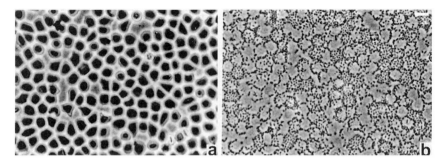

FIG. 1. Phase-contrast micrographs showing a confluent monolayer of intact RBCs (a) and a monolayer that has been incubated with wild-type EPEC for 4 hr (b). Note that all the RBCs have lysed and that numerous bacteria are bound to the resulting RBC ghost membranes. Bar: 5 μm.

Procedure

The procedure is performed at room temperature. Volumes used for all the incubations and washings are 2ml/cell culture dish and 1 ml/well of a 24-well plate. To provide a positively charged surface add the polylysine solution to the cell culture dishes/coverslips for 15 min. Meanwhile prepare a 3% suspension of washed RBCs by placing 2 ml blood in a microfuge tube and sedimenting at full speed for 1 min. Remove the plasma and white cells with a pipette and resuspend the RBCs in PBS. Wash 3× in PBS and finally resuspend the RBCs in PBS to make a 3% suspension (~20 ml from 2 ml blood). Remove the polylysine solution and wash the dishes/coverslips 3×, 1 min in PBS with gentle shaking. After the final wash cover the dishes/coverslips with the RBC suspension and allow the RBCs to settle and attach for 20 min. Wash 3×, 1 min in PBS to remove unattached RBCs. Replace the PBS with DMEM. Examination using phase optics and an inverted microscope should reveal a confluent monolayer of intact attached RBCs (Fig. 1a). Monolayers should be made and used immediately but can be stored for a short time at 4°. We routinely use human blood but we have tested sheep and horse blood and the method works effectively with these species.

Interaction of EPEC with RBC Monolayers

Once RBCs are attached to plastic/glass to form a monolayer, the interaction with EPEC can be examined in a manner identical to that used to examine EPEC interaction with epithelial cell monolayers such as Hep-2 cells in cell adhesion assays.[10]

[10] M. S. Donnenberg and J. P. Nataro, *Methods Enzymol.* **253**, 324 (1995).

Reagents

Luria broth
Stock EPEC culture
RBC monolayers in DMEM
PBS
Chemical fixatives

Procedure

The day before the experiment set up an overnight static broth culture of EPEC in Luria broth (1–2 ml). Add an aliquot of the bacterial culture (10 μl/ml) to each RBC monolayer and incubate the dishes/coverslips at 37° for up to 6 hr. At the end of the incubation period remove the DMEM, wash the monolayers 3×, 1 min. with PBS to remove nonadherent bacteria, and finally fix the cells with an appropriate fixative. Any fixation protocol can be used depending on the nature of the examination. For example, to examine the extent of hemolysis and bacterial adhesion microscopically monolayers are fixed in 4% formalin in PBS for 30 min and examined by phase contrast microscopy (Fig. 1b). Because lysed RBC ghosts are readily distinguished from intact RBCs, hemolysis can be quantified microscopically by counting a fixed number of cells in fields selected at random (e.g., 5 fields, 100 cells/field) and expressing hemolysis as a percentage (no. lysed cells/total no. cells × 100). However, hemolysis is more easily assessed by measuring hemoglobin release (see below). The extent of bacterial adhesion can also readily be determined by phase contrast microscopy (Fig. 1b). We quantified adhesion by counting the number of bacteria adhering to 100 cells and expressing adhesion as the mean number of bacteria adhering/RBC.[8] Although an inverted microscope is most convenient for examining cell culture dishes, it is worth pointing out that culture dishes can be examined using an upright microscope with nonimmersion lenses if most of the PBS covering the monolayer is removed and examination is performed quickly so as not to allow the cells to dry. Fixation in 4% formalin is also used for immunofluorescence studies whereas for electron microscopy 3% glutaraldehyde in 0.1 M phosphate buffer, pH 7.2–7.4 for 1 hr is appropriate.[11] Figure 2 illustrates EspA filament-mediated attachment of EPEC to RBCs visualized by immunofluorescence staining[7] (Fig. 2a) and by scanning electron microscopy[8] (Fig. 2b).

Quantitative Hemolysis Assay

A second quantitative method of measuring EPEC-induced hemolysis is to determine hemoglobin release. Most commercial cell culture media contains phenol

[11] S. Knutton, *Methods Enzymol.* **253**, 145 (1995).

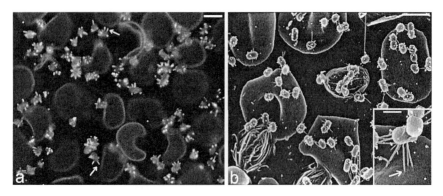

FIG. 2. Immunofluorescence (a) and scanning electron micrographs (b) following a 3 hr incubation of EPEC with RBC monolayers. Note that EspA filaments (arrows) promote attachment of EPEC to RBCs. Bars: (a,b) 1 μm; (inset) 0.2 μm.

red as a pH indicator. Because this interferes with the measurement of hemoglobin, it is important for this assay to be performed using DMEM without phenol red.

Reagents

> RBC monolayers on cell culture dishes in 2 ml DMEM without phenol red (Sigma)
> Overnight culture of EPEC in Luria broth

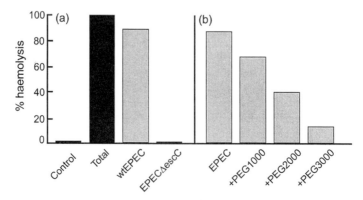

FIG. 3. (a) Hemolytic activity of wild-type EPEC and deletion mutant strain $\Delta escC$ which lacks a functional type III secretion system; (b) osmoprotection of hemolysis assessed by performing assays in the presence of 30 mM polyethylene glycol (PEG). Note the significant protection of hemolysis in the presence of >1000 molecular weight PEG.

Procedure

Each assay is performed in duplicate. Add 20 μl of the overnight EPEC broth culture to each RBC monolayer and incubate the dishes at 37° for 4 hr. We found that hemolysis reached maximal levels after 4 hr.[8] At the end of the incubation transfer the culture medium to a microfuge tube and sediment the bacteria at full speed in a microfuge for 1 min. Supernatants are monitored for hemoglobin release by measuring the optical density of the supernatants at 543 nm in a spectrophotometer. Supernatants from uninfected monolayers incubated under the same conditions are used to provide a baseline level of hemolysis (B); total hemolysis (T) is obtained from monolayers incubated with a 30-fold dilution of PBS in distilled water. Percentage hemolysis can then be calculated from the equation: $P = [(X-B)/(T-B)] \times 100$, where X is the optical density of the sample analyzed. This assay can be used to screen for the ability of wild-type and genetically modified EPEC strains to cause hemolysis. For example, EPEC strains lacking a functional TTSS are unable to induce hemolysis (Fig. 3a). The assay can also be used to assess the effect on hemolysis of adding different reagents to the culture medium. For example, we performed osmoprotection studies by adding different size sugar molecules to the DMEM; inhibition of hemolysis with high molecular weight sugars is consistent with the insertion of a hydrophilic pore into the RBC membrane.[12] (Fig. 3b).

Acknowledgment

This work was supported by the Wellcome Trust.

[12] A. Blocker, P. Gounon, E. Larquet, K. Niebuhr, V. Cabiaux, C. Parsot, and P. Sansonetti, *J. Cell Biol.* **147,** 683 (1999).

Section V

Bacterial Modification or Exploitation of Eukaryotic Signal Transduction

[26] GAP Activity of *Yersinia* YopE

By MARGARETA AILI, BENGT HALLBERG, HANS WOLF-WATZ, and
ROLAND ROSQVIST

Introduction

Yersinia pseudotuberculosis, a gram-negative enteropathogen, causes gastroin-
testinal infections in humans and animals that are characterized by diarrhea, abdominal pain, and fever. Infections in humans are usually self-limiting, but the
pathogen can cause lethal infections in rodents. Infection occurs mainly through
the oral route and the enteropathogenic *Yersinia* transit to the terminal ileum and
initiate infection by penetrating Peyer's patches. The bacteria can then colonize
the mesenteric lymph nodes and may spread to deeper tissues such as the spleen.
In rare cases, septicemia can occur.[1]

The human immune system has a highly efficient defense system that includes
an important role played by macrophages in phagocytosing invading microorganisms. However, *Y. pseudotuberculosis* and the other two pathogenic species of
Yersinia (Y. pestis and *Y. enterocolitica)* have the ability to avoid phagocytosis, remaining extracellular with the capacity to replicate and cause disease.[2] Avoidance
of host defenses is mediated by a type III secretion system (TTSS) encoded on a
70-kb virulence plasmid common to all virulent *Yersinia* strains.[2] Consisting of
approximately 20 proteins, the secretion apparatus exports several *Yersinia* outer
proteins (Yops) across the bacterial double membrane in one step.[3]

Subsequently, antihost Yop effector proteins are translocated directly into the
host cell in order to block phagocytosis and inhibit immune system activation.[4]
Thus, proteins secreted by the TTSS can be divided in two functionally distinct
groups. One group constitutes the translocated virulence effectors, each with a
specific eukaryotic target inside the host cell. The second group of proteins are
directly involved either in translocation of the effector proteins across the target
cell membrane or in regulation of the process.

The TTSS is induced *in vivo* by target cell contact.[5] This response can be
induced *in vitro* by the environmental signals, temperature, and Ca^{2+}. Maximal expression of Yops is obtained when bacteria are grown in Ca^{2+}-depleted

---

[1] T. L. Cover and R. C. Aber, *N. Eng. J. Med.* **321,** 16 (1989).

[2] G. R. Cornelis and H. Wolf-Watz, *Mol. Microbiol.* **23,** 861 (1997).

[3] C. J. Hueck, *Microbiol. Mol. Biol. Rev.* **62,** 379 (1998).

[4] M. Aepfelbacher, R. Zumbihl, K. Ruckdeschel, C. A. Jacobi, C. Barz, and J. H. Heesemann, *Biol. Chem.* **380,** 795 (1999).

[5] J. Pettersson, R. Nordfelth, E. Dubinina, T. Bergman, M. Gustafsson, K. E. Magnusson, and H. Wolf-Watz, *Science* **273,** 1231 (1996).

medium at 37°. These conditions induce abundant Yop secretion into the culture medium.[6]

To remain extracellular during infection, *Yersinia* efficiently blocks phagocytosis by macrophages. The translocated YopH and YopE effectors mediate this resistance to phagocytosis.[7] YopH is a tyrosine phosphatase responsible for dephosphorylating focal adhesion kinase (FAK) and p130^{CAS}.[8] YopE functions to disrupt actin stress fibers in infected HeLa cells. This causes these cells to round up, generating tail-like retractions of the plasma membrane.[9,10] The fact that *Y. pseudotuberculosis* strains lacking either YopH or YopE are avirulent in the mouse model indicates their importance for extracellular bacterial survival during a host infection.

Several bacterial toxins have been shown to act as GTPase activating proteins (GAPs) toward members of the small Rho GTPase family.[11–14] The Rho subfamily of small GTPases are important regulators of the actin cytoskeleton in mammalian cells. In particular, activated Rho induces stress fibers, while Rac activation leads to lamellipodia and membrane ruffle production and Cdc42 activation results in filopodia production.[15] In their active state, the Rho GTPases are bound to GTP. Inactivation occurs on hydrolysis of GTP to GDP. GAPs function to enhance the intrinsic hydrolysis of Rho GTPases resulting in their rapid inactivation. Conversely, guanine nucleotide exchange factors (GEFs) are GAP antagonists, which functions to enhance the exchange rate of GDP to GTP on activation.[16]

Common to all eukaryotic GAP proteins is the so-called arginine finger motif, which is essential for GAP activity and defines a GAP protein.[17] In recent years, the bacterial toxins SptP of *Salmonella,* ExoS of *Pseudomonas,* and YopE of *Yersinia* have been identified as GAPs.[11–14] In addition, YopE shares a high degree of homology to the N-terminal part of SptP and ExoS. However, bacterial and eukaryotic GAPs display no obvious structural similarities, except for their defining arginine

[6] T. Michiels, P. Wattiau, R. Brasseur, J. M. Ruysschaert, and G. Cornelis, *Infect. Immun.* **58,** 2840 (1990).

[7] R. Rosqvist, Å. Forsberg, M. Rimpiläinen, T. Bergman, and H. Wolf-Watz, *Mol. Microbiol.* **4,** 657 (1990).

[8] C. Persson, N. Carballeira, H. Wolf-Watz, and M. Fällman, *EMBO J.* **16,** 2307 (1997).

[9] R. Rosqvist, Å. Forsberg, and H. Wolf-Watz, *Infect. Immun.* **59,** 4562 (1991).

[10] R. Rosqvist, K.-E. Magnusson, and H. Wolf-Watz, *EMBO J.* **13,** 964 (1994).

[11] Y. Fu and J. E. Galan, *Nature* **401,** 293 (1999).

[12] U. M. Goehring, G. Schmidt, K. J. Pederson, K. Aktories, and J. Barbieri, *J. Biol. Chem.* **274,** 36369 (1999).

[13] U. von Pavel-Rammingen, M. V. Telepnev, G. Schmidt, K. Aktories, H. Wolf-Watz, and R. Rosqvist, *Mol. Microbiol.* **36,** 1 (2000).

[14] D. S. Black and J. B. Bliska, *Mol. Microbiol.* **37,** 515 (2000).

[15] A. Hall, *Science* **279,** 509 (1998).

[16] L. Van Aelst and C. D'Souza-Schorey, *Genes Dev.* **11,** 2295 (1997).

[17] K. Scheffzek, M. R. Ahmadian, and A. Wittinghofer, *Trends Biochem. Sci.* **23,** 257 (1998).

finger motif.[18,19] Therefore, in order for infecting bacteria to evade host defenses and establish colonization these bacterial toxins have mimicked the mechanism by which eukaryotic GAPs down-regulate small Rho GTPases.

We have shown that YopE has an *in vitro* GAP activity toward members of the small Rho GTPase family. YopE increases the rate of GTP hydrolysis, thereby inactivating the Rho GTPases.[13] This is consistent with the ability of translocated YopE to cause depolymerization of the actin cytoskeleton. This prevents phagocytosis of the bacteria promoting their extracellular replication. Furthermore, exchange of the essential arginine at amino acid position 144 with alanine abolished both the YopE-dependent cytotoxicity on infected HeLa cells and the *in vitro* GAP activity.[13] It follows that YopE GAP activity is required for antiphagocytic activity.[14]

This article reviews the methodology used to isolate functional YopE protein and to analyze the role of the YopE GAP activity during infection.

Expression and Purification of YopE

Yersinia grown at 37° in a Ca^{2+}-depleted medium normally result in maximal Yop expression and abundant secretion into the culture medium. This can be further exploited using a multiple Yop mutant (MYM) of *Y. pseudotuberculosis* that is defective in several Yop effectors (YopH, YopE, YopM, YopK, and YpkA) but still retains an intact regulatory circuit and secretory machinery. In particular, expression *in trans* of *yop* genes encoded on a multicopy plasmid results in specific overproduction and secretion of the corresponding protein.[8] We have taken advantage of this approach to produce large amounts of secreted YopE.

However, the solubility of secreted full-length YopE is poor, which makes purification very difficult. Because the N-terminal part of YopE causes aggregation, we considered that removal of this domain would increase solubility. Therefore, a His tag and enterokinase cleavage site are introduced between amino acids 89 and 90 of wild-type YopE. This modified YopE retains normal expression, secretion, and subsequent translocation into target cells.[13]

To purify YopE, the His-tagged variant of wt-YopE and YopE(R144A), respectively, is expressed from pUC19 under the control of its native promoter in the *Yersinia* MYM strain. An overnight culture grown in TMH medium[20] is subcultured to an OD_{600} of 0.1 in 50 ml TMH medium and grown for 30 min at 26° prior to growth at 37° for 8 hr. After 3 hr incubation at 37°, secreted YopE starts to aggregate, producing easily visible large protein filaments in the culture medium. After 8 hr the bacterial culture is drawn through a 20-μm nylon filter (type NY20,

[18] C. E. Stebbins and J. E. Galan, *Mol. Cell* **6**, 1449 (2000).
[19] M. Würtele, E. Wolf, K. J. Pederson, G. Buchwald, M. R. Ahmadian, J. T. Barbieri, and A. Wittinghofer, *Nat. Struct. Biol.* **8**, 23 (2001).
[20] S. Straley and W. S. Bowmer, *Infect. Immun.* **51**, 445 (1986).

Millipore, Bedford, MA) to collect the protein filaments. The filter is then washed three times with sterile water to remove bacteria before the protein filaments are completely dissolved in 3 M guanidine hydrochloride. The solubilized His-tagged YopE protein is incubated for 1 hr at room temperature with Ni-NTA agarose beads (Qiagen, Hilden, Germany) in the presence of 3 M guanidine hydrochloride. YopE bound to the beads is packed into a column and washed with 2 volumes each of decreasing concentrations of guanidine hydrochloride (3, 2, and 1 M, respectively), and finally with 10 volumes of enterokinase cleavage buffer (50 mM Tris-HCl, pH 8.0, 1 mM CaCl$_2$). Although YopE aggregates in the column under these conditions, the N-terminal 89 residues of YopE containing the His tag are still accessible to enzymatic digestion being cleaved with 0.3 units of the enterokinase EKMax (Invitrogen, Carlsbad, CA) per 20 μg crude protein for 16 hr at 37°. The enterokinase is then removed with one bed volume of 3 M guanidine hydrochloride. The cleaved C-terminal portion of YopE (YopE90-219) is resolubilized in 3 M guanidine hydrochloride for 10 min at room temperature prior to elution from the column. Fractions containing YopE(90-219) are pooled and dialyzed against PBS-A buffer and then concentrated, using either an Amicon (Danvers, MA) ultrafiltration device equipped with a PM10 filter or an Amicon spin column (Centricon YM10). The protein concentration is determined with Bradford reagent, using BSA as the protein standard. The purity of eluted protein is analyzed on 12% SDS–PAGE with Coomassie Brilliant Blue staining, silver staining, and by enhanced chemiluminescence (ECL). Western blots are analyzed with anti-YopE antiserum, antitotal Yops antiserum, and anti-His antibodies. No detectable contaminants are observed.[13]

GTPase Assay

Recombinant Rho GTPases are produced and purified as described.[21] Rho GTPases are loaded with [γ-^{32}P]GTP (0.88 μCi per reaction) for 5 min at 37° in loading buffer [50 mM HEPES, pH 7.3, 5 mM EDTA, 5 mg/ml bovine serum albumin(BSA)]. Cold GTP, 0.2 mM, and hydrolysis buffer [50 mM HEPES, pH 7.3, 10 mM MgCl$_2$, 1 mM dithiothreitol(DTT), 100 mM KCl, 0.1 mg/ml BSA] are added. Purified YopE(90-219) protein is added in decreasing concentrations to 25 μl of the reaction mixture containing 11 nM Rho GTPases. Incubation is performed at 16° for Cdc42 and Rac1, and 20° for RhoA. GTPase hydrolysis is stopped after 20 min by the addition of 1 ml cold stop solution (50 mM HEPES pH 7.3, 20 mM MgCl$_2$, 1 mM DTT, 10 μg/ml BSA, 0.1 mM cold GTP). The GTPase activity is subsequently analyzed by a filter-binding assay.[22] Essentially, the samples are loaded onto nitrocellulose filters [0.45 μm, type HA (Millipore)], previously

[21] A. J. Self and A. Hall, *Methods Enzymol.* **256**, 3 (1995).
[22] A. J. Self and A. Hall, *Methods Enzymol.* **256**, 67 (1995).

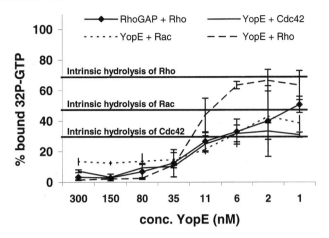

FIG. 1. Concentration-dependent *in vitro* GAP activity of purified wild-type YopE(90-219) towards RhoA, Rac1, and Cdc42 were loaded with $[\gamma\text{-}^{32}P]GTP$ for 5 min at 37°. For measurement of GTPases hydrolysis, increasing amounts of YopE were added to 11 nM RhoA (– – –), Rac1 (- - - -), and Cdc42 (——) and incubated for 20 min at 16° (Rac1 and Cdc42) and 20° (RhoA). The GAP activity of recombinant RhoGAP was measured by adding increasing amounts of RhoGAP to 11 m*M* RhoA (–◆–) and incubated for 20 min at 20°. GTPase activity was analyzed by filter binding assay as described in the text. The remaining $[\gamma\text{-}32P]$ GTP bound to RhoA, Rac1, and Cdc42 is shown as a percentage of loaded RhoA, Rac1, and Cdc42 which was kept on ice during the analysis. Each point represents the mean of three independent experiments in duplicate samples.

equilibrated in PBS-A containing 20 m*M* MgCl$_2$ and placed on a vacuum manifold (Pharmacia Biotech, Piscataway, NJ). The filters are washed 3 times with 5 ml ice-cold PBS-A containing 20 m*M* MgCl$_2$ and transferred to scintillation tubes containing 2 ml scintillation liquid (OptiPhase "HiSafe" 3, Wallac, Turku, Finland) and counted (Wallac).

To measure the intrinsic GTPase activity of Rho GTPases, one reaction is maintained on ice, while the other is incubated at the reaction temperature. The difference between reaction on ice versus reaction temperature is given as the intrinsic hydrolysis of measured GTPase (Fig. 1). The activity of the GAP protein is compared against the intrinsic hydrolysis. In all assays 100 n*M* of RhoGAP is used as the positive control.

Purified truncated YopE(90-219) shows an *in vitro* GAP activity toward all three GTPases tested. Furthermore, the results also show that a lower concentration of YopE is required to inactivate Rac than Rho or Cdc42 (Fig. 1). This indicates that *in vitro,* Rac is more susceptible to the YopE GAP activity than Rho and Cdc42.

To confirm that YopE is a GAP protein toward small Rho GTPases, the conserved arginine at position 144 is mutated to an alanine. The resulting YopE(90–219) (R144A) mutant protein is purified and tested for its ability to stimulate GTP hydrolysis under the same conditions as the purified wild-type YopE(90–219) protein.

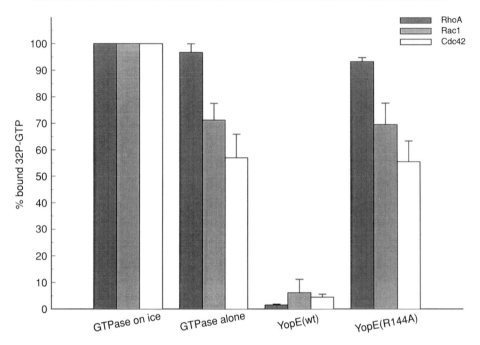

FIG. 2. Purified YopE(90-219)(R144A) has impaired GAP activity toward RhoA, Rac1, and Cdc42. Recombinant RhoA, Rac1, and Cdc42 were loaded with $[\gamma-^{32}P]GTP$ for 5 min at $37°$. For measurement of GTP hydrolysis, purified YopE (35 nM) or YopE(R144A) (35 nM) respectively, were added to 11 nM Rac1 and incubated for 20 min at $16°$. For measurement of GTP hydrolysis, purified YopE (80 nM) or YopE(R144A) (80 nM), respectively, were added to 11 nM Cdc42 or 11 nM RhoA and incubated for 20 min at $16°$ (Cdc42) or $20°$ (RhoA). As intrinsic hydrolysis control, loaded GTPases were incubated at the reaction temperature for 20 min without addition of any YopE proteins. The remaining $[\gamma-^{32}P]GTP$ is shown bound to RhoA, Rac1, and Cdc42 as a percentage of the loaded GTPases, which were kept on ice during the analysis (GTPase on ice). Each point represents the mean of three independent experiments in duplicate samples.

As expected, the addition of YopE(90–219)(R144A) has no effect on the GTP hydrolysis (Fig. 2).

Even though the recombinant YopE is capable of inactivating RhoGTPases, we wondered whether secreted full-length YopE shows the same specific activity. To test this, full-length YopE and the YopE(R144A) mutant are overproduced *in trans* in the MYM strain. The strains are grown overnight at $26°$ in brain heart infusion medium, (BHI) then subcultured to 0.1 OD_{600} in 2 ml fresh BHI and grown for 30 min at $26°$ prior to 4 hr incubation at $37°$. These conditions produce an optimal amount of soluble YopE protein, since no aggregated protein filaments are detected. Although prolonged incubation of the cultures increases the total amount of secreted YopE, a notable increase in aggregation decreases the amount of soluble YopE. Cleared culture supernatants are collected after two times 5 min centrifugation at 13,000g. The supernatant is analyzed for the presence of YopE by Western blotting. Known concentrations of YopE are compared to serial dilutions of the supernatant to provide an estimate of the YopE concentration. The YopE-containing supernatant is analyzed for *in vitro* GAP activity in the same way as the purified recombinant YopE(90-219). As a negative control, we use supernatant collected from the parental MYM strain (Fig. 3).

As expected, the secreted full-length YopE possesses an *in vitro* GAP activity toward RhoA, Rac1, and Cdc42, whereas the YopE(R144A) mutant and supernatant collected from the MYM control strain are completely devoid of any GAP

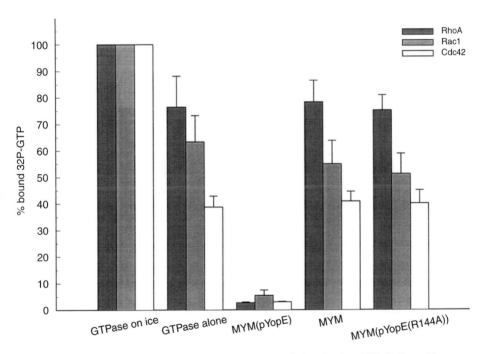

FIG. 3. Secreted wild-type YopE stimulated GAP activity of RhoA, Rac1, and Cdc42. Recombinant RhoA, Rac1, and Cdc42 were loaded with [γ-^{32}P]GTP for 5 min at 37°. For measurement of GTP hydrolysis, supernatants from MYM(pYopE), MYM, or MYM(pYopE(R144A)), respectively, were added to 11 nM Rac1, Cdc42, or RhoA, respectively, and incubated for 20 min at 16° (Rac1 and Cdc42) or 20° (RhoA). As intrinsic hydrolysis control, loaded GTPases were incubated at the reaction temperature for 20 min without addition of any YopE proteins. Shown is the remaining [γ-^{32}P]GTP bound to RhoA, Rac1, and Cdc42 as a percentage of the loaded GTPases, which had been kept on ice during the analysis (GTPase on ice). Each point represents the mean of three independent experiments in duplicate samples.

activity. Moreover, the specific activity of the full-length YopE is comparable to that of truncated YopE(90–219). Thus, this suggests that the first 89 N-terminal residues of YopE do not contribute to the GAP activity of the protein.

Allelic Exchange of Native *yopE* Gene with *yopE(R144A)* Mutation

The regulation of Yop expression and secretion in *Yersinia* is complex and involves both positive and negative elements. Furthermore, it has been observed that changes in the expression pattern of certain Yop proteins result in an altered expression level of the other Yops. This is true for overexpression of YopE, which reduces the expression of YopB, YopD, and YopH. Thus, to elucidate the importance of the YopE GAP activity during *in vivo* conditions we have exchanged the native *yopE* gene on the virulence plasmid with a 440 bp DNA fragment containing the R144A mutation. The 440 bp DNA fragment is cloned into the *Sph*I–*Xba*I-digested suicide mutagenesis vector, pDM4.[23] The subsequent plasmid pMA18 is maintained in *Escherichia coli* SY327. However, *E. coli* S17-1 is used as the donor strain in conjugal mating experiments to recombine pMA18 into the virulence plasmid of *Y. pseudotuberculosis* YPIII(pIB102).[24,25] Primary single crossover events of the suicide plasmid within the complementary *yopE* gene of pIB102 are selected using the chloramphenicol resistance (Cm^R) marker located on the suicide plasmid and growth on *Yersinia* selective agar base (Difco, Detroit, MI) containing kanamycin (50 μg/ml). To complete the allelic exchange, the integrated suicide plasmid is forced to recombine out of the chromosome using *sacB*-dependent sucrose sensitivity.[23] Three positive clones are selected and subjected to DNA sequence analysis. One of the clones having the *yopE(R144R)* mutation is selected for further work and denoted YPIII(pIB562).

Activity of YopE on HeLa Cells

In our hands, the most sensitive assay to analyze the *in vivo* activity of YopE is to study the effect of YopE on cultured HeLa cells.[9] Subconfluent monolayers of HeLa cells infected with the wild-type strain of *Y. pseudotuberculosis* show changed cell morphology over time (Fig. 4; Ref. 9). The cells round up, generating tail-like retractions of the plasma membrane on prolonged incubation (16 hr) that eventually lead to detachment of the HeLa cells from the surface. This change in morphology is dependent on the presence of translocated YopE in the cytosol of the eukaryotic cell.[7,9,10] In addition, YopE that is mechanically introduced into HeLa cells by the bead-loading technique induces a similar cytotoxic effect.[9] Using this approach, we observed that purified truncated YopE(90-219) was also

[23] D. L. Milton, R. O'Toole, P. Hörstedt, and H. Wolf-Watz, *J. Bacteriol.* **178,** 1310 (1996).
[24] M. Rimpiläinen, Å. Forsberg, and H. Wolf-Watz, *J. Bacteriol.* **174,** 3355 (1992).
[25] T. Bergman, K. Erickson, E. Galyov, C. Persson, and H. Wolf-Watz, *J. Bacteriol.* **176,** 2619 (1994).

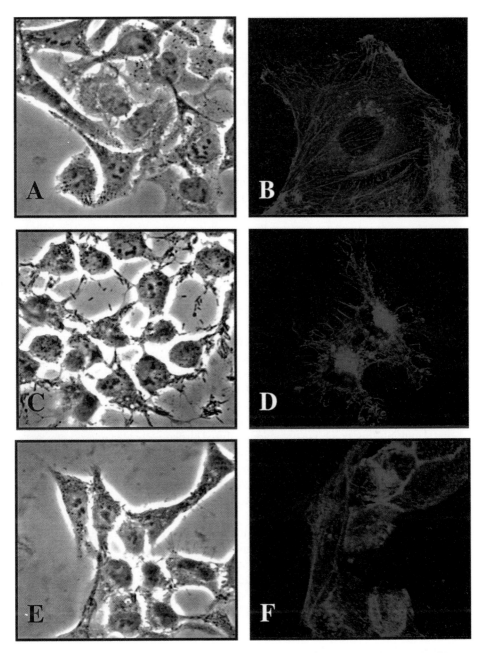

FIG. 4. YopE induced cytotoxicity in cultured HeLa cells. HeLa cells were infected with wild-type *Y. pseudotuberculosis* YPIII(pIB102) (C, D), and the YopE(R144A) mutant YPIII(pIB562) (E, F) with uninfected cells as a control (A, B). Phase contrast images (A, C, E) and actin cytoskeleton structures (B, D, F) visualized 1 hr postinfection. The actin cytoskeleton was visualized by rhodamine–phalloidin staining. Note the clearly visible retraction tails of HeLa cells infected with the wild-type strain YPIII(pIB102).

able to mediate a cytotoxic response on HeLa cells.[13] The finding that bead-loaded YopE resulted in a cytotoxic effect raises the question of whether naturally translocated variants of YopE also show a similar cytotoxic effect. In particular, can YopE(R144A), mutated in the arginine finger, induce a cytotoxic response?

HeLa cell cultures are routinely maintained in tissue culture flasks (Falcon) containing MEM (minimum essential medium, Eagle's with Earle's salt) medium supplemented with 10% heat-inactivated fetal calf serum (FCSHI), 2 mM glutamine, and 100 IU penicillin/ml and incubated at 37° in an atmosphere of 5% (v/v) CO_2 in air. The day before infection, HeLa cells are washed in PBSA and trypsinated [0.05% (w/v) trypsin and 0.02% (w/v) versenate] before the cell density is determined microscopically using a Büchner chamber. The cells are then seeded at a density of 1.6×10^5 cells per well in a 24-well tissue culture plate (15 mm diameter, Nunc, Denmark). Each well is supplied with 1 ml of medium and incubated overnight at 37°. For microscopic visualization of YopE-dependent cytotoxicity, cells are seeded on 12-mm glass coverslips placed in the tissue culture wells.

Before infection of the monolayers, penicillin is removed by washing twice with 1 ml of PBSA followed by the addition of 1 ml fresh antibiotic-free MEM containing 10% FCSHI. Overnight cultures of bacteria are diluted 1/300 in 3 ml MEM containing 10% FCSHI, then preincubated for 30 min at 26°, followed by 1 hr at 37°. To each HeLa cell monolayer, 100 μl of the bacterial suspension is added to give a multiplicity of infection of about 10 bacteria per HeLa cell. The tissue culture plates are centrifuged at 24° for 5 min at 400g to synchronize infection and facilitate contact between the bacteria and the HeLa cells. Infected cells are incubated at 37° and at multiple time points postinfection analyzed by phase-contrast microscopy. A changed morphology is visualized by rounding up of the HeLa cells and indicates a cytotoxic response (Fig. 4). At 45 min and 2 hr postinfection, the HeLa cells are washed twice by immersion in PBSA and then fixed in 2% PFA for 20 min. Following fixation, the coverslips are washed twice in PBSA before examination by phase-contrast microscopy.

Wild-type *Yersinia*, YPIII(pIB102), induces a cytotoxic effect within 45 min, visualized as rounded cells exhibiting membranous retractions on the coverslip (Fig. 4). Interestingly, the *yopE* null mutant strain YPIII(pIB522) and the GAP inactive mutant YPIII(pIB562) are unable to induce a cytotoxic response on HeLa cells (Fig. 4). This suggests that GAP activity is essential for YopE-dependent cytotoxicity on cultured HeLa cells *in vivo*.

Effect of YopE on HeLa Cell Cytoskeleton

The HeLa cell cytotoxic assay is a very sensitive assay for detecting YopE activity inside eukaryotic cells. However, change in cell morphology can only be detected about 30 min postinfection when using a multiplicity of infection of 10.

Because YopE activity must be rapid to block phagocytosis, we have investigated the YopE effect on the actin cytoskeleton at earlier time points. Cell monolayers are cultured and prepared as described above. The HeLa cells grown on coverslips are infected with different strains of *Yersinia* and fixed 15 and 60 min postinfection, respectively. The cells are permeabilized and the actin cytoskeleton stained with TRITC-conjugated phalloidin (Molecular Probes, Eugene, OR) for 30 min and mounted on microscope slides in a mounting medium containing Cityfluor as an antifading agent. The actin microfilament structure of the infected HeLa cells is analyzed by confocal microscopy.

As expected, wild-type *Yersinia* causes depolymerization of the actin cytoskeleton as early as 15 min postinfection. Moreover, at 1 hr postinfection, complete depolymerization has occurred and the F-actin is visualized as condensed aggregates (Fig. 4). The fact that cells are well rounded and leaving retraction tails is consistent with a YopE-induced cytotoxic response. On the other hand, *Yersinia* expressing YopE(R144A), YPIII(pIB562), is unable to induce actin cytoskeleton depolymerization even at 1 hr postinfection (Fig. 4). This is in good correlation with the observed noncytotoxic phenotype for this strain. These results suggest that YopE GAP activity is required for actin depolymerization and for the cytotoxic phenotype during HeLa cell infections.

YopE GAP Activity Essential for Virulence in Mice

Unlike *Y. pestis*, *Y. pseudotuberculosis* does not cause systemic infections in humans. In mice, however, *Y. pseudotuberculosis* cause symptoms that reflect *Y. pestis* infection in humans. Therefore, the mouse model of yersiniosis serves to evaluate the involvement of potential virulence factors in *Yersinia* infections.

TABLE I
VIRULENCE OF DIFFERENT *Yersinia* STRAINS IN MICE[a]

Strain	LD_{50}	Cytotoxic effect
YPIII(pIB102) (wt)	2.4×10^3	+++
YPIII(pIB522) (*yopE*)	2.6×10^7	—
YPIII(pIB562) (*yopE(R144A)*)	5.1×10^6	—

[a] C57 black mice were challenged by intraperitoneal injection with increasing doses of wild-type *Y. pseudotuberculosis* YPIII(pIB102), the *yopE* null mutant YPIII(pIB522), and the *yopE(R144A)* mutant YPIII(pIB562). LD_{50} values were calculated accordingly to the method of L. J. Reed and H. Muench, *Am. J. Hyg.* **27,** 493 (1938). Cytotoxic effect after infection of HeLa cells evaluated 2 hr postinfection. (+++) Complete depolymerization of the actin cytoskeleton; (—) no visible effect on the actin cytoskeleton.

Mice are challenged with intraperitoneal bacterial infections and the LD_{50} value is calculated for each strain. Overnight cultures of *Yersinia*, grown at 26°, are diluted in PBSA in 10-fold serial steps. Mice are challenged with bacterial concentrations of 10^3, 10^5, and 10^7. The mice are monitored daily and the LD_{50} value established accordingly to the method used in Ref. 26.

The wild-type strain YPIII(pIB102) shows full virulence in infected mice while the *yopE* null mutant, YPIII(pIB522) is avirulent. Likewise, all mice infected with the strain expressing YopE(R144A), (YPIII(pIB562), survive. Thus, the YopE GAP activity is essential for systemic infections in mice (Table I).

Acknowledgment

We thank Dr. Matthew Francis for critical review of the manuscript. This work was supported by grants from the Swedish Medical Research Council, the Swedish Foundation of Strategic Research, the Swedish Cancer Society, and the Kempe Foundation.

[26] L. J. Reed and H. Muench, *Am. J. Hyg.* **27,** 493 (1938).

[27] Tyrosine Phosphorylation of Eukaryotic Proteins and Translocated Intimin Receptor by Enteropathogenic *Escherichia coli*

By REBEKAH DE VINNEY

Introduction

Bacterial pathogens have profound effects on a number of host signaling pathways, many of which involve the modulation of protein tyrosine phosphorylation. One well-characterized example is the interaction of the human gram-negative pathogen enteropathogenic *Escherichia coli* (EPEC) with epithelial and macrophage cell lines. EPEC is a major cause of neonatal diarrhea worldwide.[1] EPEC adheres tightly to the surface of intestinal epithelial cells, which results in a dramatic remodeling of the host actin cytoskeleton, and the formation of actin pedestals beneath the adhering bacteria.[2] EPEC uses a novel mechanism to modulate the host actin cytoskeleton: it inserts its own receptor for adherence, called

[1] J. Nataro and J. Kaper, *Clin. Microbiol. Rev.* **11,** 142 (1998).
[2] H. W. Moon, S. C. Whipp, R. A. Argenzio, M. M. Levine, and R. A. Giannella, *Infect. Immun.* **41,** 1340 (1983).

Tir (translocated intimin receptor) into the host cell membrane, where it becomes tyrosine phosphorylated, binds a bacterial outer membrane ligand called intimin, and becomes the centerpiece for pedestal formation.[3] Tyrosine phosphorylated Tir is found at the tip of the EPEC pedestal, directly beneath the adherent bacteria.[3] Initially Tir was believed to be a host membrane protein called Hp90 that became tyrosine phosphorylated in response to EPEC infection.[4] Tir delivery requires the bacterial type III secretion system, and the EPEC secreted proteins EspA, EspB, and EspD, which are believed to be part of the type III translocation apparatus.[3,5–7] Once inserted into the host plasma membrane, Tir is an integral membrane protein, with the amino and carboxy termini located intracellularly, and an extracellular intimin binding domain.[3,8,9] Tir is phosphorylated on tyrosine-474, an event which is essential for pedestal formation.[10,11] Additionally, EPEC Tir is serine/threonine phosphorylated, although the function of these events is not known.

Tir as produced by EPEC migrates as a 72/78 kDa doublet on SDS–PAGE, which decreases in mobility to a 90 kDa polypeptide on insertion into the host cell membrane and phosphorylation.[3] Studies using site-directed mutagenesis have demonstrated that the mobility shift induced by Tir phosphorylation is due exclusively to serine/threonine phosphorylation.[10] A non-tyrosine-phosphorylated Tir Y474F mutant, which is still serine/threonine phosphorylated, migrates at 90 kDa. The kinetics of Tir tyrosine phosphorylation parallel that of its delivery into the host cell, which is evident as early as 90 min postinfection.[3,12] Ligand occupancy is not required for Tir tyrosine phosphorylation: strains that deliver Tir in the absence of intimin are able to stimulate Tir tyrosine phosphorylation.[4]

In addition to Tir tyrosine phosphorylation, EPEC infection stimulates both the tyrosine phosphorylation and dephosphorylation of a number of host cell proteins. In the same study that identified Hp90, Rosenshine and colleagues identified two additional epithelial cell proteins, Hp39 and Hp72, that were tyrosine phosphorylated in response to EPEC infection.[12] The kinetics of Hp39 and Hp72

[3] B. Kenny, R. DeVinney, M. Stein, D. J. Reinscheid, E. A. Frey, and B. B. Finlay, Cell **91**, 511 (1997).

[4] I. Rosenshine, S. Ruschkowski, M. Stein, D. Reinscheid, S. Mills, and B. Finlay, EMBO J. **15**, 2613 (1996).

[5] C. Wachter, C. Beinke, M. Mattes, and M. Schmidt, Mol. Microbiol. **31**, 1695 (1999).

[6] K. Taylor, C. O'Connell, P. Luther, and M. Donnenberg, Infect. Immun. **66**, 5501 (1998).

[7] C. Wolff, I. Nisan, E. Hanski, G. Frankel, and I. Rosenshine, Mol. Microbiol. **28**, 143 (1998).

[8] M. de Grado, A. Abe, A. Gauthier, O. Steele-Mortimer, R. DeVinney, and B. B. Finlay, Cell. Microbiol. **1**, 7 (1999).

[9] E. L. Hartland, M. Batchelor, R. M. Delahay, C. Hale, S. Matthews, G. Dougan, S. Knutton, I. Connerton, and G. Frankel, Mol. Microbiol. **32**, 151 (1999).

[10] B. Kenny, Mol. Microbiol. **31**, 1229 (1999).

[11] D. L. Goosney, R. DeVinney, R. A. Pfuetzner, E. A. Frey, N. C. Strynadka, and B. B. Finlay, Curr. Biol. **10**, 735 (2000).

[12] I. Rosenshine, M. S. Donnenberg, J. B. Kaper, and B. B. Finlay, EMBO J. **11**, 3551 (1992).

phosphorylation are similar to that observed for Tir. Both proteins may be associated with the host actin cytoskeleton, as they are not solubilized by Triton X-100. To date, the identities of Hp39 and Hp72 are not known. Effects on epithelial cell protein tyrosine phosphorylation occurring later in the infection time course were examined by Kenny and Finlay.[13] This resulted in the identification of a complex of proteins migrating at 150 kDa (Hp150) that were tyrosine phosphorylated in response to EPEC infection and found in the cytoplasmic and membrane-microsomal fractions. Hp150 tyrosine phosphorylation was first evident after 120 min infection (compared to 90 min for Tir), and was dependent on both intimin and the type III secretion/translocation apparatus. One member of this complex has been identified as phospholipase Cγ1 (PLCγ1), which catalyzes the hydrolysis of phosphatidylinositol 4,5-bisphosphate (PIP$_2$) into inositol trisphosphate (IP$_3$) and diacylglycerol (DAG). This may have functional significance as EPEC has been shown to modulate IP$_3$ levels in epithelial cells.[14]

EPEC infection also results in protein tyrosine dephosphorylation. In epithelial cells, EPEC infection stimulates the tyrosine dephosphorylation of a 240 kDa cytoplasmic protein and a group of 120–140 kDa cytoplasmic (series I) and membrane-microsomal (series II) proteins.[13] Tyrosine dephosphorylation occurs late in infection and initially becomes evident after 120 min. The identities of these proteins and their role in EPEC pathogenesis is unknown.

EPEC can prevent its own uptake into macrophage cell lines. This antiphagocytic effect requires the tyrosine dephosphorylation of several macrophage proteins.[15] These include proteins migrating at 50, 60, 100, and 110 kDa. If protein dephosphorylation is blocked with tyrosine phosphatase inhibitors, EPEC is phagocytosed normally. Further studies have revealed that antiphagocytosis requires phosphatidylinositol 3-kinase (PI-3K) activity, and that the tyrosine phosphorylation of proteins associating with PI-3K is blocked.[16] The identity of the macrophage proteins dephosphorylated in response to EPEC infection is not known.

In addition to Tir and its effect on host protein tyrosine phosphorylation, EPEC expresses several proteins that are substrates for protein tyrosine phosphorylation, although there is no evidence that these proteins are injected into the host cell. Rosenshine and colleagues identified three tyrosine phosphorylated EPEC proteins, Ep30, Ep51, and Ep85.[12] Further studies identified Ep85 as an EPEC-encoded tyrosine kinase, located in the bacterial inner membrane.[17] Ep85 is autophosphorylated on tyrosine and can tyrosine phosphorylate exogenous substrates as well. A second EPEC tyrosine phosphorylated protein, the 85 kDa BipA, was

[13] B. Kenny and B. Finlay, *Infect. Immun.* **65**, 2528 (1997).
[14] V. Foubister, I. Rosenshine, and B. B. Finlay, *J. Exp. Med.* **179**, 993 (1994).
[15] D. L. Goosney, J. Celli, B. Kenny, and B. B. Finlay, *Infect. Immun.* **67**, 490 (1999).
[16] J. Celli, M. Olivier, and B. B. Finlay, *EMBO J.* **20**, 1245 (2001).
[17] O. Ilan, Y. Bloch, G. Frankel, H. Ullrich, K. Geider, and I. Rosenshine, *EMBO J.* **18**, 3241 (1999).

identified by Farris and colleagues.[18] BipA is autophosphorylated on tyrosine and has GTPase activity; a *bipA* knockout mutant is unable to form pedestals within the host cell.

The following protocols describe methods that are used to detect tyrosine phosphorylated EPEC Tir and to examine the tyrosine phosphorylation profile of host proteins in response to EPEC infection. These include methods for infecting both epithelial and macrophage cell lines, preparation of bacterial contamination-free cell extracts, and anti-phosphotyrosine immunoprecipitation and immunoblotting. A method for examining the localization of Tir within pedestals formed by EPEC using immunofluorescence microscopy is also presented.

Infection of Mammalian Cells with Enteropathogenic *Escherichia coli*

A variety of epithelial and macrophage cell lines have been used to examine the effects of EPEC infection on the modulation of protein tyrosine phosphorylation. These include the human epithelial derived HeLa (ATCC CCL2, American Type Culture Collection, Manassus, VA), HEp2 (ATCC CCL23), Henle-407 (ATCC CCL6) and Caco2 (ATCC HTB37) cell lines, and mouse J774 (ATCC TIB67) and RAW 264.1 (ATCC TIB71) macrophage-derived cell lines. Cells used for EPEC infection are seeded at 3.6×10^4 cells/cm^2 16–18 hr prior to infection in the appropriate culture media, resulting in a subconfluent (80–90%) cell monolayer.[12]

Two methods for infecting eukaryotic cells with EPEC have been described. The first involves inoculating cultures of epithelial or macrophage cell lines with a standing Luria Bertani (LB) grown culture of EPEC (1 colony inoculated into 2 ml LB) at a multiplicity of infection (MOI) of 100 : 1 (100 μl/100 mm dish), and allowing the infection to proceed for 3–5 hr at 37°, 5% (v/v) CO$_2$.[12] During the first 1.5 hr of the infection period, expression of genes encoding the bacterial type III secretion apparatus and Esp-secreted proteins is induced; Tir delivery into the host cell and tyrosine phosphorylation first become evident after about 1.5 hr infection and levels off after 4–5 hr infection.[3,12] This method has been used extensively to examine Tir delivery and tyrosine phosphorylation.

A second infection method uses EPEC cultured in Dulbecco's modified Eagle's medium (DMEM) until early log phase, prior to the addition to the mammalian cell cultures.[19] During the initial culture in DMEM, expression of the type III secretion apparatus, Esp A, EspB, and EspD and Tir genes are induced, resulting in rapid adherence to the host cell, and the appearance of tyrosine-phosphorylated Tir in the mammalian cell membrane as early as 15 min postinfection. This method has been used to examine the role of macrophage protein tyrosine dephosphorylation in studies investigating EPEC antiphagocytocis.[16]

[18] M. Farris, A. Grant, T. B. Richardson, and C. D. O'Connor, *Mol. Microbiol.* **28,** 265 (1998).
[19] I. Rosenshine, S. Ruschkowski, and B. B. Finlay, *Infect. Immun.* **64,** 966 (1996).

Materials, Reagents, Equipment

Dulbecco's modified Eagle's medium (DMEM)
Standing LB overnight culture of EPEC
CO_2 incubator (37°)

Methods

A standing overnight culture of EPEC is diluted 1 : 50 into DMEM and incubated at 37°, 5% (v/v) CO_2 until the OD_{600} reaches 0.3–0.4 (early log phase, 3–4 hr). Fifteen ml of this culture is added to each 100-mm plate of mammalian cells, and cocultures are incubated at 37°, 5% CO_2 without agitation for 15–60 min. Tyrosine-phosphorylated Tir is initially evident after 15 min, and levels off after 20–30 min postinfection.[19]

Preparation of Lysates from Enteropathogenic *Escherichia coli*-Infected Cells

Lysates from EPEC-infected cells are often prepared using detergent treatment. In this method, cells are initially incubated in buffer containing saponin to release cytosolic contents and then with Triton X-100 to prepare solubilized microsomal-membrane fractions.[12] Bacteria and the host cell actin cytoskeleton are found in the detergent-insoluble pellet. Although this method has been used successfully to monitor Tir delivery and tyrosine phosphorylation, it is not without drawbacks. Both saponin and Triton X-100 have been shown to lyse EPEC resulting in the contamination of host cell fractions with detergent-released bacterial proteins.[20] As EPEC produces proteins that are substrates for tyrosine phosphorylation that may not be delivered into the host cell, it is important to minimize bacterial contamination in order to distinguish between bacterial and host tyrosine-phosphorylated proteins. Additionally, saponin treatment does not completely release the cytosolic contents, resulting in contamination of the microsomal-membrane fractions with cytosolic proteins.[20] To circumvent these problems, Gauthier and colleagues adapted a mechanical disruption/ultracentrifugation method used to fractionate mammalian cells to EPEC-host cell cocultures.[20] Both methods are presented in this chapter.

In cells infected with wild-type EPEC, both bacterial (72/78 kDa) and 90 kDa tyrosine phosphorylated Tir are found predominantly in the Triton X-100 insoluble fraction containing the host cytoskeleton and adherent bacteria, with a fraction of tyrosine phosphorylated Tir localizing with the detergent soluble membrane-microsomal fraction.[3] This is not ideal for experiments requiring soluble tyrosine-phosphorylated, Tir, such as studies involving anti-Tir immunoprecipitation. To circumvent this, infection with an EPEC intimin mutant such as CVD206 is often

[20] A. Gauthier, M. de Grado, and B. B. Finlay, *Infect. Immun.* **68,** 4344 (2000).

used. Under these conditions, tyrosine-phosphorylated Tir is predominantly found in the Triton X-100 soluble membrane-microsomal fraction, and the unphosphorylated bacterial Tir in the detergent insoluble fraction.[4] Alternatively, cells can be treated with cytochalasin D(1 μg/ml final concentration) during infection, to prevent actin polymerization and pedestal formation. This results in the majority of tyrosine-phosphorylated Tir fractionating with the membrane-microsomal proteins.[4]

Detergent Fractionation

Materials, Reagents, Equipment

Saponin lysis buffer: 50 mM Tris, pH 7.4; 0.2% saponin; 1 mM EDTA; 1 mM sodium orthovanadate; 1 mM sodium fluoride; 100 nM microcystin LR (serine/threonine phosphatase inhibitor); 100 μg/ml leupeptin; 1 μM pepstatin; 10 μg/ml aprotinin

Triton X-100 lysis buffer: 50 mM Tris, pH 7.4; 1% Triton X-100; 1 mM EDTA; 1 mM sodium orthovanadate; 1 mM sodium fluoride; 100 nM microcystin LR (serine/threonine phosphatase inhibitor); 100 μg/ml leupeptin; 1 μM pepstatin; 10 μg/ml aprotinin

Phosphate-buffered saline (PBS)

Microcentrifuge (refrigerated or at 4$°$)

Methods

Culture dishes (100 mm) containing infected cells are placed on ice, the media removed, and monolayers gently washed 3 times in ice-cold PBS to remove nonadherent bacteria. Infected cells are scraped into 1 ml ice-cold PBS and centrifuged at 5200g (7000 rpm in an Eppendorf microcentrifuge) for 2 min at 4$°$. The resulting cell pellet is resuspended in 100 μl saponin lysis buffer and incubated on ice for 5 min prior to centrifugation at 20,800g (14,000 rpm in Eppendorf microcentrifuge) for 4 min at 4$°$. The supernatant from this spin contains cytosolic proteins. The pellet is washed once in PBS (14,000 rpm, 5 min, 4$°$) prior to solubilization in 1% Triton X-100 lysis buffer. After a 5-min incubation on ice, the samples are spun for 10 min, at 14,000 rpm at 4$°$. The resulting supernatant contains solubilized membrane/microsomal proteins, and the detergent-insoluble pellet contains adherent bacteria and cytoskeletal proteins. In cells infected with wild-type EPEC, both the bacterial 72/78 kDa Tir and tyrosine-phosphorylated 90 kDa Tir are found predominantly in the Triton X-100 insoluble fraction, with a small percentage of tyrosine phosphorylated Tir found in the Triton X-100 soluble fraction. In cells infected with an EPEC intimin deletion strain, tyrosine phosphorylated Tir is found in the Triton X-100 soluble fraction, whereas bacterial (72/78 kDa) Tir is in the Triton X-100 insoluble pellet.

Mechanical Disruption/Cell Fractionation

This protocol is based on methods commonly used to mechanically disrupt mammalian cells.[20] It relies on the passage of infected cells repeatedly through a 22-gauge needle, which ruptures the cell membrane without disrupting nuclei or adhering bacteria. It is essential to handle the cells extremely gently prior to mechanical disruption to prevent premature lysis. Additionally, it is critical to monitor the degree of cell disruption microscopically. Ideally, one should observe about 80% free, intact nuclei, and very few lysed nuclei and bacteria after disruption. After cells are fractionated, it is important to assess fraction purity by monitoring for contamination by proteins released by bacterial rupture. This method is adapted from Gauthier *et al.*[20]

Materials, Reagents, Equipment

   Homogenization buffer (HB): 250 mM sucrose; 3 mM imidiazole, pH 7.4; 0.5 mM EDTA; 1 mM sodium orthovanadate; 1 mM sodium fluoride; 100 nM microcystin LR; 10 μg/ml leupeptin; 1 μM pepstatin; 10 μg/ml aprotinin
   Phosphate buffered saline (PBS)
   Low speed refrigerated centrifuge
   Tabletop ultracentrifuge (Beckman TL100/TLS55 rotor)
   15-ml Conical centrifuge tubes
   1-ml Syringe, 22-gauge needle (1-1.5 inch length)

Methods

Mechanical Disruption. Culture dishes containing infected cells are placed on ice, the media removed, and monolayers gently washed three times in ice-cold PBS to remove nonadherent bacteria. Cells are gently scraped into 2.5 ml of ice-cold PBS and transferred to a 15-ml conical centrifuge tube using a large bore pipette. A large bore pipette can be prepared by cutting off the end of a 1000-μl micropipette tip. The dishes are rinsed with 1 ml ice-cold PBS, and the rinse transferred to the conical centrifuge tube using a large bore pipette. The cells are then spun at 1000g for 5 min at 4°. The supernatant is removed, and the pellet washed with 1.5 ml HB without resuspension by gently dislodging the pellet from the bottom of the tube, and sample spun at 3000g for 10 min at 4° to pellet cells.

To mechanically disrupt cells, the pellet is gently resuspended in 300 μl HB and passed 6 times through a 22-gauge needle using a 1-ml syringe. It is essential to aspirate the cell suspension slowly into the syringe and expel it against the side of the 15-ml conical tube vigorously to adequately disrupt the cells. To assess the efficiency of cell disruption, a small aliquot (5–100 μl) of the suspension is examined using phase-contrast microscopy. If the majority of cells are not disrupted (less than 80% free, unbroken nuclei, and many intact cells), the disruption is repeated 1–3 times further. It is critical to stop disrupting cells if lysed nuclei are present.

Cell Fractionation. The disrupted cell suspension is spun at 3000g for 15 min at 4° to pellet unbroken cells, nuclei, and bacteria. This fraction contains both tyrosine-phosphorylated and unphosphorylated Tir. The supernatant is transferred into an ultracentrifuge tube and spun at 41,000g in a Beckman TL100 tabletop ultracentrifuge using a TLS55 rotor for 20 min at 4°. The supernatant contains cytosolic proteins, whereas the pellet contains the membrane-microsomal fraction, which is enriched in 90 kDa tyrosine-phosphorylated Tir. The membrane-microsomal fraction can be further solubilized in 500 μl Triton solubilization buffer as described previously in this article.

Assessment of Fraction Purity. It is important to assess both the degree of bacterial contamination and the purity of cytosol and membrane-microsomal fractions. One method involves immunoblotting with antisera against markers for bacterial cytosolic proteins and specific cellular host compartments. DnaK, an abundant bacterial cytosolic heat shock protein, has been successfully used to assess contamination due to bacterial lysis.[20] DnaK migrates as a 62 kDa polypeptide and should only be evident in samples from the pellet of unbroken cells, bacteria, and nuclei. To assess cytosolic contamination of membrane-microsomal fractions, immunoblotting with antisera against tubulin should reveal a 48-kDa band in the cytosolic and unbroken cell pellet, with no labeling evident in the membrane fraction.[20] Immunoblots probed with antisera against the integral membrane protein calnexin should reveal labeling of a 90-kDa band only in the membrane-microsomal and unbroken cell fractions.[20]

Anti-phosphotyrosine Immunoprecipitation and Immunoblotting

These two powerful techniques have been used extensively to examine Tir tyrosine phosphorylation and the effect of EPEC on host cell protein tyrosine phosphorylation. The immunoprecipitation protocol is adapted from Celli *et al.*[16]

Anti-phosphotyrosine Immunoprecipitation

Materials, Reagents, Equipment

Anti-phosphotyrosine antisera (4G10, Upstate Biotechnology, Lake Placid, NY)
Protein G-Sepharose (Pharmacia, Piscataway, NJ)
Triton lysis buffer (from detergent fractionation protocol)
10% Bovine serum albumin (BSA) in Triton lysis buffer
Cytosol or solubilized membrane-microsomal fractions prepared by either the mechanical disruption or detergent solubilization method, adjusted to 500 μl/100 mm dish infected cells with Triton lysis buffer

Laemmli SDS–PAGE sample buffer: 0.06 M Tris, pH 6.8; 2% (w/v) SDS; 10% (w/v) glycerol; 0.025% (w/v) bromphenol blue; 5% (w/v) 2-mercaptoethanol

Methods

To each 500-μl aliquot of lysate in a 1.8-ml microcentrifuge tube, add 5 μg anti-phosphotyrosine antisera and incubate for 4 hr at 4° with gentle agitation. During the incubation period, prepare BSA-blocked protein G-Sepharose beads. Blocking the beads reduces the nonspecific interaction of proteins in the lysate with the Sepharose beads. One hundred μl of a 50% slurry of protein G beads is added to 900 μl 10% BSA in Triton lysis buffer and incubated for 2 hr at 4° with gentle agitation. The beads are centrifuged at 5200g in a microcentrifuge at 4°, the BSA removed, and beads washed 3 times in 1 ml Triton lysis buffer, prior to resuspension as a 50% slurry in Triton lysis buffer. Thirty μl of the blocked beads is added to the samples and incubated for a further 2 hr, 4° with gentle agitation. Immune complexes are collected by centrifugation at 5200g in a microcentrifuge, and the supernatant saved for further analysis (unbound fraction). Beads are washed 3 times in ice cold Triton lysis buffer, and complexes dissociated by boiling for 5 min in Laemmli SDS–PAGE sample buffer.

Anti-phosphotyrosine Immunoblotting

Materials, Reagents, Equipment

SDS–PAGE and Western blot transfer apparatus
Nitrocellulose
Tris-buffered saline (TBS): 10 mM Tris, pH 7.4, 0.9% NaCl
Blocking buffer: 4% BSA/0.05% sodium azide in TBS
Primary antisera dilution buffer: 1% BSA/0.05% sodium azide in TBS
Secondary antisera dilution buffer: 1% BSA/0.1% Tween 20/TBS
Wash buffer: TBS/0.1% Tween 20
Mouse anti-phosphotyrosine antisera (4G10, Upstate Biotechnology)
Anti-mouse horeseradish peroxidase conjugate
Enhanced chemiluminescence (ECL) reagent (examples include ECL, Amersham; Lumi-Glo, NEB; SuperSignal, Pierce, Rockford, IL).
X-ray film

Methods

Proteins resolved by SDS–PAGE are transferred to nitrocellulose according to the manufacturer's instructions. For the analysis of Tir tyrosine phosphorylation, optimal results are obtained using 8% SDS–PAGE.[3] After transfer, the membranes are washed in TBS and incubated in blocking buffer for 16–18 hr at 4°. Membranes are washed once in TBS and incubated for 2 hr at room temperature in a 1 : 2000 dilution of mouse anti-phosphotyrosine antisera (4G10) in primary antisera dilution buffer. Membranes are washed three times in TBS/0.1% Tween 20, for 5 min/wash, prior to incubation for 1 hr at room temperature in anti-mouse horseradish peroxidase (1 : 1000–1 : 2000) diluted in secondary antisera dilution buffer. The blots are washed extensively TBS/0.1% Tween 20 prior to chemiluminescence detection per the manufacturer's instructions. Tyrosine phosphorylated Tir appears as a strong 90 kDa band present in lysates from EPEC infected, but not uninfected, cells.

Controls

Several controls should be performed to aid in interpretation of results. For studies examining the effects of EPEC infection on host cell protein tyrosine phosphorylation, control samples prepared from uninfected cells should be included for comparison. As effects on host cell tyrosine phosporylation often require active bacterial signaling via the EPEC type III secretion apparatus, it is important to examine the effects on cells infected with EPEC type III secretion mutant strains.[12] For studies examining Tir tyrosine phosphorylation, appropriate controls may include infection with an EPEC Tir deletion mutant strain[3] and EPEC Tir Y474F[11] or EPEC Tir Y474S[10] mutants.

Anti-phosphotyrosine Immunofluorescence Microscopy

This method directly examines the localization of tyrosine-phosphorylated Tir within the pedestals formed by EPEC. Cells are fixed and prepared for labeling with antisera against phosphotyrosine and phalloidin to label the actin pedestals, and examined using epifluorescence microscopy. Tyrosine-phosphorylated Tir is found at the tip of the EPEC pedestal and often appears horseshoe shaped.[3] A number of different fluorophore-conjugated secondary antisera are available. We commonly use Alexa series dyes (Molecular Probes, Eugene, OK) as they are more resistant to fading. Once Tir is delivered into the host cell and binds to intimin, there is no longer a requirement for live bacteria for pedestal formation.[19] Gentamicin can be added to the cultures after 3–4 hr postinfection to produce hyperelongated pedestals.[4]

Materials, Reagents, Equipment

  12-mm Round #1 glass coverslips; coverslips are washed in 70% ethanol/1% HCl, rinsed in 70% ethanol, and sterilized prior to use
  24-Well tissue culture plates
  25% Paraformaldehyde (PFA) stock solution: Dissolve 10 g paraformaldehyde in 30 ml H_2O by heating at 70–75°. Do not let temperature go above

75°. Add 5 drops 10 N NaOH to clear, then add H_2O up to 40 ml. Store at −20° in 1 ml aliquots.

2.5% Paraformaldehyde working solution: Prepare fresh for each day's experiment. Thaw an aliquot of 25% PFA by heating at 70–75°. Mix with 8 ml H_2O and 1 ml 10 × PBS. Adjust to pH 7.2–7.4 with 20% HCl. Add H_2O to 10 ml.

Mowiol mounting medium: To prepare, combine 6 g glycerol, 2.4 g Mowiol, and 6 ml H_2O and mix at room temperature for 2 hr. Add 12 ml 0.2 M Tris, pH 8.5, and heat to 50° for 10 min with occasional mixing. After the Mowiol dissolves, clarify by centrifugation at 5,000g for 15 min at room temperature, and add DABCO (1,4-diazobicyclo[2.2.2]octane) to 2.5% to reduce fading. Aliquot into airtight 1.5-ml tubes and store at −20°

Phosphate buffered saline solution, 10× and 1×

0.1% Triton X-100/Phosphate-buffered saline (PBS)

10% Normal goat serum (NGS) in PBS/0.1% Triton X-100

0.5% Triton X-100/PBS

Anti-phosphotyrosine antisera (4G10, Upstate Biotechnology)

Fluorescent labeled secondary antisera

Fluorescent labeled phalloidin: This toxin labels polymerized actin and is available from a number of suppliers including Molecular Probes. Alternatively, primary antisera against actin can be used.

Methods

Place a sterile 12-mm round coverslip into each well of a 24-well tissue culture plate. One ml cells at 2×10^4 cells/ml is seeded into each well and incubated for 16–18 hr prior to infection. Cells are infected with 1 μl of a tenfold diluted standing LB overnight culture of EPEC and incubated for 3–6 hr at 37°, 5% (v/v) CO_2. If the production of hyperelongated pedestals is desired, fresh medium containing gentamicin (100 μg/ml final concentration) is added after 4 hr infection. At the end of the infection period, the medium is removed and monolayers washed twice with warmed PBS. Cells are fixed in 0.2 ml/well 2.5% PFA (working solution) for 10 min, at 37°. Cells are washed extensively with warmed PBS (at least 5 washes) to remove residual PFA. Cells are permeabilized by the addition of 0.2 ml/well 0.5% Triton X-100/PBS and allowed to incubate for 10 min at room temperature. To decrease the nonspecific binding of the primary and secondary antisera, samples are blocked in 0.2 ml/well 10% NGS/0.1% Triton X-100/PBS for 30 min at room temperature, prior to the addition of the anti-phosphotyrosine antisera.

The mouse anti-phosphotyrosine antisera (4G10) is diluted 1 : 100 in 10% NGS/0.1% Triton X-100/PBS. Remove the blocking solution from each well and add 15 μl diluted antisera to the center of each coverslip. It is essential that the coverslips be centered and not touch the side of the well. At this stage it is important to work fairly quickly, as it is crucial that the cells do not dry out at any time as this leads to an increase in nonspecific background. Alternatively, 15 μl of antisera can be spotted onto Parafilm, and the coverslips placed cell-side-down on top of the antisera. Place the 24-well plate in a humidified chamber and incubate for 30 min at 37°. At the end of the incubation period, rinse coverslips once in PBS and wash 3 times in 0.1% Triton X-100/PBS, 5 min each wash.

During the wash steps, prepare the diluted secondary antisera. We often use goat anti-mouse Alexa 488 (Molecular Probes) as the secondary antisera, and Alexa 568 or 594 conjugated phalloidin (Molecular Probes), although other combinations of fluorophores work as well. We prepare a working solution containing goat anti-mouse Alexa 488 diluted at 1 : 400 and Alexa 568 phalloidin diluted 1 : 200, and incubate coverslips with 15 μl/well of this reagent for 30 min at room temperature in a humidified container. Again, care is taken that the coverslips are not touching the sides of the well and that the samples are not allowed to dry out. Coverslips are rinsed once with PBS, followed by 3 washes with 0.1% Triton X-100/PBS, and 1 wash in H_2O. Coverslips are mounted face down on a small (10–20 μl) drop of Mowiol mounting media, which is allowed to polymerize for 30 min prior to examination by fluorescence microscopy.

Acknowledgments

Work in R.D.'s laboratory is supported by grants from the Canadian Institutes for Health Research and the Alberta Heritage Foundation for Medical Research (AHFMR). R.D. is an AHFMR scholar.

Section VI

Type III Secretion Systems

[28] Purification and Detection of *Shigella* Type III Secretion Needle Complex

By Koichi Tamano, Shin-Ichi Aizawa, and Chihiro Sasakawa

Introduction

The type III secretion system is a highly sophisticated bacterial protein delivery system. On contact with the host cells, a set of effector proteins is delivered from the infecting bacteria to the host cells via the type III secretion system. These translocated proteins have a variety of effects on host cell function, including bacterial attachment, invasion, or avoidance from the host defense systems.[1] Although the mechanisms of protein export via the type III secretion system and biosynthesis of the secretion apparatus are still to be elucidated, some common characteristics of this secretion system have emerged.[2,3] Genetic and functional studies have indicated that the type III secretion system is encoded by more than 20 genes, including the genes encoding the secretion apparatus (called the type III secretion complex in this article), secreted proteins, chaperones, and regulators. A subset of the genes exists as a pathogenicity island (PAI), which is potentially transposed horizontally in different bacteria, thus distributing among many gram-negative animal and plant pathogenic bacteria.[2,3] In fact, there is considerable homology between the proteins of the type III secretion systems in different pathogens.[2,3] Importantly, some of the proteins of type III secretion systems also share significant similarity to the components of bacterial flagellar export machinery.[2,3] For example, some of the putative components of the type III secretion complexes such as *Salmonella typhimurium* InvA, InvB, InvC, SpaO, SpaP, SpaQ, SpaR, SpaS, and PrgK share significant similarity to FlhA, FliH, FliI, FliN, FliP, FliQ, FliR, FlhB, and FliF of the *Salmonella* flagellar export system, respectively.[3] Furthermore, the type III secretion system has been identified to be functionally and structurally similar to the flagellar export system. Secretion of a set of proteins via the type III system is dependent on the energy supply mediated by F1-type ATPase associated with the secretion apparatus.[4] The same is true for secretion of the extracellular flagellar components which form the hook, cap, and flagellar filament via the flagellar export system.[5] The supramolecular structure of the type III secretion complex

[1] J. E. Galán and A. Collmer, *Science* **284,** 1322 (1999).

[2] C. J. Hueck, *Microbiol. Mol. Biol. Rev.* **62,** 379 (1998).

[3] G. V. Plano, J. B. Day, and F. Ferracci, *Mol. Microbiol.* **40,** 284 (2001).

[4] K. Tamano, S.-I. Aizawa, E. Katayama, T. Nonaka, S. Imajoh-Ohmi, A. Kuwae, S. Nagai, and C. Sasakawa, *EMBO J.* **19,** 3876 (2000).

[5] R. M. Macnab, in *"Escherichia coli* and *Salmonella typhimurium"* (F. C. Neidhardt, ed.), p. 123. American Society for Microbiology, Washington, D.C., 1996.

of *S. typhimurium* and *Shigella flexneri* shares similarity with that of the flagellar basal body, as described below.[4,6] Finally, the expression of genes encoding the type III secretion system as well as the flagellar export system is under stringent control mediated by complicated regulatory networks in the bacteria.

Despite the structural and functional similarity of the type III secretion system in each pathogen, the proteins delivered via this system are quite diverse. For example, *S. flexneri* potentially delivers more than 20 proteins via the type III secretion system which, although some share similarity to secreted proteins from different pathogens, are unique to *S. flexneri*.[7,8] Studies of the secreted proteins from *Yersinia, Salmonella,* and *Shigella* have indicated that some of these proteins have a role in linking the secretion complex to the target host cell membrane, while others serve as effectors to modulate host cell functions.[9–11]

For analysis of the supramolecular structure, the type III secretion complexes of *S. typhimurium* and *S. flexneri* were extensively purified from the envelope fractions by adaptation of the method used for the purification of the flagellar hook–basal complex of *S. typhimurium*.[4,6,12] Although the fraction which was enriched in the type III secretion complex from *S. flexneri,* as examined by silver staining SDS–PAGE, still contained some bacterial components such as OmpC and GroEL, it seemed adequate for electron microscopic analysis or direct amino acid sequencing. The purified type III secretion complexes from *S. typhimurium* and *S. flexneri* as reported by Kubori *et al.* and Tamano *et al.,* respectively, contained four major components.[4,6] The *Salmonella* type III secretion complex contained InvG, PrgH, PrgI, and PrgK proteins, while the *Shigella* type III secretion complex contained MxiD, MxiG, MxiH, and MxiJ.[4,6,13] These type III secretion complexes have been shown to contain an additional component, PrgJ in *Salmonella* and MxiI in *Shigella*.[14,15] The supramolecular structures of the type III secretion complexes of each bacteria observed by electron microscopy are similar, being composed of two distinctive parts, the needle and basal parts (Fig. 1). The needle of the *S. typhimurium* and *S. flexneri* type III secretion complexes consists mainly of PrgI

[6] T. Kubori, Y. Matsushima, D. Nakamura, J. Uralil, M. Lara-Tejero, A. Sukhan, J. E. Galán, and S.-I. Aizawa, *Science* **280,** 602 (1998).

[7] J. E. Galán, *Mol. Microbiol.* **20,** 263 (1996).

[8] C. Buchrieser, P. Glaser, C. Rusniok, H. Nedjari, H. d'Hauteville, F. Kunst, P. Sansonetti, and C. Parsot, *Mol. Microbiol.* **38,** 760 (2000).

[9] G. R. Cornelis, *Proc. Natl. Acad. Sci. U.S.A.* **97,** 8778 (2000).

[10] J. E. Galán and D. Zhou, *Proc. Natl. Acad. Sci. U.S.A.* **97,** 8754 (2000).

[11] P. J. Sansonetti, *FEMS Microbiol. Rev.* **25,** 3 (2001).

[12] S.-I. Aizawa, G. E. Dean, C. J. Jones, R. M. Macnab, and S. Yamaguchi, *J. Bacteriol.* **161,** 836 (1985).

[13] T. Kubori, A. Sukhan, S.-I. Aizawa, and J. E. Galán, *Proc. Natl. Acad. Sci. U.S.A.* **97,** 10225 (2000).

[14] T. G. Kimbrough and S. I. Miller, *Proc. Natl. Acad. Sci. U.S.A.* **97,** 11008 (2000).

[15] A. Blocker, N. Jouihri, E. Larquet, P. Gounon, F. Ebel, C. Parsot, P. Sansonetti, and A. Allaoui, *Mol. Microbiol.* **39,** 652 (2001).

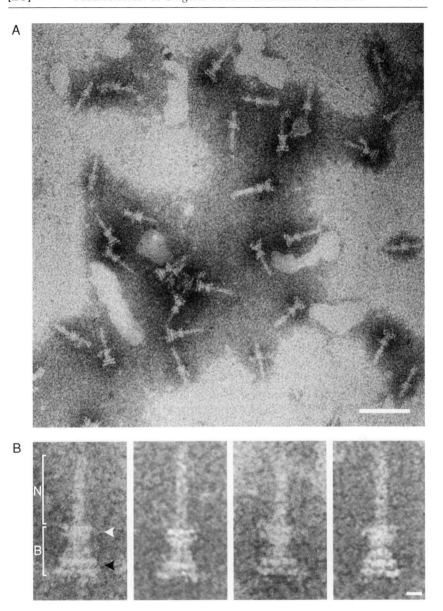

FIG. 1. Electron micrographs of purified type III secretion needle complexes from *Shigella flexneri*. (A) Purified needle complexes were negatively stained with 2% phosphotungstic acid (pH 7.0) and observed by TEM. Bar: 100 nm. (B) Needle complexes observed at high magnification. N indicates the needle, while B indicates the basal part of the complex. The open arrowhead denotes upper rings, while the closed arrowhead denotes the lower rings. Bar: 10 nm. Reproduced with permission of Oxford University Press and cited from the original paper [K. Tamano, S.-I. Aizawa, E. Katayama, T. Nonaka, S. Imajoh-Ohmi, A. Kuwae, S. Nagai, and C. Sasakawa, *EMBO J.* **19,** 3876 (2000)].

and MxiH, respectively. In addition, PrgJ in *S. typhimurium* and MxiI in *S. flexneri* are considered to be at the needle portion, forming a cap over the external needle tip.[14,15] The basal part of the *S. typhimurium* type III complex is composed of InvG and PrgH and K, while that of the *S. flexneri* complex is composed of MxiD, G and J.[4,6] Furthermore, the supramolecular structures of the basal portion of both type III complexes share significant similarity to that of the *Salmonella* flagellar basal body.[4,6]

Like the flagellar basal body, the basal part of the type III secretion complex possesses two pairs of rings, referred to as the upper and lower rings (Fig. 1B). Since the basal portion was observed to be embedded in the osmotic-shocked bacterial envelope by electron microscopy, like the flagellar basal complex, the two pairs of rings are thus assumed to be anchored to the inner and outer membranes of bacteria.[4,6] The flagellar hook forms a curved protruding structure from which a long flagellar filament is extended, while the type III secretion complex forms a straight needle protruding from the basal part. The length of the type III needle of wild-type *S. flexneri* is estimated to be 45 nm and distributed in a narrow range with a standard deviation of 3.3 nm. Interestingly, the needle length, like the flagellar hook, can be extended up to 1000 nm under certain conditions. Indeed, when MxiH was overexpressed in *Shigella* or InvJ deleted in *Salmonella,* the bacteria produced long needles.[4,13,16,17] The length of the basal body of the *Shigella* type III secretion complex is estimated to be approximately 31 nm, which is consistent with the thickness of the gram-negative bacterial envelope (approximately 25 nm). This suggests that the type III secretion complex spans both the outer and inner membranes. Although the number of needles per bacterium has not been accurately determined, based on the distribution of the type III secretion structures in a field of the osmotically shocked bacterial envelope as observed by electron microscopy, it is estimated to be at least 50 per bacterium.

Genetic and functional studies of type III secretion systems have strongly suggested that the basic morphological feature displayed by *Salmonella* and *Shigella* would be conserved among other pathogens. For example, in enteropathogenic *Escherichia coli* (EPEC) or enterohemorrhagic *E. coli* (EHEC), a long filamentous structure protrudes from the bacterium, composed of EspA which is encoded by the *espA* gene located on the locus of enterocyte effacement (LEE) pathogenicity island, downstream of the region encoding genes of the type III secretion system.[18,19] Recent study has indicated that the EspA filament extends from the

[16] K. Tamano, E. Katayama, T. Toyotome, and C. Sasakawa, *J. Bacteriol.* **184,** 1244 (2002).

[17] J. Magdalena, A. Hachani, M. Chamekh, N. Jouihri, P. Gounon, A. Blocker, and A. Allaoui, *J. Bacteriol.* **184,** 3433 (2002).

[18] S. Knutton, I. Rosenshine, M. J. Pallen, I. Nisan, B. C. Neves, C. Bain, C. Wolff, G. Dougan, and G. Frankel, *EMBO J.* **17,** 2166 (1998).

[19] F. Ebel, T. Podzadel, M. Rohde, A. U. Kresse, S. Krämer, C. Deibel, C. A. Guzmán, and T. Chakraborty, *Mol. Microbiol.* **30,** 147 (1998).

tip of the needle structure of the type III secretion complex.[20,21] This filamentous structure is essential for protein translocating events, and hence full virulence of EPEC and EHEC. In fact, the *espA* mutant of EPEC and EHEC has been shown to be deficient in forming a long filamentous structure and delivering effectors such as Tir, EspB, and EspD into the host cells, thus becoming a nonadherent mutant.[18,19] Similarly, some plant pathogens such as *Pseudomonas syringae* and *Ralstonia solanacearum* form a filamentous appendage called the Hrp pilus, which consists of HrpA in *P. syringae* or HrpY in *R. solanacearum*.[22,23] Therefore, extraction of the intact type III secretion complex or the complex together with the filamentous structure from such pathogens might need some further modification of the method used for purification of the *Shigella* type III secretion complex.

In this article, the method for purifying the type III secretion complex, tentatively named the needle complex, from *S. flexneri* 2a strain M94 will be described. Because M94 contains the intact type III secretion system required for invasion but is defective in actin-based motility in mammalian cells because of a Tn*5* insertion in the *virG* (*icsA*) gene, we used this attenuated strain for extracting the type III secretion complex for the sake of laboratory biosafety.[24] In brief, bacteria are suspended in sucrose solution supplemented with EDTA-2Na and lysozyme to allow spheroplast formation. Subsequently, bacteria are lysed by a nonionic detergent such as Triton X-100, which is followed by elimination of DNA by DNase digestion, and then the lysate is subjected to a low-speed centrifugation to remove cell debris. The intact needle complexes contained in the lysate are enriched by ultracentrifugation. The pellet is fractionated by CsCl density gradient centrifugation, and the fractions are further ultracentrifuged to sediment the extensively purified needle complex.

Purification of Type III Secretion Needle Complexes from *Shigella flexneri*

Reagents

Purification of the needle complexes from 1 liter of culture requires the following reagents:

[20] K. Sekiya, M. Ohishi, T. Ogino, K. Tamano, C. Sasakawa, and A. Abe, *Proc. Natl. Acad. Sci. U.S.A.* **98**, 11638 (2001).

[21] S. J. Daniell, N. Takahashi, R. Wilson, D. Friedberg, I. Rosenshine, F. P. Booy, R. K. Shaw, S. Knutton, G. Frankel, and S.-I. Aizawa, *Cell. Microbiol.* **3**, 865 (2001).

[22] E. Roine, W. Wei, J. Yuan, E.-L. Nurmiaho-Lassila, N. Kalkkinen, M. Romantschuk, and S. Y. He, *Proc. Natl. Acad. Sci. U.S.A.* **94**, 3459 (1997).

[23] F. V. Gijsegem, J. Vasse, J.-C. Camus, M. Marenda, and C. Boucher, *Mol. Microbiol.* **36**, 249 (2000).

[24] S. Makino, C. Sasakawa, K. Kamata, T. Kurata, and M. Yoshikawa, *Cell* **46**, 551 (1986).

100 ml Sucrose solution (0.5 M sucrose, 0.15 M Tris; pH not adjusted)

250 ml TET buffer [10 mM Tris-hydrochloride (pH 8.0), 1 mM EDTA-2Na (pH 8.0), 0.1% Triton X-100]

12.80 ml 0.1 M EDTA-2Na (pH 8.0)

2.00 ml 10 mg/ml Lysozyme (Sigma, St. Louis, MO)

1.04 ml 0.1 M Phenylmethylsulfonyl fluoride (PMSF)

11.67 ml 10% Triton X-100

0.47 ml 1 M MgCl$_2$

8 mg DNase I (powder; Sigma)

8.17 ml 5 M NaCl

1.68 g CsCl

Procedure

The procedure for purifying the type III secretion needle complexes from *Shigella* is outlined in Fig. 2.

S. flexneri 2a strain M94 (*virG*::Tn5) is used for the purification instead of wild type (YSH6000) because of its safer characteristics upon handling. Ten ml of M94 grown at 30° overnight in L-broth is used to inoculate 1 liter of L-broth, and the bacteria are grown at 37° until late logarithmic phase (approximately 1×10^9 cells/ml, OD$_{600}$ = 1.0). The bacteria are collected by centrifugation (7000 rpm, 10 min, 4°) using a fixed angle rotor (JLA-10,500, Beckman Coulter, Fullerton, CA) and suspended in 100 ml of ice-cold sucrose solution using a spatula and a pipette. The bacterial suspension is transferred to a 200-ml beaker on ice and stirred gently (~300 rpm/min). Two ml of 10 mg/ml lysozyme and 2 ml of 0.1 M EDTA-2Na (pH 8.0) are added, and the sample is incubated for 15 min at 4°. The cell suspension is transferred to a water bath and incubated for 15 min at 30° with gentle stirring. During this incubation, most of the cells are converted into spheroplasts. After incubation at 30°, spheroplast formation is complete, as confirmed by dark-field microscopy. Then, 1.04 ml of 0.1 M protease inhibitor PMSF is added to a final concentration of 1 mM. At this stage, the bacterial suspension is stirred at room temperature. The resulting spheroplasts are then lysed by adding 11.67 ml of 10% Triton X-100 (final concentration 1%). As the lysis nears completion, the lysate becomes viscous with the dispersion of the genomic DNA. For degradation of DNA, 0.47 ml of 1 M MgCl$_2$ (final concentration 4 mM) is added to activate endogenous DNase. When the viscosity was not sufficiently reduced, not more than 8 mg of DNase I is added gradually to the lysate. Viscosity is eliminated within approximately 10 min of the addition of DNase I. After the viscosity of the lysate decreases greatly, 10.8 ml of 0.1 M EDTA-2Na (pH 8.0) (final concentration 10 mM) is added, which is required for preventing reaggregation of the cell membranes and walls. Prolonged incubation of the lysate at room temperature without chelating magnesium ion (Mg^{2+}) by EDTA-2Na results in a

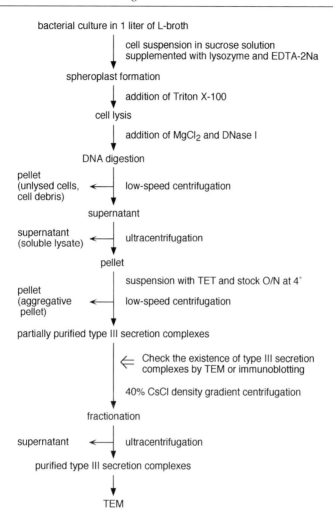

bacterial culture in 1 liter of L-broth

cell suspension in sucrose solution
supplemented with lysozyme and EDTA-2Na

spheroplast formation

addition of Triton X-100

cell lysis

addition of MgCl₂ and DNase I

DNA digestion

pellet
(unlysed cells,   ←   low-speed centrifugation
cell debris)

supernatant

supernatant
(soluble lysate)   ←   ultracentrifugation

pellet

suspension with TET and stock O/N at 4°

pellet
(aggregative   ←   low-speed centrifugation
pellet)

partially purified type III secretion complexes

← Check the existence of type III secretion
complexes by TEM or immunoblotting

40% CsCl density gradient centrifugation

fractionation

supernatant   ←   ultracentrifugation

purified type III secretion complexes

TEM

FIG. 2. Outline of the purification of type III secretion needle complexes from *Shigella flexneri*.

whitish cloudiness of the solution. Unlysed cells and cell debris are removed by two low-speed centrifugation steps (20,000g, 20 min, 4°). The cleared lysate is then transferred into another 200-ml beaker and stirred gently. Next, 8.17 ml of 5 M NaCl (final concentration, 0.3 M) is added to the lysate with gentle stirring, and the lysate is incubated at 4° for 30 min without stirring. It is worth noting that, at this stage, although alkaline treatment is recommended for extracting the needle complexes from *Salmonella* membrane, this causes extensive degradation of the needle complexes extracted from *Shigella*. Therefore this treatment is omitted in the case of *Shigella*. To collect the needle complexes, the lysate is ultracentrifuged with a fixed angle rotor such as 60TI (Beckman Coulter) at 110,000g for 60 min at 4°. After the supernatant is removed, each pellet is rinsed once with 15 ml of TET buffer, and any remaining residue of buffer on the inside surface of the centrifuge tubes is removed with tissue paper to avoid contamination with soluble proteins. Then, the pellet is suspended in 0.1–1.0 ml of TET buffer, transferred into an Eppendorf tube, and stored overnight at 4°. Insoluble materials which form a precipitate during overnight storage are removed by at least three low speed centrifugation steps (18,500g, 10 min, 4°) with a refrigerated microcentrifuge. After the centrifugation, it is recommended that the resulting supernatant, which corresponds to the partially purified needle sample, be confirmed to contain the intact needle complex by electron microscopy or immunoblotting.

For extensive purification, the partially purified needle complexes are fractionated by 40% (w/v) CsCl density gradient centrifugation. CsCl (1.68 g) is added to the partially purified needle sample, and the final volume is adjusted to 4.2 ml with TET buffer, followed by gentle mixture of the sample solution. Then, the sample is ultracentrifuged at 38,000g with a swing-out rotor such as SW50.1 (Beckman Coulter) for more than 10 hr at 24°. After ultracentrifugation, the solution is separated into six fractions. For the fractionation, a drawn-out glass pipette is moved on the surface of the sample solution, and an approximately 700 μl fraction is removed from the surface. Each of the fractions is diluted in a small amount of TET buffer (~9 ml), and fractions containing the needle complexes are ultracentrifuged again with a fixed angle rotor (70TI, Beckman Coulter) at 225,000g for 60 min at 4°. After centrifugation, the supernatant is completely removed, because the presence of small particles of remaining CsCl hinders the observation of the needle complexes by electron microscopy. The resulting pellets in one or two fractions which contain the needle complexes are suspended in 20 μl of TET buffer and stored at 4°. Figure 1 shows the purified needle complexes from *Shigella flexneri* 2a by transmission electron microscopy (TEM).

Acknowledgments

We thank Sarah Daniell for critical reading of the manuscript. We are grateful to Eisaku Katayama for close observation of the *Shigella* needle complexes by TEM. This work was supported by the Research for the Future Program of the Japanese Society for the Promotion of Science.

[29] Analysis of *Salmonella* Invasion Protein–Peptidoglycan Interactions

By M. GRACIELA PUCCIARELLI and FRANCISCO GARCÍA-DEL PORTILLO

Introduction

The peptidoglycan (murein) sacculus is a giant macromolecule that completely surrounds the bacterial cell.[1,2] This stress-bearing exoskeleton plays an essential role in ensuring bacterial viability while maintaining the high internal osmotic pressure.[2] The peptidoglycan is also involved in providing a specific cell shape. Thus, purified peptidoglycan sacculi retain the morphology of the bacteria from which they are isolated. The chemical composition of the peptidoglycan varies between bacterial genera and species, but it presents a highly conserved basic structure consisting of a glycan backbone of alternating units of N-acetylglucosamine and N-acetylmuramic acid residues (linked by β-1,4 glycosidic bonds) with short peptide chains linked to the N-acetylmuramic acid moiety. A large set of enzymes displaying transglycosylation and/or transpeptidation activities are involved in polymerizing the muropeptide units to form a net-like macromolecule entirely held by covalent linkages.[3]

Besides the relevant role of the peptidoglycan as a biopolymer ensuring physical integrity, shape, and division of bacteria, it also acts a trans-envelope barrier.[4] The effective pore size of the peptidoglycan lattice has been estimated at approximately 5 nm, a "hole" diameter that would allow the passage of globular proteins of 55 kDa of maximal size.[5] Essential processes such as protein export, conjugation, competence, and flagellar assembly rely on the passage of macromolecules through the bacterial envelope.[4] The penetration and/or assembly of these macromolecular complexes throughout the peptidoglycan layer remains largely uncharacterized. Undoubtedly, specific interactions between components of these complexes and the peptidoglycan lattice must drive these processes. On one side, the precise penetration of the peptidoglycan layer must be facilitated by autolytic peptidoglycan hydrolases. In the case of complexes that are assembled and fixed in a certain location of the cell envelope, proteins might necessarily exist that anchor the complex to the envelope, probably by maintaining stable interactions with the peptidoglycan. Finally, it should be noted that a third group of components interacting with the peptidoglycan are the lipoproteins, teichoic acids, and carbohydrates that

[1] J. V. Holtje, *Microbiol. Mol. Biol. Rev.* **62**, 181 (1998).
[2] A. L. Koch, *Crit. Rev. Microbiol.* **26**, 1 (2000).
[3] J. V. Holtje and C. Heidrich, *Biochimie* **83**, 103 (2001).
[4] A. J. Dijkstra and W. Keck, *J. Bacteriol.* **178**, 5555 (1996).
[5] P. Demchick and A. L. Koch, *J. Bacteriol.* **178**, 768 (1996).

distribute uniformly on the bacterial surface and contribute to physically link the entire membrane to the peptidoglycan layer.

Type III secretion systems (TTSS) are macromolecular complexes that span the inner and outer membranes in gram-negative bacteria.[6,7] The most intensively characterized TTSS system is the flagellar export machinery.[8] The architecture of the flagellum consists of a basal body—a rod surrounded by several rings—that is continued by a hook and a long filament.[8] Great detail has been obtained on how flagellar membrane proteins assemble to form the basal body.[9] Among these membrane proteins, motor proteins are required for proper rotation of the flagellum. Two of them, the MotA and MotB proteins, are considered "stator" proteins because of their capacity to form a complex and the possibility that MotB is attached to the peptidoglycan layer.[10,11] According to this view, the ultimate "stator" element of the flagellum would be the peptidoglycan layer. Moreover, the existence of a "P-ring," which is located in the middle part of the flagellar rod at the level of the peptidoglycan,[12] has been known for sometime. Unfortunately, no P-ring-associated protein mediating a flagellum–peptidoglycan interaction is known. Interestingly, a recent report has shown that a flagellar protein, named FlgJ, displays peptidoglycan–hydrolase activity.[13] The role of FlgJ has been proposed to be linked to the penetration of the rod part of the flagellar basal body through the peptidoglycan layer.

Virulence-related TTSS are devices used by animal and plant bacterial pathogens to inject effector proteins into their host cells.[6,14,15] The TTSS apparatus of the bacterial pathogen *Salmonella typhimurium* involved in secreting invasion factors has been visualized by electron microscopy.[16] This apparatus was termed the needle complex. Other investigators have subsequently shown the existence of a similar apparatus in *Shigella flexneri*.[17–19] Strikingly, these apparatuses have

[6] C. J. Hueck, *Microbiol. Mol. Biol. Rev.* **62,** 379 (1998).

[7] G. R. Cornelis and F. Van Gijsegem, *Ann. Rev. Microbiol.* **54,** 735 (2000).

[8] R. M. Macnab, *J. Bacteriol.* **181,** 7149 (1999).

[9] T. Minamino and R. M. Macnab, *J. Bacteriol.* **181,** 1388 (1999).

[10] R. De Mot and J. Vanderleyden, *Mol. Microbiol.* **12,** 333 (1994).

[11] R. Koebnik, *Mol. Microbiol.* **16,** 1269 (1995).

[12] M. L. DePamphilis and J. Adler, *J. Bacteriol.* **105,** 396 (1971).

[13] T. Nambu, T. Minamino, R. M. Macnab, and K. Kutsukake, *J. Bacteriol.* **181,** 1555 (1999).

[14] J. E. Galán and A. Collmer, *Science* **284,** 1322 (1999).

[15] M. S. Donnenberg, *Nature* **406,** 768 (2000).

[16] T. Kubori, Y. Matsushima, D. Nakamura, J. Uralil, M. Lara-Tejero, A. Sukhan, J. E. Galán, and S. I. Aizawa, *Science* **280,** 602 (1998).

[17] A. Blocker, P. Gounon, E. Larquet, K. Niebuhr, V. Cabiaux, C. Parsot, and P. Sansonetti, *J. Cell Biol.* **147,** 683 (1999).

[18] K. Tamano, S. Aizawa, E. Katayama, T. Nonaka, S. Imajoh-Ohmi, A. Kuwae, S. Nagai, and C. Sasakawa, *EMBO J.* **19,** 3876 (2000).

[19] A. Blocker, N. Jouihri, E. Larquet, P. Gounon, F. Ebel, C. Parsot, P. Sansonetti, and A. Allaoui, *Mol. Microbiol.* **39,** 652 (2001).

basal bodies structurally analogous to the flagellum. Considering what it is known in the flagellar system, interactions between components of the virulence TTSS and the peptidoglycan layer might occur.

The aim of the present article is to provide methods to analyze specific association of membrane proteins to the peptidoglycan layer. Although focusing mainly on the analysis of the *Salmonella* invasion TTSS, the methods can be extrapolated to determinate protein–peptidoglycan interactions in other systems, including the flagellar apparatus and virulence-related TTSS of other bacterial pathogens.

Chemistry of Peptidoglycan

Basic Methodological Principles

The characteristic chemical properties of the peptidoglycan as a large-size covalently linked polymer explain its remarkable resistance to solubilization in boiling solutions containing sodium dodecyl sulfate (SDS). Diverse protocols have been designed for isolating pure peptidoglycan from either gram-positive or gram-negative bacteria.[20] Unlike gram-negative bacteria, which harbor a peptidoglycan disposed as a monolayer or a patch-like multilayered polymer,[21] the peptidoglycan of gram-positive bacteria is multilayered and embedded in a thick wall in which other polymers such as teichoic acids and carbohydrates are covalently bound. Thus, the incubation of intact gram-positive bacteria with boiling SDS does not cause solubilization of the plasma membrane because of the inaccessibility of the detergent through the thick cell wall. Therefore, the cells need to be ruptured before the detergent treatment, a step that requires prior inactivation of autolytic enzymes.[20] In gram-negative bacteria, rapid boiling of bacteria in SDS causes "instant" cell lysis and complete inactivation of autolytic enzymes, which ensures conservation of the peptidoglycan structure during the purification process.

An additional aspect to take into consideration is the type of experimentation to be performed with the purified macromolecular peptidoglycan. Most methods have as their objective obtaining highly purified peptidoglycan as appropriate material for analysis of chemical composition i.e., the ratio of glycan to amino acid moieties[20] or alternatively, for analysis of muropeptide composition by high-performance liquid chromatography (HPLC).[22] In both cases, proteins that are either covalently bound or firmly associated to peptidoglycan are removed by protease treatment. As we describe below, the omission of the protease treatment can provide valuable information on proteins that interact with peptidoglycan.

[20] R. S. Rosenthal and R. Dziarski, *Methods Enzymol.* **235,** 253 (1994).
[21] H. Labischinski, E. W. Goodell, A. Goodell, and M. L. Hochberg, *J. Bacteriol.* **173,** 751 (1991).
[22] B. Glauner, *Anal. Biochem.* **172,** 451 (1988).

Purification of Macromolecular Peptidoglycan

Macromolecular peptidoglycan from *Salmonella typhimurium* can be purified based on the method of Glauner.[22] An experimental variable is the growth conditions at which bacteria are collected. If a major objective is to analyze interaction of *Salmonella* invasion proteins with the peptidoglycan, bacteria should be grown up to an invasion-competent phase. Some laboratories grow *Salmonella* up to exponential phase using LB–0.3 M NaCl medium with shaking.[23] Our experience shows that bacteria grown overnight in LB medium under static conditions are fully competent for invasion of eukaryotic cells.[24]

Day 0. In 1000 ml flasks inoculate 200 ml of LB medium with bacteria from a single colony previously grown on LB-agar plates. Incubate the culture overnight at 37° under nonshaking conditions.

Day 1

1. Monitor the bacterial growth by measuring the optical density at 550 nm (OD_{550}). Under the growth conditions used, the overnight culture reaches an $OD_{550} = 1.0$ [approximately 10^9 colony forming units(cfu)/ml]. Thaw the culture rapidly by inserting the flask in a bath containing a mixture of ice and NaCl.

2. Harvest bacteria by centrifugation ($12,000g$, 15 min, 4°). Suspend the bacterial pellet in 1/10 volume of phosphate-buffered saline (PBS), pH 7.4. Repeat the centrifugation step and suspend bacteria to a final volume of 3 ml of cold PBS buffer, pH 7.4.

3. Boil 3 ml of a 8% SDS solution in a tightly closed glass tube to avoid evaporation. Insert a magnetic stir rod. Add 3 ml of bacterial suspension dropwise to the boiling SDS solution. Maintain vigorous stirring and boiling of the solution for a minimum of 4 hr. Complete solubilization of membranes and DNA is obtained under these conditions. Keep the solution at 80° overnight.

Day 2. Centrifuge the sample at $300,000g$, 20 min, 30°. Resuspend the pellet containing insoluble macromolecular peptidoglycan, in warm distilled water (60°). Repeat the washing process at least four times to eliminate SDS completely. This is easily monitored by the loss of bubbling on pellet suspension. The pellet should have a quasi-transparent appearance and it is finally resuspended in water or the appropriate buffer (see below) and stored frozen ($-20°$).

Quantification of Peptidoglycan

The most widely used method to determine the amount of peptidoglycan purified from gram-negative bacteria relies on the estimation of the relative amount of diaminopimelic acid (DAP). This diamino acid can be quantified by a colorimetric

[23] C. M. Collazo, M. K. Zierler, and J. E. Galán, *Mol. Microbiol.* **15,** 25 (1995).
[24] F. García-del Portillo and B. B. Finlay, *Infect. Immun.* **62,** 4641 (1994).

assay based on the method originally described by Work.[25] Briefly, the macro-molecular peptidoglycan is hydrolyzed in 6N HCl for 14 hr at 95°. This treatment causes the rupture of the peptide bonds. The next day, samples are dried by evaporation and resuspended in distilled water. A volume of the sample is mixed with an equal volume of glacial acetic acid and ninhydrin reagent (250 mg of ninhydrin in 4 ml of 0.6 M phosphoric acid and 6 ml of glacial acetic acid). DAP gives a yellow color and thus the peptidoglycan concentration is determined by measuring absorbance at 436 nm and comparing the absorbance against samples containing defined concentrations of the diamino acid. The typical yield for *S. typhimurium* is about 6–10 mg of pure peptidoglycan per 200 ml of overnight culture harvested at an OD$_{550}$ of 1.0. Loss of macromolecular peptidoglycan due to the numerous centrifugation steps included in the purification protocol can be traced by radioactive labeling of the peptidoglycan.[20] The amino sugars and amino acids present in the peptidoglycan can be analyzed and quantified by acid hydrolysis of macromolecular peptidoglycan and further HPLC or standard ion-exchange chromatography.[26]

Analysis of Muropeptide Composition of Peptidoglycan by HPLC

In many instances it is necessary to monitor whether structural alterations occur in the peptidoglycan. This is the case in phenotypic analysis of mutations in genes encoding enzymes of peptidoglycan metabolism, in changes in growth parameters (temperature, nutrient composition, phase of the cell cycle, etc.), or in the effect of antibiotics that block peptidoglycan metabolism. High-performance liquid chromatography (HPLC) has been used to define the number and relative amount of the complex mixture of muropeptide species that can form the peptidoglycan.[27] The application of novel and powerful mass spectrometric methods such as the plasma desorption time-of-flight mass spectrometry has been used to discern the exact chemical structure of muropeptides.[27,28] The material to be used in these techniques requires prior digestion of the macromolecular peptidoglycan with a specific set of enzymes that can remove both high molecular weight carbohydrates trapped in the peptidoglycan lattice and covalently bound proteins. Our procedure is based on the method described by Glauner.[22]

Day 1

1. Resuspend the pellet containing purified macromolecular peptidoglycan in a solution of 10 mM NaCl–10 mM Tris-HCl pH 7.0. Treat with α-amylase

[25] E. Work, *Biochemistry* **67**, 416 (1957).

[26] B. Glauner, J. V. Holtje, and U. Schwarz, *J. Biol. Chem.* **263**, 10088 (1988).

[27] G. Allmaier and E. R. Schmid, *in* "Bacterial Growth and Lysis" (M. A. de Pedro, J. V. Höltje, and W. Löffelhardt, eds.). Plenum Press, New York and London, 1993.

[28] M. Roos, E. Pittenauer, E. Schmid, M. Beyer, B. Reinike, G. Allmaier, and H. Labischinski, *J. Chromatogr. B. Biomed. Sci. Appl.* **705**, 183 (1998).

(100 μg/ml, 2 hr, 37°) to remove high molecular weight glycogen that could be trapped in the peptidoglycan lattice.

2. Add an appropriate volume of a concentrated pronase E solution (10 mg/ml to a final concentration of 100 μg/ml). Incubate for 1 hr at 60°. The pronase E enzyme is a protease that has to be preactivated by incubation of the concentrated solution at 60° for 1 hr. The pronase E treatment causes the digestion of any protein covalently or firmly associated to the peptidoglycan.

3. Inactive the pronase E enzyme by adding SDS to a 4% final concentration. Incubate the sample at 100° for 30 min.

4. Remove the SDS by successive centrifugations (300,000g, 20 min, 30°) until bubbling due to the detergent is no longer observed.

5. Resuspend macromolecular peptidoglycan in muramidase buffer (50 mM sodium phosphate buffer, pH 4.9) and add Cellosyl [muramidase from the fungus *Chalaropsis* (Hoechst, Frankfurt, Germany)] to a final concentration of 20 μg/ml. Incubate for at least 14 hr at 37°. Similar to lysozyme, the *Chalaropsis* muramidase cleaves the glycosidic bond between the C-1 atom of *N*-acetylmuramic acid and the C-4 atom of *N*-acetylglucosamine.

Day 2

1. Stop the *Chalaropsis* muramidase treatment by boiling the sample for 10 min.

2. Separate the muramidase digestion products (muropeptides) from undigested macromolecular peptidoglycan by centrifugation (15,000g, 15 min, room temperature). The amount of muropeptides that are typically obtained from a 200 ml bacterial culture (OD$_{550}$ = 1.0) is 0.4–0.5 mg. The supernatant containing the muropeptides is further lyophilized, dissolved in 250 μl of Milli-Q water (Millipore, Bedford, MA) and stored frozen (−20°).

3. To avoid the different anomeric configurations of muropeptides with a free reducing sugar and therefore the appearance of two to four peaks per compound in the chromatographic analysis, the muramic acid is reduced to corresponding sugar alcohol (muramitol). To achieve this reduction, a 1 : 10 volume of 1 M borate buffer pH 9.0 is added to the solution containing soluble muropeptides. Then sodium borohydride (10 mg/ml final concentration) is immediately added. Samples are incubated for 30 min at room temperature. The reaction is stopped with phosphoric acid (of highest purity grade) and adjusted to pH 4.0.

4. Prior to the HPLC analysis, the muropeptide solution is filtered through Millipore filters (SJHV, 0.45 μm pore size) to eliminate insoluble impurities. Reduced muropeptide-containing samples can be stored at −20° until HPLC chromatography.

We perform the HPLC analysis in Spectra Physics equipment (model SP8700), connected to a Waters detector (model 481), and a Hewlett-Packard printer (model

1050). The type of reversed-phase column used is Hypersil RP18 (250 mm by 4 mm, 3 μm particle size). Elution buffers are 50 mM sodium phosphate, pH 4.35 (buffer A), and 15% (v/v) methanol in 75 mM sodium phosphate, pH 4.95 (buffer B). The elution conditions are 7 min of isocratic elution in buffer A; 115 min of linear gradient to 100% of buffer B; and 30 min of isocratic elution in buffer B. The flow rate is adjusted to 0.5 ml/min and the temperature of the column to 40°. Muropeptides are detected by measuring the absorbance at 204 nm. The HPLC analysis allows resolution of a complex mixture of muropeptides in the peptidoglycan of *Escherichia coli*.[22,26] About 30–50 μg of total muropeptides is used per chromatographic separation yielding a clear peak pattern. This experimental approach allowed us to discern subtle differences between the peptidoglycan structure in laboratory-grown *S. typhimurium* versus the same bacteria growing within cultured mammalian cells.[29]

Analysis of Role of Peptidoglycan in Functionality of *Salmonella* Invasion Apparatus

The *S. typhimurium* invasion apparatus is assembled in the bacterial cell envelope as a macromolecular complex that spans both the inner and outer membranes.[16,30,31] Like flagella, the existence of interactions between components of the invasion apparatus and the peptidoglycan layer is conceivable. If this hypothesis is true, structural modifications in the peptidoglycan should affect the capacity of bacteria to invade eukaryotic cells. To introduce changes in the peptidoglycan structure we routinely use a method first described for *Escherichia coli*, which relies on the incubation of bacteria in medium containing specific D-amino acids such as D-methionine (D-Met) or D-phenylalanine (D-Phe).[32] Under these conditions the peptidoglycan is modified by live bacteria, which incorporate these D-amino acids into the peptidoglycan. This reaction takes place by competition of the D-amino acid versus the natural D-Ala substrate. Interestingly, a maximum of peptidoglycan modification (about 25% of total peptidic side chains) can be achieved in *E. coli* under growth conditions in which neither morphology nor cell integrity is affected.[32] HPLC analysis of muropeptides obtained from these bacteria denotes the appearance of new muropeptides containing the corresponding D-amino acids.[32] In the case of *S. typhimurium*, an overnight growth of bacteria in LB medium containing 5 mg/ml of either D-Met or D-Phe causes a negligible loss of viability (<5%). The normal bacillar morphology and motility are also retained (M. G. Pucciarelli, C. Quintela, M. A. de Pedro, and F. García-del Portillo,

[29] J. C. Quintela, M. A. de Pedro, P. Zollner, G. Allmaier, and F. García-del Portillo, *Mol. Microbiol.* **23**, 693 (1997).

[30] T. G. Kimbrough and S. I. Miller, *Proc. Natl. Acad. Sci. U.S.A.* **97**, 11008 (2000).

[31] A. Sukhan, T. Kubori, J. Wilson, and J. E. Galán, *J. Bacteriol.* **183**, 1159 (2001).

[32] M. Caparrós, A. G. Pisabarro, and M. A. de Pedro, *J. Bacteriol.* **174**, 5549 (1992).

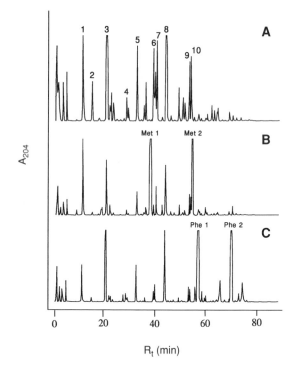

FIG. 1. Chromatographic HPLC analysis that demonstrates the *in vivo* incorporation of D-Met and D-Phe into the *S. typhimurium* peptidoglycan. Bacteria were grown overnight in: (A) LB medium; (B) LB medium containing 5 mg/ml D-Met; or, (C) LB medium plus 5 mg/ml D-Phe. Muropeptides were obtained after muramidase treatment of purified macromolecular peptidoglycan (see text for details), fractionated in a HPLC column and simultaneously detected by a UV detector set at 204 nm. Muropeptides marked from 1 to 10 are the major muropeptides present in *E. coli* peptidoglycan [B. Glauner, *Anal. Biochem.* **172**, 451 (1988)]. Identification of D-amino acid-containing muropeptides was made by comparison with modified muropeptides of known structure in D-amino acid-treated *E. coli* [M. Caparrós, A. G. Pisabarro, and M. A. de Pedro, *J. Bacteriol.* **174**, 5549 (1992)].

unpublished results, 2000). To discern whether the modification occurs at the same extent as has been previously reported for *E. coli,* the peptidoglycan is purified and processed for HPLC analysis of the muropeptide composition. The appearance in *S. typhimurium* of a new characteristic group of muropeptides containing the D-amino acid confirms the alteration of peptidoglycan structure (Fig. 1).

Once it has been confirmed that the peptidoglycan of *S. typhimurium* is modified by the incorporation of D-amino acids, two functional assays indicative of the functionality of the invasion apparatus can be performed. First, Western analysis of the levels of effector proteins secreted by the apparatus can be tested in the presence of D-amino acids. Prepare samples from bacteria grown in the presence of the respective L-isomers (L-Met, L-Phe) or the D-amino acid D-proline. None of these latter amino acids incorporate into macromolecular peptidoglycan,[32] thus serving as appropriate controls in the experiment. The secretion of effector invasion proteins, such as Sips, depends on the functionality of the *Salmonella* invasion apparatus,[33] and it can be monitored in bacteria grown in the presence of these amino acids. The extract containing secreted proteins is prepared following the method described by Kaniga *et al.* based on filtration of the culture supernatant, trichloroacetic protein precipitation, and acetone washing.[34] If secretion is impaired because of the modification of the peptidoglycan structure, lower levels of extracellular Sip proteins will be observed. A complementary assay is the determination of the invasion rate of cultured epithelial cells by *S. typhimurium* harboring a modified peptidoglycan. In this case, bacteria grown overnight in the presence of D-Met or D-Phe (5 mg/ml) are used to infect cultured nonphagocytic cells such as the HeLa human epithelial cell line following standardized methods.[35,36] Under these conditions, it is essential to confirm that the viability of D-amino acid-treated bacteria used to infect the eukaryotic cells remains minimally altered. We use a series of viability tests, including: (i) the determination of colony forming units (cfu) by plating in LB-agar plates; (ii) the estimation of cell lysis by detecting in the culture supernatant cytoplasmic proteins as the ribosomal elongation factor EF-Tu or the chaperonin GroEL; and (iii) the microscopic observation of D-amino acid-treated bacteria resuming growth in fresh medium lacking the D-amino acid but containing the β-lactam antibiotic furazlocillin (1.5 μg/ml), which blocks specifically cell division. This last assay is probably the most reliable since it can detect loss of viability without concomitant cell lysis and bacteria are not subjected to extensive handling. If bacterial viability has not been affected by the D-amino acid treatment, the entire bacterial population is visualized by microscopy as elongated bacteria after growing at 37° for 5 hr. Recording the percentage of "nonfilamented" bacteria provides a direct quantification of nonviable cells. Our experience is that overnight growth of *S. typhimurium* at 37° in a static nonshaking LB medium containing 5 mg/ml of either D-Met or D-Phe causes a negligible loss of viability ($\leq$5%).

Analysis of Association of *Salmonella* Invasion Proteins to Peptidoglycan

Membrane proteins that interact with peptidoglycan have been characterized biochemically using methods that rely in subcellular fractionation and identification of proteins that withstand solubilization in solutions containing SDS.

[33] J. E. Galán, *Curr. Opin. Microbiol.* **2,** 46 (1999).
[34] K. Kaniga, S. Tucker, D. Trollinger, and J. E. Galán, *J. Bacteriol.* **177,** 3965 (1995).
[35] M. G. Pucciarelli and B. B. Finlay, *Methods Enzymol.* **236,** 438 (1994).
[36] R. Ménard and P. J. Sansonetti, *Methods Enzymol.* **236,** 493 (1994).

Membrane proteins such as the Braun's (murein) lipoprotein of *E. coli,*[37] or the major outer membrane protein of *Legionella pneumophila,*[38] are not completely extracted from macromolecular peptidoglycan by boiling in 4% SDS. In this case, the fraction of the protein that remains insoluble is covalently bound to peptidoglycan. On the other hand, proteins that associate to peptidoglycan in a noncovalent form are solubilized by SDS at temperatures above 56°–60°, as occurs with the major outer membrane protein A (OmpA) of *E. coli.* Other investigators have cataloged peptidoglycan-associated proteins as those resisting solubilization in 2% SDS at 30°.[39,40] In addition, membrane proteins have also been considered as associated to peptidoglycan if they are efficiently cross-linked to macromolecular peptidoglycan with low molecular weight dithiol-containing cross-linker reagents.[41] In this latter case, the extremely short distance between the groups that are cross-linked (estimated as ≤1.2 nm for this group of reagents)[41] is considered optimal for promoting protein–peptidoglycan interactions. Finally, the existence of a specific protein–peptidoglycan association will gain further support if the protein under study is directly visualized in purified peptidoglycan sacculi.[42,43] All these methods can be successfully applied to determine whether membrane proteins of TTSS interact with the peptidoglycan.

Subcellular Fractionation: Preparation of Membrane and PG Fractions

Traditionally, subcellular fractionation of bacteria involves the rupture of the cells to obtain cytosolic and cell envelope fractions. In gram-negative bacteria, the cell envelope material is further treated with nonionic detergents (Triton X-100, sodium *N*-lauroylsarcosinate) to partition the sample into two fractions: one enriched in inner membrane proteins (material soluble in the detergent) and other containing mostly outer membrane proteins (material insoluble in the detergent). The ease of extraction by the detergent often relates to the relative size of the protein not inserted in the membrane. Thus, many outer membrane lipoproteins, which are anchored to the membrane only by their lipid portion, can be substantially extracted with nonionic detergents, as has been shown for the *Salmonella* outer membrane lipoprotein InvH, involved in bacterial invasion.[44,45] Subsequently, the soluble and insoluble membrane fractions are boiled in sample buffer (4×: 200 m*M*

[37] V. Braun, *Biochim. Biophys. Acta* **415**, 335 (1975).
[38] C. A. Butler and P. S. Hoffman, *J. Bacteriol.* **172**, 2401 (1990).
[39] J. C. Lazzaroni and R. Portalier, *Mol. Microbiol.* **6**, 735 (1992).
[40] E. Bouveret, H. Benedetti, A. Rigal, E. Loret, and C. Lazdunski, *J. Bacteriol.* **181**, 6306 (1999).
[41] M. Leduc, K. Ishidate, N. Shakibai, and L. Rothfield, *J. Bacteriol.* **174**, 7982 (1992).
[42] B. Walderich and J. V. Holtje, *J. Bacteriol.* **173**, 5668 (1991).
[43] Z. Li, A. J. Clarke, and T. J. Beveridge, *J. Bacteriol.* **178**, 2479 (1996).
[44] A. M. Crago and V. Koronakis, *Mol. Microbiol.* **30**, 47 (1998).
[45] S. Daefler and M. Russel, *Mol. Microbiol.* **28**, 1367 (1998).

Tris-HCl pH 6.8, 8% SDS, 8% 2-mercaptoethanol, 40% glycerol, and 0.04% bromphenol blue) for 10 min and loaded on gels. We routinely separate proteins by SDS–polyacrylamide gel electrophoresis (8 or 10% acrylamide gels) using the Tris–Tricine system described by Schägger and von Jagow.[46] Following the standardized method of membrane fraction preparation, the macromolecular peptidoglycan remains insoluble and is removed by centrifugation before sample loading. Obviously, when discarding macromolecular peptidoglycan no information is obtained on proteins covalently or firmly associated that are nonextracted by boiling SDS. Considering this relevant aspect, we have optimized a method to analyze the proteins present in the SDS-insoluble material. This fraction has been termed the peptidoglycan or PG fraction. The method has the following steps.

Day 0. In 1000-ml flasks inoculate 200 ml of LB medium with bacteria from a single colony previously grown on LB-agar plates. Incubate the culture overnight at 37° under nonshaking conditions.

Day 1

1. Monitor the bacterial growth by measuring the optical density at 550 nm (OD_{550}). Under the conditions used, the overnight culture reaches an $OD_{550} = 1.0$ (approximately 10^9 colony-forming units/ml). Thaw the culture in a mixture of ice and NaCl.

2. Harvest bacteria by centrifugation ($12,000g$, 15 min, 4°). Suspend the bacterial pellet in 1/10 volume of phosphate-buffered saline (PBS), pH 7.4. Repeat the centrifugation step and suspend bacteria in the same final volume of PBS buffer, pH 7.4. Separate 0.4 ml for preparation of membrane material. The rest of the sample is used to obtain the peptidoglycan (PG) fraction (see below).

Preparation of membrane fractions

1. Rupture cells by sonication (3 pulses, 20 sec). Remove unbroken cells by centrifugation ($5,000g$, 5 min, 4°).

2. Perform a high-speed centrifugation ($200,000g$, 20 min, 4°) to obtain cell envelope material.

3. Resuspend envelope material in 100 μl of phosphate-buffered saline (PBS) pH 7.4 buffer containing 0.4% Triton X-100. Incubate at 4° for 3 hr.

4. Centrifuge the sample ($15,000g$, 15 min, 4°). Remove supernatant cautiously so as not to take any material present in the pellet. This fraction contains Triton-soluble membrane material, enriched with inner membrane proteins.

5. Wash the pellet with 1 ml PBS pH 7.4 buffer containing 0.4% Triton X-100. Repeat the centrifugation ($15,000g$, 15 min, 4°).

6. Resuspend the pellet in 100 μl of PBS pH 7.4 buffer containing 0.4% Triton X-100. This fraction contains the Triton-insoluble material enriched with outer membrane proteins.

---

[46] H. Schägger and G. von Jagow, *Anal. Biochem.* **166**, 368 (1987).

7. Add to the detergent-soluble and -insoluble membrane fractions an appropriate volume of concentrated Tris–Tricine sample buffer. Boil the samples for 10 min. Centrifuge in a table-top centrifuge (15,000g, 15 min, room temperature) to remove macromolecular peptidoglycan and load the samples in gels.

Preparation of peptidoglycan (PG) fraction

1. Centrifuge bacteria and suspend in cold PBS buffer, pH 7.4 to a final volume of 3 ml. Add bacteria to an equal volume of boiling 8% SDS and obtain pure macromolecular peptidoglycan as described (see the purification of macromolecular peptidoglycan).

2. Treat with *Chalaropsis* spp. muramidase as described above (see the analysis of muropeptide composition of the peptidoglycan by HPLC). Note that the α-amylase and protease treatments are omitted to preserve integrity of proteins that associate with peptidoglycan.

3. Stop the muramidase treatment by heating at 100° for 10 min. Under these conditions, proteins released from macromolecular peptidoglycan precipitate as a result of acid pH (pH 4.9 of muramidase buffer) and heat (100°, 10 min) treatments.

4. Centrifuge the sample at 15,000g, 15 min, room temperature. If desired, store the supernatant to test whether proteins are still present.

5. Resuspend the pellet in 50 μl of PBS pH 7.4 buffer. Add an appropriate volume of concentrated Tris–Tricine sample buffer. Boil the sample for 10 min. Centrifuge in table-top centrifuge (15,000g, 15 min, room temperature) to remove any undigested macromolecular peptidoglycan. Load the supernatant on gels. This fraction contains exclusively proteins that associate to peptidoglycan either by covalent linkages (like the Braun murein lipoprotein) or by strong noncovalent interactions not disrupted by extensive boiling in 4% SDS.

Salmonella membrane proteins that form part of the invasion apparatus can be analyzed for their relative distribution in these envelope fractions by Western analysis using specific antibodies. Among these components, lipoproteins such as PrgK and InvH are *a priori* putative components that interact with peptidoglycan. Besides their relative small size, PrgK form part of the rod portion of the basal body of the needle complex,[30,31] whereas InvH contributes to an efficient assembly process of the apparatus.[31] Using this method, we have been able to detect both lipoproteins in the PG fraction (M. G. Pucciarelli, C. Quintela, M. A. de Pedro, and F. García-del Portillo, unpublished results, 2000). IagB is another *Salmonella* invasion protein that should be tested for association to PG since it presents similarities to members of a family of lysozyme-like virulence factors.[47] Include as controls in Western analysis proteins expected to be present in the PG fraction,

[47] A. R. Mushegian, K. J. Fullner, E. V. Koonin, and E. W. Nester, *Proc. Natl. Acad. Sci. U.S.A.* **93**, 7321 (1996).

such as the Braun murein lipoprotein and other membrane proteins not known to interact with the peptidoglycan. These controls serve to confirm the purity of the PG fraction and to demonstrate the absence of contaminating proteins that could eventually be trapped in the peptidoglycan lattice.

Cross-Linking of *Salmonella* Invasion Proteins to Peptidoglycan

Membrane proteins that interact with peptidoglycan can be identified with bi-functional dithiol-containing compounds that cross-link amino groups that lie less than 1.2 nm apart.[41] The rationale is that if a certain protein is in high proximity to the peptidoglycan, it should be effectively cross-linked with these bifunctional cross-linker reagents. The work of Leduc *et al.* showed that most of the cell envelope lipoproteins interact with the peptidoglycan.[41] This type of assay can be used to test whether components of the *Salmonella* invasion apparatus are cross-linked to macromolecular peptidoglycan. Whereas the method described by Leduc *et al.* for *E. coli* shows no difference when using either DSP (dithiobissuccinimidylpropionate) or DTBP (dimethyl-3,3'-dithiobispropionimidate) as cross-linkers,[41] our experience with *S. typhimurium* has provided successful results only with the DSP reagent. The exact reasons for these differences are unknown, although they might rely on a different penetration rate of the reagents through the lipopolysaccharide outer layer present in virulent isolates of *S. typhimurium,* a molecule that lacks the O-antigen part in most of the *E. coli* laboratory strains. The main steps of the method we routinely used with *S. typhimurium* are as follows.

1. Grow bacteria overnight in 200 ml of LB medium under nonshaking conditions at 37°.
2. Harvest bacteria by centrifugation (12,000g, 15 min, 4°) and suspend in 1 ml of 50 mM sodium phosphate pH 7.4 containing 20% sucrose to plasmolyze the cells.
3. Add 125 μl of a DSP stock solution [80 mg/ml in dimethyl sulfoxide (DMSO)]. Incubate for 30 min at room temperature.
4. Remove the DSP reagent by centrifugation (15,000g, 15 min, 4°). Resuspend the bacterial pellet in cold PBS pH 7.4 buffer.
5. Purify macromolecular peptidoglycan by adding bacteria to an equal volume of boiling 8% SDS (see the purification of macromolecular peptidoglycan). Leave the peptidoglycan intact by omitting the α-amylase/pronase and muramidase treatments.
6. Resuspend the macromolecular peptidoglycan in a suitable volume (50 μl) of PBS pH 7.4 buffer. Release cross-linked proteins by adding an appropriate volume of solubilization buffer containing 1% SDS and 1.5 M 2-mercaptoethanol.[41] Heat the sample for 10 min at 100°. Remove macromolecular peptidoglycan by centrifugation (15,000g, 15 min, room temperature) and load the supernatant onto gels.

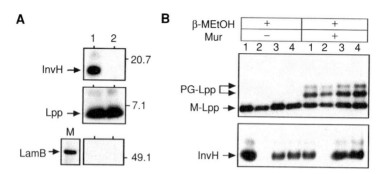

FIG. 2. (A) Cross-linking of the *S. typhimurium* invasion protein InvH to the peptidoglycan layer. Isogenic wild-type (lane 1) and *invH* mutant (lane 2) strains of *S. typhimurium* were treated with the cross-linker reagent dithiobissuccinimidylpropionate (DSP) (see text). After peptidoglycan purification, proteins cross-linked directly or indirectly to the peptidoglycan were released with 1% SDS and 1.5 *M* 2-mercaptoethanol. The extract containing these proteins was tested for the presence of InvH, the murein lipoprotein (Lpp), and the outer membrane protein LamB. Note the efficient cross-linkage of both the invasion lipoprotein InvH and Lpp. By contrast, none of the membrane-associated LamB (marked as M) is cross-linked to peptidoglycan in this assay. (B) Distinction between the Lpp molecules that are covalently (PG-Lpp) and noncovalently (M-Lpp) associated to the peptidoglycan in different *S. typhimurium* strains: wild type (1), *invH* (2), *invA* (3), and *invJ* (4). Note the visualization of both forms when disulfide bonds are disrupted and the peptidoglycan is digested with muramidase. In the case of InvH, the same forms are observed irrespective of the muramidase treatment.

The sample containing the proteins that were cross-linked either directly or indirectly (via an intermediate protein) to peptidoglycan can be analyzed by Western assays for the presence of *Salmonella* invasion proteins. Thus, when samples are prepared from isogenic wild-type and *invH* mutant strains of *S. typhimurium,* the efficient cross-linking of the lipoprotein InvH to macromolecular peptidoglycan is evident (Fig. 2A). When these samples are tested for the presence of cross-linked murein lipoprotein (Lpp), an equal amount of this protein is cross-linked in both strains (Fig. 2A). The Lpp molecules that are released on rupture of the disulfide bonds belong to the 70% fraction that is not covalently bound to peptidoglycan. The Lpp molecules that are covalently bound can be detected if a muramidase digestion is included in the protocol before the 2-mercaptoethanol treatment (Fig. 2B). In this case, the Lpp cross-linked to macromolecular peptidoglycan and the covalently bound form migrate differently and are detected simultaneously in the gel (Fig. 2B). When this assay is performed to detected InvH, no differences in the migration of the protein are observed irrespective of the muramidase digestion (Fig. 2B).

Electron Microscopy Analysis of Peptidoglycan-Associated Proteins

If biochemical evidence has been obtained for a firm association between the peptidoglycan and the protein of interest (e.g., resistance to total extraction by boiling in 4% SDS), a further assay to visualize the protein in purified peptidogly-can sacculi is recommended. Although not broadly used, the method is simple and has provided indications for strong protein–peptidoglycan associations in the case of the soluble transglycosylase of *E. coli*[42] and a major autolysin of *Pseudomonas aeruginosa*.[43] The immunodetection of proteins in purified peptidoglycan sacculi requires a few modifications in the standard protocol. First, no enzymatic treatment is used to preserve protein integrity. Second, the peptidoglycan sacculi are repeat-edly boiled and washed in solutions containing SDS to improve purity of the sacculi for the electron microscopy analysis.[48] Carbon-coated copper grids (200 mesh) are glow discharged (10 min) and the purified peptidoglycan sacculi adsorbed to the grid for 15 min at room temperature. A solution containing 0.5% bovine serum al-bumin (BSA)–0.2% gelatin–PBS pH 7.4 is used as blocking and antibody buffer. Preblocking is performed for 15 min and incubation with the specific primary antibody is carried out for 60 min. As secondary antibody, anti-species-specific antibody conjugated to gold particles (6 or 10 nm) is used. Alternatively, when using rabbit antibody as primary antibody, protein A conjugated to gold particles can be used. The peptidoglycan sacculi are finally stained with 1% (w/v) uranyl acetate in water, washed in water, and air dried. The sample can be observed in a Jeol1200EX electron microscope at an acceleration voltage of 60 kV. Diverse controls should be included in the immunodetection assays such as peptidoglycan sacculi incubated only with secondary antibody (or protein A) and, if possible, label peptidoglycan sacculi of a mutant bacterial strain lacking the protein under study. To get a rough estimation of the relative amount of protein bound to pepti-doglycan, parallel labeling of peptidoglycan sacculi with anti-murein lipoprotein is recommended. Obviously, this complementary assay relies on the availability of specific anti-murein lipoprotein antibodies. It has been estimated that 2×10^5 molecules of murein lipoprotein are covalently bound to peptidoglycan per *E. coli* cell.[37,49] This number can be useful to evaluate the relative amount of the protein of interest that is associated to peptidoglycan.

Concluding Remarks

The peptidoglycan has unique chemical characteristics and a very precise lo-cation in the bacterial cell envelope. Therefore, this polymer constitutes a foremost candidate to serve as a molecule through which large macromolecular complexes assemble and freeze their relative position in the cell envelope. The methods described in this article, which are summarized in Fig. 3, offer the opportunity to demonstrate the existence of specific interactions between members of TTSS and the peptidoglycan layer. Even in the case of the flagella, which is perhaps

[48] M. A. de Pedro, J. C. Quintela, J. V. Holtje, and H. Schwarz, *J. Bacteriol.* **179**, 2823 (1997).
[49] M. Inouye, J. Shaw, and C. Shen, *J. Biol. Chem.* **247**, 8154 (1972).

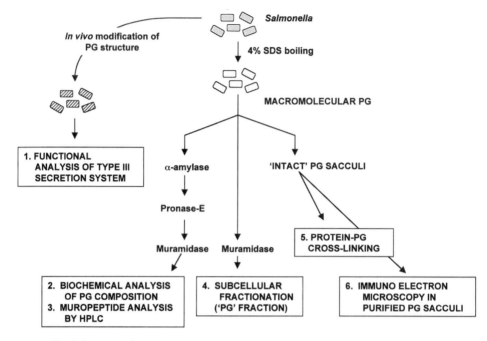

FIG. 3. Summary of the diverse methods described in this article. Note the extensive enzymatic treatment of the macromolecular peptidoglycan required for muropeptide analysis. The existence of protein–peptidoglycan interactions can be analyzed by diverse methods including subcellular fractionation, protein cross-linking, and electron microscopy analysis of purified peptidoglycan sacculi.

the surface-located macromolecular complex most profoundly studied to date, a biochemical proof for the proposed protein–peptidoglycan interactions is still lacking. The assays described here, which focus fundamentally on the analysis of the proteins remaining associated to macromolecular peptidoglycan on extensive boiling in SDS, will certainly provide the opportunity to obtain this valuable information. In the case of virulence-associated TTSS, our work with the *S. typhimurium* invasion apparatus indicates that the peptidoglycan is required for functionality of the apparatus and that distinct components of the apparatus might have strong interactions with the peptidoglycan. In this respect, the methods reported for isolation of *Salmonella* and *Shigella* needle complexes include treatment with a peptidoglycan-digesting enzyme such as lysozyme.[16–19,30,50] Further experimentation is required to determine whether this muramidase treatment may have some consequences in removing accessory proteins mediating interactions with the peptidoglycan layer. The isolation of needle complexes without peptidoglycan enzymatic digestion is without any doubt a major methodological challenge.

[50] T. Kubori, A. Sukhan, S. I. Aizawa, and J. E. Galán, *Proc. Natl. Acad. Sci. U.S.A.* **97,** 10225 (2000).

Additional areas of interest are the analysis of protein–peptidoglycan interactions in bacterial mutants harboring partial or fully assembled needle complexes as well as the identification of the protein domain(s) mediating their association to the peptidoglycan.

Acknowledgments

We thank Miguel A. de Pedro, Carlos Quintela, and Jorge E. Galán for helpful discussions on protein–peptidoglycan interactions. Work in our laboratory is supported by a grant from "Comunidad de Madrid" (08.2/0045. 1/2000). M.G.P. is supported by a postdoctoral fellowship from the Consejería de Educación de la Comunidad de Madrid.

Section VII

Quorum Sensing and Gene Regulation

[30] RP4-Based Plasmids for Conjugation between *Escherichia coli* and Members of the Vibrionaceae

By ERIC V. STABB and EDWARD G. RUBY

Introduction

The Vibrionaceae family includes both important pathogens and useful model organisms in the study of processes ranging from the regulation of light production by quorum sensing to the establishment of mutualistic animal–bacteria associations. Genetics has proven a powerful discipline in the study of *Vibrio* species, and the refinement of genetic tools for use with these important and interesting bacteria will facilitate future studies of their biology.

The self-transmissible broad host-range IncPα plasmid RP4[1] provides the means for genetic manipulation of many diverse bacteria, including members of the Vibrionaceae. Derivatives of RP4 facilitate the conjugal transfer of vectors engineered using *Escherichia coli* as a host into other bacterial species of interest. As a general strategy, the *cis*-acting origin of transfer, *oriT* or "mob," is cloned into vectors, which can then be mobilized if *trans*-acting *tra*-and *trb*-encoded RP4 transfer functions are provided. The *trans*-acting transfer functions are typically provided by an RP4-derived helper plasmid such as pRK2013[2,3] or by strains such as β2155, S17-1, or SM10, which have chromosomally integrated helper plasmids.[4,5] The RP4-based helpers pRK2013, S17-1, and SM10, together with the pSUP series of *oriT*-containing plasmids,[5,6] have been the foundation for genetic manipulations in many gram-negative bacteria including the Vibrionaceae. We have improved the utility of RP4-based conjugation for use with the Vibrionaceae (and other gram-negative bacteria) by modifying conjugal tools to meet three criteria.

Criterion 1: An ideal conjugal helper should not introduce insertion elements into recipient cells, except where that function is intended. If this criterion is not met, transposition of insertion elements from the donor into recipient DNA could complicate interpretation of experimental data. Helper plasmids and the chromosomes of helper strains contain *oriT* and are themselves self-transmissible; their use can result in unintended transfer of transposable elements into recipients.

[1] W. Pansegrau, E. Lanka, P. T. Barth, D. H. Figurski, D. G. Guiney, D. Haas, D. R. Helinski, H. Schwab, V. A. Stanisich, and C. M. Thomas, *J. Mol. Biol.* **239,** 623 (1994).
[2] D. H. Figurski and D. R. Helinski, *Proc. Natl. Acad. Sci. U.S.A.* **76,** 1648 (1979).
[3] G. Ditta, S. Stanfield, D. Corbin, and D. R. Helinski, *Proc. Natl. Acad. Sci. U.S.A.* **77,** 7347 (1980).
[4] C. Dehio and M. Meyer, *J. Bacteriol.* **179,** 538 (1997).
[5] R. Simon, U. Priefer, and A. Pühler, *Biotechnology* **1,** 784 (1983).
[6] R. Simon, M. O'Connell, M. Labes, and A. Pühler, *Methods Enzymol.* **118,** 640 (1986).

For example, SM10, S17-1, and β2155 can transfer dozens of insertion elements native to *E. coli* (e.g., IS*1*, IS*2*, IS*3*, IS*4*, IS*5*, IS*30*, IS*150*, and IS*186*) into recipients. Each of these strains also carries transposable elements on their integrated helper plasmids. For example S17-1 contains a chromosomal copy of RP4-2-Tc::Mu-Km::Tn7 harboring Tn7 and Mu, both of which have been shown to transpose in *Vibrio* species. Furthermore, we found that S17-1 can transfer Tn7 to *Vibrio fischeri* recipients. Helper plasmid pRK2013 lacks the transposable elements IS*21* and Tn*1* of its parent vector, RP4; however, an examination of pRK2013's construction,[7–11] and Southern blotting (unpublished data) revealed that pRK2013 carries Tn*903*. Although unintended transposition of Tn*903* or Tn7 into recipients can often be readily detected by screening for drug resistances encoded by these transposons, the presence of other elements (e.g., Mu, IS*1*, IS*903*) can only be screened by laborious PCR- or hybridization-based methods. Accidental incorporation of foreign insertion elements into recipient strains can be virtually eliminated by the use of a conjugal helper plasmid lacking insertion elements. We therefore constructed three such helper plasmids (pEVS101, pEVS103, and pEVS104) that lack insertion elements.

Criterion 2: For many purposes, a conjugal helper that replicates in the donor, but not in the recipient, is desirable. pRK2013 contains the ColE1 plasmid origin which allows its replication in certain recipients, particularly members of the Enterobacteriaceae and Vibrionaceae. Although pRK2013 can be readily lost from *Vibrio* strains during nonselective growth, its retention under selective pressure can be problematic in two ways. First, using pRK2013 we could not select for mobilization of constructs conferring kanamycin resistance (KnR), because this marker is found on pRK2013 and kanamycin (Kn) selection identifies, with high frequency, *Vibrio* transconjugants retaining pRK2013. Second, when the goal is to mobilize suicide plasmids for marker exchange or transposon mutagenesis, the homologous *oriT* regions of the helper and the suicide plasmid it is mobilizing can recombine, and the resulting product can be replicated by the helper's ColE1 origin. Thus, plasmid–plasmid recombinants may be the unintentional products of a mating intended to produce chromosomal insertions. We therefore constructed helper plasmid pEVS104 such that it contains only the R6Kγ origin of replication,[12] which is not maintained in *Vibrio* recipients.

Criterion 3: Optimally, mobilizable vectors should contain only a defined, minimal *oriT*. The *cis*-acting *oriT* region in many mobilizable (e.g., pSUP) plasmids

[7] S. N. Cohen, A. C. Y. Chang, H. W. Boyer, and R. B. Helling, *Proc. Natl. Acad. Sci. U.S.A.* **70**, 3240 (1973).

[8] M. A. Lovett and D. R. Helinski, *J. Bacteriol.* **127**, 982 (1976).

[9] A. Oka, H. Sugisaki, and M. Takanami, *J. Mol. Biol.* **147**, 217 (1981).

[10] R. P. Silver and S. N. Cohen, *J. Bacteriol.* **110**, 1082 (1972).

[11] T. Watanabe, C. Ogata, and S. Sato, *J. Bacteriol.* **88**, 922 (1964).

[12] R. Kolter, M. Inuzuka, and D. R. Helinski, *Cell* **15**, 1199 (1978).

is undefined[5,6] and includes *tra* genes flanking the *oriT* (i.e., *traJ* and *traK*) that are nonessential for *cis*-acting transfer function. The expression of these genes by mobilized plasmids is unnecessary and could present a burden on recipient cells making plasmid carriage unfavorable. Furthermore, the presence of this extraneous DNA can unnecessarily complicate cloning strategies. Plasmids with minimal *oriT* regions have been generated and were conjugally functional.[13,14] We have extended this approach by constructing a series of minimal *oriT* cassettes flanked by convenient restriction sites, facilitating the subcloning of *oriT* into other vectors. Using these cassettes we constructed plasmids with utility for allelic exchange or complementation in *Vibrio* species.

The new RP4-based constructs described here are at least as efficient as existing tools for mediating conjugal transfer between *E. coli* and members of the Vibrionaceae. Helper plasmid pEVS104 mediated conjugation into seven Vibrionaceae species as efficiently as did pRK2013, but, unlike pRK2013, did not result in recipients bearing helper-plasmid-encoded KnR. Shuttle vector pEVS78 was transferred to *V. fischeri* 100-fold more efficiently than pSUP102, and CC118λpir pEVS104 transferred pEVS78 to most *Vibrio* recipients more efficiently than did strain S17-1. Advantages of these new tools include: (i) conjugal helpers that do not introduce transposable elements into recipients, (ii) a conjugal helper, pEVS104, that does not replicate in *Vibrio* recipients, (iii) convenient *oriT* cassettes that are defined and free of extraneous gene sequences, and (iv) mobilizable cloning, marker exchange, and shuttle vectors incorporating this *oriT* cassette. The tools and methods described here should also be useful for mediating conjugation between *E. coli* and many other gram-negative bacteria. Helper plasmid pEVS104 will be particularly advantageous for recipients, such as the Enterobacteriaceae, that replicate ColE1-based vectors or support transposition of IS*903*, Mu, Tn*7*, or *E. coli* IS elements.

General Methods for Cloning and Conjugation

Bacterial strains used in conjugation experiments, and their origins, are listed in Table I. *E. coli* strains S17-1, DH5α, or DH5α-32 are used as the host for plasmids with either ColE1 or p15A origins of replication. Strain BW23474 serves as the host for plasmids with the R6Kγ origin of replication, with the exception of plasmid pEVS104, which is maintained in strain CC118λpir. When added to LB medium[15] for selection of *E. coli,* tetracycline (Tc), ampicillin (Ap), chloramphenicol (Cm), and Kn are used at concentrations of 10, 100, 20, and 40 μg ml^{-1}, respectively. For selection of *E. coli* with erythromycin (Em), 150 μg ml^{-1} is added

---

[13] M. F. Alexeyev and I. N. Shokolenko, *Gene* **160,** 59 (1995).

[14] A. Schafer, A. Tauch, W. Jager, J. Kalinowski, G. Thierbach, and A. Pühler, *Gene* **145,** 69 (1994).

[15] J. H. Miller, "A Short Course in Bacterial Genetics." Cold Spring Harbor Laboratory Press, Cold Spring Harbor, NY, 1992.

TABLE I
BACTERIAL STRAINS, PLASMIDS, AND OLIGONUCLEOTIDES

Strain, plasmid, or oligonucleotide	Relevant characteristics[a]	Source or reference[b]
Bacterial strains		
E. coli		
DH5α	F^- *Φ80dlacZΔM15 Δ(lacZYA-argF)U169 deoR supE44 hsdR17 recA1 endA1 gyrA96 thi-1 relA1*	(1)
DH5α-32	DH5α, demi-Tn5EmR::*trpA*	(2)
BW23474	Δ*lac-169 robA1 creC510 hsdR514 uidA (ΔMlu*I*)::pir-116 endA (BT333) recA1*	(3, 4)
S17-1	*thi pro hsdR hsdM recA* RP4 2-Tc::Mu-KnR::Tn7 (TpR, SpR, SmR)	(5)
CC118λ*pir*	Δ*(ara-leu) araD ΔlacX74 galE galK phoA20 thi-1 rpsE rpoB argE*(Am) *recA1*, lysogenized with λ*pir*	(6)
TH1191	*dam⁻ dcm⁻*	K. Visick
Vibrionaceae		
ES114	*V. fischeri;* wild-type *Euprymna scolopes* light organ isolate	(7)
ESR1	Spontaneous RfR derivative of ES114	(8)
ATCC33934	*V. orientalis*	ATCC
ATCC33869	*V. splendidus*	ATCC
KNH1	*V. parahaemolyticus*	(9)
BB120	*V. harveyi*	B. Bassler
KNH6	*P. leiognathi*	This study
HM21R	*A. veronii;* spontaneous RfR derivative of wild-type isolate HM21	(10)
Plasmids		
pACYC184	p15A ori; CmR, TcR	(11)
pSUP102	RP4 *oriT*-containing, ~2 kb, *Sau*3AI (partial digest) fragment in *Bcl*I site pACYC184; CmR, TcR	(12)
pSUP202	RP4 *oriT*-containing, ~2 kb, *Sau*3AI (partial digest) fragment in *Sau*3AI (partial digest) site pBR325; ColE1 ori, ApR, CmR, TcR	(5)
pBluescriptKS+	ColE1 ori, MCS::*lacZα*; ApR	Stratagene
pBCSK+	pBluescriptSK+, *cat* replaces *bla;* CmR	Stratagene
pCR2.1	ColE1 ori, MCS::*lacZα*, PCR-product cloning vector; ApR KnR	Invitrogen
pUC4K	ColE1 ori, KnR cassette; ApR KnR	(12)
pRK2013	ColE1 ori, RP4 *oriT, trb* and *tra*, R6-5 Tn*903* fragment; KnR	(13)
pKV38	oriR6K, EmR	K. Visick
pEVS70	*oriT* PCR product (from primers EVS49 and EVS50) in pCR2.1	This study
pEVS71	pEVS70 *oriT Xba*I fragment, and *Xba*I-*Xba*I MCS fragments, in pBCSK+ *Spe*I site, *lacZα⁺*	This study
pEVS72	pEVS70 *oriT Xba*I fragment in pBCSK+ *Spe*I site, *lacZα⁻*	This study
pEVS73	pEVS70 *oriT Xba*I fragment, and *Xba*I-*Xba*I MCS fragments, in pBluescriptKS+ *Spe*I site, *lacZα⁺*	This study
pEVS75	*oriT* PCR product (from primers EVS58 and EVS59) in pCR2.1	This study
pEVS76	pEVS75*Alu*I *oriT* fragment in pUC4K *Stu*I site	This study
pEVS77	pEVS71 *Bam*HI *oriT* fragment in pACYC184 *Bcl*I site	This study

TABLE I (*continued*)

pEVS78	pEVS77 Δ*Eco*RV (ΔTcR)	This study
pEVS79	pEVS75 *Alu*I *oriT* fragment in pBCSK+ *Xmn*I site	This study
pEVS94	pEVS72 *Xba*I-*Bam*HI *oriT* fragment in *Xba*I-*Bam*HI digested pKV38; artificial MCS in *Xba*I site	This study
pEVS94S	5′-AATTGGTCGACC-3′ self-annealed and ligated into *Eco*RI site of pEVS94	This study
pEVS99	pKV38 *Xba*I digested, in pRK2013 *Avr*II site	This study
pEVS101	pEVS99 Δ*Stu*I (Δ Tn903)	This study
pEVS103	pEVS76 KnR *oriT Sal*I cassette in pEVS101 *Xho*I site	This study
pEVS104	pEVS103 Δ*Eco*RI (Δ ColE1)	This study
pEVS113	pUC4K KnR *Pst*I fragment in pEVS94S *Pst*I site, Δ*Eco*RV (Δ EmR)	This study
pEVS114	CmR *Pst*I fragment in pEVS94S *Pst*I site, Δ*Eco*RV (Δ EmR)	This study
Oligonucleotides		
EVS49	5′GGA TCC TCT AGA CTG GAA GGC AGT ACA CCT TGA TAG 3′	This study
EVS50	5′GGA TCC TCT AGA TTC CTG CAT TTG CCT GTT TCC AG 3′	This study
EVS58	5′ CAT GAT CGA GCT TAA TTC TGG AAG GCA GTA CAC CTT GAT AG 3′	This study
EVS59	5′ CAT GAT CGA GCT TAA TTC CTG CAT TTG CCT GTT TCC AG 3′	This study

[a] ATCC, American Type Culture Collection (Manassus, VA); MCS, multiple cloning site; ApR, ampicillin resistance; CmR, chloramphenicol resistance; EmR, erythromycin resistance; KnR, kanamycin resistance; RfR, rifampicin resistance; SpR, spectinomycin resistance; SmR, streptomycin resistance; TcR, tetracycline resistance; TpR, trimethoprim resistance.

[b] Key to references: (1) D. Haldiman, *J. Mol. Biol.* **100**, 357 (1983); (2) E. V. Stabb, E. G. Ruby, "American Society for Microbiology, 98th General Meeting," p. 375. American Society for Microbiology, Washington, D.C., 1998; (3) A. Haldimann, M. K. Prahalad, S. L. Fisher, S. K. Kim, C. T. Walsh, and B. L. Wanner, *Proc. Natl. Acad. Sci. U.S.A.* **93**, 14361 (1996); (4) W. W. Metcalf, W. Jiang, and B. L. Wanner, *Gene* **138**, 1 (1994); (5) R. Simon, U. Priefer, and A. Pühler, *Biotechnology* **1**, 784 (1983); (6) M. Herrero, V. De Lorenzo, and K. N. Timmis, *J. Bacteriol.* **172**, 6557 (1990); (7) K. J. Boettcher and E. G. Ruby, *J. Bacteriol.* **172**, 3701 (1990); (8) J. Graf, P. V. Dunlap, and E. G. Ruby, *J. Bacteriol.* **176**, 6986 (1994); (9) S. V. Nyholm, E. V. Stabb, E. G. Ruby, and M. J. McFall-Ngai, *Proc. Natl. Acad. Sci. U.S.A.* **97**, 10231 (2000); (10) J. Graf, *Infect. Immun.* **67**, 1 (1999); (11) A. C. Y. Chang and S. N. Cohen, *J. Bacteriol.* **134**, 1141 (1978); (12) J. Vieira and J. Messing, *Gene* **19**, 259 (1982); (13) D. H. Figurski and D. R. Helinski, *Proc. Natl. Acad. Sci. U.S.A.* **76**, 1648 (1979).

to BHI medium (Difco, Sparks, MD). With the exception of *Aeromonas veronii,* which is grown in LB, Vibrionaceae strains are grown in LBS medium.[16] When added to LBS or LB medium for selection of Vibrionaceae strains, rifampicin (Rf), trimethoprim (Tp), Cm, streptomycin (St), spectinomycin (Sp), and Kn are used

[16] E. V. Stabb, K. A. Reich, and E. G. Ruby, *J. Bacteriol.* **183**, 309 (2001).

at concentrations of 100, 5, 2, 200, and 100 μg ml^{-1}, respectively. Strain KNH6 was isolated from near-shore seawater in Kaneohe Bay, Oahu, Hawaii, and is designated *Photobacterium leiognathi* based on: (i) luciferase enzyme kinetics; (ii) galactose utilization; (iii) nonutilization of maltose, trehalose, and sucrose; and (iv) growth at 37°.[17,18]

Except in conjugations with *A. veronii,* donor and recipient cells are grown to stationary phase in LB or LBS, respectively, with appropriate antibiotics, and 100 μl of each culture is combined in a microfuge tube. The cells are pelleted by centrifugation, washed in fresh antibiotic-free LBS, repelleted, suspended in 10 μl of fresh LBS, dropped onto the surface of fresh LBS agar medium, incubated for 16 hr at 28°, resuspended in 750 μl of LBS, serially diluted, and plated. Matings with *A. veronii* are performed similarly, except that the recipient culture is grown in LB to an OD$_{600}$ between 0.2 to 0.3, 800 μl of this *A. veronii* culture is combined with donor cells, and matings are incubated on LB agar medium at 37°. Selection, or strong enrichment, against *E. coli* donor strains is accomplished using Rf for matings with *A. veronii,* or by plating at 22° for *V. orientalis, V. splendidus, V. parahaemolyticus, V. fischeri,* and *P. leiognathi.* When *V. harveyi* BB120 is the recipient, plated mating mixtures are incubated at 28°, and recipient colonies are easily distinguished by their bioluminescence.

Construction of New Conjugal Helper Plasmids

Plasmids pEVS101, pEVS103, and pEVS104 are derivatives of pRK2013 that retain the *tra* and *trb trans*-acting functions necessary for conjugal transfer, but lack Tn*903* (Fig. 1 and Table I). Plasmids pEVS101, pEVS103, and pEVS104 each function as a conjugal helper plasmid. Helpers pEVS101 and pEVS103 contain both a ColE1 plasmid origin and an R6Kγ origin of replication, whereas pEVS104 contains only the R6Kγ origin and therefore requires the presence of the R6K *pir* gene in host strains for plasmid maintenance.[12]

Construction of Minimal oriT Cassettes

We use PCR (polymerase chain reaction) to amplify a 400-bp region of RP4 surrounding *oriT* (bp 51033 to 51432), and through cloning and subcloning of these amplification products we have generated small cassettes containing the *oriT* region flanked by convenient restriction sites in different orientations (Fig. 2). These cassettes contain the recognition, binding, and *nic* sites for TraI and TraJ, as well as the nearby bent DNA region that serves as the target for TraK. However, the

[17] J. J. Farmer III and F. W. Hickman-Brenner, *in* "The Prokaryotes," 2nd Ed. (A. Balows, H. G. Trüper, M. Dworkin, W. Harder, and K.-H. Schleifer, eds.), Vol. 3, p. 2952. Springer-Verlag, New York, 1992.

[18] K. H. Nealson, *Methods Enzymol.* **57,** 153 (1978).

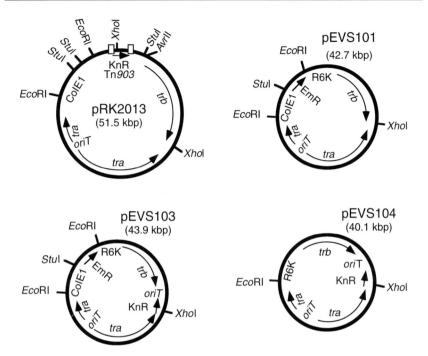

FIG. 1. New conjugal helper plasmids. Positions of plasmid origins, transfer origins, drug resistance genes, and selected restriction sites and transcripts (e.g., *tra* and *trb*) are indicated. Not all genes are shown (e.g., *kor* genes). *Stu*I sites blocked by *dcm* methylation are not shown. Plasmid construction is outlined in Table I.

constructs lack the *traJ* and *traK* genes flanking *oriT*.[1] We select for the presence of *oriT* cassettes in vectors based on conferred mobilizability from strain DH5α to DH5α-32. Oligonucleotide primer sequences and plasmids used to generate and isolate the *oriT* cassettes are listed in Table I. The sequence of each *oriT*-cassette-bearing plasmid is known and available on request.

The *oriT* cassettes in pEVS70, pEVS71, pEVS72, pEVS73, pEVS75, and pEVS76 are flanked by recognition sequences for common restriction enzymes, facilitating the subcloning of *oriT* into other vectors. The *oriT* cassette in pEVS75 is flanked on either side by the 4-bp recognition sequences for *Nla*III, *Sau*3AI, *Taq*I, *Alu*I, *Mse*I, and *Tsp*509I, and digestion with these enzymes results in DNA ends that are compatible with those generated by a number of 6-bp recognizing enzymes (e.g., *Eco*RI, *Cla*I, *Sph*I, *Bam*HI, and *Ase*I). This allows the pEVS75 *oriT* cassette to be cloned into a variety of sites without the introduction of additional 6-bp enzyme recognition sites that might otherwise complicate further cloning strategies (e.g., see the generation of pEVS79, below). The minimal *oriT* present on an *Alu*I fragment of pEVS75 is subcloned into the *Stu*I site in pUC4K (purified from *dcm*−

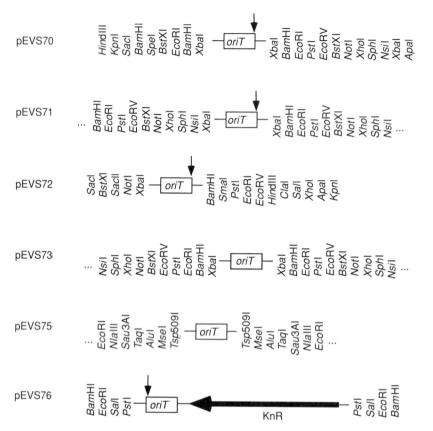

FIG. 2. Restriction maps of *oriT* cassettes. "..." indicates that the cassette shown is embedded within a multiple cloning site that extends beyond the restriction sites shown. Enzymes that cut more than once, and on both sides of the *oriT* cassette, are highlighted with bold type. During conjugation, the site of nicking occurs near the center of *oriT* and, in asymmetric constructs, an arrow indicates the end of *oriT* first transferred to a recipient. The construction of these cassettes is described in Table I, and plasmid sequences are available on request.

strain TH1191 to allow *Stu*I digestion), generating pEVS76. pEVS76 contains a KnR/*oriT* cassette flanked on either end by *Pst*I, *Sal*I, *Bam*HI, *Hinc*II, and *Eco*RI. KnR provides an antibiotic selection to facilitate cloning this cassette into other vectors making them both KnR and mobilizable. The cassette in pEVS71 contains a direct repeat of several restriction enzyme recognition sequences flanking *oriT*. Digestion of this plasmid with either *Bam*HI or *Nsi*I releases a cassette with *oriT* adjacent to a multiple cloning site, and incorporating this cassette into vectors can simultaneously confer mobilizability and a convenient multiple cloning site (e.g., see the generation of pEVS77 and pEVS78 below).

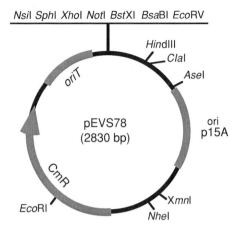

Fig. 3. Map of shuttle vector pEVS78. The *oriT*-containing *Bam*HI fragment of pEVS71 was cloned into the *Bcl*I site of pACYC184 generating pEVS77, and pEVS78 was generated by digestion of pEVS77 with *Eco*RV followed by self-ligation.

Construction of Shuttle Vectors Stable in Vibrio Recipients

We use the *oriT*-MCS cassette from pEVS71 to generate shuttle vector pEVS77 and its Δ*tetA* derivative pEVS78 (Table I). These vectors contain the p15A plasmid origin of replication, which is stably maintained in *V. fischeri*.[19,20] In triparental conjugation from *E. coli* to *V. fischeri*, we have found that pEVS78 (Fig. 3) is stably transferred 100-fold more efficiently than pEVS77 (see below), making it a more promising shuttle vector for use with *Vibrio* recipients. pEVS78 is small and contains several unique restriction sites, which should facilitate the cloning of genes of interest into this shuttle vector. To demonstrate the broader utility of pEVS78, we have conjugally transferred this plasmid to six other members of the Vibrionaceae: *V. orientalis, V. splendidus, V. harveyi, V. parahaemolyticus, A. veronii,* and *P. leiognathi.* The complete sequences of pEVS77 and pEVS78 are available on request.

Construction of Vectors for Allelic Exchange

We have also used the minimal *oriT* cassettes described above to generate mobilizable vectors for cloning and allelic exchange. The *oriT*-containing *Alu*I fragment of pEVS75 is cloned into the *Xmn*I site of pBCSK+ (Stratagene, La Jolla, CA) generating pEVS79 (Table I, Fig. 4). The introduction of the *oriT* fragment introduces only one restriction site (*Acc*I) also present in the multiple cloning site. Thus,

[19] K. L. Visick and E. G. Ruby, *in* "Bioluminescence and Chemiluminescence" (J. W. Hastings, L. J. Kricka, and P. E. Stanley, eds.), p. 119. John Wiley and Sons, New York, 1997.
[20] K. M. Gray and E. P. Greenberg, *J. Bacteriol.* **174,** 4384 (1992).

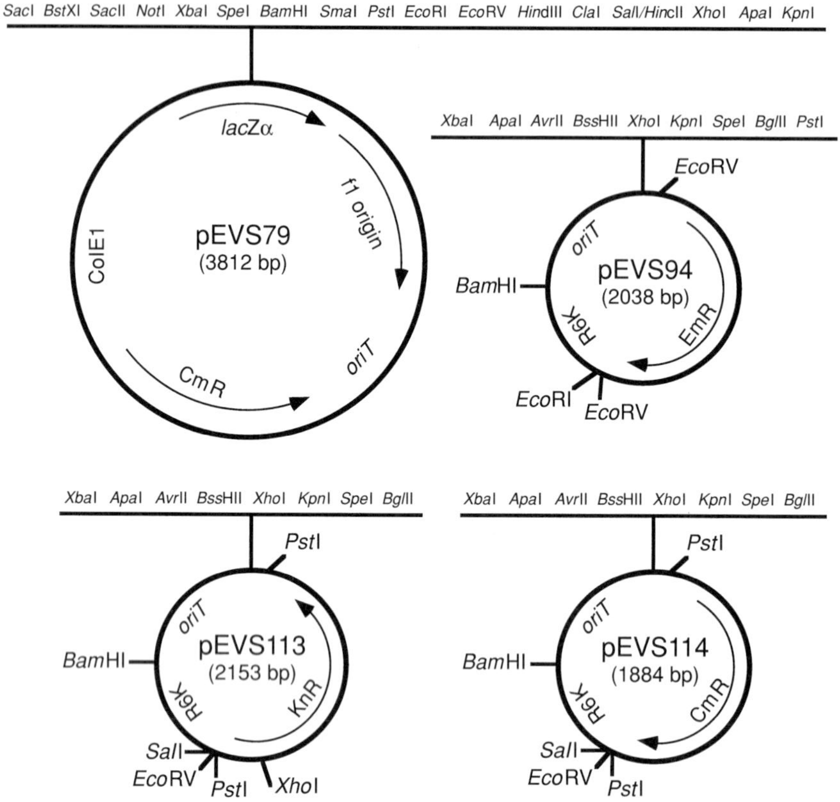

FIG. 4. Maps of cloning and allelic exchange vectors pEVS79, pEVS94, pEVS113, and pEVS114. The *oriT*-containing *Alu*I fragment of pEVS75 was cloned into the *Xmn*I site of pBCSK+ (Stratagene) generating pEVS79. The *oriT*-containing *Xba*I–*Bam*HI fragment of pEVS72 and an artificial MCS were cloned into pKV38 to generate pEVS94. The unique *Eco*RI site in pEVS94 was replaced with a *Sal*I site, the EmR-containing *Eco*RV fragment deleted, and KnR or CmR cloned into the *Pst*I site to generate pEVS113 and pEVS114, respectively.

pEVS79 retains the utility of pBCSK+ as a cloning vector (e.g., blue/white screening, universal primer sites), while also being conjugally mobilizable. We and others (K. Visick, P. Fidopiastis, C. Lupp, D. Millikan; personal communications, 2002) have used pEVS79 to clone *V. fischeri* DNA, engineer mutations in it, and mobilize these mutant alleles back into the *V. fischeri* chromosome. Although pEVS79 contains the ColE1 origin and replicates as a plasmid in *V. fischeri* (and presumably in other *Vibrio* species), it is sufficiently unstable that recombinants, which are more stable than free plasmid, can be identified following nonselective growth. Interestingly, such single recombinants appear to resolve as double recombinants at an unusually high frequency, possibly because leaky replication of the plasmid origin amplifies the duplicated region surrounding the site of the initial recombination event, providing additional homologous targets for double recombination. Replication of pEVS79 as a plasmid seems to be disfavored at lower temperature, and plating the results of matings selectively at 22°, rather than 28°, enriches for recombinants.

Another mobilizable vector suitable for cloning and marker exchange is pEVS94 (Fig. 4), which contains the minimal *oriT* cassette from pEVS72, a gene encoding EmR, and the R6Kγ origin of replication. Several unique restriction sites in the MCS of pEVS94, together with the small size of this vector (2038 bp), facilitate the cloning of fragments into pEVS94. Because the R6Kγ origin of replication does not replicate in *Vibrio* species, plasmid maintenance following selection for the vector-encoded resistance marker is not a concern, as it is for pEVS79 above. To obtain high yields of pEVS94 DNA, we used host strain BW23474,[21] which contains a chromosomal integrant of the *pir-116* allele and thereby maintains plasmids with the R6Kγ origin of replication in high copy.[22] The unique *Eco*RI site in pEVS94 was replaced with a *Sal*I generating pEVS94S. KnR and CmR determinants were cloned into the *Pst*I site of pEVS94S, and the EmR *Eco*RV fragment deleted in each case, generating plasmids pEVS113 and pEVS114 (Fig. 4). The complete sequences of pEVS79, pEVS94, pEVS94S, pEVS113, and pEVS114 are available on request.

Mobilization and Reconstruction of Transposon Insertions

The KnR/*oriT* cassette from pEVS76 can be used in conjunction with Tn*10* derivatives that contain the R6Kγ origin of replication to quickly reconstruct transposon-insertion mutations in different *V. fischeri* genetic backgrounds (Fig. 5) as follows: (i) mini-Tn*10*Cm transposons containing the R6Kγ origin of replication and conferring CmR are used to mutagenize the *V. fischeri* chromosome; (ii) in a mutant of interest, the Tn insertion is cloned directly, by digesting chromosomal DNA with an enzyme that does not cut in the transposon, self-ligating this CmR- and ori R6Kγ-containing fragment, and transforming an *E. coli* strain that contains the *pir* gene; (iii) the resulting plasmid is relinearized with the same enzyme used to excise it from the *V. fischeri* chromosome and ligated to the KnR/*oriT* cassette from pEVS76; (iv) the resulting construct is mobilized into another *V. fischeri* strain, selecting for KnR and CmR single recombinants; (v) screening for CmR, Kn-sensitive double recombinants allows the regeneration of the Tn::chromosome mutation in a different strain. This technique has been used to move a mini-Tn*10*Cm::*hadA* allele, initially isolated in *V. fischeri* strain KV150

[21] A. Haldimann, M. K. Prahalad, S. L. Fisher, S. K. Kim, C. T. Walsh, and B. L. Wanner, *Proc. Natl. Acad. Sci. U.S.A.* **93**, 14361 (1996).

[22] W. W. Metcalf, W. Jiang, and B. L. Wanner, *Gene* **138**, 1 (1994).

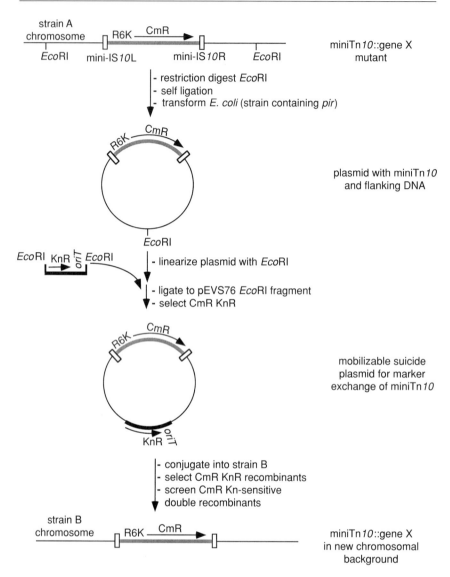

FIG. 5. Schematic representation of method for reconstruction of transposon insertions in new strain background.

(RfR, *lux*⁻), into wild-type strain ES114 (B. Feliciano, personal communication, 2000). In this scheme for mobilization and reconstruction of transposon insertions, linearized suicide vectors such as pEVS94 could be used in much the same way as the KnR/*oriT* cassette from pEVS76.

TABLE II
CONJUGAL TRANSFER FROM *E. coli* TO *V. fischeri* STRAIN ES114

Mating	Donor strain(s)	Frequency of recipients	
		CmR	KnR
A	DH5α pEVS77	$<10^{-8}$	$<10^{-8a}$
B	CC118λpir pRK2013 and DH5α pEVS77	2×10^{-5}	2×10^{-2}
C	CC118λpir pEVS103 and DH5α pEVS77	3×10^{-5}	2×10^{-2}
D	CC118λpir pEVS104 and DH5α pEVS77	4×10^{-5}	$<10^{-8a}$
E	CC118λpir pEVS104 and DH5α pSUP102	3×10^{-5}	nd[b]
F	CC118λpir pEVS104 and DH5α pACYC184	$<10^{-8}$	nd
G	CC118λpir pEVS104	$<10^{-8}$	nd
H	CC118λpir pEVS104 and DH5α pEVS78	7×10^{-3}	nd
I	CC118λpir pEVS104 pEVS78	3×10^{-2}	nd
J	S17-1 pEVS78	4×10^{-3}	nd

[a] Very small KnR colonies arose at a frequency between 10^{-5} and 10^{-6}.
[b] nd, Not determined.

Comparison of Conjugation Methods

In many studies, spontaneous antibiotic-resistant mutant strains are used as conjugal recipients, providing a counterselection against donor *E. coli* cells. However, such marked strains are not truly wild type and may display unexpected mutant phenotypes. For example, RfR and naladixic acid-resistant mutants display pleiotropic patterns of gene expression,[23–26] including a reduced ability to compete with wild-type cells for colonization of host tissue.[27] Because we and others are interested in studying the interactions between *Vibrio* strains and their animal hosts, or the regulatory patterns of *Vibrio* strains, we have experimented, successfully, with the use of low growth temperature (22°) as an enrichment against *E. coli* donor cells following conjugal transfer to true wild-type *Vibrio* isolates. We have found that low-temperature incubation is an effective enrichment against *E. coli,* indistinguishable from rifampicin counterselection, and have used this approach to isolate transconjugants of wild-type strains of *V. fischeri* (Table II), *V. parahaemolyticus, V. splendidus, V. orientalis,* and *P. leiognathi.*

In triparental matings between *E. coli* and *V. fischeri* strain ES114, we have found that pEVS103 and pEVS104 are as efficient as pRK2013 at mediating

[23] L. Gutmann, R. Williamson, N. Moreau, M. D. Kitzis, E. Collatz, J. F. Acar, and F. W. Goldstein, *J. Infect. Dis.* **151,** 501 (1985).
[24] D. J. Jin and C. A. Gross, *J. Bacteriol.* **171,** 5229 (1989).
[25] A. Blanc-Potard, E. Gari, F. Spirito, N. Figueroa-Bossi, and L. Bossi, *Mol. Gen. Genet.* **247,** 680 (1995).
[26] C. Yanofsky and V. Horn, *J. Bacteriol.* **145,** 1334 (1981).
[27] J. Björkman, D. Hughes, and D. I. Andersson, *Proc. Natl. Acad. Sci. U.S.A.* **31,** 3949 (1998).

the stable transfer of pEVS77, as measured by the frequency of CmR recipients (Table II; matings B, C, and D). Control experiments (Table II; matings A and G) have demonstrated that spontaneous CmR is absent (or rare) in *V. fischeri,* indicating that the frequency of CmR is a good measure of conjugal transfer. Although each helper plasmid confers KnR (Fig. 1), only the use of pEVS101 (data not shown), pEVS103, or pRK2013 gives rise to KnR recipients through plasmid maintenance or possibly, in the case of pRK2013, transposition of Tn*903* (Table II; matings B and C). In contrast, use of pEVS104, which contains only the R6Kγ plasmid origin of replication, does not result in levels of KnR recipients that are above background (Table II; matings A and D). The minimal *oriT* cassette present in pEVS77 mediates conjugal transfer as efficiently as the larger undefined *oriT*-containing fragment present in the otherwise isogenic pSUP102, while the *oriT*-lacking parent plasmid, pACYC184, is not transferred (Table II; matings D, E, and F). Interestingly, pEVS78, a Δ*tetA* derivative of pEVS77, is stably transferred 100-fold more efficiently than pEVS77 (Table II; matings D and H), and may therefore represent a more generally useful shuttle vector. We also have found that CC118γ*pir* pEVS104/pEVS78 transfers this shuttle vector 10-fold more efficiently than does S17-1 pEVS78 (Table II; matings I and J). Triparental mating using the combination of CC118λ*pir* pEVS104 and DH5α pEVS78 as donors is as efficient as biparental mating with donor S17-1 pEVS78 at transferring shuttle vector pEVS78 (Table II; matings H and J).

Similar results have been obtained in matings between *E. coli* and six other members of the Vibrionaceae: *V. orientalis, V. splendidus, V. harveyi, V. parahaemolyticus, A. veronii,* and *P. leiognathi* (data not shown). We have found that pEVS103 and pEVS104 are as efficient as pRK2013 at mediating the stable transfer of pEVS78. Also, for three species tested (*V. orientalis, V. harveyi,* and *V. parahaemolyticus*), the use of pRK2013 gives rise to KnR recipients through plasmid maintenance or transposition of Tn*903,* whereas use of pEVS104 does not result in KnR above background levels. In general, biparental matings result in more efficient transfer of pEVS78 than triparental matings, although for the recipients *A. veronii* and *P. leiognathi* triparental mating is similar to biparental mating using strain S17-1. In biparental matings we have found that CC118λ*pir* pEVS104/pEVS78 transfers the shuttle vector roughly 10-fold more efficiently than does S17-1 pEVS78, although these donors are roughly equivalent when *P. leiognathi* or *V. splendidus* serve as the recipient.

Acknowledgments

We thank Todd Vas-Dias for technical assistance, B. Bassler, J. Graf, J. Messing, and K. Visick for sharing plasmids and strains, and J. Flory, M. O. Martin, and D. Millikan for insightful comments on the manuscript. This work was supported by the National Institutes of Health Grant RR12294 to E. G. Ruby and M. McFall-Ngai, and by National Science Foundation Grant IBN-9904601 to M. McFall-Ngai and E. G. Ruby. E.V.S. was supported by a National Research Service Award, F32 GM20041, from the National Institutes of Health.

[31] Role of Autoinducers in Gene Regulation and Virulence of *Pseudomonas aeruginosa*

By LUCIANO PASSADOR

Populations of bacterial cells exhibit and maintain the capacity for cooperative behavior. Organisms such as *Proteus* spp. and the myxobacteria manifest these "multicellular" interactions macroscopically in the form of swarming motility and fruiting body formation, respectively. An important component of this prokaryotic behavior is the utilization of chemical signals in cell–cell communication mechanisms to allow the bacteria to monitor their environment. One such intercellular communication mechanism has been termed "quorum sensing."[1] Quorum sensing allows bacterial populations to respond to fluctuations in their cell density by coordinating the expression of various gene products. Studies of such systems were initiated several decades ago with the studies of the autoinduction processes exhibited by the marine organism *Vibrio fischeri*. During growth *V. fischeri* can synthesize and elaborate a signal molecule termed the autoinducer (AI). When *V. fischeri* cells are present at high cell densities, the concentration of AI rises above a threshold level resulting in the stimulation of several genes required for bioluminescence. This AI signal was subsequently identified as a member of a class of molecules known as acylated homoserine lactones (acyl-HSLs).

Over the past decade it has become increasingly clear that quorum sensing is not confined to *V. fischeri* and its close relatives. Indeed, many gram-negative bacteria have been shown to possess the components for such a regulatory system.[1] Although the mechanistic details can vary slightly between organisms, in general, gram-negative bacteria accomplish this regulation through the production and sensing of acyl-HSL molecules. General reviews on the topic of gram-negative quorum sensing are plentiful[1–3] and the reader is encouraged to refer to these for a more detailed treatment of aspects of quorum sensing that lie outside the scope of the present discussion.

The main focus of this article will be an overview of the AI molecules produced by the opportunistic human pathogen *Pseudomonas aeruginosa*, and their involvement in gene regulation and virulence of this organism. *P. aeruginosa* is a pathogen of individuals in which the normal immune defenses have been compromised because of a genetic predisposition, medical treatment, or some form of trauma. Once established, *P. aeruginosa* can cause a wide variety of acute or

---

[1] N. A. Whitehead, A. M. Barnard, H. Slater, N. J. Simpson, and G. P. Salmond, *FEMS Microbiol. Rev.* **25**, 365 (2001).

[2] T. R. de Kievit and B. H. Iglewski, *Infect. Immun.* **68**, 4839 (2000).

[3] M. B. Miller and B. L. Bassler, *Ann. Rev. Microbiol.* **55**, 165 (2001).

0076-6879/02 $35.00

chronic infections.[4] These infections can be difficult to treat and/or eradicate because of the innate resistance of *P. aeruginosa* to many antibiotics and its ability to produce a large array of both cell-associated and secreted virulence products. Although these traits make *P. aeruginosa* a formidable pathogen, they also provide investigators with a valuable model system with which to study bacterial pathogenesis. Over the past decade *P. aeruginosa* has become one of the most intensively studied quorum sensing model systems. The resulting studies have provided much information with respect to not only the molecular interactions involved in quorum sensing but also its role in the production of disease.

In the published literature the term AI is often used to refer to the acyl-HSL molecules which are required as cognate signals for quorum sensing to occur. However, studies have also suggested the involvement of other, non-acyl-HSL compounds in *P. aeruginosa* quorum sensing. Currently, no conclusive evidence exists for interaction of the non-acyl-HSL compounds with the R-proteins required for quorum sensing. Nevertheless, for the purposes of this discussion, the term AI will be used to refer to both acyl-HSL and non-acyl-HSL signal compounds.

Quorum Sensing Paradigm

In reality, quorum sensing is a relatively simple process and, in its most fundamental form, requires only two components: a transcriptional activator protein (R-protein) and its cognate acyl-HSL signal molecule. The current paradigm proposes that the R-protein remains inactive until it becomes complexed with its cognate acyl-HSL. Once formed, the acyl-HSL/R-protein complex can function as a transcriptional regulator by binding upstream of the target gene and interacting with RNA polymerase to affect gene expression.

The link between the expression of target genes and cell density is based on the observation that the acyl-HSL molecule(s) produced by the bacterium can traverse the bacterial cell wall. Hence, the concentration of acyl-HSL reflects the number of bacterial cells and a group of cells can measure their population density simply by monitoring the concentration of acyl-HSL molecules present. Thus, during periods of low cell density, the concentration of acyl-HSL molecules is also low, resulting in a limitation on the numbers of acyl-HSL/R-protein complexes that can form. As the density of the population increases, there is a concomitant increase in acyl-HSL levels both internal and external to the bacterial cell. On reaching a threshold concentration of acyl-HSL, sufficient acyl-HSL/R-protein complexes are formed that the bacterial population senses it has achieved "quorum" and the appropriate response can ensue.

---

[4] M. Pollack, *in* "Principles and Practices of Infectious Diseases" (G. L. Mandell, J. E. Bennett, and R. Dolin, eds.), p. 1980. Churchill-Livingstone, 1995.

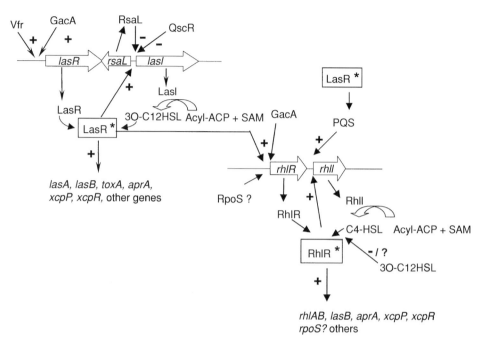

FIG. 1. Quorum sensing circuits in *P. aeruginosa*. The genes and products involved in the *las* and *rhl* quorum sensing circuits of *P. aeruginosa* are shown. Both circuits are interconnected via the regulation of the *rhl* system by LasR as well as a number of shared target genes. Boxes and asterisks (∗) indicate the respective R-protein/acyl-HSL complex. Plus (+) and minus (−) signs indicate positive and negative effects, respectively. Processes which have been suggested but which have come under some doubt are indicated by a question mark (?). Details of the illustrated interactions are given in the text.

Pseudomonas aeruginosa Quorum Sensing Circuits

Perhaps as an indication of its pathogenic versatility, *P. aeruginosa* presents a much more complicated scenario than that described above in that it contains at least two, linked, quorum sensing circuits[5] (see Fig. 1). The first identified circuit was shown to positively regulate the expression of the *lasA* and *lasB* elastase genes and thus was termed the *las* system. This circuit consists of the LasR transcriptional activator protein encoded by *lasR* and its cognate AI *N*-3-oxododecanoylhomoserine lactone (3O-C12HSL), whose synthesis is directed by the *lasI*-encoded AI synthase. The expression of *lasI* is exquisitely sensitive to the presence of LasR and 3O-C12HSL.[6] In fact, *lasI* is the most tightly regulated LasR/3O-C12HSL-dependent gene identified to date. This control provides an

[5] T. R. de Kievit and B. H. Iglewski, *Sci. Med.* **6**, 42 (1999).

[6] P. C. Seed, L. Passador, and B. H. Iglewski, *J. Bacteriol.* **177**, 654 (1995).

autoregulatory loop in which the expression of *lasI* as well as that of other LasR/3-O-C12HSL-dependent target genes is tightly regulated by the concentration of 3O-C12HSL present and thus by cell density. The expression of *lasR* also increases with cell density and reaches maximal levels during stationary phase of growth when cell densities are high.[7,8] However, it is not clear that *lasR* expression is directly up regulated by the presence of LasR/3O-C12HSL complexes.

Subsequent studies of the *las* system indicate that the *las* circuit actually regulates the expression of a large number of genes many of which encode products that will eventually be secreted from the cell.[5] However, a key target gene is *rhlR*, which along with the *rhlI* gene encodes homologs of LasR and LasI and forms the components of the second quorum sensing circuit termed the *rhl* system. The *rhlRI* genes were initially identified via their being required for the expression of *rhlAB* which encodes products involved in the production of the biosurfactant/ciliotoxin known as rhamnolipid. Analogous to the *las* system, RhlI is the synthase responsible for the synthesis of *N*-butanoyl-L-homoserine lactone (C4-HSL), the second major AI of *P. aeruginosa* and the cognate signal for RhlR. The *rhl* system also regulates a variety of genes not the least of which is *rhlI*.[5] Hence the *rhl* system also contains an autoregulatory loop linking expression of the target genes to cell density via the presence of C4-HSL.

In addition to regulation by the *las* system, the two systems are also linked by their regulation of at least several target genes in common. Among this subset of genes are those for elastase (*lasB*), alkaline protease (*aprA*), and some components of the protein secretion apparatus (*xcpR* and *xcpP*). Although the reason for this sharing of target genes is unclear it may reflect a situation in which different environmental signals are used to preferentially stimulate a specific circuit. In that regard it must be noted that the quorum sensing circuits are also regulated by various other global regulatory factors including Vfr, the two-component regulatory system GacA/LemA, and the stress sigma (σ) factor RpoS.[1]

Adding a further level of complexity is the fact that there appears to be a hierarchy of expression for many of the quorum sensing-regulated genes.[6] Studies using transcriptional gene fusions of *lasI* and *lasB* in an *Escherichia coli* lysogen system indicate that the expression of some quorum sensing target genes is more responsive to the presence of LasR/3O-C12HSL than is that of others. It is clear that the presence of two distinct but interconnected quorum sensing circuits provides *P. aeruginosa* with a complex method of gene regulation. This likely allows the organism to respond to a wide variety of environmental signals in addition to, and perhaps concomitant with, cell density. The importance of this regulatory system

[7] A. M. Albus, E. C. Pesci, L. J. Runyen-Janecky, S. E. West, and B. H. Iglewski, *J. Bacteriol.* **179,** 3928 (1997).

[8] E. C. Pesci, J. P. Pearson, P. C. Seed, and B. H. Iglewski, *J. Bacteriol.* **179,** 3127 (1997).

is underscored by the proposal that as many as 3–4% of all *P. aeruginosa* genes may be regulated by quorum sensing.[9]

The *las* and *rhl* systems may not be the sole quorum sensing circuits present in *P. aeruginosa*. Protein homology searches of the information encoded by the *P. aeruginosa* genome indicate that other R-protein homologs may exist. One of these, QscR, has already been described, but to date no cognate AI synthase homolog has been identified.[10] As an added twist, it appears that QscR may function as a negative regulator in contrast to the positive effect of both LasR and RhlR.

Given their role as signal molecules and ligands of the R-proteins, it is clear that AI molecules are critical components of the quorum sensing mechanism. In *P. aeruginosa* AIs and quorum sensing are important for the expression of virulence genes and by implication, critical determinants of the ability of this organism to elicit disease. With regards to disease, the purpose of linking virulence factor production to cell density is not clear. Maintaining production and secretion of various products tightly regulated may prevent early detection by the host and may allow *P. aeruginosa* to unleash its arsenal so as to overwhelm the host immune response.

Pseudomonas aeruginosa Acyl-HSL Structure and Synthesis

As might be expected, all the acyl-HSL molecules identified from various bacteria have similar structures (Table I). In general these molecules consist of a homoserine lactone ring moiety to which is attached an acyl side chain. All the molecules contain the homoserine lactone ring in common, but may differ in the length of the acyl side chain, the degree to which the acyl chain is saturated, and the specific substitutions present on the chain. It is this differing length and the presence of various moieties on the chain that confer specificity to each quorum sensing system. Furthermore, as a result of the variety of side chains some of the molecules are predicted to differ with respect to their overall hydrophobicity, a characteristic that may affect their movement through the lipid bilayers of the bacterial membranes. Thus far all the acyl-HSL molecules identified have side chains of between 4 and 14 carbons in length.

P. aeruginosa produces two acyl-HSL compounds, 3O-C12HSL and C4-HSL, that appear to function as the major signals in quorum sensing (Fig. 2). However, *P. aeruginosa* has also been shown to also synthesize minor amounts of other acyl-HSL compounds, *N*-3-oxohexanoyl-L-homoserine lactone (3O-C6HSL)

[9] M. Whiteley, K. M. Lee, and E. P. Greenberg, *Proc. Natl. Acad. Sci. U.S.A.* **96**, 13904 (1999).
[10] S. A. Chugani, M. Whiteley, K. M. Lee, D. D'Argenio, C. Manoil, and E. P. Greenberg, *Proc. Natl. Acad. Sci. U.S.A.* **98**, 2752 (2001).

TABLE I
ACYL-HSL SIGNALS PRODUCED BY BACTERIA

Major signal molecule	Produced by organism
C4-HSL	*Pseudomonas aeruginosa*
	Aeromonas hydrophila
	Aeromonas salmonicida
	Serratia liquifaciens
3-Hydroxy-C4-HSL	*Vibrio harveyi*
C6-HSL	*Pseudomonas aureofaciens*
	Erwinia chrysanthemi
	Chromobacterium violaceum
	Rhizobium leguminosarum
	Yersinia enterocolitica
3-Oxo-C6-HSL	*Vibrio fischeri*
	Erwinia carotovora subsp. *carotovora*
	Erwinia chrysanthemi
	Erwinia stewartii
	Enterobacter agglomerans
7-*cis*-C14-HSL	*Rhodobacter sphaeroides*
3-Hydroxy-7-*cis*-C14-HSL	*Rhizobium leguminosarum*
C8-HSL	*Rhizobium leguminosarum*
	Burkholderia cepacia
	Ralstonia solanacearum
	Yersinia pseudotuberculosis
3-Oxo-C8-HSL	*Agrobacterium tumefaciens*
3-Oxo-C12-HSL	*Pseudomonas aeruginosa*

and *N*-hexanoyl-L-homoserine lactone (C6HSL).[1] It is of interest to note that 3O-C6HSL is also the cognate AI molecule for LuxR in *V. fischeri* (Table I). The role of the minor compounds has not been clearly defined. They may, under specific conditions, function as additional signals for LasR, RhlR, or as yet unidentified R-proteins. This appears to be unlikely at least for LasR as experimental evidence suggests that 3O-C6HSL is not capable of functioning as an AI for LasR.[11] A survey of identified AI structures also reveals that a number of organisms produce the same acyl-HSL (Table I). This finding has suggested the possibility of communication between various organisms.

The biosynthesis of the *P. aeruginosa* AIs has not been intensively studied. However, a mechanism likely to be common to the majority of AI synthases has been derived from studies of both *V. fischeri* LuxI and *Agrobacterium tumefaciens* TraI. Studies involving the expression of various LuxI homologs in heterologous

[11] L. Passador, K. D. Tucker, K. Guertin, M. Journet, A. S. Kende, and B. H. Iglewski, *J. Bacteriol.* **178,** 5995 (1996).

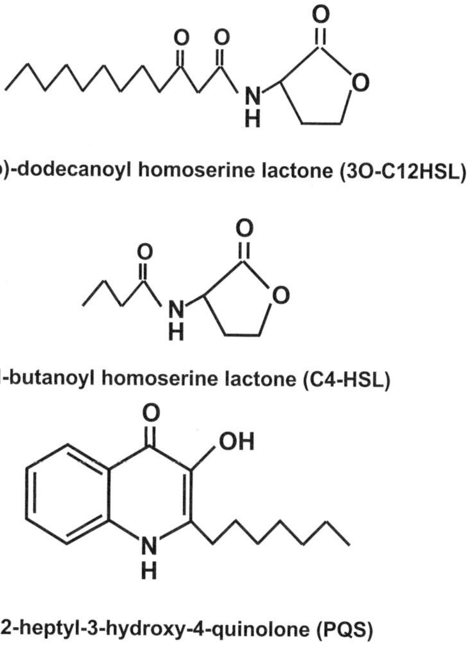

N-3-(oxo)-dodecanoyl homoserine lactone (3O-C12HSL)

N-butanoyl homoserine lactone (C4-HSL)

2-heptyl-3-hydroxy-4-quinolone (PQS)

FIG. 2. Autoinducers of *P. aeruginosa*. The structure and nomenclature of both acyl-HSL compounds and the PQS molecule are illustrated. Their role in gene regulation is described in the text.

hosts have clearly demonstrated that these proteins were indeed involved in the synthesis of the AI molecules. However until recently the mechanism of the biosynthesis and direct evidence for the enzymatic activity of the synthase protein remained elusive. Based on the structure of the acyl-HSL molecule, it was hypothesized that the acyl side chain would be derived from fatty acid metabolism and that the ring moiety would most likely come from modification of an amino acid. Indeed, a number of studies of LuxI and its *Agrobacterium tumefaciens* TraI homolog supported this hypothesis by indicating that *S*-adenosylmethionine was the amino acid substrate which would give rise to the homoserine lactone ring moiety.[1]

Studies of 3O-C6HSL synthesis in *V. fischeri* suggested that the most likely candidates to function as donors of the acyl side chain were coenzyme A (CoA) acetylated with a 3-oxohexanoyl moiety or the acyl carrier protein (ACP) conjugated with 3-oxohexanoic acid.[12,13] Studies of *E. coli*[13] and *P. aeruginosa* fatty acid mutants[14] suggested that acylated ACP pools served as the major substrates

[12] A. Eberhard, T. Longin, C. A. Widrig, and S. J. Stranick, *Arch. Microbiol.* **155,** 294 (1991).
[13] D. L. Val and J. E. Cronan, Jr., *J. Bacteriol.* **180,** 2644 (1998).
[14] T. T. Hoang and H. P. Schweizer, *J. Bacteriol.* **181,** 5489 (1999).

for acyl-HSL synthesis. These findings were corroborated by studies indicating that both LuxI and TraI will use appropriately acylated ACP but not the acylated CoA as substrate.[15,16]

Interestingly, *P. aeruginosa* RhlI appears to be able to use a CoA derivative as a donor of the acyl side chain but the acylated-ACP is the preferred substrate.[17-19] RhlI appears to synthesize C4-HSL via a sequential ordered pathway in which *S*-adenosylmethionine initiates the binding of butyryl-ACP. An amide bond is formed to link the acyl chain to the *S*-adenosylmethionine with the resultant release of the holo-ACP. The homoserine ring is lactonized and the process is completed with the release of the butyrylhomoserine lactone.[19]

The vast majority of acyl-HSL synthases exhibit a large degree of similarity both in structure and function. The proteins thus far identified range from 194 to 226 amino acids in length and as expected demonstrate a significant number of similar as well as identical residues when their protein sequences are aligned. The amino acid conservation is most evident in the amino-terminal half of the proteins suggesting that this portion of the protein may contain the active site of the molecule. Comparison of the carboxy-terminal half of the identified synthases indicates that this region exhibits significant divergence. Thus it has been suggested that the carboxy terminus may be involved in the recognition of the acyl-ACP substrate.[20] Mutational studies of the RhlI protein support the proposed roles for the various regions of the protein. Studies have identified eight residues within the amino-terminal portion of RhlI required for synthesis of the acyl-HSL.[21] Interestingly seven of these residues corresponded to LuxI residues which, when mutated, resulted in a decrease in enzyme activity.

Some organisms are capable of synthesizing a range of acyl-HSL molecules which are cognate ligands for quorum sensing systems of other species.[1] In addition, there exist a number of studies demonstrating that a given R-protein can recognize and interact, to varying degrees of effectiveness, with noncognate acyl-HSL structures. Given these observations, it is exciting to speculate that interspecies communication can occur. Such a proposal is not infeasible given that many of the organisms that inhabit similar ecological niches utilize acyl-HSL based quorum

[15] M. I. More, L. D. Finger, J. L. Stryker, C. Fuqua, A. Eberhard, and S. C. Winans, *Science* **272**, 1655 (1996).

[16] A. L. Schaefer, D. L. Val, B. L. Hanzelka, J. E. Cronan, Jr., and E. P. Greenberg, *Proc. Natl. Acad. Sci. U.S.A.* **93**, 9505 (1996).

[17] Y. Jiang, M. Camara, S. R. Chhabra, K. R. Hardie, B. W. Bycroft, A. Lazdunski, G. P. Salmond, G. S. Stewart, and P. Williams, *Mol. Microbiol.* **28**, 193 (1998).

[18] T. T. Hoang, Y. Ma, R. J. Stern, M. R. McNeil, and H. P. Schweizer, *Gene* **237**, 361 (1999).

[19] M. R. Parsek, D. L. Val, B. L. Hanzelka, J. E. Cronan, Jr., and E. P. Greenberg, *Proc. Natl. Acad. Sci. U.S.A.* **96**, 4360 (1999).

[20] C. Fuqua and A. Eberhard, *in* "Cell-Cell Signaling in Bacteria" (G. M. Dunny and S. C. Winans, eds.), p. 211. ASM Press, Washington, D.C., 1999.

[21] M. R. Parsek, A. L. Schaefer, and E. P. Greenberg, *Mol. Microbiol.* **26**, 301 (1997).

sensing mechanisms to regulate gene expression. In such a process, one organism could elaborate an acyl-HSL that can not only serve as a signal for its own gene expression but also alter the gene expression of different species that can recognize and respond to the acyl-HSL. Such interactions could be used as positive regulation, such as inducing the expression of enzymes in a responding species to provide nutrients in a form that the organism expressing the signal cannot provide for itself. Alternatively such interactions could be used to negatively regulate the production of products that might be toxic to a given species. Regardless, this area of investigation is gaining momentum and should provide fascinating insights as to how bacteria may interact in naturally occurring mixed populations, be they located in an environmental setting or in an infected host. Two studies involving acyl-HSL signaling in *Pseudomonas* spp. have been reported. One study demonstrated that *P. aeruginosa* AIs were capable of significantly stimulating production of siderophores, lipase, and proteases by *Burkholderia cepacia*.[22] This finding has special relevance because these two organisms are both important lung pathogens in individuals afflicted with cystic fibrosis. A second study indicated that the production of acyl-HSLs by one population of *P. aureofaciens* could stimulate production of phenazine by a second distinct population.[23] The findings of these studies strongly support the existence of communication between distinct populations of bacteria that happen to coreside in a given environment.

Novel Acyl-HSL Synthases

There are a number of published reports describing novel acyl-HSL synthases that do not exhibit sequence homology to the LuxI family of AI synthases.[24–26] These proteins include the AinS protein of *V. fischeri,* the LuxM protein of *V. harveyi,* and the HtdS protein of *Pseudomonas fluorescens*. Interestingly, although AinS and LuxM exhibit similarity to each other HtdS appears to be unrelated to any identified acyl-HSL synthase. The AinS and LuxM proteins are responsible for the synthesis of *N*-octanoyl-L-homoserine lactone and *N*-3-hydroxybutanoyl-L-homoserine lactone, respectively. Although similar to the LuxI type synthases in that it can utilize *S*-adenosylmethionine and the appropriate acyl-ACP for AI synthesis, AinS differs in that it can also use the appropriate acyl-CoA substrate. In this regard it is similar to the *P. aeruginosa* RhlI protein. Acyl-HSL synthesis by both AinS and RhlI can be inhibited by *S*-adenosylhomocysteine and holo-ACP

[22] D. McKenney, K. E. Brown, and D. G. Allison, *J. Bacteriol.* **177,** 6989 (1995).
[23] D. W. Wood, F. Gong, M. M. Daykin, P. Williams, and L. S. Pierson III, *J. Bacteriol.* **179,** 7663 (1997).
[24] B. L. Bassler, M. Wright, R. E. Showalter, and M. R. Silverman, *Mol. Microbiol.* **9,** 773 (1993).
[25] L. Gilson, A. Kuo, and P. V. Dunlap, *J. Bacteriol.* **177,** 6946 (1995).
[26] B. E. Laue, Y. Jiang, S. R. Chhabra, S. Jacob, G. S. Stewart, A. Hardman, J. A. Dowine, F. O'Gara, and P. Williams, *Microbiology* **146(Pt 10),** 2469 (2000).

suggesting that both enzymes may utilize similar enzymatic mechanisms. The substrate requirements for acyl-HSL synthesis by HtdS and LuxM have not yet been elucidated. It is quite likely that LuxM will exhibit similarities to the LuxI family of synthases based on its similarity to AinS.

HtdS appears to be involved in the synthesis of *N*-3-hydroxy-7-*cis*-tetradecenoyl-L-homoserine lactone. It is interesting to note that this molecule is also produced by *Rhizobium* spp. which like *P. fluorescens* are plant-associated bacteria. In the case of *Rhizobium* spp. the molecule has been designated a bacteriocin. The description of HtdS in another pseudomonad suggests that the possibility of identifying a novel acyl-HSL synthase in *P. aeruginosa* cannot be dismissed.

Other Signaling Molecules Involved in *Pseudomonas aeruginosa* Quorum Sensing

Two new classes of molecules that appear to play a role in *P. aeruginosa* quorum sensing have been identified. The synthesis of the *Pseudomonas* quinolone signal (PQS; 2-heptyl-3-hydroxy-4-quinolone) appears to occur following the exponential growth phase of liquid cultures and it requires the presence of LasR.[27,28] Studies demonstrate that the expression of *lasB* and *rhlI* can be induced by the exogenous addition of PQS. In addition, PQS appears to have the ability to stimulate expression of *lasR* and *rhlR* albeit to a much lesser degree than that seen for *lasB* and *rhlI*. The exact role of this molecule in quorum sensing and overall *P. aeruginosa* physiology is not clear. It has been proposed that it may function as a means to further stimulate *rhl* quorum sensing circuit during stationary phase growth, perhaps in an attempt to obtain further supplies of nutrients. Whatever its function, PQS provides another connection between the *las* and *rhl* quorum sensing circuits and a further level of complexity to an already complex regulatory mechanism.

Holden and colleagues[29] have reported on the production of a class of cyclic dipeptides known as diketopiperazines by *P. aeruginosa* and several other gram-negative bacteria. It has been proposed that these molecules may interact with the R-protein to directly affect, both positively and negatively, the quorum sensing systems of several bacteria. However, the concentration of these compounds that must be used to exert the effect is greatly in excess of the acyl-HSL concentration required for a similar effect. This finding has led to some doubt as to whether

[27] E. C. Pesci, J. B. Milbank, J. P. Pearson, S. McKnight, A. S. Kende, E. P. Greenberg, and B. H. Iglewski, *Proc. Natl. Acad. Sci. U.S.A.* **96,** 11229 (1999).

[28] S. L. McKnight, B. H. Iglewski, and E. C. Pesci, *J. Bacteriol.* **182,** 2702 (2000).

[29] M. T. Holden, S. Ram Chhabra, R. de Nys, P. Stead, N. J. Bainton, P. J. Hill, M. Manefield, N. Kumar, M. Labatte, D. England, S. Rice, M. Givskov, G. P. Salmond, G. S. Stewart, B. W. Bycroft, S. Kjelleberg, and P. Williams, *Mol. Microbiol.* **33,** 1254 (1999).

diketopiperazines actually would interact with quorum sensing systems in those organisms that actually produce acyl-HSLs.

Regulation of *Pseudomonas aeruginosa* Autoinducer Synthesis

Given the importance of the AIs in the quorum sensing mechanism it is not surprising that *P. aeruginosa* regulates their synthesis. This regulation ensures that the concentrations of the AIs are adjusted to reflect the physiological state of the population. This is important in order to prevent premature stimulation of target gene expression, which may have deleterious effects on cell growth and survival. *P. aeruginosa* appears to maintain a number of mechanisms to both positively and negatively regulate the synthesis of AI.

In *P. aeruginosa,* the expression of both *lasI* and *rhlI* is positively affected by the presence of the LasR/3O-C12HSL and RhlR/C4-HSL complexes, respectively.[1] The placement of *lasI* and *rhlI* in such an autoregulatory loop ensures that production of AI is tightly linked to its own levels in the local environment. Thus during periods of low cell density, when expression of quorum sensing regulated genes is quiescent, AI expression can be maintained at basal levels. Conversely, it can be synthesized rapidly during periods of high cell density allowing for a rapid response and adaptation to the given situation.

Evidence suggests that expression of *lasI* may be negatively regulated by two, distinct negative regulatory proteins. Work by deKievit and colleagues[30] identified a small 10 kDa protein termed RsaL that appears to inhibit the transcription of *lasI*. The exact mechanism by which the negative regulation occurs has not been elucidated. RsaL does not appear to exhibit homology to any identified protein. Interestingly, the expression of RsaL requires the presence of LasR and 3O-C12HSL thereby linking it to the availability of AI. The expression of *rsaL* is very sensitive to the presence of AI (P. Kiratisin and L. Passador, unpublished results, 2001) leading to the suggestion that it may play a role in maintaining AI concentrations at basal levels within the cell. This state would occur until the appropriate conditions exist such that the negative effect of RsaL could be neutralized by some as yet undeciphered mechanism.

Work by Chugani *et al.*[10] has identified a protein, QscR which appears to function by repressing *lasI* expression. A null mutant of QscR produces 3O-C12HSL earlier in growth suggesting that the protein functions as a repressor, and it has been proposed that the role of QscR may be to ensure that genes regulated by quorum sensing are activated only when necessary and appropriate for a specific environment. In contrast to RsaL, QscR is a homolog of LasR and RhlR but no cognate AI molecule or synthase has been identified. Although most R-proteins function as transcriptional activators, the finding that QscR can function as a repressor is

[30] T. de Kievit, P. C. Seed, J. Nezezon, L. Passador, and B. H. Iglewski, *J. Bacteriol.* **181,** 2175 (1999).

not necessarily novel. Studies in the plant pathogen *Erwinia stewartii* indicate that EsaR represses the expression of genes involved in exopolysaccharide synthesis.[1] It is proposed that EsaR keeps target gene expression repressed until binding of its cognate acyl-HSL occurs. In support of this is the finding that EsaR null mutants constitutively express the exopolysaccharide genes.

In addition to the regulation by RhlR and its cognate AI, expression of the *P. aeruginosa rhlI* gene appears to be positively impacted by the presence of PQS quinolone molecule. However, neither the mechanism nor the role that this regulation plays in the quorum sensing circuitry has been elucidated.

Given that quorum sensing is most often seen during the stationary phase of growth of cultures when nutrients become limiting, it is not surprising that nutritional stresses may affect its function. Inorganic polyphosphate is purported to function by allowing bacteria to adapt to situations of environmental and nutritional stress. The enzyme PPK, which catalyzes the synthesis of the polyphosphate, or polyphosphate itself can affect quorum sensing in *P. aeruginosa*.[31] The exact mechanism by which the expression of synthases was reduced has not been deciphered but it is clear that in a *ppk* mutant AI levels were reduced to 50% of those seen in the wild type. In other studies, expression of *lasI* was significantly stimulated when *P. aeruginosa* was grown under iron-limited conditions.[32] This correlates well with evidence that quorum sensing regulates expression of the *P. aeruginosa* siderophore pyoverdin.[33] Finally, both erythromycin[34] and azithromycin[35] have been postulated to affect AI synthesis. Neither the mechanisms nor the significance of the findings is clear.

Methods to Detect, Identify, and Characterize acyl-HSL Molecules

A number of methods have been developed to identify the production of various acyl-HSL compounds by bacteria including *P. aeruginosa*. These methods range from simple bioassays to more rigid biochemical and biophysical characterizations. Therefore, a brief discussion might be useful to those readers who may wish to incorporate any number of these methods in their own studies. A detailed description of each method is out of the scope of this paper and the reader is referred to an excellent overview of the methods.[36]

---

[31] M. H. Rashid, K. Rumbaugh, L. Passador, D. G. Davies, A. N. Hamood, B. H. Iglewski, and A. Kornberg, *Proc. Natl. Acad. Sci. U.S.A.* **97,** 9636 (2000).

[32] N. Bollinger, D. J. Hassett, B. H. Iglewski, J. W. Costerton, and T. R. McDermott, *J. Bacteriol.* **183,** 1990 (2001).

[33] A. Stintzi, K. Evans, J. M. Meyer, and K. Poole, *FEMS Microbiol. Lett.* **166,** 341 (1998).

[34] D. Sofer, N. Gilboa-Garber, A. Belz, and N. C. Garber, *Chemotherapy* **45,** 335 (1999).

[35] K. Tateda, R. Comte, J. C. Pechere, T. Kohler, K. Yamaguchi, and C. Van Delden, *Antimicrob. Agents Chemother.* **45,** 1930 (2001).

[36] M. Camara, M. Daykin, and S. R. Chhabra, *in* "Bacterial Pathogenesis" (P. Williams, J. Ketley, and G. P. Salmond, eds.), Vol. 27, p. 319. Academic Press, San Diego, 1998.

The detection of the presence of various acyl-HSL compounds can be undertaken using any one of a number of simple assays. Gene fusion constructs have been used as the basis of "biosensor" strains that can detect a wide variety of acyl-HSL compounds as well as in studies of very specific targets. Such strains usually harbor either plasmid-borne or chromosomally integrated genetic constructs in which a fragment of a known target gene, including its promoter and relevant upstream region, is cloned upstream of a gene whose product is easily assayed and which serves as a reporter for the system. In addition to the reporter system, the strain also carries the structural gene for an R-protein capable of stimulating expression of the target gene fusion when a suitable acyl-HSL is present. Source material in the form of either cell-free culture supernatants or solvent extracts of such supernatants can then be added to the biosensor strain and the production of the reporter product is monitored. Commonly used reporter systems include β-galactosidase (LacZ), which is easily measurable in a colorimetric assay, or *luxAB*, which encodes the luciferase required for light production. More recently, fusions or target genes to the gene for green fluorescent protein (GFP) have been employed.[37,38] Such fusions are especially useful in single-cell analysis or *in vivo* infection studies since they can be followed using fluorescent or confocal laser scanning microscopy.

The use of reporter systems can be cumbersome if a large number of samples are to be examined, especially if the assay requires specific instrumentation. These disadvantages can be overcome by using a simpler agar plate based system to detect the presence of acyl-HSLs. This assay is based on the requirement for *N*-hexanoyl-L-homoserine lactone in the production of a purple pigment, violacein, by *Chromobacterium violaceum*. *C. violaceum* CV026 is a violacein nonproducing strain in which production of violacein can be restored only by the addition of exogenous acyl-HSLs. The strain can be grown on plates in the presence of source material suspected to contain acyl-HSLs and observed for production of the purple pigment. The one limitation of the assay is that compounds with acyl side chains of 10 to 14 carbons are not capable of inducing pigment production. However, an ingenious assay has also been developed to exploit this shortcoming by testing such compounds in the presence of an acyl-HSL known to activate pigment production and measuring the ability of the source material to compete with the activating acyl-HSL. In this assay, inhibition of pigment production will result in colonies that exhibit a white halo against a purple background.

With regard to *P. aeruginosa,* C4-HSL can stimulate the expression of violacein whereas 3O-C12HSL cannot.[36] However, 3O-C12HSL can function as an inhibitor

[37] H. Wu, Z. Song, M. Hentzer, J. B. Andersen, A. Heydorn, K. Mathee, C. Moser, L. Eberl, S. Molin, N. Hoiby, and M. Givskov, *Microbiology* **146(Pt 10),** 2481 (2000).
[38] J. B. Andersen, A. Heydorn, M. Hentzer, L. Eberl, O. Geisenberger, B. B. Christensen, S. Molin, and M. Givskov, *Appl. Environ. Microbiol.* **67,** 575 (2001).

of pigment production.[36] Of the two minor acyl-HSLs produced by *P. aeruginosa*, C6-HSL is a much stronger inducer of violacein production than is 3O-C6HSL. Neither is able to function in the inhibitory assay.

Of course detection is only the initial step in the identification of the acyl-HSL. A number of methods can be used to separate and identify acyl-HSLs produced by bacteria including *P. aeruginosa*. Such methods can provide clues as to the type of compounds and their relative abundance in a given sample. One of the most commonly used methods is reversed-phase HPLC. One disadvantage of this method is the particular attention that must be paid with respect to the acquisition and preparation of the sample material. If one only needs a more general notion of the types of compounds present then a thin-layer chromatography system is much simples and forgiving than HPLC. Shaw and colleagues[39] developed a system that was capable of resolving acyl-HSLs with chains from 4 to 12 carbons in length. Cell-free culture supernatants are simply solvent extracted and the resulting material is spotted on C_{18} reversed phase silica plates and chromatography is performed. Known acyl-HSLs serve as a set of standards for identification of the resulting spots on the developed chromatogram. To visualize the acyl-HSL spots, the chromatogram is overlaid with a thin agar film containing any of the biosensor strains described above. In the initial description of the method, a strain of *Agrobacterium tumefaciens* carrying a *traG::lacZ* fusion was used. TraR, a homolog of LasR and LuxR, regulates the expression of *traG*. The incorporation of 5-bromo-4-chloro-3-β-D-galactopyranoside in the film caused the acyl-HSLs to appear as blue spots. If *lux* reporters are used, then autoradiography will be necessary for detection. The spots can be scraped from the plates and subjected to further analysis for more complete characterization.

For more rigid chemical characterization, methods such as mass or NMR (nuclear magnetic resonance) spectroscopy and fast atom bombardment become necessary.[36] Of course the ultimate proof is the chemical synthesis of the acyl-HSL for use in specific bioassays.

Movement of *Pseudomonas aeruginosa* AIs across the Cell Wall

The initial studies of the *V. fischeri* AI 3O-C6HSL suggested that the molecule was freely diffusible across the bacterial cell membranes.[40] An interesting development is the report by Welch and co-workers,[41] who observed that the ability of the *Erwinia carotovora* subsp. *carotovora* CarR to be stimulated was dependent

[39] P. D. Shaw, G. Ping, S. L. Daly, C. Cha, J. E. Cronan, K. L. Rinehart, and S. K. Farrand, *Proc. Natl. Acad. Sci. U.S.A.* **94,** 6036 (1997).

[40] H. B. Kaplan and E. P. Greenberg, *J. Bacteriol.* **163,** 1210 (1985).

[41] M. Welch, D. E. Todd, N. A. Whitehead, S. J. McGowan, B. W. Bycroft, and G. P. Salmond, *EMBO J.* **19,** 631 (2000).

on the ability of 3O-C6HSL to avoid aggregation in the bacterial cell membrane, suggesting that the acyl-HSL was quite hydrophobic. Given that *E. carotovora* subsp. *carotovora* utilizes 3O-C6HSL as its AI, this appears to cloud the interpretation of the 3O-C6HSL diffusion studies performed in *V. fischeri*. Resolution of this conflict awaits further studies.

The identification of acyl-HSL molecules that contained longer side chains, such as the 3O-C12HSL of *P. aeruginosa,* raised questions as to whether diffusion was used for influx/efflux of all AI molecules. It was predicted that those molecules with longer side chains would have greater difficulty crossing membranes because they were more hydrophobic in nature. As a result, the existence of transport systems for the influx and/or efflux of some acyl-HSLs were hypothesized. The first clues that this might be true came from studies indicating that mutants which overexpressed the MexAB-OprM antibiotic efflux pump exhibited decreased levels of 3O-C12HSL which correlated with a decline in levels of several quorum sensing dependent virulence factors in these strains.[42] These results suggested that 3O-C12HSL might be a substrate for the MexAB-OprM pump. A subsequent, more rigorous kinetic study indicated that this indeed was the case.[43] This study demonstrated that while 3O-C12HSL was actively effluxed by MexAB-OprM, C4-HSL appeared to diffuse across the cell wall.

Overexpression of a second *P. aeruginosa* drug efflux pump, MexEF-oprN, has been shown to result in decreased production of those virulence products regulated by the *rhl* system.[44] The findings of this study suggested that the *rhl*-specific effects are due to the role of MexEF-OprN in regulation of PQS levels. Recall that PQS has been shown to stimulate expression of *rhlI,* the C4-HSL synthase. Consistent with the hypothesis is the finding that mutants that overexpress MexEF-OprN also exhibit decreased *rhlI* expression and C4-HSL production. It is interesting to note that the PQS molecule consists of a quinolone ring moiety to which is attached a 7-carbon acyl side chain which presumably confers a hydrophobic character on the molecule. An additional finding of this study is that MexEF-OprN may contribute to the secretion of the 3O-C12HSL compound. Thus it appears that long-chain acyl-HSLs and other hydrophobic quorum sensing signal compounds utilize transport systems to move in and out of the cell while less hydrophobic signal molecules can simply diffuse across the membranes. It should be noted that to date no system involved specifically in the internalization of either 3O-C12HSL or PQS has been identified. Furthermore, the use of such systems in the efflux of AIs has only been demonstrated for *P. aeruginosa.*

[42] K. Evans, L. Passador, R. Srikumar, E. Tsang, J. Nezezon, and K. Poole, *J. Bacteriol.* **180,** 5443 (1998).

[43] J. P. Pearson, C. Van Delden, and B. H. Iglewski, *J. Bacteriol.* **181,** 1203 (1999).

[44] T. Kohler, C. van Delden, L. K. Curty, M. M. Hamzehpour, and J. C. Pechere, *J. Bacteriol.* **183,** 5213 (2001).

Interactions of Acyl-HSLs with R-Proteins

The main mechanism by which AI molecules affect the expression of genes, while admittedly indirect, is via the interaction with their cognate R-protein. Given that until recently no R-proteins had been purified in an intact form, it is no surprise that there is a paucity of studies looking at the molecular interactions of acyl-HSL molecules with their cognate receptor protein. This is especially true of the *P. aeruginosa* LasR and RhlR proteins. Nonetheless, clues to the interactions that may occur are becoming available from a number of studies of various bacterial systems.

The R-proteins involved in gram-negative quorum sensing belong to a family of transcriptional activators named after the archetypal LuxR protein.[45] The proteins within this family generally range from 235 to 260 amino acid residues in length and, not surprisingly, exhibit significant homology to each other at the amino acid level. This homology is particularly high in two regions of the protein. The first region lies within the carboxy-terminal portion (roughly residues 180–240) of the protein and represents that segment of the protein believed to be involved in the binding of the DNA element. The presence of a helix–turn–helix motif within this region, as well as mutational studies, supports its proposed function. The second region resides within the amino-terminal portion of the protein (roughly residues 65–140) and is believed to bind the acyl-HSL.

The R-proteins are believed to exist in an inactive form until they bind their respective acyl-HSL signal. Most of the early evidence for interaction of an AI with an R-protein came from experiments indicating a requirement for the presence of both acyl-HSL and its cognate R-protein to elicit a specific phenotypic response. More direct evidence comes from studies of radiolabeled acyl-HSL interactions with strains that overproduce the cognate receptor protein. In *P. aeruginosa*, tritiated 3O-C12HSL was shown to associate with cells overexpressing LasR.[11] Similar studies indicated the same was true for RhlR and C4-HSL.[8] One interesting finding of such studies is the observation that interaction of LasR with 3O-C12HSL did not increase significantly when the GroEL/ES chaperone proteins were concomitantly overexpressed.[11] This is in contrast to studies of LuxR[46] and RhlR[8] which suggested that GroEL/ES aided in the proper folding of the R-protein. The implications of these contrasting observations for LasR and RhlR are not immediately clear but indicate that the two homologs may require different accessory factors for their optimal function.

The strongest support for the binding of acyl-HSLs to R-proteins comes from studies of purified *E. carotovora* CarR and *A. tumefaciens* TraR.[41,47] In both of

[45] A. M. Stevens and E. P. Greenberg, *in* "Cell-Cell Signalling in Bacteria" (G. M. Dunny and S. C. Winans, eds.), p. 231. ASM Press, Washington, D.C., 1999.
[46] B. L. Hanzelka and E. P. Greenberg, *J. Bacteriol.* **177**, 815 (1995).
[47] J. Zhu and S. C. Winans, *Proc. Natl. Acad. Sci. U.S.A.* **96**, 4832 (1999).

these studies, the receptor proteins were shown to associate with their cognate acyl-HSL in a molar ratio of 1 : 1.

To study interactions of acyl-HSL with their receptor proteins in *P. aeruginosa,* a number of studies have utilized AI analog compounds.[11,48] Studies of the *P. aeruginosa* 3O-C12HSL indicated that analogs carrying acyl side chains of 8, 10, or 14 carbons in length were capable of activating LasR albeit less efficiently than the native AI. Compounds with side chains of six carbons or fewer were incapable of activating LasR. In addition analogs which did not carry the 3-oxo moiety, or which contained various substitutions at this position, were also less efficient at inducing LasR activity. From these findings it is clear that the length of the acyl side chain and the constituents present on it are crucial for efficient activation of LasR. This result is not surprising given that quorum sensing systems must exhibit specificity at the R-protein/acyl-HSL interaction, especially in organisms such as *P. aeruginosa* which produce a number of acyl-HSL molecules. It is also interesting to note that a number of studies have shown that the binding of acyl-HSL by an R-protein does not appear to destroy the signal molecule.[1] In those studies biologically active acyl-HSL has been recovered following interaction with R-proteins.

On binding of the acyl-HSL, it is hypothesized that the R-protein undergoes a conformational change that allows subsequent binding of the target DNA element and subsequent interaction with RNA polymerase. Certainly the requirement for AI in R-protein–DNA interactions have been convincingly shown in quorum sensing systems other than *P. aeruginosa,* most notably the TraR and CarR systems.[41,49] However, evidence that this is true for *P. aeruginosa* comes from one study which described the ability of a glutathione *S*-transferase–LasR protein fusion to bind upstream of the *lasB* promoter when 3O-C12HSL was present.[50]

It is also becoming evident that interaction with the acyl-HSL molecule promotes the multimerization of the R-protein. This hypothesis was strongly supported by the identification of dominant negative alleles of both LuxR and TraR.[51,52] Much more conclusive evidence has come from studies of preparations of purified TraR that was shown to dimerize in the presence of its cognate AI.[51,52] In the case of *P. aeruginosa* the lack of a purified form of LasR has hampered such studies. Studies by Kiratisin and Passador (manuscript submitted) have utilized a LexA protein interaction assay to demonstrate that LasR does indeed multimerize in the presence of 3O-C12HSL. In the same study a truncated LasR molecule was also shown to function as a dominant negative allele, lending further support to the

[48] T. Kline, J. Bowman, B. H. Iglewski, T. R. de Kievit, Y. Kakai, and L. Passador, *Bioorg. Med. Chem. Lett.* **9,** 3447 (1999).

[49] J. Zhu and S. C. Winans, *Proc. Natl. Acad. Sci. U.S.A.* **98,** 1507 (2001).

[50] Z. You, J. Fukushima, T. Ishiwata, B. Chang, M. Kurata, S. Kawamoto, P. Williams, and K. Okuda, *FEMS Microbiol. Lett.* **142,** 301 (1996).

[51] S. H. Choi and E. P. Greenberg, *Mol. Mar. Biol. Biotechnol.* **1,** 408 (1992).

[52] Z. Q. Luo and S. K. Farrand, *Proc. Natl. Acad. Sci. U.S.A.* **96,** 9009 (1999).

multimerization hypothesis. Furthermore, various truncated forms of LasR could only activate the expression of target genes when they were able to interact with AI and multimerize.

Interestingly, the quorum sensing systems of the *Erwinia* spp. and the closely related *Pantoea stewartii* appear to have various twists with respect to the outcome of R-proteins binding AI.[1] The *E. carotovora* CarR and *Pantoea stewartii* EsaR proteins appear to dimerize in the absence of their cognate AIs. Indeed, CarR can form even larger multimers when its AI is present. Although CarR can recognize target sequences regardless of the presence of AI, it is proposed that EsaR functions as a repressor in the absence of AI and that AI binding relieves the repression.

Competition between Acyl-HSL Molecules and Posttranslational Control

As mentioned previously, *P. aeruginosa* produces a number of different acyl-HSL molecules. In addition, it has been shown that some synthases including RhlI can produce acyl-HSLs with different chain lengths if given the appropriately charged ACP as a substrate.[19] These findings raise the question of why an organism would produce a variety of acyl-HSL molecules and whether all of them are important for quorum sensing signals or other metabolic processes. The answer to these questions is not known but it is conceivable that the additional acyl-HSLs may play a role as antagonists of the quorum sensing mechanism.

Through the use of radiolabeled compounds, the *P. aeruginosa* 3O-C12HSL AI has been shown to compete with C4-HSL for binding of RhlR.[8] In an *E. coli*-based bioassay, this competition was manifested as decreased expression of a *rhlA::lacZ* gene fusion as concentrations of 3O-C12HSL were increased. These findings suggested that 3O-C12HSL could exert a level of posttranslational control on the activation of RhlR. This hypothesis has been brought into question by the finding that expression of the *P. aeruginosa lecA* gene, which encodes a lectin, was indeed reduced in the presence of 3O-C12HSL when assayed in an *E. coli* host but not when assayed in *P. aeruginosa*.[53] Thus, although the hypothesis is intriguing, more work will be needed to determine whether such a posttranslational control mechanism does actually exist. To date no data exist with regard to possible functions of the minor acyl-HSLs that *P. aeruginosa* produces. It is unlikely that one of them, 3O-C6HSL, could function as an antagonist at least for LasR based on previously published data indicating that this acyl-HSL cannot activate LasR or compete for 3O-C12HSL binding to LasR.[11] This is not to say that it may not play a role as an agonist or antagonist for RhlR or as yet unidentified R-protein homologs of *P. aeruginosa*.

[53] K. Winzer, C. Falconer, N. C. Garber, S. P. Diggle, M. Camara, and P. Williams, *J. Bacteriol.* **182**, 6401 (2000).

Detection of Autoinducers in *Pseudomonas aeruginosa* Infections

A role for AIs in the virulence of *P. aeruginosa* is supported by a variety of studies, many of which are outlined below. Most of these studies implicate an indirect involvement for AIs via the demonstration that functional quorum sensing is required for maximal virulence in various animal models. It is becoming widely accepted that bacteria, including *P. aeruginosa,* exist in the form of biofilms and that these biofilm structures are important and relevant to disease. As a result the involvement of quorum sensing in biofilm formation and persistence has become an area of intensive study. The concept of AIs acting directly as virulence factors has also been supported by a number of studies that are described further below.

Many of the findings with respect to *P. aeruginosa* quorum sensing are derived from studies using synthetic media and standard laboratory culture conditions which may not necessarily translate to the *in vivo* situation. As a result there have been questions raised as to whether quorum sensing actually functions in an infection and whether AIs are actually synthesized in such an environment. Studies have begun to address both questions.

In the past few years, the production of AIs during infection has been demonstrated by a number of investigators. Wu and co-workers[37] utilized mice infected with *E. coli* harboring various quorum sensing dependent reporter constructs as a monitor for acyl-HSL synthesis by *P. aeruginosa* during pulmonary infection. The findings demonstrated that acyl-HSL production did occur *in vivo*. In fact, signal production was most evident in those lung tissues demonstrating the most severe pathological changes, suggesting a correlation between production of acyl-HSL signals and pathogenesis by *P. aeruginosa*. A separate study[54] made use of reporter constructs and thin-layer chromatography to demonstrate the production of acyl-HSLs by *P. aeruginosa* isolates from cystic fibrosis (CF) patients. Finally, analysis of sputum from CF individuals detected the presence of both major AIs from *P. aeruginosa*.[55]

Evidence of Role for Quorum Sensing
in *Pseudomonas aeruginosa* Virulence

The role of quorum sensing in *P. aeruginosa* virulence has been studied most extensively through the use of mutants lacking one or both of the quorum sensing circuits of this organism.

An early study examining the effect of various *P. aeruginosa* gene products on immortalized airway epithelial cells, was the first to link the production of

[54] O. Geisenberger, M. Givskov, K. Riedel, N. Hoiby, B. Tummler, and L. Eberl, *FEMS Microbiol. Lett.* **184**, 273 (2000).

[55] P. K. Singh, A. L. Schaefer, M. R. Parsek, T. O. Moninger, M. J. Welsh, and E. P. Greenberg, *Nature* **407**, 762 (2000).

3O-C12HSL to stimulation of interleukin-8 (IL-8) production.[56] When such cells were exposed to *P. aeruginosa* strain PAO-R1, a LasR null mutant, IL-8 production was not evident. This finding was borne out in a later study aimed at examining the role of various virulence determinants in respiratory tract infections.[57] In that study a number of defined mutants, including *P. aeruginosa* PAO-R1, were used to infect neonatal mice. The data demonstrated that even though the LasR-deficient mutant was able to colonize the respiratory tract it was essentially avirulent when compared to the parent strain. It was hypothesized that the ability of this strain to colonize might be due to the decreased levels of IL-8 produced. Thus, taken together with the knowledge that strain PAO-R1 produces drastically reduced levels of 3O-C12HSL and C4-HSL, these findings provided a link between the production of AIs and the production of IL-8. Another study[58] provides further evidence of a role for quorum sensing in *P. aeruginosa* pathogenesis. The virulence of parent PAO1 strain as well as that of defined *lasI, rhlI,* or *lasI* and *rhlI* mutants was examined in a neonatal mice model. When compared to the parent strain, all mutants demonstrated a decreased lethality and ability to cause pneumonia and bacteremia. The strongest decrease was seen with the *lasIrhlI* double mutant. Complementation studies using plasmid-borne *lasI, rhlI,* or *lasI* and *rhlI* genes exhibited enhanced virulence, thereby confirming the requirement for these gene products in infection.

Interestingly, a study by Sawa and co-workers[59] appeared to dispute the findings of a link between quorum sensing and virulence. Using an acute lung infection model to examine the role of quorum sensing in *in vitro* cytotoxicity as well as *in vivo* virulence, that study suggested an inverse correlation between quorum sensing and virulence. However, a number of differences in the methods and bacterial strains used may explain the discrepancy. First, there were differences in the route of administration of the *P. aeruginosa* into the mice. In contrast to the other studies, which used intranasal inoculation of bacteria, the study by Sawa used direct instillation of the *P. aeruginosa* into the lungs of adult mice. In addition, the strains that were compared in the Sawa study were not isogenic and thus a truly clear distinction could not be made with regard to virulence in general. The most striking finding of the Sawa study was that strain PA103 which does not express LasR was actually more virulent than the PAO1 parent strain. However, it should be noted that strain PA103 produces a cytotoxin, ExoU, which strain PAO1 does not. What is clear from the Sawa study is that the cytotoxicity of *P. aeruginosa* does not

[56] E. Dimango, H. J. Zar, R. Bryan, and A. Prince, *J. Clin. Invest.* **96**, 2204 (1995).

[57] H. B. Tang, E. DiMango, R. Bryan, M. Gambello, B. H. Iglewski, J. B. Goldberg, and A. Prince, *Infect. Immun.* **64**, 37 (1996).

[58] J. P. Pearson, M. Feldman, B. H. Iglewski, and A. Prince, *Infect. Immun.* **68**, 4331 (2000).

[59] T. Sawa, M. Ohara, K. Kurahashi, S. S. Twining, D. W. Frank, D. B. Doroques, T. Long, M. A. Gropper, and J. P. Wiener-Kronish, *Infect. Immun.* **66**, 3242 (1998).

appear to be controlled by quorum sensing. Finally, the *lasIrhlI* double mutant has been shown to exhibit decreased virulence in a rat model of chronic infection.[60]

Although many studies of *P. aeruginosa* virulence target respiratory infections, most likely as a result of the great impact of CF and *P. aeruginosa* pneumonias, it must be remembered that *P. aeruginosa* is capable of causing a wide variety of infections. Preston and co-workers[61] investigated a murine model of corneal infection comparing the virulence of various elastase-deficient mutant strains as well as that of the LasR mutant strain PAO-R1 to the parent PAO1 strain. Interestingly, both the PAO-R1 and the *lasB* mutant strains appeared to be as virulent as the parent, whereas *lasA* and *lasAB* double mutants appeared to exhibit a reduced virulence. These findings clearly suggested that LasA plays a significant role in corneal infections. However, neither the exogenous addition of purified LasA nor complementation with a functional *lasA* gene could restore virulence to the mutant. In addition, the finding that PAO-R1 was virulent raised a number of questions since expression of both *lasA* and *lasB* has been shown to be under the control of quorum sensing.[62] However, this might be explained by the finding that expression of *lasA* is not as tightly regulated by the *las* system as is that of *lasB*. Furthermore, the authors of the study speculated that the *lasA* mutant might have other additional mutations. Although these data are difficult to correlate and explain, the fact remains that the loss of the *las* system did not appear to affect virulence in corneal infections. Strains deficient in the *rhl* system have not been studied in this system and thus it is unknown whether the *rhl* system may be involved. The authors of the study also raised the interesting possibility that LasR may be a tissue-specific regulator. Most importantly, this study suggests that quorum sensing may not be critical for all types of *P. aeruginosa* infection.

The *lasI, rhlI,* and *lasIrhlI* double mutant have also been studied in a murine thermal injury model intended to represent *P. aeruginosa* infections of burn wounds. The data convincingly demonstrate that these mutants are significantly less virulent than the wild type and that virulence can be restored to the mutants by complementation with the affected genes. A study[63] has also shown that strains which lack either the *lasI* or both the *lasI* and *rhlI* genes were greatly diminished in their capacity to colonize the lungs in a pneumonia model using adult mice.

Storey *et al.*[64] were able to demonstrate a correlation in the accumulation of *lasA* and *lasB* transcripts with those of *lasR* in direct examination of CF sputum. This finding suggests that quorum sensing is active during infection of the CF

[60] H. Wu, Z. Song, M. Givskov, G. Doring, D. Worlitzsch, K. Mathee, J. Rygaard, and N. Hoiby, *Microbiology* **147**, 1105 (2001).
[61] M. J. Preston, P. C. Seed, D. S. Toder, B. H. Iglewski, D. E. Ohman, J. K. Gustin, J. B. Goldberg, and G. B. Pier, *Infect. Immun.* **65**, 3086 (1997).
[62] D. S. Toder, M. J. Gambello, and B. H. Iglewski, *Mol. Microbiol.* **5**, 2003 (1991).
[63] R. S. Smith, S. G. Harris, R. Phipps, and B. H. Iglewski, *J. Bacteriol.* **184**, 1132 (2002).
[64] D. G. Storey, E. E. Ujack, H. R. Rabin, and I. Mitchell, *Infect. Immun.* **66**, 2521 (1998).

lung and importantly was the first to begin to look at the role of quorum sensing in a human scenario. CF sputum transcript analysis[65] has identified that *migA* expression, which is under the control of the *rhl* quorum sensing system, is also increased in CF individuals. MigA is a putative glycosyltransferase proposed to modify low molecular mass lipopolysaccharide.

Infection of the nematode *Caenorhabditis elegans* with various *P. aeruginosa* mutants indicated that the *lasR* and *gacA* genes were important for virulence. Also, some virulence mutants were identified that exhibited decreased production of pyocyanin, whose expression is regulated by the RhlR/RhlI system. A number of these mutants have now been shown to be avirulent in mouse and plant model systems, suggesting that there may be conserved pathways for virulence which are common to infections across kingdoms. In addition, a study[66] reported killing of *C. elegans* by a diffusible factor that required both the *las* and *rhl* systems for its production. The identity of the factor is not yet known but it does not appear to be either of the *lasA* or *lasB* elastases or exotoxin A.

Pseudomonas aeruginosa Quorum Sensing and Biofilms

In any environment, bacteria may exist either as free-living planktonic cells or as communities which become attached to a solid surface and are known as biofilms.[67] These biofilm structures usually arise from small groups of cells known as microcolonies that are formed when planktonic cells adhere to a surface and begin to multiply. Eventually the population grows to the point of becoming a biofilm consisting of microcolonies embedded in an exopolysaccharide matrix of bacterial origin. A series of fluid-filled channels separate the large clusters of bacteria and are believed to act as conduits to provide nutrients and remove waste. The diversity of microenvironments found within a biofilm lead to the formation of a heterogeneous population of bacterial cells. The biofilm structures can form on virtually any surface including the epithelial cells of the lung and the surface of indwelling medical devices. Biofilms are extremely important in disease as they may provide resistance to antimicrobials and to mechanisms of the immune response. Given their persistence and the fact that cells may leave the biofilm and colonize new surfaces, biofilms can act as important sources of bacteria for spread of an organism.

One of the most intensively studied models for biofilm development is that of *P. aeruginosa*. The high population densities observed in biofilm structures naturally led to the postulation of a linking of biofilm structure to quorum sensing. The importance of quorum sensing in biofilm formation was clearly corroborated in a

[65] H. Yang, M. Matewish, I. Loubens, D. G. Storey, J. S. Lam, and S. Jin, *Microbiol.* **146(Pt 10)**, 2509 (2000).

[66] C. Darby, C. L. Cosma, J. H. Thomas, and C. Manoil, *Proc. Natl. Acad. Sci. U.S.A.* **96**, 15202 (1999).

[67] N. Hoiby, H. K. Johansen, C. Moser, Z. Song, O. Ciofu, and A. Kharazmi, *Microbes Infect.* **3**, 23 (2001).

study[68] which demonstrated that LasI mutants strains of *P. aeruginosa* produced biofilms that were thinner, lacked the organized structure seen in parental biofilms, and were susceptible to disruption by sodium dodecyl sulfate. The exogenous addition of 3O-C12HSL restored wild-type biofilm formation and resistance. This initial study has been supported by studies of biofilms on submerged stones[69] and on urethral catheters[70] which were shown to produce acyl-HSL molecules. A report by De Kievit and colleagues[71] demonstrated that *lasI* expression in *P. aeruginosa* biofilms was widespread but tended to decrease over time. In contrast the expression of *rhlI* remained relatively constant but occurred in a lower percentage of cells within the biofilm. Further, it was shown that quorum sensing mutants were significantly impaired in their ability to attach to a surface.

Direct Effects of Autoinducers

The role of AIs as cell-to-cell signal molecules is firmly established. More recent evidence suggests that the AI molecules themselves may act directly as virulence factors. A number of publications have surfaced which demonstrate that the *P. aeruginosa* AI molecules can affect expression of eukaryotic processes. Indeed, one mechanism by which AIs could act as virulence factors is by modulating the human immune response.

One of the hallmarks of *P. aeruginosa* lung infection is the recruitment of neutrophils to the affected site. The neutrophils are most likely responding to the presence of a chemoattractant chemokine identified as IL-8.[57,72]

Early studies demonstrated that epithelial cells incubated in the presence of 3O-C12HSL produced significant amounts of IL-8.[56] Similar experiments found that a *P. aeruginosa* LasR null mutant, which produces negligible amounts of 3O-C12HSL, was not able to stimulate IL-8 production.[57] A more recent study by Smith *et al.*[73] examined the ability of human lung cells to produce IL-8. Their findings suggested that the stimulation of IL-8 production in the presence of 3O-C12HSL was due to a stimulation of the transcription factors NF-κB and activator protein AP2. In an *in vitro* study, Telford *et al.*[74] demonstrated that 3O-C12HSL could inhibit the proliferation of lymphocytes and the production of tumor necrosis

[68] D. G. Davies, M. R. Parsek, J. P. Pearson, B. H. Iglewski, J. W. Costerton, and E. P. Greenberg, *Science* **280**, 295 (1998).

[69] R. J. McLean, M. Whiteley, D. J. Stickler, and W. C. Fuqua, *FEMS Microbiol. Lett.* **154**, 259 (1997).

[70] D. J. Stickler, N. S. Morris, R. J. McLean, and C. Fuqua, *Appl. Environ. Microbiol.* **64**, 3486 (1998).

[71] T. R. De Kievit, R. Gillis, S. Marx, C. Brown, and B. H. Iglewski, *Appl. Environ. Microbiol.* **67**, 1865 (2001).

[72] P. G. Jorens, J. B. Richman-Eisenstat, B. P. Housset, P. P. Massion, I. Ueki, and J. A. Nadel, *Eur. Respir. J.* **7**, 1925 (1994).

[73] R. Smith, E. R. Fedyk, T. A. Springer, N. Mukaida, B. H. Iglewski, and R. P. Phipps, *J. Immunol.* **167**, 366 (2001).

[74] G. Telford, D. Wheeler, P. Williams, P. T. Tomkins, P. Appleby, H. Sewell, G. S. Stewart, B. W. Bycroft, and D. I. Pritchard, *Infect. Immun.* **66**, 36 (1998).

factor α by stimulated macrophages. This same AI was also shown to decrease the production of IL-12 and inhibit antibody synthesis by splenic cells. The effects on tumor necrosis factor α (TNFα) and IL-12 were proposed to lead to a shift from a host Th1 response to a Th2 response, which would be more favorable to the bacterium. This would then be aided by the decreased antibody production affected by the 3O-C12HSL. It is proposed that the alterations in the immune response would allow the bacteria to persist and escape the immune response. Injection of 3O-C12HSL directly into the skin of mice resulted in a significant increase in the mRNA for the cytokines IL-la and IL-6 as well as for a number of chemokines and cyclooxygenase-2.[63] In apparent contrast to the findings of Telford and colleagues described above, this publication reported the activation of T-cells to produce the inflammatory cytokine interferon gamma, which would promote a Th1 response. Why the two studies produced disparate observations is not clear.

In another study, acyl-HSLs were shown to inhibit the production of an antibacterial factor in human CF tracheal cells but not in normal tracheal cells.[75] The ability of 3O-C12HSL to inhibit muscle contraction has also been demonstrated.[76] This has been suggested as a mechanism by which *P. aeruginosa* could increase the supply of nutrients to the site of infection and bypass the immune response.

Concluding Remarks

The AIs used in *P. aeruginosa* quorum sensing are critical for the expression of virulence products and in the pathogenic potential of the organism. Even though only a relatively small amount is known about AI properties and interactions, the studies that have surfaced to date have revealed many interesting data regarding their role in the quorum sensing regulatory mechanism and in the promotion of pathogenesis. One of the most interesting facets of these molecules is the potential that they provide for therapeutic measures. Although AIs are crucial components of quorum sensing, ironically they may prove to be the Achilles heel of the mechanism. The most direct therapy would be the design and identification of AI molecule analogs that could act as true antagonists. Thus far only two studies using a limited number of acyl-HSL analogs have been reported for *P. aeruginosa*.[11,48] Although neither study has identified an antagonist they have provided information regarding moieties of the molecule that may hold promise for alteration. However, the design of an antagonist acyl-HSL analog may be complicated by the fact that *P. aeruginosa* contains at least two interconnected circuits. Hence, finding an antagonist for one circuit may not necessarily eliminate the

---

[75] A. Saleh, C. Figarella, W. Kammouni, S. Marchand-Pinatel, A. Lazdunski, A. Tubul, P. Brun, and M. D. Merten, *Infect. Immun.* **67**, 5076 (1999).

[76] R. N. Lawrence, W. R. Dunn, B. Bycroft, M. Camara, S. R. Chhabra, P. Williams, and V. G. Wilson, *Brit. J. Pharmacol.* **128**, 845 (1999).

activity of the other. The identification of suppressor mutations[77] may also hinder progress. Some promise for the inhibition of R-protein/acyl-HSL interaction lies in the identification of various halogenated furanone analogs of the marine alga *Delisea pulchra* as potential inhibitors.[78] These compounds have been shown to be able to displace the acyl-HSL signal from its receptor R-protein. Aside from the therapeutic use in humans such antagonists would be useful in industrial and medical device applications. For example, the finding that *lasI* and *rhlI* expression in biofilms appears to occur primarily at the substratum[71] suggests that the application of antagonist molecules to any desired surface may provide a means to controlling or even eliminating biofilm formation.

The inhibition of R-protein/acyl-HSL interaction described above is only one method by which quorum sensing mechanisms may be interfered with to prevent bacterial growth. A second target would be the inhibition of the synthesis of acyl-HSLs altogether. In this regard two potential targets can be considered given the presently available data. The inhibition of FabI, a protein involved in *P. aeruginosa* fatty acid biosynthesis, by administration of the drug triclosan was shown to efficiently inhibit production of C4-HSL.[14] The identification of novel FabI inhibitors or inhibitors of other proteins in fatty acid biosynthesis could prove beneficial. A second potential approach would be to design inhibitors of the acyl-ACP/SAM interaction. While the cell uses SAM in a variety of processes, it is clear that the LuxI homologs utilize SAM in a novel manner. The identification of compounds that would inhibit the interaction of SAM with the acyl-ACP would effectively eliminate AI synthesis entirely.

It is a logical progression that if the acyl-HSL signal cannot escape from a cell or enter another cell, quorum sensing would be inhibited. Thus interference with the efflux and import systems for various AIs may hold some promise as a control measure. Lastly, given its high degree of conservation PPK might provide a useful target for inhibiting quorum sensing.

The elucidation of AI interactions within and between bacteria has provided much information as to how and why bacteria behave as they do. The realization that these molecules can communicate with eukaryotic cells is opening a new frontier. Such interactions have already provided insight into how various eukaryotic pathways may function. It is clear that modulation of the immune response is a major effect of AI/eukaryotic cell interactions. Perhaps future studies will allow for the identification of novel therapeutic reagents for a wide variety of ailments not connected to bacterial infection at all.

---

[77] C. Van Delden, E. C. Pesci, J. P. Pearson, and B. H. Iglewski, *Infect. Immun.* **66,** 4499 (1998).

[78] M. Manefield, R. de Nys, N. Kumar, R. Read, M. Givskov, P. Steinberg, and S. A. Kjelleberg, *Microbiol.* **145,** 283 (1999).

[32] Quorum-Sensing System of *Agrobacterium* Plasmids: Analysis and Utility

By STEPHEN K. FARRAND, YINPING QIN, and PHILIPPE OGER

Introduction

Agrobacterium tumefaciens has served as a model organism for studies concerning a number of important biological phenomena and systems including pathogen–host signaling, trans-kingdom gene transfer, type IV macromolecular secretion, and, most recently, quorum sensing. In this article we will present strategies and protocols for studying the quorum-sensing system used by *A. tumefaciens* to regulate conjugal transfer of its Ti plasmids from one bacterium to another as well as the applications of these protocols to quorum-sensing systems of other bacteria.

The System

Agrobacterium tumefaciens induces tumorous growths, called crown galls, on susceptible host plants. The galls result from a genetic transformation event; the bacterium transfers to the plant cells a specific fragment of DNA, called the T-DNA, which integrates into the nuclear genome of the infected host cell. Certain genes on the integrated T-DNA express in the plant cell, leading to the tumorous phenotype. Additional T-DNA genes code for enzymes that condense intermediary metabolites in the crown gall tumor cell into novel compounds generically called opines. The opines, in turn, are released into the soil where they can be taken up and catabolized by the agrobacteria that induced the tumor (Fig. 1).

Most of the genetic determinants, including the T-region, required for virulence of *A. tumefaciens* reside on large extrachromosomal elements called Ti plasmids. The Ti plasmids also code for the uptake and catabolism of the opines produced by the crown gall tumors. These plasmids code for two independent transfer systems; one, called *vir,* is responsible for processing and transfer of the T-region from the bacterium to host plant cells. The second, called *tra,* confers transfer of the entire Ti plasmid from the host donor bacterium to a recipient bacterium by conjugation. Both transfer systems are regulated at the transcriptional level by signal-sensing systems. The *vir* regulon is controlled by a classical two-component signal transduction system which senses and responds to plant-produced factors that signal a suitable infection site. The *tra* system responds to factors produced by the crown gall tumors that signal an environment suitable for propagation of the bacteria that serve as hosts for the Ti plasmid itself.

FIG. 1. Biology of the *Agrobacterium*–plant interaction. Soil-living strains of *A. tumefaciens* are attracted to wound sites where (1) specific signal compounds released by the plants (AS) activate expression of the Ti plasmid *vir* regulon via VirA–VirG, a two-component signal transduction system. Once activated, the Vir functions produce a copy of the T-region (T-strand) and export this DNA molecule to the host plant cell via the type IV VirB secretion apparatus (2). The T-strand migrates to the nucleus where it integrates in a semirandom fashion into the nuclear genome of the plant cell (3). Expression of the *onc* genes on the integrated T-DNA leads to the transformation of the plant cell to a crown gall tumor cell (3). The tumor arises by the uncontrolled proliferation of these cells. Additional genes on the T-DNA direct the synthesis by the tumor cells of low molecular weight carbon compounds called opines (Op). These compounds are released by the tumor cells into the environment where they are taken up and catabolized by the inducing agrobacteria via opine transport and catabolism functions also coded for by the Ti plasmid (4). The opines provide an advantage to the catabolizers when in competition with other bacteria that cannot utilize these substrates (5). The opines also serve to induce the conjugal transfer system (*tra*) of the Ti plasmid, allowing this element to transfer itself to other agrobacteria and related organisms in the vicinity of the crown gall tumors (6). On acquisition of a Ti plasmid, these bacteria gain the capacity to utilize the opines produced by the plant tumors (7). Reprinted with permission from S. K. Farrand, *in* "Conjugal Transfer of *Agrobacterium* Plasmids" (D. B. Clewell, ed.), Chap. 10. Plenum Press, 1993.

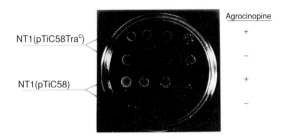

FIG. 2. Opines induce conjugal transfer of Ti plasmids. A spot-plate mating in which the Ti plasmid-less recipient, *A. tumefaciens* C58C1RS, has been spread over the surface of a plate of media selective for transconjugants that have acquired a Ti plasmid. Five-μl volumes of 10-fold dilutions of donors, grown with (+) or without (−) the conjugal opine, agrocinopine, were spotted onto the surface of the plate which was incubated for 48 hr at 28°. Transconjugants appear as small colonies within the confines of the donor spots. Donor strain NT1(pTiC58) harbors a wild-type Ti plasmid, pTiC58 while donor strain NT1(pTiC58tra^c) harbors an opine-independent transfer-constitutive derivative of pTiC58.

Regulation of Conjugal Transfer by Hierarchical System Involving Opines and Quorum Sensing

Almost 30 years ago two groups first reported that transfer of the Ti plasmids from one bacterium to another by conjugation is strongly induced by specific opines produced by the crown gall tumors[1,2] (Fig. 2). Moreover, the inducer, called the conjugal opine, is specific for a particular Ti plasmid. For example, octopine, an imine opine, specifically induces transfer of octopine/mannityl opine-type Ti plasmids[2] while agrocinopines A + B, a pair of sugar phosphodiester opines, stimulate transfer of the nopaline–agrocinopine-type Ti plasmids.[3] Octopine does not induce transfer of the latter, and the agrocinopine opines do not induce transfer of the former. These observations led to a model proposing that for each Ti plasmid a central regulatory element, responsive to these tumor-produced substrates, coregulated expression of genes involved in opine metabolism and conjugal transfer. This model is, in part, correct. For nopaline/agrocinopine-type Ti plasmids, conjugal transfer and catabolism of the agrocinopine opines is coregulated by AccR, a transcriptional repressor specifically responsive to agrocinopines A and B.[4] Similarly, transfer and catabolism of octopine in octopine/mannityl opine-type Ti plasmids is coregulated by OccR, a transcriptional activator that is responsive to octopine.[5,6]

[1] A. Kerr, P. Manigault, and J. Tempé, *Nature (London)* **265,** 560 (1977).
[2] C. Genetello, N. van Larebeke, M. Holsters, A. De Picker, M. van Montagu, and J. Schell, *Nature (London)* **265,** 561 (1977).
[3] J. G. Ellis, A. Kerr, A. Petit, and J. Tempé, *Mol. Gen. Genet.* **186,** 269 (1982).
[4] S. Beck von Bodman, G. T. Hayman, and S. K. Farrand, *Proc. Natl. Acad. Sci. U.S.A.* **89,** 643 (1992).
[5] L. Habeeb, L. Wang, and S. C. Winans, *Mol. Plant–Microbe Interact.* **4,** 379 (1991).
[6] C. Fuqua and S. C. Winans, *J. Bacteriol.* **176,** 2796 (1994).

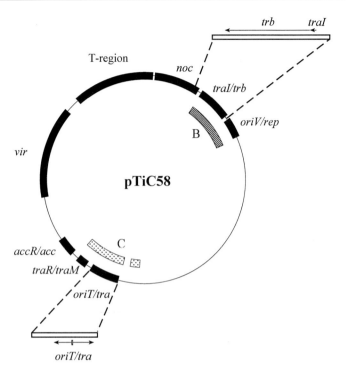

Fig. 3. Physicogenetic map of a Ti plasmid. These elements, usually between 150 and 250 kb in size, are divided into functional segments including: *oriV/rep*, replication; T-region; *vir* region; opine catabolism (*noc*, nopaline catabolism and *acc*, agrocinopine catabolism); *traI/trb* and *oriT/tra*, the conjugal transfer (*tra*) regulon. The *tra* regulon is composed of three operons, *traI-trb* that encodes production of the acyl-HSL quormone and also the type IV secretion apparatus and two *tra* operons, *traAFB* and *traCDG* which are oriented divergently from an interoperonic region containing the *oriT* site. The shaded arcs inside the map represent two of the four regions that are strongly conserved between many Ti plasmid types. Reproduced with permission from S. K. Farrand, *in* "The *Rhizobiacea*: Molecular Biology of Model Plant-Associated Bacteria" (H. P. Spaink, A. Kondorosi, and P. J. J. Hooykaas, eds.), Chap. 10. Kluwer Academic Publishers, 1998.

The genes of the *tra* regulon are organized into three highly conserved operons located on these virulence elements[7] (Fig. 3). One large operon, *trb,* is closely linked to the plasmid replication region and codes for the type IV macromolecular transporter of the mating system. The two remaining operons, *traAFB* and *traCDG,* are arranged as divergently oriented gene sets separated by a short noncoding region

---

[7] S. K. Farrand, *in* "The Rhizobiaceae. Molecular Biology of Model Plant-Associated Bacteria" (H. P. Spaink, A. Kondorosi, and P. J. J. Hooykaas, eds.), p. 199. Kluwer Academic Publishers, Dordrecht, The Netherlands, 1998.

containing the origin of conjugal transfer, *oriT,* and promoter elements responsible for regulation of their expression. These genes code for the DNA processing functions of conjugation, and also for proteolytic processing of the prepilin protein coded for by *trbC* of the *trb* operon. So, expression of the genes responsible for conjugation requires no fewer than three regulated promoters. While AccR and OccR regulate expression of their corresponding opine catabolism gene systems directly, they only indirectly control expression of the genes responsible for conjugal transfer of the Ti plasmids. Expression of the three target *tra* operons of both types of Ti plasmids is controlled directly by a highly conserved transcriptional activator, TraR, the gene for which also is located on the Ti plasmid. At its discovery, TraR was identified as a member of the LuxR family of quorum-sensing transcription factors.[8] This was a particularly intriguing observation since it had been reported earlier that conjugal transfer of Ti plasmids required, in addition to the conjugal opine, a diffusible factor, called CF (conjugation factor), produced by the donor bacteria themselves.[9] CF later was shown to be an acylhomoserine lactone, *N*-(3-oxooctanoyl)-L-homoserine lactone.[10] Synthesis of this quormone is coded for by the *luxI* homolog *traI,* which is the first gene of the Ti plasmid *trb* operon.[11]

Although TraR directly activates expression of the three *tra* operons, the conjugal opines and their associated transcriptional control factors are required to initiate the process. Dependence on opine induction derives from a regulatory linkage between *traR* expression and opine-mediated regulation; in all cases examined to date, the *traR* gene is a member of an operon the expression of which is controlled by the transcription factor responsive to the conjugal opine. For example, in the case of the nopaline/agrocinopine-type Ti plasmids such as pTiC58, *traR* is a member of a five-gene operon, called *arc*[12] (Fig. 4A). Expression of *arc* is controlled by AccR, the repressor responsible for coregulation of agrocinopine catabolism and conjugal transfer. Similarly, in the case of the octopine/mannityl opine-type Ti plasmids such as pTiR10, *traR* is the last gene of a 14-gene operon controlled by OccR, the activator responsible for coregulation of octopine catabolism and conjugal transfer of this Ti plasmid type[6] (Fig. 4B).

This conserved gene organization whereby *traR* is a member of an opine-regulated gene set results in a hierarchical system of signaling in which the tumor-produced substrates are at the apex. In the absence of these compounds *traR* is not expressed and the quorum-sensing system is inactive. However, because the *trb* operon, and consequently *traI,* is expressed at a low basal level, the acyl-HSL quormone is synthesized and accumulates at a slow rate even in the absence of the conjugal opine. When the opine signal is present, the quorum-sensing system is

[8] K. R. Piper, S. Beck von Bodman, and S. K. Farrand, *Nature (London)* **362,** 448 (1993).

[9] L.-H. Zhang and A. Kerr, *J. Bacteriol.* **173,** 1867 (1991).

[10] L.-H. Zhang, P. J. Murphy, A. Kerr, and M. E. Tate, *Nature (London)* **362,** 446 (1993).

[11] I. Hwang, P.-L. Li, L. Zhang, K. R. Piper, D. M. Cook, M. E. Tate, and S. K. Farrand, *Proc. Natl. Acad. Sci. U.S.A.* **91,** 4639 (1994).

[12] K. R. Piper, S. Beck von Bodman, I. Hwang, and S. K. Farrand, *Mol. Microbiol.* **32,** 1077 (1999).

A. The *acc-traR* Control Region

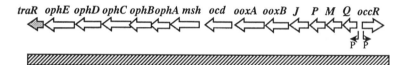

B. The *occ-traR* Control Region

FIG. 4. The conjugal control regions of two Ti plasmids in which transfer is regulated by different opines. (A) In the nopaline/agrocinopine A + B type Ti plasmids *traR* is a member of the five-gene *arc* operon divergently expressed from the *acc* operon that codes for uptake and catabolism of the conjugal opine. Expression of the two operons is regulated by repression by AccR. (B) In the octopine/mannityl opine-type Ti plasmids *traR* is a member of the 14-gene *occ* operon that codes for, among other functions, the uptake and catabolism of octopine and its related opines. Expression of *occ* is regulated by activation by OccR. Except for *traR,* which is 98% identical at the nucleotide sequence level, none of the genes in the control region of one plasmid are related to those in the control region of the other. But in both cases *traR* is a member of an opine-regulated operon. Reproduced with permission from S. K. Farrand, *in* "The *Rhizobiacea:* Molecular Biology of Model Plant-Associated Bacteria" (H. P. Spaink, A. Kondorosi, and P. J. J. Hooykaas, eds.), Chap. 10. Kluwer Academic Publishers, 1998.

activated through induction of *traR*. Activation of the *tra* regulon then occurs, but only after the acyl-HSL has accumulated to some critical level that corresponds to the preset donor quorum[13] (Fig. 5).

In all Ti plasmids examined to date *traR* is located adjacent and in opposite orientation to the *traAFB* operon. However, the upstream opine-controlled gene systems with which *traR* is associated differ depending on the particular conjugal opine. For example, with the exception of *traR,* the genes of the *arc* operon of pTiC58 all are unrelated to those of the *occ* operon of pTiR10.[7,12] These observations suggest that in each case dependence on opine induction results from the fortuitous association of *traR* with a set of genes that are themselves regulated by an opine system. Moreover, several such associations have occurred independently during the evolution of the Ti plasmids, each resulting in a different opine

[13] K. R. Piper and S. K. Farrand, *J. Bacteriol.* **182,** 1080 (2000).

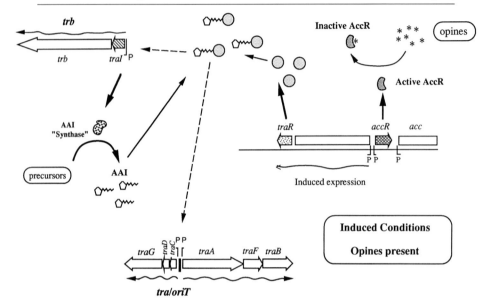

FIG. 5. Hierarchical regulation of conjugal transfer by opines and quorum sensing. In the absence of agrocinopines A + B, AccR of the nopaline/agrocinopine A + B-type Ti plasmid represses expression of *acc*, and also *arc*. Under these conditions, *traR* is not expressed at levels sufficient to activate the *tra* regulon. However, *traI* is expressed at a low basal level, resulting in the production of small amounts of the acyl-HSL quormone (AAI), and the signal slowly accumulates as the population of cells increases. In the presence of the conjugal opine, AccR no longer represses expression of *acc* or *arc*, and expression of *traR* is induced. When the quormone has accumulated to its critical level, the signal-bound dimer form of TraR activates expression of the three *tra* operons, the conjugal apparatus is constructed, and the donors can transfer the Ti plasmid to recipient bacteria. Modified with permission from K. R. Piper and S. K. Farrand, *J. Bacteriol.* **182,** 1080 (2000).

serving as the master signal.[7,12,14,15] The fact that in all Ti plasmids examined to date opines regulate conjugation suggests that it is important to the biology of these elements that their transfer be directly linked to the presence of these tumor-produced substrates.

Molecular Basis of Acyl-HSL-Mediated Quorum Sensing

Much is known concerning the biology of quorum-regulated transfer of the Ti plasmids, and studies focusing on the molecular mechanism of the process have established TraR, and its associated quorum-sensing system, as a paradigm for this regulatory mechanism. Several characteristics specific and unique to TraR have contributed to the development of this system. First, the activity of TraR is

[14] P. Oger and S. K. Farrand, *Mol. Microbiol.* **41,** 1173 (2001).
[15] P. Oger and S. K. Farrand, *J. Bacteriol.* **184,** 1121 (2002).

easily detected by assays of conjugal transfer. Since under opine-induced, quorum-sensing conditions, the Ti plasmid transfers at a frequency of around 10^{-2} per input donor with a limit of detection of one event in 10^8, TraR activity can be assayed with a sensitivity spanning six orders of magnitude. In addition direct gene expression assays based on reporter gene fusions[6,8] and on direct measurements of RNA transcripts[12,16] have been developed. Several of the genetic constructs useful in such studies are minimalistic, comprising only the *traR* gene and a reporter gene directly driven by a TraR-dependent promoter.[17,18] Systems also are available in which the expression of *traR* itself can be strongly repressed and then up-regulated to produce increasing amounts of the activator protein.[18] Second, *Agrobacterium* is amenable to sophisticated genetic analysis, and good shuttle vectors exist for moving genes between *A. tumefaciens* and *Escherichia coli*. Third, genetic constructs have been developed by which the DNA binding properties of TraR can be examined independent of its function as a transcriptional activator.[17] This development in particular has led to the isolation of positive control mutants of TraR which, while retaining DNA binding activity, are specifically defective in activator function.[17] Fourth, TraR is the only member of the LuxR family that has been purified in its full-sized active form.[19] This remarkable achievement has opened the system to biochemical analysis which, when coupled with the powerful genetic tools, allows for a detailed dissection of the functional characteristics of the protein. Finally, the crystal structure of TraR in combination with its quormone ligand and its DNA substrate has been determined at 1.6 Å resolution.[20] Aside from its intrinsic value to understanding structure–function relationships of TraR, the structure will be useful as a predictive model to guide the genetic studies of other members of the LuxR family.

Given these tools, studies with TraR have yielded an understanding of the role of the quormone in transcriptional activation as well as the structure and functional domains of the protein itself. TraR activates expression of gene systems by interacting with a specific promoter element, the *tra* box.[17,19] This *cis*-acting 18 bp inverted repeat invariably is located directly upstream of the −35 elements of the target promoters. Genetic studies suggest that the acyl-HSL quormone is required for TraR to bind the *tra* box,[17] a conclusion confirmed by biochemical analyses.[19,21] Purified active TraR binds DNA, footprinting precisely to the *tra* box sequence.[19] Moreover, when provided with RNA polymerase holoenzyme

[16] C. Fuqua and S. C. Winans, *J. Bacteriol.* **178,** 435 (1996).

[17] Z.-Q. Luo and S. K. Farrand, *Proc. Natl. Acad. Sci. U.S.A.* **96,** 9009 (1999).

[18] Z.-Q. Luo, Y. Qin, and S. K. Farrand, *J. Biol. Chem.* **275,** 7713 (2000).

[19] J. Zhu and S. C. Winans, *Proc. Natl. Acad. Sci. U.S.A.* **96,** 6036 (1999).

[20] Z. Rong-guang, T. Pappas, J. L. Brace, P. C. Miller, T. Oulmassov, J. M. Molyneaux, J. C. Anderson, J. K. Bashkin, S. C. Winans, and A. Joachimiak, *Nature (London)* **417,** 971 (2002).

[21] Y. Qin, Z.-Q. Luo, A. J. Smyth, P. Gao, S. Beck von Bodman, and S. K. Farrand, *EMBO J.* **19,** 5212 (2000).

(RNAP), the bound TraR activates expression from the *tra* box promoter.[19] Interestingly, however, although TraR will bind linear DNA, to activate transcription in conjunction with RNAP the DNA must be in circular, supercoiled form; purified TraR will not activate transcription from linear templates.[19] Active TraR contains 1 mol of the acyl-HSL bound per mole of protein.[19] The ligand is bound very tightly and its removal from TraR by extensive dialysis results in the loss of DNA binding activity[19,21] and the subsequent activation of RNAP-mediated transcription.[19]

Active TraR is soluble and exists in the dimer form; when purified from cells of *E. coli* or *A. tumefaciens* grown with the acyl-HSL, TraR elutes upon size-exclusion chromatography as a single homogeneous peak with a size of 52 kDa, precisely that expected of a homodimer.[21] Moreover, no additional peaks corresponding to monomers or to higher order multimers of TraR are detectable in such preparations. Genetic analyses using fusions of TraR to the DNA binding domain of λcI repressor indicate that conversion from the monomer to dimer form *in vivo* is entirely dependent on exposure of the cells to the acyl-HSL ligand.[21] Consistent with this role for the quormone, removing the acyl-HSL by extensive dialysis results in the conversion of the dimer to monomer form[21] and, as described above, concomitant loss of activator and DNA binding activities.[19]

Studies with LuxR suggest that this activator and, by association, other members of the LuxR family is composed of two domains, a C-terminal DNA binding domain and an N-terminal signal-binding/dimerization domain.[22-24] Consistent with this hypothesis, extensive genetic and biochemical analyses located the dimerization domain of TraR to the N-terminal half of the protein,[25] a conclusion entirely consistent with the crystal structure. A region within this domain also is responsible for binding the acyl-HSL ligand.[25] The C-terminal domain contains a helix–turn–helix (H–T–H) motif and mutations in this domain abolish DNA binding activity.[8,17] The X-ray structure of the TraR–DNA complex predicts that the C-terminal end interacts with DNA and that elements of the H-T-H motif make direct contacts with critical, conserved target bases of the *tra* box element.[20] Consistent with the structural predictions, mutations in a subset of these residues and their target bases abolish the ability of TraR to bind the *tra* box.[17]

From these studies, supplemented by the extensive literature of LuxR, emerges a model in which the active form of TraR exists as a ligand-bound homodimer. Moreover, it is only in the dimer form that TraR can bind the *tra* box, its cognate promoter recognition element. In turn dimerization of TraR is absolutely dependent upon binding the acyl-HSL signal. These functional considerations now can be placed into the context of the structure of the TraR dimer as determined by X-ray

[22] J. Slock, D. Van Riet, D. Kolibachuk, and E. P. Greenberg, *J. Bacteriol.* **172,** 3974 (1990).
[23] S. H. Choi and E. P. Greenberg, *J. Bacteriol.* **174,** 4064 (1992).
[24] S. H. Choi and E. P. Greenberg, *Mol. Marine Biol. Biotech.* **1,** 408 (1992).
[25] Z.-Q. Luo, P. Gao, Y. Qin, and S. K. Farrand, submitted, 2002.

crystallography. The predictions from the crystal structure, in turn, can be tested, both *in vivo* and *in vitro,* by the analysis of directed and random-selected amino acid substitution mutations introduced into the protein itself.

Although this model may be transferable to other LuxR-like transcription factors, it is not at all certain that it is applicable to all members of the family. LuxR binds its quormone,[26] and genetic analysis using the λcI fusion system suggests that acyl-HSL binding drives dimerization of this activator in a manner analogous to TraR.[21] However, EsaR, the quorum-sensing transcription factor from *Pantoea stewartii* multimerizes even in the absence of its quormone.[27] Similarly, CarR, the LuxR homolog from *Erwinia carotovora,* purifies as a multimer from cells grown in the absence of quormone.[28] Moreover, both EsaR and CarR can bind DNA, both *in vivo* and *in vitro,* in the absence of their acyl-HSL ligands.[27,28] Both proteins also bind acyl-HSLs but with different consequences. Ligand-bound EsaR loses affinity for its promote,[27] whereas CarR alters its oligomeric state and the nature of its interaction with target promoters.[28] These results, in comparison to those obtained with TraR and LuxR, suggest that the family of LuxR-like transcription factors may divide into at least two classes based on structural and functional considerations.

Usefulness of TraR-Mediated Quorum-Sensing System

The quorum-sensing system of the Ti plasmid conjugal transfer apparatus, as well as the components of this system, has provided broadly applicable models and tools for the study of this gene regulatory paradigm in other bacteria. As described above, physiological, genetic, biochemical, and structural studies of TraR and the target genes it regulates by quorum sensing have contributed to our understanding of the nature of these systems as well as the molecular mechanisms by which acylhomoserine lactone signals are translated to gene regulation through LuxR-like transcription factors. Biological reporter systems based on TraR have proven useful in surveys for the production of acyl-HSL signals by other bacteria,[29] as well as for the characterization of these compounds present in complex mixtures.[30]

To date, TraR is the only member of the LuxR family that can be purified in its full-sized, biologically active form; that is, in a form that binds specifically to

[26] B. L. Hanzelka and E. P. Greenberg, *J. Bacteriol.* **177,** 815 (1995).

[27] T. D. Minogue, M. Wehland-von Trebra, F. Bernhard, and S. Beck von Bodman, *Mol. Microbiol.* **44,** 1625 (2002).

[28] M. Welch, D. E. Todd, N. A. Whitehead, S. J. McGowan, B. W. Bycroft, and G. P. C. Salmond, *EMBO J.* **19,** 631 (2000).

[29] C. Cha, Y.-C. Chen, P. D. Shaw, and S. K. Farrand, *Mol. Plant–Microbe Interact.* **11,** 1119 (1998).

[30] P. D. Shaw, G. Ping, S. L. Daly, C. Cha, J. E. Cronan, Jr., K. L. Rinehart, and S. K. Farrand, *Proc. Natl. Acad. Sci. U.S.A.* **94,** 6036 (1997).

its DNA recognition element and can activate transcription *in vitro* in the presence of RNAP. In this regard a number of tools have been developed for or have been adapted to studies with TraR that should be applicable to the study of other members of the LuxR family. These include genetic tools and biochemical approaches designed to assess the capacity of these proteins to bind DNA and to form dimers or higher order homomultimers in the presence or absence of their cognate acyl-HSL signals.

In this section we will describe a number of these tools and assays. In each case we will present a brief rationale for the assay or technique, and then outline in general form the method itself. We will indicate the general applicability of, and, where appropriate, the strengths, limitations, and pitfalls of each test.

Assays to Assess TraR-Mediated Gene Expression in Agrobacterium

There are two general classes of assays to assess TraR activity in *A. tumefaciens:* one measures the frequency of Ti plasmid conjugal transfer, the second measures activation of a *tra* gene using reporter fusions or direct detection of transcription products.

Conjugal Transfer Assays. As described above, Ti plasmid transfer is controlled by quorum sensing via TraR and its acyl-HSL signal, *N*-(3-oxooctanoyl)-L-homoserine lactone. Quorum sensing, in turn is controlled by opine availability. Dependency on the conjugal opine can be dispensed with by using strains harboring Ti plasmids that constitutively express their opine system.[31] These strains regulate transfer of their Ti plasmids solely by TraR and its quormone ligand.

Conjugal transfer frequencies can be measured over six orders of magnitude, giving a large range for assessing TraR activity. However, transfer is an indirect measure of TraR activity, being dependent on expression of the *tra* genes, assembly and activation of the conjugal transfer system, the formation of a mating pair with an appropriate recipient, and transfer to and successful establishment of the Ti plasmid in the recipient.[13] The limitations imposed by the latter two requirements can be minimized by conducting matings under conditions in which recipients are in excess.

In any bacterial mating, conditions must be established by which transconjugants are selected and both donors and unmated recipients are counterselected. Most Ti plasmid matings use recipients that express resistance to two antibiotics to which the donor is susceptible. Medium containing these two antibiotics, most often rifampicin and streptomycin,[1,31] counterselects the donors. Opine catabolism, which is encoded by the particular Ti plasmid, or resistance to antibiotics in cases in which the Ti plasmid is marked with an appropriate cassette or transposon can be used to select for transfer of the element. Whereas the opine-initiated regulatory

---

[31] S. Beck von Bodman, J. E. McCutchan, and S. K. Farrand, *J. Bacteriol.* **171,** 5281 (1989).

cascade occurs in donor cells grown on solid or in liquid media, mating pairs form efficiently only on solid surfaces.[32] Thus donor and recipient cultures can be grown in liquid or on solid media, but matings must be conducted on surfaces such as micropore membranes or an agar-based medium.

Given these considerations, matings routinely are conducted using either of two methods.

Standard Filter Matings

1. Liquid cultures of donor and recipient are grown overnight with shaking at 28° in L broth or in a suitable *Agrobacterium* minimal medium such as AB[33] or AT[34] with mannitol as carbon source. The next morning each culture is diluted into fresh medium and grown as above to late exponential phase corresponding to 5×10^8 to 10^9 colony-forming units (cfu) per ml. In cases in which the donors contain wild-type Ti plasmids and must be induced with the appropriate opine, both donors and recipients should be grown in minimal mannitol medium with that of the donor supplemented with the appropriate conjugal opine.

2. A sample of the donor and recipient cultures is removed and suitably diluted in series, and volumes of the appropriate dilutions plated on nonselective medium to determine viable cell titers at the time of the initiation of the mating.

3. At the same time a 0.1 ml volume of the donor culture is mixed with a 1.0 ml volume of the recipient culture and the mixture is collected onto a sterile micropore membrane by vacuum filtration. Similar volumes of donor and recipient cultures are collected separately on membrane filters as controls. Alternatively, 10–20 μl volumes of the mixture and of the donor and recipient cultures can be spotted, using a micropipette, directly onto the surface of micropore membrane filters. Each filter is placed bacteria-side-up on the surface of a plate containing fresh agar medium, either L broth- or minimal medium-based, depending on the medium used to grow the parent cultures. As above, if the donor requires opine induction, a minimal medium containing mannitol and the conjugal opine should be used for the mating.

4. The plates are incubated at 28° for a minimum of 2 hr. Plates can be incubated for longer periods, but care must be taken when comparing results for matings conducted for different times; growth of the transconjugant on the filters can influence the apparent transfer frequencies.

5. At the end of the mating period the filters are transferred to sterile tubes containing 1 ml of a solution of 0.9% NaCl and the cells are removed by vortexing or vigorous agitation.

[32] K. R. Piper and S. K. Farrand, *Appl. Environ. Microbiol.* **65,** 2798 (1999).
[33] M.-D. Chilton, T. C. Currier, S. K. Farrand, A. J. Bendich, M. P. Gordon, and E. W. Nester, *Proc. Natl. Acad. Sci. U.S.A.* **71,** 3672 (1974).
[34] A. Petit and J. Tempé, *Mol. Gen. Genet.* **167,** 147 (1978).

6. Decade dilutions are made from each of the three cell suspensions, the mating mix, and the donor and recipient controls, and volumes, typically 100 μl, of appropriate dilutions are spread onto plates containing the selective medium. Generally, plating volumes of undiluted and 1/10 diluted suspensions of the donor and recipient preparations are sufficient to control for mutation to the selective agents while the mating mix should be diluted and plated over a range of at least 5 to 6 logs.

7. The selection plates are incubated at 28° for 48 to 72 hr, the numbers of colonies are counted, and the titers of transconjugants in the 1 ml volumes of the resuspension solution are calculated from the dilution factor corrected for the appearance of selection-resistant donors and recipients. Frequencies of transfer, expressed as transconjugants recovered per input donor, can be calculated by dividing the titer of the recovered transconjugants by the titer of the donor culture determined at the beginning of the mating.

The effects of agents to be tested on TraR-mediated *tra* gene activation easily can be assessed using this technique. Generally the agent to be tested should be included in the medium in which the donors are grown prior to mating, as well as in the medium in the plates on which the mating filters are incubated.

Spot Plate Matings. In this technique suspensions of donor cells are spotted onto the surface of agar selection medium previously seeded with a culture of the recipient.[9] The technique is fast, technically simple, and economical; several independent matings can be conducted on the same plate. Moreover, because each transconjugant colony that appears derives from a single mating the method yields a direct measure of initial transfer events.

1. Cultures of the donor and recipient strains are grown overnight in the appropriate medium, diluted, and grown to mid- to late-exponential phase as described above. Treatments of the donor, including opine induction or addition of the acyl-HSL signal, must be done at this stage of the mating.

2. Volumes of 100 μl of the recipient are spread evenly over the surface of the selection medium. Make sure that no moisture remains on the surfaces of these plates.

3. A dilution series of the donor is prepared in a 0.9% solution of NaCl, and 5 μl volumes of a suitable set of dilutions are spotted onto the lawn of recipient cells. Volumes of each dilution can be spotted in multiples to facilitate statistical analysis. Similar 5 μl volumes of the undiluted donor cell suspension are spotted onto the surface of a plate containing selection medium only to assess for mutation to resistance to the counterselection agents. There is no need to spot the recipient to such plates; its presence as a lawn on the surface of the mating/selection plates serves as such a control. The titer of the donor culture is determined by dilution plating as described for the standard filter mating technique.

4. The plates are incubated for 24 to 72 hr and the numbers of transconjugant colonies arising within the confines of each site at which donors were spotted is determined with the aid of a dissecting stereomicroscope if necessary (Fig. 2).

5. As above, the transfer frequency, expressed as transconjugants appearing per input donor, can be calculated from the titer of the donor culture determined just prior to spotting onto the mating plates.

In both techniques antibiotics to which the recipient is resistant are used to select against the donor. Counter to intuition, in the spot plate assay antibiotics such as rifampicin and streptomycin do not inhibit transfer of the plasmid from the susceptible donors. Transconjugants can be selected for their capacity to utilize opines associated with the Ti plasmid being transferred using AB or AT minimal agar media, or for resistance to an antibiotic associated with an appropriate resistance determinant carried by the conjugal element. In the latter case, a rich medium such as L agar or nutrient agar supplemented with the appropriate antibiotics can be used.

Gene Expression Assays

REPORTER FUSIONS. Assays for expression of genes regulated by TraR are based on fusions of *lacZ* or *uidA* to an appropriate *tra* gene expressed from a *tra* box-dependent promoter. Many biovar 1 strains of *A. tumefaciens* lack endogenous β-galactosidase and β-glucuronidase activities making these reporters particularly useful. Colonies can be examined on medium containing X-Gal or X-Glu, and the two enzymes can be assayed quantitatively by standard methods.[35] Several such reporter constructs exist.[16–18] In some the reporter fusion is carried on one recombinant plasmid while TraR is provided either by a second recombinant plasmid[8] or by a Ti plasmid itself.[6] In others, *traR* and the reporter fusion are carried on the same recombinant plasmid.[29] Alternatively, reporter fusions can be introduced into the Ti plasmid as double crossover marker exchanges or single crossover Campbell-like insertions using standard genetic techniques.[14,15] In such cases, control of expression of the *tra* regulon, as assessed by the *in situ* reporter fusion, can be monitored within the context of the Ti plasmid itself. In cases in which the reporter strain lacks *traI,* the acyl-HSL signal must be added exogenously to the growth medium.

The *tra* box controlling expression of the divergent *traAFB* and *traCDG* operons is of particular interest since a single DNA recognition element serves to bind the transcription factor responsible for activating expression from promoters oriented in both directions. The *tra* box itself is situated in the canonical site, just upstream of the −35 elements of the two divergent operons.[6,36] We have constructed a double reporter vector, pDCAC2b, to monitor simultaneously the

[35] J. Sambrook, E. F. Fritsch, and T. Maniatis, "Molecular Cloning: A Laboratory Manual," 2nd Ed. Cold Spring Harbor Laboratory Press, Cold Spring Harbor, NY, 1989.
[36] S. K. Farrand, I. Hwang, and D. M. Cook, *J. Bacteriol.* **178**, 4233 (1996).

TABLE I
TraR-Mediated Expression from Divergent *tra*-Box Promoter System[a]

| Plasmid | Construct | β-Galactosidase activity[b] | | β-Glucuronidase activity[f] | |
		$-$TraR[c]	$+$TraR[d]	$-$TraR[c]	$+$TraR[d]
None	None	N.T.[e]	2.7	N.T.	1.7
pRG970b	Vector	4.4	4.4	3.0	5.0
pDCAC2b	Intact *tra* box	8.4	1535.5	2.5	18.1
pDCAC2bF	Mutant *tra* box	5.0	2.7	N.T.	1.7

[a] Located between the *tra* operons of pTiC58.
[b] Activity of the *traA::lacZ* fusion expressed as units of β-galactosidase activity per 10^9 viable cells.
[c] Tested in *A. tumefaciens* strain NT1 which lacks a Ti plasmid.
[d] Tested in *A. tumefaciens* strain NT1(pTiC58ΔaccR) which harbors a Ti plasmid that constitutively expresses *traR* and the quorum-sensing system.
[e] N.T., Not tested.
[f] Activity of the *traC::uidA* fusion expressed as units of β-glucuronidase activity per 10^9 viable cells.

activation of expression of these two operons by TraR.[37] The vector, which is based on pRG970,[38] contains a 251 bp fragment corresponding to the intergenic region between *traAFB* and *traCDG* cloned between divergently oriented promoterless copies of *lacZ* and *uidA*. Expression of the former can be measured by assessing β-galactosidase activity, while that of the latter can be measured as β-glucuronidase activity.

The utility of this vector is shown in Table I. The activity of both enzymes was assayed in extracts from a derivative of *A. tumefaciens* harboring pDCAC2b and either lacking or containing a Ti plasmid constitutive for expression of *traR*. The results show that both operons are expressed, but only in a strain that produces TraR and its acyl-HSL ligand. Moreover, a reporter plasmid, pDCAC2bF, in which the *tra* box was altered by the addition of two bases at the center of the inverted repeat, failed to express either fusion even in a strain providing TraR.

Measurements of *tra* gene expression by monitoring fusions is subject to all of the advantages and pitfalls of the technique itself. Expression levels may depend on whether the fusion is transcriptional or translational, and for these reasons one cannot meaningfully compare levels of expression of two different genes of the regulon. Moreover, since the technique measures enzyme activity, a function of a protein, assessing fusions is not a reliable method to determine levels of message, the direct outcome of transcriptional activation. Finally, β-galactosidase is a

[37] D. M. Cook, Ph.D. dissertation. University of Illinois at Urbana-Champaign, 1995.
[38] G. Van den Eede, R. Deblaere, K. Goethals, M. van Montagu, and M. Holsters, *Mol. Plant–Microbe Interact.* **3,** 228 (1992).

relatively stable enzyme, making fusions to this protein not well suited for steady-state expression assays or for assessing mechanisms by which gene expression is down-regulated.

MESSENGER RNA ASSAYS. Expression of *traR,* as regulated by opines, or of the *tra* regulon, as controlled by TraR, also can be assessed by techniques that detect specific mRNA populations including primer extension, RNase protection, and RT-PCR (reverse transcriptase–polymerase chain reaction) assays. Methods have been developed to isolate high-quality RNA from *A. tumefaciens* strains suitable for use in such assays.[12,16] The detection techniques are at least semiquantitative and have the advantage that they can be used to measure directly transcriptional activation as well as message stability. These techniques also can be used to assess conditions that lead to down-regulation of the *tra* regulon or its controlling elements.

Strategies and Assays for Opine-Regulated Control of traR Expression. Opines control conjugal transfer by regulating expression of *traR,* and subsequently, the TraR-dependent quorum-sensing system. Of the eight known chemical classes of opines only two, the arginyl opines octopine and nopaline, and the sugar phosphodiester opines agrocinopines A + B and C + D are known to induce conjugal transfer.[2,3] Moreover, new Ti plasmids are described on a regular basis, but usually no efforts are made to determine if these elements are conjugal and, if so, whether transfer is regulated by a hierarchical opine quorum-sensing cascade.

There are relatively simple ways to determine if one or more of the opines found in tumors induced by a given strain of *A. tumefaciens* initiates induction of conjugal transfer. We consider two cases. In the first, the opines produced by the tumor have been identified. In the second, either none of the opines known to be present in a tumor induce transfer, or the tumors do not produce a known opine. Case 1 defaults to case 2 should none of the known opines present in such tumors activate transfer of the Ti plasmid present in the inducing bacterium. For each, we provide two strategies: one designed to determine if the opine, known or unknown, induces conjugal transfer, the second to determine if the opine induces a component of the quorum-sensing system.

CASE 1: TUMORS CONTAIN ONE OR MORE KNOWN OPINES. In this case, the opines, if available, can be purchased commercially. Octopine, nopaline, and mannopine are available from several chemical supply companies. If the opines are not commercially available, they can be synthesized by published methods, or isolated to any state of purity from crown gall tumors. The reader is directed to an excellent review by Dessaux *et al.*[39] describing in detail methods for detecting, purifying, and characterizing known opines present in tumorous tissues induced by strains of *A. tumefaciens.* The review also describes the classic strategies for detecting the presence of previously unidentified opine types.

---

[39] Y. Dessaux, A. Petit, and J. Tempé, *in* "Molecular Signals in Plant–Microbe Communications" (D. P. S. Verma, ed.), p. 109. CRC Press, Boca Raton, FL, 1992.

CASE 2: TUMORS CONTAIN KNOWN OPINES THAT DO NOT INDUCE TRANSFER OR DO NOT CONTAIN A KNOWN OPINE. In both cases, the goal first is to determine, using the tests listed below, whether the Ti plasmid present in the inducing strain has a transfer system that is inducible by an unknown opine in the tumor. In the simplest sense, one can test extracts from tumors for their ability to induce a reporter/detection system. However, the problem is complicated by the observation that some tumor extracts contain substances that are inhibitory to the induction of transfer, thus necessitating at least a partial purification of any unknown compounds that might have inducing activity.

A two-step procedure can be used in the initial purification of opines from tumor tissues. The methodology is described by Dessaux et al.[39] in detail; we will present here only an outline of the strategy.

1. Aqueous or water–ethanol extracts are prepared from minced or finely ground crown gall tumor tissue induced on any susceptible plant by the A. tumefaciens strain in question.

2. The extract is subjected to a "biological purification" step. The extract is inoculated with a strain of A. tumefaciens that lacks a Ti plasmid or known opine catabolic element such as NT1 or C58C1. This culture is incubated at 28° with shaking for 48–72 hr. Since virtually all opine catabolism systems are associated with Ti, Ri, or opine catabolic plasmids, the bacterial strains will catabolize many small organic components present in the extract, but leave the opines intact. However, there is some danger in this technique. At least one opine, deoxyfructosylglutamine (dfg), is catabolized by genes present in strains NT1 and C58C1 as well as in several other Ti-plasmidless isolates.[40]

3. Following biological purification, the extract is taken to dryness, redissolved in distilled water, and subjected to a separation technique, most often high voltage paper electrophoresis. In this step, the extract is applied to the paper as a long strip across the baseline. An extract prepared from normal tissue of the plant type on which the tumor was induced is prepared and applied to a second sheet of paper as a comparative control. The samples are subjected to electrophoresis using one or more of the buffers described by Dessaux et al.[39]

4. Following electrophoresis, the paper is cut into long strips along the direction of electrophoresis. Each strip can be stained with a reagent designed to visualize classes of organic compounds, alkaline silver nitrate for α-diols, ninhydrin for amino acids, the Pauli reagent for imidazole-containing opines, and phenanthrene quinone for the arginyl opines, for example.[39] Opines of various classes migrate as a function of their electrical charge which depends on the pH of the buffer used for the electrophoresis. Strips containing extracts from normal and tumorous tissues

[40] V. Vaudequin-Dransart, A. Petit, W. S. Chilton, and Y. Dessaux, Mol. Plant–Microbe Interact. 11, 583 (1998).

are compared with compounds appearing in the extracts of the latter but not the former being noted as candidate opines.

5. An additional 1-cm-wide unstained strip is cut width-wise yielding a series of 1-cm squares. These squares will be used to test for conjugation-inducing activity.

Tests for Conjugal Opine Activity

INDUCTION OF CONJUGAL TRANSFER. Conjugal transfer assays, conducted as described above, can be used to assess which, if any, opines present in the tumors induces transfer of a particular Ti plasmid. In case 1, purified opines, purchased or synthesized, are included in the premating medium in which the donor is grown. The bacteria should be grown only in minimal medium using mannitol or glucose as the primary carbon source. Rich media should be avoided; in at least one case components of such media inhibit opine-mediated induction of conjugation.[41] In case 2 the unknown compounds identified as being specific to the tumor extract can be tested for induction of transfer. The unknown compound can be eluted from the paper electrophoretogram strip into distilled water. A portion of this solution then is added to the minimal medium in which the donor will be grown. Alternatively, a section of the strip can be placed into the tube containing the minimal medium in which the donor will be grown. Following growth in this medium, the donor can be mated with a suitable recipient as described in the section on matings.

Detecting transfer of the Ti plasmid can be problematic; most such elements do not code for easily selectable phenotypes. However, opine catabolism can be a useful selection marker. For example, in case 1, the known opines associated with the Ti plasmid can be tested for induction of transfer, but they also can be used as selection agents following the mating itself. In these cases, AB or AT agar minimal medium containing the opine as sole carbon source, and antibiotics counterselective to the donor, can be used to select transconjugants as described in the section on mating techniques. In case 2, if the tumors produce known opines that do not induce transfer, these compounds can, at the least, be used to select for transconjugants in matings with donors induced by the candidate unknown compounds identified by the electrophoresis screens.

Difficulties arise in the second iteration of case 2 where the tumors do not contain a known opine but do contain candidate opines of unknown structure. In this case, generally these candidate agents are not available in amounts sufficient for direct genetic selections. It is possible to introduce selectable markers, such as an antibiotic resistance gene, onto the Ti plasmid, but this can be a tedious and difficult exercise in bacterial genetics. Fortunately, alternative strategies for detecting opine-mediated induction of the transfer system exist.

[41] P. J. J. Hooykaas, C. Roobol, and R. A. Schilperoort, *J. Gen. Microbiol.* **110,** 99 (1979).

INDUCTION OF QUORUM-SENSING SYSTEM. In the uninduced state strains harboring Ti plasmids show little expression of any of the three *tra* operons. Moreover, the strains produce only barely detectable levels of the acyl-HSL quormone.[6,11,29] On opine-mediated induction, however, the quormone is produced in high amounts, and expression of the *tra* regulon is induced to easily detectable levels. Either phenotype can be used to test whether a given opine serves as the conjugal signal.

DETECTING INCREASES IN PRODUCTION OF ACYL-HSL SIGNAL. *traI,* the gene responsible for production of the Ti plasmid mating quormone, N-(3-oxooctanoyl)-L-HSL, is the first gene of the Ti plasmid *trb* operon, and its expression is dependent on TraR.[11] In turn, expression of *traR* requires the conjugal opine; in its absence *traR* is not expressed at functional levels, *traI* is expressed at a very low basal level, and subsequently, the acyl-HSL signal accumulates to only a barely detectable level. In the presence of the conjugal opine, TraR is produced, and it, in conjunction with the slowly accumulating acyl-HSL signal, activates expression of *traI*. This condition leads to a rapid postinduction accumulation of the quormone to high levels. Given these considerations, the signaling activity of a given opine can be assessed by determining its affect on the production of the quormone. It should be noted that these assay can be used in case 1 and in case 2 in lieu of the mating tests.

To conduct this assay:

1. Two cultures of the *A. tumefaciens* strain to be tested are established at low population density (ca. 10^5 cfu) in a small volume (ca. 1 ml) of AB or AT minimal medium containing glucose or mannitol as primary carbon source. A volume of the opine or candidate opine is added to one culture, and the two are incubated in parallel with shaking at 28°.

2. When the cultures have reached late exponential phase, the cells are removed by centrifugation and the culture supernatants are mixed with equal volumes of acidified ethyl acetate (0.1 ml of glacial acetic acid per liter of solvent) to extract the acyl-HSL.

3. Following phase separation, which can be facilitated by low speed centrifugation, the ethyl acetate phase is removed to a clean vessel.

4. The culture supernatant is extracted again, the phases are allowed to separate, and the organic phase of the second extraction is combined with that of the first.

5. Residual water in the pooled ethyl acetate phase is removed by addition of a small amount of anhydrous magnesium (or sodium) sulfate, and the organic phase is transferred to a clean vessel.

6. The organic solvent is evaporated to dryness under a stream of dry nitrogen or argon, and the remaining residue is taken up in a small, measured volume of ethyl acetate.

7. The amount of acyl-HSL present in this concentrated extract is determined by bioassay as described in a later section.

Alternatively, the effect of the opine in question on activation of a standard TraR-dependent *tra::lacZ* reporter fusion can be tested. This strategy assumes that the opine regulates expression of the Ti plasmid *traR* gene, which, in turn, activates expression of the reporter fusion. The test also assumes that the TraR of the Ti plasmid being tested recognizes the *tra* box of the standard *tra* fusion element being used as the reporter. While such conservation has proven the case among the *Agrobacterium* plasmids tested to date, this by no means ensures universal adherence to the rule. Finally, the assay assumes that the isolate being tested does not produce its own β-galactosidase.

To conduct this assay:

1. Introduce by mating from *E. coli* or by electroporation[42] a standard *tra::lacZ* reporter plasmid such as pZLb251[17] into the strain of *A. tumefaciens* being tested. This broad host range plasmid contains a fusion of *lacZ* to *traG* driven by the native *traCDG* TraR-dependent promoter. Screen this reporter strain for the plasmid and also for expression of β-galactosidase activity on medium supplemented with X-Gal but lacking the opine to ensure that the reporter is not expressed in the absence of a tumor-specific signal.

2. The influence of the opine on expression of the fusion can be assessed qualitatively on a minimal agar medium containing X-Gal. The opine can be included into the medium, or small paper disks saturated with the opine can be place on top of a lawn of, or adjacent to a colony of the bacterium being tested.

The influence of the opine on induction of the reporter can be quantified as follows:

1. Establish two cultures of the strain harboring the reporter plasmid in AB or AT minimal medium, one of which is supplemented with the opine being tested.

2. Incubate each culture with shaking at 28° until the cells reach late exponential phase, harvest the cells and quantify the β-galactosidase activity using any adaptation of the Miller assay.

If the opine activates conjugation, the level of β-galactosidase activity should be substantially higher in cells from the culture grown with the signal as compared to the culture grown in the unsupplemented medium. The results from such experiments can be verified by assaying the supernatants from the two cultures for levels of the acyl-HSL as described above.

[42] G. A. Cangelosi, E. A. Best, G. Martinetti, and E. W. Nester, *Methods Enzymol.* **204,** 384 (1991).

Assays for Interaction of TraR with Its DNA Target. Having purified TraR allows for direct, *in vitro* assays for the interaction of this protein with its DNA recognition site. Readers are directed to the work by Zhu and Winans[19] in which standard techniques including gel mobility shift assays and DNA footprinting are used to demonstrate and map the site of the interaction between TraR and its recognition element.

Interactions of TraR and other members of the LuxR family with their cognate DNA recognition elements also can be assessed using genetic techniques. Expression of an activatable reporter fusion is not a suitable assay for DNA binding since activation requires at least two events, protein–promoter interaction, and activation of RNAP-mediated transcription. However, DNA binding by these proteins can be detected using an assay based on repression of expression of a reporter gene. One first must construct a reporter system in which the DNA recognition element is located within the promoter such that binding by the LuxR-like transcription factor will interfere with the initiation or extension of transcription by RNAP. For example, Luo and Farrand[17] constructed a *lacZ* reporter in which the *tra* box is located directly over the −10 element of the *lac* promoter. TraR repressed expression of this reporter, but only when the cells were grown with quormone. Similarly, Egland and Greenberg[43] constructed a reporter in which the *lux* box, the 20 bp LuxR recognition element, is located between the −10 and −35 hexamers of a promoter driving expression of *lacZ*. LuxR repressed expression of this reporter, again in a quormone-dependent manner. Mutations in the DNA recognition element, as well as in the C-terminal DNA binding domain of the protein, abolished repression in both cases.[17,43]

This strategy should be applicable to any of the members of the LuxR family. The reporter can be used to assess DNA binding by the transcription factor and the dependence of this binding on the quormone signal. The assay also can be used to isolate mutations in the protein or in the *cis*-acting binding site that interfere with DNA recognition and binding. The repressor system, when combined with a second screen using a standard activatable promoter, can be used to isolate positive control (PC) mutants: those that retain the capacity to bind DNA but have lost activation functions. Such PC mutants of TraR and of LuxR have been obtained using this strategy.[17,43]

Assays for Dimerization of TraR. Because it can be purified in active form, TraR is an ideal target for studies designed to determine how these proteins dimerize in combination with their ligand, how they interact with their target DNA, and how they activate transcription. However, several of the techniques used to assess the dimeric properties of active TraR and the role of the signal in the process of dimerization do not depend on prior purification of the protein and should be applicable to studies of other members of the family.

[43] K. A. Egland and E. P. Greenberg, *J. Bacteriol.* **182,** 805 (2000).

USING FUSIONS TO λCI REPRESSOR. James Hu has constructed a multimerization reporter system based on the cI repressor of phage λ.[44,45] The cI protein is composed of two functional regions, an N-terminal DNA binding domain and a C-terminal multimerization domain. Stable, high affinity binding of cI to its promoter recognition site requires the protein be in a dimeric or tetrameric form. The reporter system consists of a strain stably lysogenized with a derivative of λ containing a *lacZ* gene expressed from the cI-repressible P_R promoter and a plasmid containing a truncated form of the *cI* gene, *cI'*, coding for only the N-terminal DNA binding domain of the repressor. This gene, which is expressed from the *lac* promoter, is followed directly by a multiple cloning site (mcs). Since the N-terminal peptide lacks a dimerization domain it will not stably bind the P_R promoter and the reporter is expressed at high levels.

The gene for the protein of interest is cloned into the multiple cloning site to create a translational fusion containing at the N terminus the cI DNA binding domain and, at the C terminus, the protein to be tested. If the protein being tested multimerizes, the chimeric protein, expressed following induction by IPTG, can bind the P_R promoter tightly, thereby repressing expression of the *lacZ* reporter. A positive control in the form of a cI'::LeuZip-GCN4 fusion protein is available[44] and should be included in all such assays. In addition, the *in vivo* stability of the fusion protein should be monitored, especially in cases in which no repression is observed.

COAFFINITY DISPLAY SYSTEMS. Affinity-tagged chromatography can be used to display proteins that interact with a target protein *in vivo*. Although most often used to detect interactions between heterologous proteins, the technique also can be used to probe homomeric interactions. In the general case, a strain is constructed that expresses two forms of the protein, one tagged with an affinity epitope, the other wild-type. The affinity tag serves two purposes: it provides a way to isolate the protein in its dimeric form, and it alters the size of the protein in comparison to the wild-type form. Soluble extracts from the strain are chromatographed on the appropriate affinity matrix, and, following extensive washing, the bound species are collected by elution with the proper displacing agent. The eluted affinity-tagged protein then is analyzed by denaturing sodium dodecyl sulfate–polyacrylamide gel electrophoresis (SDS–PAGE). If the protein does not dimerize, only the affinity-tagged version of the protein will be retained on the matrix, and a single band, corresponding to this species, will be detectable on the gel. However, if the protein forms homomultimers, three species will form in the cell: homo-tagged dimers, homo-untagged dimers, and heterodimers composed of one affinity-tagged protomer and one wild-type protomer. Assuming that, in the pseudoheterodimeric form, the epitope tag is still displayed, two bands will appear, one corresponding to the affinity-tagged protein, the other to the coretained untagged species

[44] J. C. Hu, *Structure* **3**, 431 (1995).
[45] J. C. Hu, E. K. O'Shea, P. S. Kim, and P. T. Sauer, *Science* **25**, 1400 (1990).

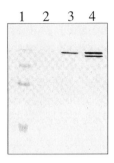

FIG. 6. Demonstration of TraR dimerization by affinity trapping chromatography. Strains express-ing wild-type *traR* (lane 2), *his-traR* (lane 3), or both genes (lane 4) were grown in medium supple-mented with *N*-(3-oxooctanoyl)-L-HSL to late exponential phase. The cells were harvested and broken by French press treatment, and the lysates were clarified by centrifugation. The cleared lysates were subjected to affinity chromatography on NTA columns and the fractions eluted using buffer containing high concentrations of imidazole were collected and subjected to SDS–PAGE. TraR and its His-tagged derivative were visualized by Western analysis using anti-TraR antibody. Note that no TraR is detected in lane 2 indicating that the wild-type protein is itself not retained on the NTA column. However, when the two proteins are coexpressed, wild-type TraR (lower band in lane 4) is retained because it forms a pseudoheterodimer with the His-tagged form (upper band in lanes 3 and 4) of the activator. Lane 1 contains a set of molecular weight marker proteins.

(Fig. 6). Because only small amounts of protein may be recovered, sensitive de-tection methods must be available. The proteins can be radiolabeled *in vivo* and detected following SDS–PAGE by autoradiography. Alternatively, the proteins can be detected using immunological methods assuming that specific antiserum, re-active against both forms of the protein, is available. Any affinity tag should be useable as long as the epitope does not interfere with multimerization.

A general protocol might proceed as follows:

1. Plasmids coding for an epitope-tagged protein such as TraR and for its wild-type counterpart are introduced into an appropriate bacterium.

2. Two cultures are established, one of which is supplemented with the acyl-HSL signal. The cultures are grown in parallel to late exponential phase. As con-trols, cultures of strains expressing only one of each of the proteins are prepared and processed in parallel.

3. The cells of each culture are harvested and lysed, and debris is removed by centrifugation.

4. The cleared lysates from each culture are subjected to chromatography on the appropriate affinity resin. The resin is washed repeatedly with a buffer allowing retention of the epitope-tagged protein.

5. The resin is washed with a buffer containing the appropriate displacement agent for the epitope tag being used. The eluates are retained for analysis by SDS–PAGE.

6. Each eluate is concentrated if necessary, and the proteins are separated by SDS–PAGE under reducing conditions.

7. The separated proteins are visualized by staining or by Western analysis.

Attention must be paid to several considerations with this technique. First, both species of the protein must be stable and expressed at similar levels *in vivo*. Second, although not essential, results from these types of experiments are most convincing in cases where the affinity-tagged protein retains biological activity. Third, the dimers must be sufficiently stable to remain intact during chromatography. Fourth, although the technique can detect multimerization, it cannot reliably differentiate between dimerization and the formation of higher order multimers.

Two important controls are required. First, extracts prepared from cultures expressing only one or the other of the two proteins should be analyzed in parallel; following SDS–PAGE the sample containing the epitope-tagged protein should yield a single band while that containing the untagged species should be free of the protein. Second, in the special case in which dimerization may be dependent on some condition, it is important to test extracts from cells grown under treated and nontreated conditions. This control is of particular importance for studies concerning acyl-HSL-mediated dimerization. Furthermore, extracts from cells expressing one or the other of the two proteins, when mixed together following lysis, should yield only the epitope-tagged protein following chromatography.

Qin *et al.*[21] used cI' fusions to show that TraR and also LuxR form stable dimers, and that dimerization is dependent on growth with the appropriate acyl-HSL signal. They also used the coaffinity display system to confirm the results of the cI' fusion experiments and to map the region of TraR required for dimerization. In this case, strains were constructed that coexpressed His-tagged TraR and ordered sets of N- and C-terminal deletion derivatives of the activator. The N- and C-terminal borders of the dimerization region were delineated by determining whether a given deletion derivative formed a heterodimer with His-tagged TraR.[21]

General Assays Utilizing TraR System. In addition to studies concerning the biology and molecular biology of quorum sensing in *A. tumefaciens,* TraR has proven to be a valuable tool to those interested in quorum sensing in other bacteria. Over the past few years literally hundreds of bacterial strains have been screened for the production of compounds with acylhomoserine lactone activity,[29,46,47] with the production of such compounds taken to signify the presence of a quorum-sensing system in the tested bacterium. Acyl-HSLs are difficult to detect by standard

[46] S. Swift, M. K. Winson, P. F. Chan, N. J. Bainton, M. Birdsall, P. J. Reeves, C. E. D. Rees, S. R. Chhabra, P. J. Hill, J. P. Throup, G. P. C. Salmond, P. Williams, and G. S. A. B. Stewart, *Mol. Microbiol.* **10,** 511 (1993).

[47] M. Elasri, S. Delorme, P. Lemanceau, G. Stewart, B. Laue, E. Glickmann, P. M. Oger, and Y. Dessaux, *Appl. Environ. Microbiol.* **67,** 1198 (2001).

chemical tests. These compounds react with very few indicator reagents and the molecules do not contain strongly absorbing chromophores. Compounding the problem of detection, these signal molecules generally are produced at very low levels, usually in the nanomolar to micromolar range.[48]

Traditionally, quorum-sensing signals have been detected using sensitive bioreporter systems. In the general sense, these reporter bacteria contain a member of the LuxR family and a gene coding for a detectable phenotype, the expression of which is driven by a promoter activatable by the LuxR-like transcription factor. Moreover, the reporter strains do not produce their own signal, but can respond to acyl-HSLs added to the culture medium. A remarkably large number of bioreporter strains have been constructed from systems including *lux* of *V. fischeri, las* of *P. aeruginosa, tra* of *A. tumefaciens,* and *Chromobacterium violaceum.* Detection phenotypes include bioluminescence,[48] pigment production,[49] GFP-mediated fluorescence,[50] and β-galactosidase activity.[29,51] Each reporter phenotype has its strengths and weaknesses. Moreover, each activator–reporter system exhibits a characteristic specificity for signal structure. Most of the reporters respond to a range of signals with varying side-chain lengths and different substitutions at the C-3 position, but in general, they respond with considerably lesser sensitivity to noncognate signals.[29] No one reporter responds to all of the known acyl-HSL signals. Thus, in screens designed to detect these compounds two or more reporters that, in combination, cover the spectrum of known signals should be used.

TraR has proven particularly valuable in screens for acyl-HSLs produced by other bacteria. Among the LuxR-like reporters, TraR detects the widest range of signal molecules, being sensitive to 3-oxo- and 3-hydroxy-substituted acyl-HSLs with chain lengths ranging from six to 14 carbons.[29,30] The activator also responds reasonably well to alkanoyl-HSLs with chain lengths ranging from eight to 12 carbons. Of the known activators TraR exhibits the highest level of sensitivity to the 3-oxo- and 3-hydroxy-substituted signals. However, TraR is less sensitive to the alkanoyl-HSLs, especially those with side chains of six carbons or shorter and does not respond to any C_4 analogs.

Our standard TraR-based detector strain, *A. tumefaciens* NTL4(pZLR4),[29] lacks a Ti plasmid and is nonpathogenic. The strain also lacks *traI* and does

[48] L. Ravn, A. B. Christensen, S. Molin, M. Givskov, and L. Gram, *J. Microbiol. Meth.* **44,** 239 (2001).

[49] J. P. Throup, N. J. Bainton, B. W. Bycroft, P. Williams, and G. S. A. B. Stewart, *in* "Bioluminescence and Chemiluminescence: Fundamental and Applied Aspects" (A. K. Campbell, L. J. Kricka, and P. E. Stanley, eds.), p. 89. Wiley, Chichester, UK, 1995.

[50] J. B. Andersen, A. Heydorn, M. Hentzer, L. Eberl, O. Geisenberger, B. B. Christensen, S. Molin, and M. Givskov, *Appl. Environ. Microbiol.* **67,** 575 (2001).

[51] J. P. Pearson, K. M. Gray, L. Passador, K. D. Tucker, A. Eberhard, B. H. Iglewski, and E. P. Greenberg, *Proc. Natl. Acad. Sci. U.S.A.* **91,** 197 (1994).

not produce any compounds with detectable acyl-HSL activity. Plasmid pZLR4 contains a copy of *traR* and a fusion of *lacZ* to *traG* which is expressed from a TraR-dependent promoter. In the absence of an active signal, this strain does not produce significant amounts of β-galactosidase and forms white colonies on medium containing X-Gal. When exposed to optimum amounts of an active signal, the strain produces on the order of hundreds of units of β-galactosidase activity and forms strongly blue colonies when grown on medium containing the chromogenic substrate. The strain is extremely sensitive to its cognate signal, *N*-(3-oxooctanoyl)-L-HSL, yielding half-maximal induction of the reporter fusion at quormone concentrations as low as 1–2 n*M*. On plates, the strain will detect femtomolar amounts of *N*-(3-oxo-octanoyl)-L-HSL and nanomolar to micromolar amounts of noncognate analogs.[30]

Assays for Detecting Acyl-HSL Signals. Two types of assays can be used to detect acyl-HSL signal molecules. The qualitative assays detect only the presence of bioactive compounds, whereas the analytical assays can provide a minimum estimate of how many chemically different acyl-HSLs are present in a given sample.

QUALITATIVE ASSAYS. Samples can be assayed for compounds with acyl-HSL activity simply by spotting them onto the surface of an agar-base medium on which a culture of the bioindicator strain is growing as a confluent lawn. In its simplest version, colonies of the bacterium to be tested are inoculated onto an overlay of NTL4(pZLR4) on suitable medium containing X-Gal. Alternatively, culture supernatants of the bacteria to be tested, or ethyl acetate extracts of such supernatants, can be spotted onto the overlays. The plates then are incubated overnight at 28°. If the tested culture or sample yields a compound with acyl-HSL activity a diffuse blue zone will appear in the overlay around the site of application. Eight to 10 such samples can be tested on a standard 100-mm diameter petri dish; a correspondingly larger number can be tested on 150-mm diameter plates or on square antibiotic assay plates.

The technique is rapid, sensitive, technically undemanding, inexpensive, and amenable to processing large numbers of samples in parallel. However, given the insensitivity of the *A. tumefaciens* reporter to shorter chain alkanoyl-HSLs, we generally conduct a second set of assays using *C. violaceum* CV026blu[49] as the indicator. This strain responds to exogenous acyl-HSLs by producing an intense purple pigment and detects specifically alkanoyl-HSLs with side chains ranging from four to eight carbons.[29,49]

Test sample preparation. Several options are available. Colonies of the bacteria to be tested can be patched directly onto the soft agar overlays. This is a useful and rapid test but only when the bacteria to be tested can grow on AB mannitol medium. Alternatively, small volumes of culture supernatants or concentrated ethyl acetate extracts of culture supernatants can be spotted onto the overlay. The bacteria to be tested should be grown in several media as signal production may be medium dependent.

1. Grow the bacteria to be tested in 5–10 ml volumes of the appropriate medium at the appropriate temperature to late exponential or early stationary phase.

2. For each culture, remove the cells by centrifugation and transfer the culture supernatant to a clean 25-ml Erlenmeyer flask. Volumes of this culture supernatant can be spotted directly to the surface of the indicator plates. If the bacterium to be tested produces its own β-galactosidase, it is useful to sterilize the culture supernatant by filtration.

Alternatively:

3. Add an equal volume of acidified ethyl acetate and mix well. Allow the phases to separate and transfer the organic phase to a clean 25-ml Erlenmeyer flask.

4. Repeat step 3 at least once, combining the organic phases.

5. Remove residual water by addition of a small amount of anhydrous magnesium (or sodium) sulfate, transfer the extract to a clean test tube, and take the sample to dryness under a stream of dry nitrogen or argon gas at room temperature.

6. Redissolve the residue in 1/10 the volume of the sample extracted of ethyl acetate. Use immediately, or store at $-20°$ in vapor-tight glass sample tubes.

Preparation of the A. tumefaciens reporter overlay

1. The day before the assay is to be performed inoculate 5 ml of AB mannitol medium supplemented with 30 μg/ml of gentamicin with a colony of NTL4(pZLR4) taken from a fresh plate. Incubate the culture with shaking at $28°$ overnight.

2. The day of the assay, inoculate a fresh culture of AB mannitol medium containing 3 μg/ml of gentamicin with a 1/20 volume of the overnight preculture. Grow to late exponential phase, 6–8 hr, as above. The volume of this culture depends on how many assay plates will be prepared.

3. For each petri dish to be prepared seed a 5 ml volume of melted 0.7% water agar containing 40 μg/ml X-Gal with about 0.5 ml of the indicator culture. Mix well by gentle vortexing to prevent foam and bubbles. It is not necessary to include antibiotics in the overlay medium. The melted soft agar should be kept at a temperature no higher than $45°$ and, once inoculated, used as rapidly as possible; *A. tumefaciens* will not survive at this temperature for any length of time.

4. Immediately pour the inoculated soft agar over the surface of a standard petri dish containing a base of 25 ml of AB mannitol medium. Rock and tilt the plate to evenly distribute the overlay over the entire surface.

5. Allow about 5 min at room temperature for the overlay to set before spotting on the samples to be tested.

6. Colonies can be patched onto the surface of the agar overlays using an inoculating loop, or more conveniently, the flat end of a sterile wood toothpick. Liquid samples can be applied in small volumes (1–5 μl) using glass capillary micropipettes or standard micropipettors.

For careful analyses, we prefer a somewhat more complicated assay. Ethyl acetate extracts are prepared from spent culture medium in which the bacterium of interest has been grown as described above. Small volumes, usually 2–3 μl, of these organic extracts are arrayed as spots onto the surface of a 20 × 20 cm C_{18} reversed-phase thin-layer chromatography plate. These samples can be applied as a set of serial dilutions to assess relative amounts of activity in a given sample. In all cases fixed volumes of a dilution series of a pure standard acyl-HSL should be applied as a positive control to gauge the response of the indicator strain. This technique is best suited to organic extracts as aqueous samples tend to bead up on the highly hydrophobic C_{18} matrix. Following sample application the TLC plates are overlaid with the indicator strain as follows:

1. Place the 20 × 20-cm glass-backed C_{18} reversed-phase TLC plate onto which samples have been applied and allowed to dry into the Plexiglas jig (the construction of which is described later).

2. For each such plate to be prepared, seed a 50 ml volume of the indicator culture into 100 ml of 0.7% agar-based AB mannitol medium containing 40 μg/ml X-Gal held at 45°. Mix well.

3. Pour a 150 ml volume of the inoculated soft agar medium onto the surface of the TLC plate. Quickly, using a Plexiglas comb, evenly spread the melted soft agar over the surface of the plate.

4. Allow 5–10 min at room temperature for the overlay to set, score the agar around the edge of the plate with a clean knife or spatula, and remove the plate from the jig.

The Plexiglas jig and spreader comb should be disinfected by wiping with 70% (v/v) ethanol prior to each use.

Incubation and analysis. The cultures are incubated at 28° for 18–24 hr. The overlaid TLC plates should be incubated in covered clear plastic boxes lined on the bottom with autoclaved paper towels dampended with sterile water to prevent the agar layer from drying out. The boxes should be wiped with 70% ethanol between uses.

Following growth and color development, the petri dishes and TLC plates can be photographed using conventional or digital cameras. The agar layer on the TLC plates can be dried to a thin film in a fume hood, and the dried plates can be stored in flexible plastic sleeves for several months. Images of the dried plates can be archieved by photography or by digital color scanning (Fig. 7A).

Controls. Several controls should be included in all assays:

1. A positive control, either a colony of a bacterium known to produce a detectable acyl-HSL or an aqueous or organic sample containing a known, active acyl-HSL, should be included on all plates.

2. As negative controls, colonies known not to produce a detectable signal should be tested. When liquid samples are being tested, volumes of uninoculated media used to grow the test bacteria, or ethyl acetate extracts of such media, should be spotted to the plates to control for components in these media that might gratuitously activate the reporter. In our hands, highly concentrated ethyl acetate extracts of L-broth contain an unknown compound with weak activity.

3. In both types of assays it is important to prepare a duplicate plate of samples in which the overlay contains X-Gal, but no indicator bacteria. We have received reports of false-positive reactions in which blue zones form around some tested bacteria even in the absence of the reporter strain. Presumably during growth these bacteria release into the medium some activity that hydrolyzes X-Gal.

ANALYTICAL ASSAYS. Bacteria that use acyl-HSL-mediated quorum sensing usually produce more than one form of the quormone. Traditionally, the number of discrete signal molecules produced by a bacterium has been determined by combining separation using HPLC on C_{18} reversed-phase columns with detection using one or more of the available bioreporters.[52] Thin-layer chromatography, again coupled with biodetection systems, has been used widely in such studies.[30] The technique is simple, rapid, and can handle many samples in parallel. In addition, the developed plates are visually appealing (Fig. 7B).

TLC Analysis

1. Spot 1–2 μl volumes of concentrated ethyl acetate extracts along a baseline 1–1.5 cm from the bottom of the plate. Up to eight such samples can be applied on a single 20 × 20 cm TLC plate. Spot an appropriate volume of a mixture of standard acyl-HSLs on the same line at each edge of the plate, and also in the middle.

2. Place the plate in a leveled TLC tank containing methanol–water, 60 : 40 (v/v), as the mobile phase. Ensure that the interior atmosphere is saturated with the solvent by lining three sides of the tank with solvent-soaked filter paper.

[52] A. L. Schaefer, B. L. Hanzelka, M. R. Parsek, and E. P. Greenberg, *Methods Enzymol.* **305,** 288 (2000).

A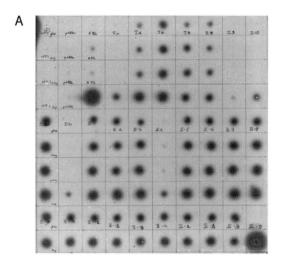

B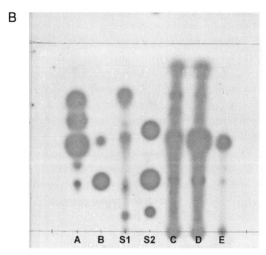

A   B   S1   S2   C   D   E

FIG. 7. TraR-based bioassays for acyl-homoserine lactones. (A) Spot plate assay. Two-μl volumes of ethyl acetate extracts of culture supernatants, and of serial dilutions of these extracts were spotted onto the surface of a C_{18} reversed-phase TLC plate. After the samples had dried, the plate was placed in a Plexiglas jig and overlaid with a suspension of the indicator strain NT1(pZLR4) in a soft agar-based minimal medium containing X-Gal and the plate was incubated overnight at 28°. The next morning the plate was examined, the agar layer was allowed to dry to a thin film, and the image of the plate was archived by digital scanning. (B) Thin-layer chromatography. Two-μl volumes of ethyl-acetate extracts from culture supernatants were spotted at the origin of a C_{18} reversed-phase TLC plate and the plate was chromatographed in the ascending mode against a solvent of methanol–water, 60/40 (v:v). Following completion of the chromatography, the plate was removed from the tank, dried, and overlaid with the indicator strain as described above. The plate was incubated, examined, and archived also as described above. Lanes S1 and S2 contain a mix of standards of 3-oxo, and alkanoyl-HSLs with different chain lengths, respectively. Lanes A–E contain samples from cultures of *Pseudomonas fluorescens, Ralstonia solanacearum, A. tumefaciens, Erwinia carotovora,* and *Rhizobium leguminosarum,* respectively.

3. Cover the tank with a vapor-tight top and allow the solvent to migrate to within 1 cm of the top of the plate. Remove the plate from the tank, mark the solvent boundary with a pencil line, and dry the plate in a fume hood.

4. Place the dried TLC plate in the Plexiglas jig and overlay with a soft agar suspension of the bioreporter strain as described in the section on the preparation of the *A. tumefaciens* overlay.

5. When the overlay has gelled, carefully score the agar around the edges of the plate using a clean knife or spatula and remove the plate from the jig.

6. Incubate the plate in a clear plastic box as described above.

7. Examine the plate, record by photography, or dry down the overlay and record the results by digital scanning as described above.

Data interpretation. Each acyl-HSL migrates with a characteristic mobility and spot shape. 3-Oxoacyl-HSLs produce tear-shaped spots, whereas alkanoyl and 3-hydroxy forms migrate as well-defined circles. Distance of migration (R_f value) is dependent on side-chain length with shorter chain species of each class migrating faster than their longer chain analogs. The methanol–water solvent system reproducibly resolves acyl-HSLs with chain lengths ranging from 4 to at least 14 carbons.

As noted above, while the *A. tumefaciens* reporter detects a large number of acyl-HSLs, it is insensitive to signals with four-carbon side chains. We suggest chromatographing a set of the same samples on a second plate using the *C. violaceum* reporter in the overlay.

Applications. The TLC method is most useful as an analytical tool to assess the number of acyl-HSL signals produced by a given bacterium. The technique can be used as a preparative step to produce enough partially purified sample for structural studies.[30] However, HPLC on C_{18} reversed-phase columns coupled directly to the mass spectrometer is a far more efficient method for determining the structures of these signal molecules.[52] Given its high resolving power and the remarkable sensitivity of the reporter strains, TLC is useful for monitoring purification of the acyl-HSLs by other methods.[10]

The technique also can be used to quantify each of the separated components. Spot size is proportional to the amount of signal over a wide concentration range.[29] Accurate quantitation requires that the structure of the acyl-HSL corresponding to the spot be known, and that a set of standards of known concentrations of that acyl-HSL also be included on the plate. A standard curve is constructed relating spot area or diameter to amount of standard spotted.[29] The area of the spot from the experimental sample is then determined and compared to the standard curve for that acyl-HSL. As long as the value for the unknown lies within the nonsaturated range of values for the standard, the signal can be quantified with reasonable accuracy.

Limitations and considerations. Most importantly, the TLC method cannot be used to assign structure. Moreover, the appearance of a blue zone in the indicator

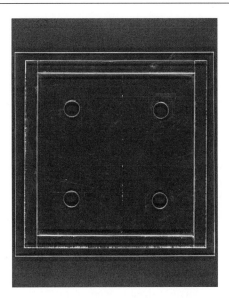

FIG. 8. The Plexiglas Jig. The jig shown holds a single 20 × 20-cm TLC plate or one or more smaller plates using appropriate spacers. The construction and use of the jig are described in the text.

layer does not prove that the active agent is an acyl-HSL; the bioindicators are known to respond to compounds that are not acyl-HSLs.[53]

Since the reporter responds to each analog with a different sensitivity, spot intensity is not a measure of relative abundance among a set of signals with different acyl chain lengths in a given sample. As an example, the *A. tumefaciens* reporter responds with maximum sensitivity to its cognate signal, *N*-(3-oxooctanoyl)-L-HSL. At nonsaturating levels, roughly 1000 times as much *N*-(3-oxohexanoyl)-L-HSL is required to give a spot equal in intensity to that of the cognate signal. However, the intensity of spots corresponding to the same acyl-HSL species in different samples can be used to compare their relative amounts as long as all samples are chromatographed in parallel on the same plate.

Miscellaneous Information

Standards. At the time of writing a limited number of pure acyl-HSLs were available from laboratory chemical supply companies or from companies that specialized in quorum-sensing systems. Should these no longer be available, the alkanoyl- and 3-oxoacyl-HSLs can be synthesized with relative ease using procedures recently described by Eberhard and Schineller.[54] Alternatively, ethyl acetate

[53] M. Teplitski, J. B. Robinson, and W. D. Bauer, *Mol. Plant–Microbe Interact.* **13,** 637 (2000).
[54] A. Eberhard and G. B. Schineller, *Methods Enzymol.* **305,** 301 (2000).

extracts of culture supernatants from bacteria known to produce a characteristic suite of different acyl-HSL species can be used, with appropriate caution, as second-derivative standards.

The Plexiglas Jig. The jig (Fig. 8) is designed to hold a standard 20 × 20-cm glass-backed TLC plate in a well that is 2–3 mm deeper than the thickness of the plate. This configuration allows the soft agar suspension of the reporter bacterium to be spread over the surface of the plate in an even layer. To construct this jig:

1. Cut a base plate of 3/8-inch-thick Plexiglas measuring about 30 × 30 cm.

2. Glue to the surface of this plate four 15-mm-wide strips of 4-mm-thick Plexiglas to form a square defining an interior well with dimensions of 20.2 × 20.2 cm. Make sure that the TLC plate will fit within this square with about a 1-mm space on each edge before gluing down the strips.

3. Drill four 3/4- to 1-inch-diameter holes in the base plate within the area formed by the strips. These holes ease removal of the plate from the well after the overlay has gelled.

4. The jig can be made to accommodate 5 × 20 or 10 × 20-cm TLC plates by manufacturing appropriately sized spacers from 4-mm-thick Plexiglas sheets. The smaller TLC plate is placed in the jig up against three edges. The appropriate spacer then is inserted to occupy the remaining space in the well. Alternatively, custom jigs can be manufactured for each size of TLC plate.

Acknowledgments

Studies in the author's labortatory that contributed to this article have been funded by Grant No. R01 GM 52465 from the NIH, and Grant No. IDACF 01I-3-3 CS from C-FAR, State of Illinois to S.K.F. We thank Ryan M. Farrand for excellent assistance with graphics.

Author Index

Numbers in parentheses are footnote reference numbers and indicate that an author's work is referred to although the name is not cited in the text.

A

Aballay, A., 3, 14
Abalos, R. M., 31
Abdalla, A. M., 320, 322, 323(6; 9)
Abe, A., 371, 389
Aber, R. C., 359
Abshire, K. Z., 109, 120(3), 228, 229, 234(6)
Abu Kwaik, Y., 335
Acar, J. F., 425
Acheson, S. W. K., 151
Achtman, M., 188, 204, 206
Adams, J. M., 335
Adams, L. G., 148
Adams, M. D., 165, 291
Adler, J., 234, 394
Aebersold, R., 208, 243
Aebischer, A., 307, 316(4)
Aepfelbacher, M., 359
Afarwal, B., 320
Afsar, A., 135
Ahmad, M., 335, 342(12)
Ahmadian, M. R., 360, 361
Ahmer, B. M., 73
Ahringer, J., 25, 26(46)
Aili, M., 359
Aizawa, S.-I., 385, 386, 386(4), 387,
     388(4; 6; 13), 389, 394, 399(16), 408,
     408(16; 18)
Akerley, B. J., 100, 101, 102, 107, 107(8), 151
Akins, D. R., 165, 166
Aktories, K., 360, 361(13), 362(13), 368(13)
Ala'Aldeen, D. A., 207
Alaniz, G. R., 75
Albus, A. M., 430
Aldovini, A., 71
Alexeyev, M. F., 415
Ali, A., 151
Allaoui, A., 386, 388(15), 394, 408(19)
Allaoul, K., 350

Allaway, D., 66
Allison, D. G., 435
Allmaier, G., 397, 399
Allocati, N., 326
Alnemri, E. S., 335, 342(12)
Alon, U., 205
Altieri, M., 151
Altschul, S. F., 121, 199
Ambros, V., 24
Ameisen, J. C., 335
Andersen, J. B., 46, 439, 445(37), 476
Andersen, P., 248
Andersen, S. J., 75
Anderson, D. E., Jr., 165
Anderson, J. C., 459
Anderson, K. V., 10
Anderson, N. L., 254
Anderson, P., 20
Andersson, D. I., 425
Andrews, H. L., 350
Ang, S., 198
Appel, R. D., 228
Appleby, P., 449
Aravind, L., 19
Ares, M., Jr., 205
Arfin, S. M., 178, 182(3), 186(3), 189, 191(6),
     196(6), 198, 205(6)
Argenzio, R. A., 370
Armitige, L. Y., 75
Arnold, C. N., 198
Arnold, F. H., 62
Aronson, J. D., 32
Arrington, J. B., 35
Attia, L., 320
Attree, I., 345
Auffray, Y., 257
Ausubel, F. M., 3, 4, 4(1; 5–8), 9, 9(1), 13,
     13(5), 14, 15, 16(15), 21(11), 27(12),
     114, 195
Autret, N., 75

Subject Index

A

Acylated homoserine lactones, *see also* Quorum
  sensing
assays
  Agrobacterium tumefaciens TraR system
    Agrobacterium tumefaciens compound
      detection, 470–471
    controls, 480
    detection from other bacteria, 475–484
    qualitative assays, 477–480
    reporter overlay preparation, 478–479
    thin-layer chromatography, 480,
      482–484
  Pseudomonas aeruginosa, 438–440
  competition between molecules and
    posttranslational control, 444
  R-protein interactions, 442–444, 451
  structures and synthesis, 431–435
  synthases, 435–436
Agrobacterium tumefaciens
  conjugal opines, 454, 458
  crown gall induction, 452
  plant interaction biology, 452–453
  quorum sensing, molecular mechanisms,
    458–461
  Ti plasmid
    conjugation regulation, 454–458
    functions, 452
    map, 455–456
  TraR
    conjugal opine regulation
      assays, 467–469
      overview, 456–457
    dimerization assays
      coaffinity display systems, 473–475
      fusions to λcI repressor, 473
      overview, 472
    DNA binding and assay, 460, 472
    operon regulation, 456, 459
    quorum sensing system
      acylated homoserine lactone detection,
        470–471, 475–484

applications, 461–462
conjugal transfer assay, 459, 462–463,
  469
filter mating assay, 463–464
induction, 470
messenger RNA assays of gene
  expression, 467
rationale for development, 459
reporter fusion assays of gene
  expression, 465–467
spot plate mating assay, 464–465
structure, 459–461
Allelic exchange, *see Mycobacterium
  tuberculosis;* RP4-based plasmids; YopE
Antisense, *see Staphylococcus aureus*
Apoptosis
  caspase-independent mechanisms, 335–336
  Chlamydia modulation, *see Chlamydia,*
    apoptosis modulation
  Helicobacter pylori induction in gastric
    epithelium, *see Helicobacter pylori*
  morphological features, 334–335

B

Bacillus Calmette–Guerin
  proteome analysis with two-dimensional gel
    electrophoresis and mass spectrometry
    bacillus Calmette–Guerin genome
      comparison with *Mycobacterium
      tuberculosis,* 244–247
    cellular protein preparation, 249–250
    culture, 249
    culture supernatant preparation, 250–251
    MALDI-MS peptide mass fingerprinting
      data acquisition, 255
      database searching, 255–256
      principles, 243–244
      sample preparation, 254–255
    materials, 248–249
    membrane fraction preparation, 251
    mycobacteria of phagosomal origin,
      preparation, 251–253

515

ribosomal loci blocking on chromosomal
    DNA, 117
self-annealing of complementary DNA,
    117–118
Shigella flexneri, type III secretion system
    proteins, 386
    purification, 389–392
    structure, 386, 388
Signature-tagged mutagenesis
    Mycobacterium tuberculosis, 98
    principles, 141–142
    transposome mutagenesis, 138–139
Staphylococcus aureus, inducible antisense
        RNA for essential gene identification
    chromosomal DNA random library
        construction, 124–125
    quantitative titration of essential genes
        in vitro, 127
        in vivo, 127–128
    rationale, 123
    screening growth defect and lethal colonies,
        126–127
    tet regulatory system, 123–124
STM, *see* Signature-tagged mutagenesis

T

Terminal deoxyuridine nick end labeling,
    Helicobacter pylori apoptosis induction in
        gastric epithelium
    embedding and sectioning, 321–322
    fixation, 321
    incubation conditions, 322
    materials, 320–321
    quantification considerations, 322–323
    rationale for *in vivo* studies, 319–320
Thin-layer chromatography, acylated
        homoserine lactone assay
    applications, 482
    development, 480, 482
    interpretation, 482
    limitations, 482–483
    plexiglass jig, 484
    standards, 483–484
Tir, tyrosine phosphorylation
    assays
        detergent fractionation, 375
        immunofluorescence microscopy, 379–381
        immunoprecipitation, 377–378
        infection of mammalian cells, 373–374

lysate preparation from infected cells,
    374–375
mechanical disruption and cell
    fractionation, 376–377
Western blot analysis, 378–379
overview, 372–373
TLC, *see* Thin-layer chromatography
Translocated intimin receptor, *see* Tir
Transposome
    definition, 128–129
    efficiency of transposition, 128–129
    electroporation
        advantages over other delivery methods,
            129, 140
        conditions, 134
        electrocompetent cell preparation,
            133–134
    formation reaction, 130
    insertion site localization
        cloning of insertions, 131–132
        genomic DNA isolation, 130–131
        rescue using conditioned origin of
            replication, 132–133
        sequencing, 133
    prospects for virulence gene studies, 139
    Proteus vulgaris mutagenesis and phenotypic
        changes, 138
    signature-tagged mutagenesis, 138–139
    transposons for exported or membrane
        peptide mutagenesis
        EZ::Tn transposon, 135–136
        insertion clones, 136–137
        Tn*PhoA* limitations, 134–135
    Xylella fastidiosa mutagenesis, 137
Transposon footprinting
    advantages, 152
    bacteria species applicability, 151–152
    cloning and sequencing of significant DNA
        fragments, 147
    gel electrophoresis, 147
    polymerase chain reaction
        gel fragment amplification, 147
        retrieval of target mutants, 148
        template preparation, 146, 149, 151
        transposon-flanking sequences, 146–147
    pools of transposon mutants, preparation, 145
    principles, 142–143
    Salmonella typhimurium
        transposon mutagenesis, 143–145
        virulence gene identification, 148–149

ISBN 0-12-182261-3

90051

9 780121 822613